Modern Physics

In the photograph facing the title page we observe the effect of a single plane of atoms with the light microscope. Chemical etching produced these pyramidal pits in the surface of a copper crystal. Each pyramid was created by the preferential etching wherever the perfect order of the crystal was interrupted by a dislocation, an extra half-plane of atoms. The microscopic technique makes use of interference to distinguish regions of different slope by providing a different hue for each slope. This etch-pit method is one of the ways the world of atomic sizes can be studied, even though individual atoms cannot be seen directly.

Dr. F. W. Young, Jr., and Dr. L. D. Hulett of the Oak Ridge National Laboratory made this photograph, which is reproduced here at a magnification of 800X. For additional analysis of such photographs, see Secs. 10-2e and 10-2f.

Modern Physics

The quantum physics of atoms, solids, and nuclei

SECOND EDITION

Robert L. Sproull

PROFESSOR OF PHYSICS, CORNELL UNIVERSITY

John Wiley & Sons, Inc., New York · London

Preface

An analytical study of the quantum physics of atoms, molecules, solids, and nuclei is presented in this textbook. This area of physics, usually but not altogether appropriately called "modern physics," is frequently taught in a largely descriptive fashion in an introductory college course and in a highly mathematical fashion in a graduate course. My aim in this book has been to cultivate a middle ground, in which a quantitative understanding of atomic, molecular, solid-state, and nuclear physics is attained through the application of elementary quantum mechanics. Quantum mechanics, once thought to be too mysterious for undergraduates, is presented here at an introductory level with emphasis on its concepts and methods, which become familiar to the student through analysis of one-dimensional problems. Application of these concepts and methods permits answering the most intriguing questions of modern physics—questions like: What holds matter together? How does the variety of chemical properties of different elements arise? How do electrons move through solids? Why do the nuclei occurring in nature have only certain combinations of protons and neutrons?

The first edition of this book developed from a course for undergraduate engineering students, and the book was therefore addressed to engineers. Quantum physics was rarely taught in engineering curricula at the time of that edition, and I included many examples of applications in an attempt to persuade engineering students that this area of physics was an appropriate part of their professional training. Such encouragement seems out of place today, since engineers are taking quantum physics in their stride. Furthermore, it soon developed that many instructors of courses for majors in physics, chemistry, and other sciences welcomed the analytical approach to atomic physics provided by the first edition, despite its subtitle referring to engineers.

This second edition is intended as a text for a course following an introductory general physics course and is addressed to both engineering and science majors. Engineering and science students will probably build on the subject matter of this book in different ways. For example, the electrical engineer may combine the knowledge gained here with work in circuit theory to prepare himself for conducting original work in solid-state devices or nuclear particle detectors; the physics or chemistry major may use the knowledge gained here in his intermediate theory and laboratory courses and later return to quantum mechanics in graduate-level courses before doing original research. Although students' paths will diverge after studying this book, it seems to me that the aims of engineering and science majors in studying modern physics at the level presented here should be basically the same, namely, to understand as fully and as quantitatively as possible the fundamental processes of quantum physics.

In keeping with the analytical character of this book, I have placed considerable emphasis on the solution of problems by the student. It is not much of an exaggeration to say that no topic is introduced in this text unless it is possible for the student to solve meaningful problems after studying that topic. The number of problems has been more than doubled in the new edition, largely in order to make more choice available to the instructor who is using the book year after year. The order of presenting problems at the end of each chapter parallels the order of presentation of the relevant topics in the text. The relative difficulty of problems varies widely; some merely demonstrate orders of magnitude or simple applications, but many constitute substantial extensions of the text, frequently guided by appropriate hints. Answers to problems marked by asterisks are given in Appendix H, primarily as a help to readers who are studying this book without the benefit of an instructor.

The second edition retains the basic organization and approach of the first, but many sections have been rewritten and some have been shortened. New sections on the Stern-Gerlach experiment, superconductivity, diffusion, thermoelectricity, and thermionic energy conversion have been added. Chapter 5 has been rewritten and expanded. The most substantial change is that Chapter 13 has been replaced by two longer chapters on nuclear physics that treat this subject more extensively and more analytically.

The order in which modern physics is discussed here remains the same as in the first edition. The first chapter presents elementary descriptions of the particles composing atoms and nuclei, the ingredients of modern physics. Chapter 2 develops the concept of a distribution function, needed later; it also adduces evidence on the size of atoms and shows that any system in thermal equilibrium, such as an atom or a solid, will be in nearly the lowest possible energy state for that system. Chapter 3 presents the demonstration that atomic physics can be split into two parts, the study of the extranuclear structure of the atom and the study of the nucleus. The separate treatments are possible because of the small size and large binding energy of the nucleus relative to the scales of sizes and energies involved in experiments on the extranuclear structure of the atom. These three chapters may be review for many students, since the material contained in them (and even much in Chapter 4) is now commonly included in a first course in physics.

The heart of modern physics is quantum mechanics, and the central part of this book is devoted to it. Chapter 4 presents the experiments that show the *necessity* for quantum theory, that produce much interesting information about atoms, and that provide hints of the form quantum theory must take. Chapter 5 describes quantum mechanics and some simple, artificial examples that illustrate the *method* of using it. The *application* of quantum mechanics begins in Chapter 6 and continues throughout the remainder of the book.

Attention could be turned again to nuclear physics after Chapter 6, but the development of molecular and solid-state physics follows more naturally (Chapters 7 through 11). Chapter 12 on physical electronics is also interposed, and then the subject of nuclear physics is taken up again at the point where it was left at the end of Chapter 3. The discussion of nuclear physics in Chapters 13 and 14 can thus use not only quantum theory but also several topics from Chapters 10 through 12 that help to explain the instrumentation of nuclear experiments. The teacher who objects to the dispersal of nuclear physics may decide to interpose Chapters 13 and 14 between Chapters 6 and 7.

A detailed picture of the structure of this book can be obtained by reading the "Introduction" sections of every chapter. The student would be well advised to read all these sections before embarking upon the study of this book and to read all of them again after completing it.

An attempt has been made to raise the level of difficulty gradually from an easy start in Chapter 1 to the more difficult portions of Chapter 8. After Chapter 8, the level of difficulty should be sensibly constant.

References are provided at the end of each chapter to aid the student studying on his own and to provide an entry into the literature if additional topics are to be studied in a higher-level course.

Relatively little attention is paid in this book to the history of the development of twentieth century physics, for three principal reasons. First, each year quantum physics becomes less of a novel curiosity and more of a vital part of the core of physics; and it seems to me that its parentage should naturally begin to recede into the background, just as the development of Newtonian mechanics now claims little space in a classical mechanics textbook. Furthermore, lectures (rather than a textbook) are probably a more suitable medium for presenting the fascinating history of the ideas of quantum physics. And finally, there are already excellent books available that feature the historical approach.

The units used throughout are rationalized mksa units, except for frequent substitutions of the electron volt as a unit of energy and the angstrom as a unit of length. Use of these two special units permits a better visualization of quantities of atomic size. Perhaps more frequent use of grams, centimeters, millimeters, microns, fermis, and even inches would also help in visualization. I have usually avoided such use, however, in order to make the strictly mechanical part of solving problems as straightforward as possible and thus to permit the student to concentrate on the physics. Appendix D presents simple instructions for converting mksa to cgs units, examples of conversions, and a table of conversion factors.

I acknowledged gratefully in the first edition indispensable help from Drs. E. M. Pell, A. R. Moore, and R. L. Pritchard; Professors D. R. Corson, J. W. DeWire, H. F. Newhall, H. S. Sack, and L. P. Smith; and especially Professor J. A. Krumhansl. The revision in no way lessens my debt to them. The second edition has profited greatly from study and suggestions by Dr. Pell, Dr. P. J. Leurgans, Dr. D. S. Billington, Professor DeWire, Professor Jay Orear, and a reviewer chosen by the publisher. Substantial as their contributions were, they were surpassed by the criticism, suggestions, and encouragement provided by Professor C. S. Smith of Case Institute of Technology, whose participation throughout the planning and writing of the second edition was uniquely valuable. I am grateful also to Nancy, Robert F., and Mary Sproull for their help in preparing the manuscript. Finally, I should like to acknowledge with gratitude the comments and criticisms of many students and instructors who have given me the benefit of their experience with the first edition.

R. L. SPROULL

Ithaca, New York
February, 1963

Contents

xi

Modern Physics

1

Fundamental Particles

1-1 INTRODUCTION

An analytical introduction to the basic physics developed in the twentieth century is presented in this book. The new physics has been of great intrinsic interest, almost a new science in itself, and in addition has provided applications in engineering that are already considerable and are expanding rapidly. The study of modern physics leads to new devices and energy sources, to more convenient and accurate instruments, to new materials of construction, and to a clearer understanding of existing materials.

This book is primarily concerned with physical laws and processes, but applications in other sciences and in engineering will be described frequently. Television camera tubes, transistors, nuclear reactors, and other devices will be analyzed as part of the application of the basic physics; but most of the applications will be found in other science and engineering courses and in engineering practice.

In this book we shall first describe the elementary particles that are the *ingredients* of modern physics. The variation of mass with velocity and the famous Einstein $E = Mc^2$ relation (basic to the whole field of atomic energy) will be presented in conjunction with the properties of these particles in the present chapter. In Chapters 2 and 3 we shall present an introduction to the interactions of

1

these particles, and we shall present evidence that our study can be divided into two nearly separate studies of atoms and of nuclei. Then (Chapters 4, 5, and 6) we shall show that new physical laws govern the behavior of atoms and nuclei, and we shall state and illustrate these *quantum physics* laws. At the end of Chapter 6 we shall be in a position to apply the new understanding of quantum physics to molecules, solids, and nuclei. As we carry out this application in Chapters 7 to 14 we shall find that the quantum physics provides explanations of many phenomena of engineering interest, such as the properties of solids, the characteristics of electron tubes, and the operation of transistors. Furthermore we shall learn that the applications in many areas, such as metallurgy and nuclear energy, are developing rapidly and that much rewarding work remains for new generations of scientists and engineers.

Many students will find the first chapters largely a review of earlier work. In these chapters, it will frequently be necessary to assert some properties of particles and to assume the existence of sources and detectors, assertions and assumptions that will be justified only much later in the book. In the later chapters, however, we shall attempt a uniformly *analytical* approach, that is, we shall make assertions or conclusions only by logical argument based on experiments or on theories well tested by experiments.

1-2 THE ELECTRON

(a) Source

The usual source of electrons in the laboratory is a hot filament in a vacuum tube. The physical process of emission of electrons from this source will be discussed in Chapter 12. Such emission, called thermionic emission, produces electrons with small initial kinetic energies.

Another source of electrons to be discussed also, but later, is the emission from radioactive nuclei. Such electrons are emitted with a wide range of energies extending up to very high energies. Since this source was discovered before the emitted particles were identified, they were not called electrons but were called β-*rays* and later β^-- or e^--particles.

(b) Size

No experiments capable of measuring the size or shape of the electron have been performed. An *upper limit* to the size of the electron can be obtained, however, from experiments in which electrons at very high energies are used as projectiles and nuclei are used as targets. The *maximum* size that an electron could have and be consistent with these

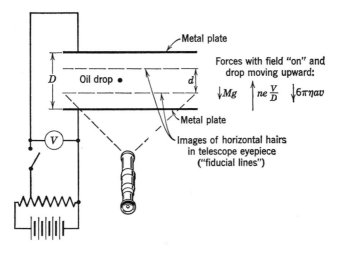

Fig. 1-1. Geometry of the oil-drop experiment for determining e. The forces diagrammed are on the assumption that the charge on the drop is negative.

experiments is about 10^{-14} m.* (distance across the electron, or diameter if it were a sphere). This distance is so small compared to the other distances we shall be concerned with that we can consider the electron as a *mass point* with zero extension in space. Such a mass point is called a *particle* in mechanics.

(c) Charge

The charge of the electron is negative, and its magnitude is

$$e = 1.602 \times 10^{-19} \text{ coulomb}$$

Throughout this book we use the symbol e for the absolute magnitude of the electronic charge. The charge of the electron is therefore $-e$.

The electronic charge can be measured by the *Millikan oil-drop experiment*. The apparatus used is shown schematically in Fig. 1-1. A pair of horizontal, parallel condenser plates is mounted inside an enclosure. The enclosure prevents drafts and permits varying the pressure. Except in very precise work (where the pressure must be varied in order to determine small corrections), the chamber is filled with ordinary air at atmospheric pressure. An atomizer permits spraying fine drops of a non-evaporating oil into the space between the plates. A telescope with horizontal hairs permits the observation of a single drop and the measure-

---

* "m." will be used as an abbreviation for meter; the symbol m will be used later for the mass of the electron.

ment of the vertical velocity of a drop. The velocity is determined by measuring the time required for the drop to rise or fall the fixed distance d between the images of the hairs. This distance is usually considerably less than D, the spacing between the plates.

A source of ionizing radiation that can be turned on or off is provided. This source can be an X-ray tube or an ultraviolet arc. The *process* of ionization will be considered in detail in later chapters; at this point, all we need to know is that it is possible to remove an electron from an atom by X-rays or by ultraviolet light. If this atom is a gas atom, a positive ion and an electron are provided, either one of which may be captured by the oil drop. If this atom is one of the atoms making up the oil drop, the oil drop will attain a positive charge. Thus, while the ionizing radiation is turned on, the oil drop can have its charge either increased or decreased but always by an *integral number of electronic charges*.

A falling drop of the size used in this experiment reaches its *terminal velocity* very quickly ($\ll 1$ sec). When the drop moves at this velocity the resistance of the air to the motion of the drop equals the negative of the applied force. In other words, the drag caused by the viscosity of air is equal in magnitude and opposite in sign to the other forces (gravitational or gravitational plus electrical) acting on the drop. For spherical drops this viscous force has been found by ordinary hydrodynamics experiments to be

$$F = -6\pi\eta a v \tag{1-1}$$

where η is the viscosity of the medium, a is the radius of the sphere, and v is its terminal velocity. Equation 1-1 is called *Stokes' law* (it requires a

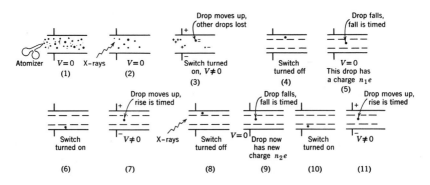

Fig. 1-2. Procedure of the oil-drop experiment. Steps 1–3 produce many drops and select a suitable drop. Steps 4–11 are repeated many times with the same drop.

correction of a few per cent for the very small drops used to measure e, but this correction can be found by varying the pressure of the air in the chamber). Drops of the size encountered in this experiment are spherical because of the action of the surface-tension force.

The procedure of this experiment is illustrated in Fig. 1-2. A drop is selected, its time t_0 of fall through a distance d is measured with no electric field, and its terminal velocity is then determined from the relation $v_0 = -d/t_0$ (velocities upward are considered positive). Since the drop is not accelerating, the sum of the forces on it must be zero. Therefore the sum of the gravitational force $-Mg$ (downward) and the viscous force F (upward) must be zero, and

$$Mg = -6\pi\eta a v_0 = 6\pi\eta a (d/t_0) \tag{1-2}$$

where g is the acceleration of gravity (9.80 m./sec^2) and M is the mass of the drop. Since the drop is spherical,

$$M = \tfrac{4}{3}\pi a^3 \rho \tag{1-3}$$

where ρ is the density of the oil. Equations 1-2 and 1-3 are two equations with two unknowns (M and a), and so M and a can be determined from them. Because the drop is so small, neither M nor a can be measured directly; the order of magnitude of a is 10^{-6} m.

Next a short burst of X-rays produces some charge on the drop, the electric field V/D is applied in the correct sense to move the drop upward, and the time t_1 of rise is measured. The velocity of rise is d/t_1, and, if there are n_1 electronic charges on the drop,

$$n_1 e(V/D) - Mg = 6\pi\eta a(d/t_1) \tag{1-4}$$

The drop is again allowed to fall without an electric field, another burst of ionization changes the charge to $n_2 e$, and a new time t_2 of rise is observed:

$$n_2 e(V/D) - Mg = 6\pi\eta a(d/t_2) \tag{1-5}$$

The subtraction of eq. 1-4 from eq. 1-5 yields

$$(n_2 - n_1)e = \frac{6\pi\eta a D d}{V}\left(\frac{1}{t_2} - \frac{1}{t_1}\right)$$

This procedure is repeated over and over. By making the bursts of ionization short enough, the differences $(n_2 - n_1)$, $(n_3 - n_2)$, etc., can be kept small. These differences are therefore small integers (like $3, -2, 4, 1, \cdots$). A table of values $(n_{i+1} - n_i)e$ can be prepared, and the integers and e can be determined from this table. Hundreds of rise times have been measured for a single drop, and never has a change in charge smaller than 1.6×10^{-19} coulomb been observed. This fact constitutes evidence that the fundamental quantity of charge is the charge of the electron, and that all electrical processes (e.g., ionization) involve the transfer of an integral number of electronic charges.

It should be noted that the Millikan oil-drop experiment permits the determination of a quantity of *atomic size* (the charge of the electron) by measurements of quantities of ordinary *laboratory size* (lengths of a few millimeters, times of a few seconds, and potential differences of a few volts). This result is accomplished by the use of an oil drop that is large enough to be seen and to move slowly, yet is small enough so that its motion is appreciably affected by a change in charge of only 1.6×10^{-19} coulomb. It should also be noted that e can be measured indirectly by combining the results of other experiments.

(d) e/m

The ratio of the electron's charge to its mass m is

$$e/m = 1.759 \times 10^{11} \text{ coulombs/kg}$$

and therefore the mass m is

$$m = 9.11 \times 10^{-31} \text{ kg}$$

We shall see in Sec. 1-7 that at velocities near the velocity c of light the mass of a particle depends on its velocity. The values of e/m and of m given above are the values for velocities very much less than c and are, strictly speaking, the values e/m_0 and m_0 appropriate to zero velocity. Since in almost all this book we shall be dealing with velocities very much less than c, we shall usually omit the subscript. Where both m and m_0 appear in an equation, m will be the actual mass and m_0 will be the mass of the electron at rest.

The e/m of the electron is much larger than the similar ratio for any other particle or aggregate of particles. This fact and the relative ease with which electrons can be emitted from solids are responsible for the great usefulness of electrons in vacuum-tube devices. Particles with smaller charge-to-mass ratios are more sluggish in electric and magnetic fields. If such particles were used in electron-tube devices, the tubes could be used only at low frequencies and the space-charge-limited currents would be much smaller (see prob. 1-11).

The e/m of the electron can be measured by the deflection of an electron beam in electric and magnetic fields or by the use of an electric field and a measurement of a time of flight. There is a wide variety of possible ways of making such measurements. We can illustrate the principles of possible measurements by an example.

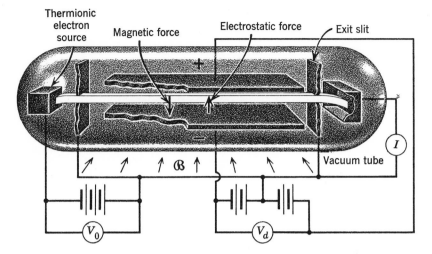

Fig. 1-3. Vacuum tube for measuring e/m. There is a magnetic induction ℬ directed as shown by the arrows.

A vacuum tube (Fig. 1-3) constructed of non-ferromagnetic materials is placed in a uniform magnetic field with the magnitude of the magnetic induction equal to ℬ. A thin, flat beam of electrons is emitted from the hot cathode and enters the deflection region at the center of the tube. The force on a charge $-e$ moving with velocity **v** (a vector) is *

$$\mathbf{F} = -e\mathbf{v} \times \mathbf{ℬ} \qquad (1\text{-}6)$$

The direction of this force is at right angles to both **v** and ℬ, and its sense is illustrated in Fig. 1-3. Since **v** is perpendicular to ℬ in this experiment, the magnitude F of the force is

$$F = |\mathbf{F}| = ev\mathbf{ℬ}$$

where v is $|\mathbf{v}|$ (the *speed*) and ℬ is $|\mathbf{ℬ}|$. This force produces a deflection toward the lower plate; hence the electron beam does not pass through the exit slit at the right, and no current is measured in the meter I.

Next an electrostatic deflection is produced by introducing the potential difference V_d, and V_d is adjusted until the beam passes through

---

* The vector cross product **v** ✕ ℬ is defined as follows. Its *magnitude* is vℬ sin θ, where v and ℬ are the magnitudes of **v** and ℬ, respectively, and where θ is the smaller of the two angles between **v** and ℬ. Its *direction* is at right angles to the plane of **v** and ℬ and pointed in the direction a right-hand screw would travel if turned from **v** to ℬ through the angle θ.

the exit slit and a current I is observed.   The electrostatic force eV_d/d must be just equal in magnitude (but opposite in direction) to the magnetic force:

$$eV_d/d = ev\mathcal{B} \tag{1-7}$$

This vacuum tube is therefore a *velocity selector*, since only electrons with a velocity satisfying eq. 1-7 can traverse the tube.

The potential difference V_0 through which the electrons have been accelerated is also measured.   The work done by this potential difference on one electron equals the increase in kinetic energy of the electron (which started with nearly zero kinetic energy at the cathode):

$$eV_0 = \tfrac{1}{2}mv^2 \tag{1-8}$$

The combination of eqs. 1-7 and 1-8 gives

$$\frac{e}{m} = \frac{v^2}{2V_0} = \frac{V_d{}^2}{2V_0 d^2 \mathcal{B}^2} \tag{1-9}$$

Therefore e/m can be determined from the observable quantities· This method is capable of high precision, but careful attention must be paid to edge effects of the deflecting plates and to precision of construction.

Measurements of e/m have been very useful in identifying electrons, since no particle has been discovered with an e/m at all close to the electron's.   In the experiments to be discussed in the following chapters any doubt as to the nature of the particles participating can usually be removed by measuring their charge-to-mass ratios.

(e) Other properties

The behavior of an electron is such that it must possess a definite angular momentum.   It acts as if it were spinning about its center with this angular momentum.   The principal experiments indicating this fact are observations of details of the line spectra of light emitted by atoms, which will be described briefly in Chapter 6.   This property is usually referred to as the *spin angular momentum* (or just *spin*) to distinguish it from any angular momentum the electron may have because of its motion from one point in space to another.   (The angular momentum of a *rigid body* is represented as a vector that can be computed from the angular velocity vector and the moments and products of inertia. Since the size and shape of an electron are not properties that can be found by experiment, it is useless to inquire about the angular velocity and moments of inertia of an electron.)

The electron is also known to have a definite magnetic moment. This fact, too, is demonstrated chiefly by experiments on line spectra. The magnetic moment of a bar magnet is usually thought of as the product of pole strength and pole separation, but these quantities have limited physical significance even for the bar magnet. Like angular velocity and moment of inertia, these quantities have no significance for the electron.

These two properties (spin and magnetic moment) are not so obvious or so often mentioned as the charge and mass, because they do not play a dominant role in experiments like those already described. The forces between electrons (or between electrons and external magnetic fields) because of their magnetic moments are usually much less than the forces present because of the charge and velocity of the electron. Spin and magnetic moment are nevertheless important attributes of the electron, as we shall discover when considering electrons in atoms and in solids.

1-3 THE PROTON

(a) Source

Protons are usually obtained by ionizing hydrogen atoms, each of which consists of a proton and an electron, and are accordingly also called hydrogen nuclei or hydrogen ions. When an energetic electron in a hydrogen gas discharge collides with a hydrogen atom it can remove the electron and thereby produce a free proton.

(b) Size

An effective proton radius of about 3×10^{-15} m. can be determined from experiments in which high-energy protons collide with other protons. This size is so small that we can consider the proton to be a particle (that is, a mass point) until nuclear physics will be discussed in detail in Chapter 13.

(c) Charge

The charge of the proton is positive and exactly equal in magnitude to the electron's charge e. This statement follows from the fact that the hydrogen atom is electrically neutral.

(d) Ratio of charge to mass

This ratio for the proton is

$$e/M = 9.58 \times 10^7 \text{ coulombs/kg}$$

and leads to a mass M of the proton:

$$M = 1.673 \times 10^{-27} \, \text{kg}$$

that is 1836 times the electron mass. (The symbol M is not reserved exclusively for the proton but is used for any mass other than the electron mass.)

The methods described in Sec. 1-2d could be used to measure e/M of the proton, but they would not be quite so accurate. The difficulty is that practical sources provide protons with an appreciable spread of energies. Therefore eq. 1-8 does not hold accurately, since it assumes that the initial energy of the particle is practically zero. A more precise method is to utilize the principle of the cyclotron, which is explained in Sec. 1-9. The cyclotron itself could be used for this measurement, but it is more convenient and accurate to use a special tube. The measurement of e/M is reduced to the measurement of a magnetic field strength and of the frequency of a radio-frequency oscillator. As in the case of an electron, measurement of the e/M usually identifies a particle as a proton.

(e) Other properties

The proton has a spin angular momentum equal to that of the electron. Its magnetic moment is about $1/600$ of the electron magnetic moment.

1-4   THE NEUTRON

(a) Source

Neutrons cannot be obtained so easily as electrons or protons. They are produced as the results of nuclear reactions, and there is no other source. Illustrations of the production of neutrons will be given in Chapters 13 and 14.

(b) Size

The statements of Sec. 1-3b about the proton size apply also to neutrons.

(c) Charge

The neutron has no charge. Neutrons can penetrate large thicknesses of matter because they have no electrical interactions with the electrons and nuclei in matter. The lack of electrical charge makes it impossible to detect neutrons in the same way protons and electrons are detected. Some methods of detection will be considered in Chapter 14.

(d) Mass

The mass of the neutron is

$$M = 1.675 \times 10^{-27}\ \text{kg}$$

which is only a little larger than the proton mass. Since the charge is zero, e/M measurements cannot be performed to determine the mass. The neutron mass is measured indirectly, as will be explained in Sec. 3-4.

(e) Other properties

The neutron has a spin angular momentum equal to that of the proton. Its magnetic moment is somewhat smaller than the proton's and is in the opposite direction relative to its spin.

1-5 OTHER PARTICLES

There are many kinds of particles other than electrons, protons, and neutrons, but the others will not be discussed until Chapter 13 since they do not enter in an essential way into the understanding of atomic, molecular, and solid-state physics.

1-6 ENERGY IN ELECTRON VOLTS

The joule is so large that it is inconvenient as a unit for the energy of a single electron. A much easier unit to use (and to visualize the magnitude of) is the *electron volt*, abbreviated to e.V. The electron volt is defined as the kinetic energy of an electron accelerated from rest through a potential difference of 1 volt. The work done by the electric field in this case is:

$$(1\ \text{volt}) \times (1.602 \times 10^{-19}\ \text{coulomb}) = 1.602 \times 10^{-19}\ \text{joule}$$

Therefore

$$1\ \text{e.V.} = 1.602 \times 10^{-19}\ \text{joule} \qquad (1\text{-}10)$$

For example, an electron in a radio tube may be accelerated through a potential difference of 200 volts; its kinetic energy is hence 200 e.V. The energies we actually measure in the laboratory are always very large numbers of electron volts, since very large numbers of electrons are participating in laboratory-scale experiments. An electron volt is therefore *not* a convenient unit to use in measuring, for example, the energy taken from a battery when a charge of 1 coulomb has passed. The electron volt *is* a convenient and appropriate unit throughout atomic and nuclear physics, since we are frequently concerned with an individual electron or positive ion. It must be kept in mind that a *volt* is a unit of *potential difference*. An *electron volt* is a unit of *energy*.

1-7 DEPENDENCE OF MASS ON VELOCITY

All the experiments we have considered up to this point have involved particle velocities less than about 10^7 m./sec. In this velocity range, the e/m of the electron and the e/M of the proton are found experimentally to be independent of velocity, at least within the limits of error of the experiments. Of course, laboratory-size objects moving at laboratory-size velocities (a few meters per second) also have masses independent of velocity and, if charged, charges independent of velocity.

At higher velocities, m and all other masses vary appreciably with velocity. The experiments described below actually prove only that the ratio of charge to mass varies with velocity. We ascribe the variation to the mass (rather than to e) for two reasons: (1) There are experiments that show that e is constant even at very high velocities; one of these (the Duane-Hunt limit experiment) will be described in Sec. 4-6. (2) The theory of relativity predicts that mass should change with velocity in just the way m/e is observed to change, and this theory agrees with experiment in many other ways.

The earliest experiments that show the way m varies with velocity were performed by Bucherer, Wolz, and Neumann. Neumann's apparatus is shown schematically in Fig. 1-4. There are two parallel plate electrodes, arranged like those in Fig. 1-3 but very close together (0.251-mm separation in the original apparatus). A small quantity of radium emits β^--particles (high-energy electrons). A uniform magnetic field is provided by a large solenoid. A photographic plate is placed a distance a from the edge of the electrodes, and the apparatus is evacuated.

Electrons are emitted by the radium in all directions and with all energies up to 1.2 million electron volts (M.e.V.). We shall be concerned with only those electrons that move in the plane $AOBC$ and that are emitted toward the right. When there is no electric or magnetic field, they strike the photographic plate at O, and an exposure of several hours is made under these conditions. Then electric and magnetic fields are applied in the directions indicated in Fig. 1-4. The magnetic and electrostatic forces are oppositely directed, and hence we have again a *velocity selector* as explained in conjunction with eq. 1-7. Any electron that traverses the selector must have a velocity

$$v = V_d/\mathscr{B}d \qquad (1\text{-}11)$$

relative to the apparatus.

After leaving the velocity selector the electrons are acted upon by only the magnetic force, which has the constant magnitude

$$F = ev\mathcal{B}$$

and acts in the direction at right angles to the velocity **v**. Therefore the acceleration F/m is constant in magnitude and at right angles to **v**. This is precisely the condition that the orbit of the electron be an arc of a circle. Motion in a circle of radius r requires a centripetal acceleration v^2/r, constant in magnitude and everywhere perpendicular to the circular path. Hence:

$$mv^2/r = F = ev\mathcal{B}$$

$$mv = \mathcal{B}re \tag{1-12}$$

A region of constant magnetic field and zero electric field is therefore a *momentum selector*, since particles with different values of the momentum describe circles of different radii. An exposure of several hours is made with the electric and magnetic fields "on." From the position B of the exposed spot on the photographic plate we can measure y and hence r.

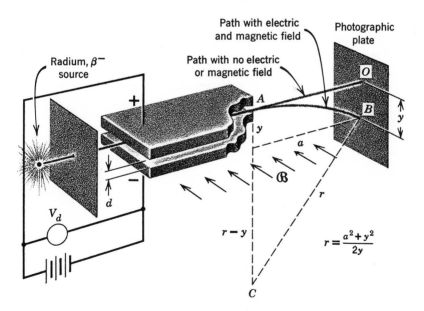

Fig. 1-4. Apparatus for measuring the dependence of mass on velocity. The apparatus is evacuated, and there is a magnetic induction $\mathcal{B}$ perpendicular to the plane $AOBC$. [G. Neumann, *Ann. Physik*, **45**, 529 (1914).]

As mentioned above, we assume that e is constant, independent of v. By measuring V_d, $\mathcal{B}$, and d we obtain v from eq. 1-11, and by also determining r we find mv from eq. 1-12. A series of exposures with different values of V_d therefore permits measurement of m as a function of v.

The results of many photographs for values of v from $0.4c$ to $0.8c$, where c is the velocity of light (3.00×10^8 m./sec), are all consistent with the expression

$$m = \frac{m_0}{(1 - v^2/c^2)^{\frac{1}{2}}} \tag{1-13}$$

m_0 is called the *rest mass*, the value of m when $v^2 \ll c^2$. This expression also follows from the Einstein theory of relativity. (As mentioned in Sec. 1-2*d*, it was really m_0 that was measured in the experiment described there.)

The validity of eq. 1-13 is not confined to the electron. We see in Sec. 1-9 how proof of this relation for protons and heavier particles is provided as a by-product of the use of high-energy accelerators.

It should be noted that we could not expect to prove or disprove eq. 1-13 by performing mechanics experiments such as pendulum or inclined-plane experiments, since the velocities used are so small compared to c that an impossible precision would be required. Figure 1-5 illustrates the velocity scale involved. For example, pendulum experiments in the laboratory adequately verify Newton's laws but could never tell us whether Newton's laws were still valid in some widely different range of masses or velocities. As experiments open up new ranges of any parameter (e.g., the velocity), we must test our familiar, laboratory-scale

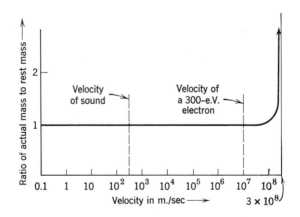

Fig. 1-5. Dependence of mass on velocity.

mechanics in this range. In most experiments Newtonian mechanics works well in the new range. But a situation may be found (e.g., v near c) where a new, more general physics must replace the old physics in order to obtain agreement with experiment. Of course the new laws must reduce to the old laws in the regions (e.g., low velocity) where the old laws have been observed to be adequate. Thus eq. 1-13 reduces to a practically constant mass when $|v|/c$ is less than 0.01.

The preceding paragraph has presented a rather obvious situation in some detail. The reason for the present discussion is that we shall find a very similar situation to the variation of mass with velocity when we discuss the new wave mechanics in Chapters 4 and 5. There, too, when experiments entered new ranges (in that case, the very small lengths and momenta necessary to describe electrons in atoms) the old laws were found to be invalid in these ranges. New laws were found that are valid in the new ranges of parameters and reduce to the old laws for laboratory sizes.

1-8 MASS-ENERGY RELATION

In elementary particle mechanics, (force) = (mass) × (acceleration), and (force) = (rate of change of momentum). These statements say the same thing if the mass is constant, since then $\dfrac{d}{dt}(Mv) = M\dfrac{dv}{dt}$. These statements are incompatible in the region of high velocities, where M is a function of v; since v usually is a function of t, we cannot treat M as a constant in the differentiation process. Which one of these statements (if either) are we to use when v is appreciable compared to c? The theory of relativity answers this question in favor of the momentum statement. The paths and times of flight of particles in high-energy accelerators (see Sec. 1-9) confirm this theory. We therefore write

$$F = d(Mv)/dt \qquad (1\text{-}14)$$

which holds for *all* velocities, even velocities near c.

It is instructive to consider a simple, one-dimensional problem of the acceleration of a particle of rest mass M_0 and charge e along the direction of x. The particle is being accelerated from rest in a constant electric field of magnitude $\mathcal{E}$, and this field is in the $+x$-direction. The force on the particle is, of course, $e\mathcal{E}$. Therefore:

$$e\mathcal{E} = \frac{d}{dt}(Mv) = \frac{d}{dt}\left\{\frac{M_0 v}{(1 - v^2/c^2)^{1/2}}\right\} = \frac{M_0(dv/dt)}{(1 - v^2/c^2)^{3/2}} \qquad (1\text{-}15)$$

We should like to obtain an equation for the kinetic energy K. K is defined, as in Newtonian mechanics, as the work done by the field, since the initial kinetic energy is zero. This work is $\int F\,dx = \int e\mathcal{E}\,dx$. The simplest approach is to multiply both sides of eq. 1-15 by $dx = v\,dt$ and to integrate:

$$K = \int_0^x e\mathcal{E}\,dx = \int_0^v \frac{M_0 v\,dv}{(1 - v^2/c^2)^{3/2}}$$

$$K = eV_0 = \frac{M_0 c^2}{(1 - v^2/c^2)^{1/2}} - M_0 c^2 \qquad (1\text{-}16)$$

$$K = eV_0 = Mc^2 - M_0 c^2 \qquad (1\text{-}17)$$

Here V_0 has been used for $\int_0^x \mathcal{E}\,dx$, the potential difference through which the particle has been accelerated. It should be noted that as the acceleration proceeds the velocity approaches the velocity c of light but never reaches or exceeds it (see prob. 1-19).

Equation 1-17 states that the increase in kinetic energy is equal to the product of c^2 and the increase in mass. We have proved a very special case of a general law: Energy and mass are merely two different ways of describing the same thing, and

$$K = (\Delta M)c^2 \qquad (1\text{-}18)$$

Thus, for any change in mass of a particle or a system, there is a change in kinetic energy. A particle of rest mass M_0 at rest in the laboratory has a *rest energy* equal to $M_0 c^2$. If it is given a kinetic energy, its new total energy (kinetic plus rest energy) is Mc^2. The rest energy is thus an "available" energy, similar to the potential energy. The familiar conservation of energy must therefore be generalized to become the *conservation of mass-energy*. Equation 1-18 is a consequence of the theory of relativity, but we shall make no attempt to give a general proof of this relation. The nuclear binding energy experiments described in Chapter 3 and the work of Chapters 13 and 14 will give experimental confirmation of this relation.

It is easy to show that eq. 1-16 or 1-17 is not in conflict with Newton's laws in the region $v^2 \ll c^2$, where these laws should apply. We can do this by expanding eq. 1-16 by means of the binomial expansion:

$$(1 + u)^n = 1 + nu + \frac{n(n-1)u^2}{2!} + \frac{n(n-1)(n-2)u^3}{3!} + \cdots$$
$$(1\text{-}19)$$

This expression is *valid* when $u^2 < 1$ and *useful* when $u^2 \ll 1$.

We set $n = -\frac{1}{2}$ and $u = -v^2/c^2$, and obtain from eq. 1-16

$$K = M_0 c^2 \left\{ \left(1 - \frac{v^2}{c^2} \right)^{-\frac{1}{2}} - 1 \right\} = M_0 c^2 \left\{ 1 + \frac{v^2}{2c^2} + \frac{3\,v^4}{8\,c^4} + \cdots - 1 \right\}$$

When $(v^2/c^2) \ll 1$, the terms in v^4/c^4 and higher powers of v/c can be neglected, and then

$$K = M_0 v^2/2 \tag{1-20}$$

in agreement with Newton's laws.

This section and Sec. 1-7 have described the consequences of relativity physics that are most important for our work. We have based our discussion on the Neumann *experiment*. It would be more satisfying, but it would require an extensive digression from our study of atomic physics, to give here a systematic account of the Einstein *theory* of relativity. This theory developed from the postulate that the laws of physics are the same in all reference systems moving with constant velocity with respect to each other. The implications of this postulate are far-reaching and surprising, requiring time ("time dilation") and dimensions ("Lorentz contraction") to have different values in different reference systems moving with respect to one another. The theory of relativity is a fascinating study, and the books cited at the end of this chapter give introductory accounts of it.

1-9 HIGH-ENERGY-PARTICLE ACCELERATORS

In order to investigate the properties of atoms and especially of nuclei it is necessary to have particles with very large kinetic energies. Radioactive elements (e.g., radium) provide energetic particles, but their energies are never larger than a few million electron volts. Furthermore, the particles are emitted in all directions, and the rate of emission cannot be controlled. A *beam* of charged particles with known kinetic energy is what is desired for convenience and precision in most experiments.

Energies up to a few million electron volts can be obtained by direct acceleration through the required potential difference. Insulation and corona problems prevent the application of this method to energies much above 10 M.e.V. The commonest type of such an accelerator is the Van de Graaff accelerator, in which the potential difference is obtained by an electrostatic belt generator working on the same principle as the early electrostatic machines. The Van de Graaff accelerator can be applied to the acceleration of electrons, protons, or other ions. The beam is homogeneous in direction and energy. For very precise work a velocity selector can be added at the output end of the machine. With such a selector, eq. 1-16 can be verified for electrons, since m_0, e, and c are known and V_0 and v can be measured.

All accelerators for higher energies have this in common: Acceleration occurs in a large number of small steps or continuously at a slow rate, rather than in one large step. Thus only moderate voltages are required, but very high kinetic energies of the particles are attained. Of all the types of accelerators we discuss only two, the cyclotron and the synchrotron. References that describe the betatron, linear accelerator, synchrocyclotron, and other accelerators are listed at the end of the chapter.

The *cyclotron* is an accelerator for protons or heavier particles. It is illustrated schematically in Fig. 1-6. An evacuated box (the tank) is situated between the cylindrical pole pieces of a large magnet. Inside the vacuum tank there are two electrodes, called "D's" because of their shape. These are hollow, open, copper boxes each with a cylindrical side and semicircular ends. Near the center of the chamber is an ion source. If protons are to be accelerated, the source is an electric arc in a hydrogen gas atmosphere. If doubly charged helium ions (α-particles) are to be accelerated, the gas is helium. In any case, gas is admitted in a

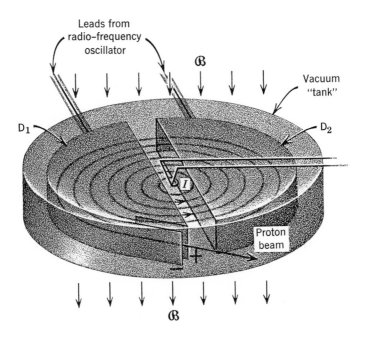

Fig. 1-6. The vacuum tank, "D's," ion source I, and beam trajectory in a cyclotron.

fine stream at the arc and is pumped rapidly away by the pumps evacuating the tank.

The D's are connected to a radio-frequency oscillator. Thus a proton emitted when D_2 is negative with respect to D_1 will be at first accelerated toward the right. Since it is in a magnetic field it will be deflected into an approximately circular path as demonstrated in connection with eq. 1-12. If the frequency of the oscillator is correct, when the ion returns to the gap between the D's one-half cycle of the radio frequency will have elapsed and D_1 will be negative with respect to D_2. Thus the particle will again be accelerated. This process can be repeated over and over, and the required frequency does not have to change as the radius of the proton's orbit increases.

The operation of the cyclotron rests on the fact, which we now prove, that the time taken by the particle to make one revolution is independent of the radius of its orbit. In order to show this, we approximate a section of the spiral path by a circle of radius r. Then the time for one revolution is

$$T = 2\pi r/v$$

and eq. 1-12, but with the mass M in place of the electron mass m, gives the result

$$T = 2\pi M/\mathfrak{B}e \tag{1-21}$$

This expression contains only constants, and therefore the particle can be accelerated by a constant-frequency a-c field each time it crosses the gap between the D's. As the radius of the particle s orbit grows, its velocity increases proportionally, and therefore the time for one revolution is constant.

Some typical dimensions and other parameters for a cyclotron are given in prob. 1-25. It is apparent from this example that the proton makes hundreds of revolutions in its spiral path, and therefore the circle we assumed in the preceding paragraph is a good approximation to a segment of the path. The maximum $\mathfrak{B}$ we can readily obtain (limited by saturation in the iron pole pieces) is about 1.5 webers/m.^{2} This means that, in order to obtain large energy, and hence large momentum, we must use large values of r (see again eq. 1-12). The pole pieces, electromagnet, D's, and vacuum tank must then be large enough to accommodate the large radius.

Up to this point we have treated M as a constant, but clearly continued acceleration of the particle eventually increases M because of the dependence of mass on velocity at high energies. The ordinary cyclotron therefore cannot be used to produce arbitrarily high energies. In order to attain higher energies, a modification of the cyclotron called the *synchrocyclotron* is used.   In this machine the protons are emitted in pulses. The oscillator frequency is varied during the acceleration of each pulse in order to accommodate the changing mass.   Magnet designs more elaborate than the simple design with $\mathcal{B}$ independent of position are also required to attain very high energies.   Energies approaching the proton rest energy $M_0 c^2 = 938$ M.e.V. have been obtained in this way.   A cyclotron is not useful as an *electron* accelerator since the electron's energy when its mass starts to change is so low.

Another very useful device is the *synchrotron*, which can accelerate either electrons or protons.   Proton synchrotrons with energies up to tens of billions of electron volts have been built.   The machine described below is a 1500-M.e.V. electron synchrotron; it is illustrated in Fig. 1-7.   It consists principally of four quadrants in each of which a magnetic field is created by electromagnets with C-shaped cross sections. In the magnet gap is the glass vacuum tube.   The quadrants are separated by short straight sections where there is no magnetic field.   All

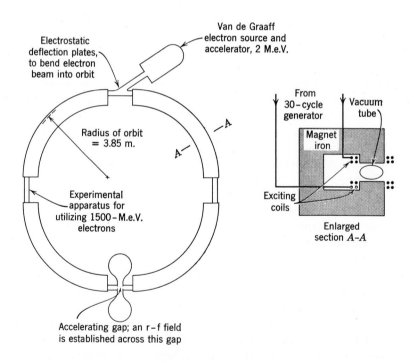

Fig. 1-7.   A 1500-M.e.V. electron synchrotron.

these sections are exploited to connect pumps and instruments to the accelerator tube, and two of them are used for injection and acceleration of the electron beam.

Electrons are injected in short pulses into the vacuum tube by a Van de Graaff accelerator and electrostatic deflection system. Since the electron's energy on entering the synchrotron is 2 M.e.V. (which is about four times the electron's rest energy), the electron is already moving at a velocity near c. At the time of injection the magnetic field is very weak (0.0021 weber/m.^{2}). Once each revolution, the pulse of electrons passes the accelerating gap and receives energy from the radio-frequency field created in the gap by an oscillator. The period of this oscillator must be the same as the period of revolution of the electron in its nearly circular orbit with a 3.85-m. radius. Thus the oscillator frequency must change with time during the early part of the acceleration cycle. Soon, however, it can be constant since the electron moves with almost exactly the speed of light. The electron's energy is increasing because its mass is increasing; its velocity remains nearly constant. While this is going on, the magnetic field is also increasing with time in order to keep the electrons moving in a circle with the same radius. The magnet is excited by a 30 cycle/sec motor-generator set. Electrons are injected each $\frac{1}{30}$ sec at a phase such that the magnetic field is nearly zero and is increasing. At the end of $\frac{1}{4}$ cycle the magnetic field is a maximum (1.5 webers/m.^{2}). The electrons have then reached their maximum energy and are used in experiments.

The advantage of the synchrotron over the cyclotron for high energies is that the magnet required is much cheaper. The overall diameter required in either machine is the same for the same energy. But in the synchrotron the magnetic field is required only over a thin annulus of this diameter, whereas in the cyclotron the magnetic field must extend over *all* values of r from zero up to the maximum. The cyclotron has a varying radius of orbit and constant magnetic field strength during acceleration; the synchrotron has a constant radius of orbit and varying magnetic field strength.

In this discussion we have not considered the trajectories of the particles in detail. In order to achieve an appreciable beam current great attention must be paid to possible motion in the vertical plane, as well as in the horizontal plane, and to the stability of the orbits. The acceleration cycle requires very long paths of the particles (see prob. 1-27), and they must never touch the walls. The synchrotron focusing theory employs eqs. 1-13 and 1-14; the fact that its results agree with experiment is further evidence of the variation of mass with velocity.

REFERENCES

Relativity

J. Orear, *Fundamental Physics*, Wiley, New York, 1961, Chapter 11.
R. M. Eisberg, *Fundamentals of Modern Physics*, Wiley, New York, 1961, Chapter 1.
J. L. Synge and B. A. Griffith, *Principles of Mechanics*, McGraw-Hill, New York, 3rd Ed., 1959, Chapter 18.

Accelerators

D. Halliday, *Introductory Nuclear Physics*, Wiley, New York, 2nd Ed., 1955, Chapter 12.
E. M. McMillan in *Experimental Nuclear Physics*, Vol. III, edited by E. Segrè, Wiley, New York, 1959, pp. 639–785.
M. S. Livingston and J. P. Blewett, *Particle Accelerators*, McGraw-Hill, New York, 1962.

PROBLEMS

1-1.* In a Millikan oil-drop experiment, the condenser plates are separated by 0.0160 m., the distance of rise or fall is 0.0060 m., the potential difference between the plates is 4550 volts, and the density of the oil used is 858 kg/m.^{3} at 25°C. The experiment is carried out at 25°C, at which temperature the viscosity η of air is 1.83×10^{-5} kg m.$^{-1}$ sec^{-1}. The average of the times of fall (no electric field) is 21.2 sec. The following times of rise are observed: 46.1, 15.6, 28.0, 13.0, 45.2, and 20.1 sec. Compute e. (A somewhat better value of e would have been obtained if we had used the corrections to Stokes' law applicable for very small drops.)

1-2. With experimental conditions identical with those of prob. 1-1, another drop is studied and the average of the times of fall (no electric field) is 14.7 sec. The following times of rise are observed: 42.7, 14.3, 9.7, 17.2, 12.3, 9.6, 14.6, and 11.0 sec. Compute e.

1-3.* Compute the charge-to-mass ratio of the oil drop in prob. 1-2 at the time of the first rise. Compare with the e/m of an electron. (Note how small the charge-to-mass ratio must be compared to e/m in order that gravitational forces be comparable to electrostatic forces.)

1-4. What difficulty is caused by using too long or too intense a burst of X-rays between observations of rise times in the oil-drop experiment? *Hint:* What would happen if the integers n_1, n_2, etc., in probs. 1-1 or 1-2 were 10 or 100 times their values in those problems?

1-5.* In order to see the order of magnitude of quantities in Stokes' law, apply it to raindrops of radius 0.4 mm. Compute the terminal velocity by using η from prob. 1-1. In order to check this velocity with your experience, also compute the

* Answers to the numerical parts of problems marked by an asterisk appear in Appendix H.

angle the track of such a raindrop makes with the vertical when there is a 30-mph (13.4-m./sec) wind blowing.

1-6. A student has measured four rise and four fall times of a single oil drop using good equipment. He reports that his value of e is $(3.1 \pm 0.2) \times 10^{-19}$ coulomb. If he has made no computational errors, what is probably his trouble, and what should he do?

1-7. Show that after a few rises and falls of a drop in the oil-drop experiment there will be only a single drop in the field of view, even though hundreds of drops may have been present initially.

1-8.* The vacuum tube illustrated in Fig. 1-3 has a deflection plate spacing of 0.015 m., V_0 is set at 300 volts, $\mathcal{B}$ is 0.0012 weber/m.^{2}, and V_d is 184 volts. Calculate e/m.

1-9.* An electron has been accelerated from rest through a potential difference V_0, where V_0 is small enough for m to be practically equal to m_0. Give an expression for the velocity v in terms of V_0. (This expression will be useful in many later calculations.)

1-10. Devise a vacuum tube for measuring e/m by a time-of-flight method, without the use of a magnetic field. Develop the equations connecting observable quantities with e/m. Your method should use sinusoidal radio-frequency voltages. Compute the frequency required for your particular tube dimensions and for electrons with 100-e.V. energy.

1-11.* The space-charge-limited current density of electrons between parallel planes a distance d apart with a potential difference V_0 between them is

$$J = \frac{4\epsilon_0}{9} \sqrt{\frac{2e}{m}} \frac{V_0^{3/2}}{d^2} \text{ amp/m.}^2$$

Compute the space-charge-limited current for an area of 10^{-4} m.^{2} with $d = 10^{-3}$ m. and $V_0 = 300$ volts. Make the similar computation but assume that the particles had the e/M characteristic of protons.

1-12.* The force of gravitational attraction between two particles each of mass M is $F = -GM^2/r^2$, where G is the gravitational constant (6.66×10^{-11} joule m./kg^2) and r is the distance between them. Compare this force with Coulomb's law, and obtain the ratio of the electrostatic force between two protons to the gravitational force between them (for the same value of r).

1-13. Make the calculation comparable to prob. 1-12 but for electrons instead of protons.

1-14. Electrons leave the cathode of a plane-parallel vacuum tube with zero kinetic energy (let the potential energy equal zero at the cathode) and are accelerated toward the plate which is at a potential of 300 volts. Make a sketch of the kinetic energy, the potential energy, and the total energy of an electron as a function of distance from the cathode. Express the ordinate scale both in joules and in electron volts.

1-15. The following data are from an experiment like Neumann's experiments on the variation of electron mass with velocity: $d = 2.51 \times 10^{-4}$ m., $\mathcal{B} = 0.01772$

weber/m.^{2}, and $a = 0.0247$ m. These values of V_d and y were observed:

V_d, volts	y, meters
530	0.0082
652	0.0060
813	0.0043
930	0.0033
1060	0.0025

Calculate v/c and m for each line; plot m vs. v/c; plot eq. 1-13 on the same scale, using $m_0 = 9.11 \times 10^{-31}$ kg.

1-16.* Calculate the rest energy of an electron and of a proton. Express in millions of electron volts.

1-17. What is v/c for an electron such that $K = m_0c^2$?

1-18. Plot $\log_{10} K$ vs. $\log_{10}(v/c)$ for both protons and electrons on the same plot. The range of v/c values for each should extend from 0.10 to 0.99.

1-19.* What is the ratio v/c for an electron with a kinetic energy of: (a) 50,000 e.V.; (b) 500,000 e.V.; (c) 5,000,000 e.V.?

1-20. An electron is accelerated from rest through a potential difference of 300 volts. What is the ratio of its kinetic energy to its rest energy? If eq. 1-8 with $m = m_0$ is used to calculate its velocity v instead of eq. 1-16, what percentage error in v is made? *Hint:* Let $\beta = v/c$; let $u = 300/m_0c^2$; compare the values of β^2 calculated in the two ways and then calculate the error in β.

1-21.* An electron in a television projection cathode-ray tube is accelerated through a potential difference of 80 kilovolts. What is the ratio m/m_0 of its actual mass after acceleration to its rest mass? What is the percentage error we would make in computing v if we used eq. 1-8 with $m = m_0$ instead of eq. 1-16?

1-22. We might think that the Newtonian expression $K = \frac{1}{2}Mv^2$ could be preserved at high velocities merely by inserting M from an equation like eq. 1-13. Show that this is *not* true, since this process does *not* give eq. 1-17, which was derived from the definition of K.

1-23. In deriving eq. 1-12, the tacit assumption was made that m was constant during the acceleration of the particle by the magnetic force. Show that m is constant during such a process and that eq. 1-12 follows from the application of eq. 1-14 to this problem.

1-24.* An electron beam from a 2.0-M.e.V. Van de Graaff enters at right angles to a uniform magnetic field of 0.05 weber/m.^{2} What is the radius of curvature of the path of the electrons? What would have been calculated if we had (incorrectly) used the non-relativistic eq. 1-8 with $m = m_0$ to calculate this radius?

1-25. A cyclotron is to be constructed to accelerate protons to an energy of 10 M.e.V. The maximum $\mathfrak{B}$ attainable in the iron used for the d-c magnet is 1.55 webers/m.^{2}, and this value is therefore also the maximum $\mathfrak{B}$ possible in the accelerating chamber. What must the diameter of the "D's" be? The root mean square (rms) value of the radio-frequency voltage between the D's is 20 kilovolts. How many revolutions does each proton make?

1-26.* Protons are accelerated to 740 M.e.V. in the Berkeley synchrocyclotron, which has a $\mathfrak{B} = 2.3$ webers/m.^{2} What is the oscillator frequency at the time a pulse of protons is injected at the center ($M = M_0$)? What is the oscillator frequency at the time the protons have reached the rim with $K = 740$ M.e.V.?

1-27. In the synchrotron described in the text, the time during which a pulse is in the machine is $\frac{1}{120}$ sec (one-fourth cycle of the 30-cycle frequency, the frequency of the motor-generator set powering the magnets). The electrons are injected with 2-M.e.V. energy and attain an energy of 1500 M.e.V. What is the average increase in kinetic energy per revolution? How many miles does an electron travel during this time (1 mile $= 1609$ m.)?

1-28.* Use the data given in the text for the electron energy at the time of injection in the 3.85-m. radius synchrotron to compute v/c and m/m_0 of the electrons at the time of injection.

1-29. Assume that classical mechanics (eq. 1-20) applies to 2-M.e.V. electrons. Calculate the magnetic induction required to accelerate such electrons in an orbit of 3.85-m. radius. Compare with the observed value of $\mathfrak{B}$ (0.0021 weber/m.^{2}) for 2-M.e.V. electrons and this radius. (Since $\mathfrak{B}$, r, and the electron's energy can be measured, the failure of eq. 1-20 at high speeds is thus experimentally proved.)

1-30.* The CERN (European Center for Nuclear Research) proton synchrotron has a maximum energy of 25,000 M.e.V. and an orbit radius of 100 m. What is the frequency of revolution of the protons when they are injected at an energy of 50 M.e.V.? What is the frequency after acceleration to 25,000 M.e.V.?

1-31. Classical electrodynamics predicts that an accelerated electric charge radiates energy. For an electron of kinetic energy K in a circular orbit of radius r, the radiation is $(\mu_0 e^2 c^2/3r)(K/m_0 c^2)^4$ joules per revolution. Calculate the radiation loss in electron volts per revolution if $r = 3.85$ m. and $K = 1500$ M.e.V. (The r-f oscillator must provide this amount of energy to the electrons each revolution just to prevent their deceleration.)

2

Assemblies of Particles

2-1 INTRODUCTION

We considered in Chapter 1 the properties of individual particles. We shall next study some of the properties of an assembly with a large number of particles. The most familiar example of such an assembly is the ideal gas, and the laws of its behavior will be used as the starting point here. In Sec. 2-2 the average energy of the random motion of a gas molecule will be computed. This kinetic energy of random motion is a universal property of assemblies of particles. Section 2-3 will present the distribution of energies and velocities about this average value, the Maxwell distribution. In Sec. 2-4 we shall study the distribution in space of an assembly of particles subject to a force that is a function of position, the Boltzmann distribution of potential energy.

These distributions of molecular energies and velocities are called *classical statistics*. Although they will be developed here for an ideal gas, they are applicable to any assembly of particles that is sufficiently dilute. That is, they apply to any system where the particles are relatively far apart and interact only rarely, like the molecules in an ideal gas. Both the Maxwell and the Boltzmann distributions can be derived from a very few general assumptions in a branch of physics called statistical mechanics. In the following sections a completely different and more restricted approach is followed: Special cases of these distributions are demonstrated by experiments, and the general situation is stated without proof.

These two distributions are presented for four reasons: (1) They are of interest in themselves, especially for such problems as the motion of electrons and ions in gas discharges. (2) They illustrate how average quantities (e.g., the average energy) can be computed when the distribution in energy is known; this technique will be useful in much of the later work. (3) They show that the energy state of a system (e.g., an electron in an atom) that will be observed in nature at ordinary temperatures is always within a few tenths of an electron volt of the *lowest* possible energy state for that system; this result will be useful for identifying the normal states of nuclei and atoms in subsequent chapters. (4) They indicate the extent of *fluctuations* about average values, fluctuations that persist in the *quantum statistics* to be studied in Chapters 6 and 9.

In Sec. 2-5 the motion of an individual gas molecule will be examined, and the frequency of its collisions with other molecules will be studied. From this study we obtain a knowledge of the sizes of atoms and molecules that will be useful for the separation of nuclear physics from atomic physics.

2-2 ENERGY OF RANDOM MOTION

Before we study the energy of the particles in a gas it is useful to review the relation between macroscopic, laboratory-size magnitudes and atomic magnitudes. These magnitudes are related by Avogadro's number $N_0 = 6.02 \times 10^{26}$ molecules/kilomole. (One kilomole is the amount of a substance whose weight expressed as a number of kilograms equals its molecular weight.) Avogadro's number is also the number of *atoms* per kilogram-*atomic* weight.

Avogadro's number N_0 is, of course, a very large number. If it were a great deal smaller, the "graininess" of matter, which is not continuous but built up of a finite number of molecules, would be visible with a microscope. The use of Avogadro's number enables us to translate any measured quantity of laboratory size (e.g., the mass of a kilomole of hydrogen) into the comparable quantity of atomic size (e.g., the mass of a hydrogen molecule).

Electrolysis experiments provide a method of measuring the product $N_0 e$. For example, the amount of silver plated on the cathode of an electrolytic cell can be determined by weighing the cathode before and after a measured charge (product of electrical current and time) has passed through the cell. Suppose that 1 kilogram-atomic weight

(107.87 kg) has been plated. We assume that we know from auxiliary chemical experiments that each silver ion has a charge e, that is, that each has lost a single electron. Therefore the total charge transported must have been equal to $N_0 e$. N_0 can also be determined from X-ray measurements of the spacing between atoms in crystalline solids as will be explained in Sec. 4-4.

The *ideal gas law* states that

$$pV_m = \mathsf{R}T \qquad (2\text{-}1)$$

Here p is the pressure in newtons/meter^2, V_m is the volume in meters^3 of 1 kilomole of the gas, R is the gas constant per kilomole, and T is the absolute temperature. R is a universal constant, the same for all gases, and its value is 8314 joules/deg. (We can express p in atmospheres if we set R equal to 8.21×10^{-2} m.^{3} atmospheres/deg.) One kilomole of gas, containing N_0 molecules, occupies a volume of 22.4 m.^{3} at atmospheric pressure and 0°C. The restriction to ideal gases means that the motions of the molecules must be nearly independent of one another, and this condition is well satisfied by gases at low pressures since the molecules are usually relatively far apart. The ideal gas law is therefore obeyed by all gases at low enough pressures and is an excellent approximation to gases at atmospheric pressure.

The average energies of the molecules or atoms in a gas can be determined from the ideal gas law. (We shall henceforth assume a gas of molecules in order to avoid repeating "molecules or atoms" at each step.) The pressure p of a gas is caused by the motion of the molecules; if the gas molecules in a rectangular box were at rest, there would be only a very small pressure on the floor of the box and no pressure at all on the sides and top. We shall now compute the pressure on a wall of a rectangular box in terms of the mass and speeds of the molecules. We shall assume that the wall and the gas are at the same temperature so that, on the average, molecules striking the wall neither gain nor lose energy. Therefore the collision of a molecule with the wall is (at least on the average) perfectly *elastic*. We shall first compute the change in momentum, on the average, per molecule incident on the wall. From this result and from a knowledge of the average number of molecules striking the wall per unit time we shall be able to compute the time-average force, which is equal to the time-average rate of change of momentum supplied by the wall.

A molecule of mass M that approaches the wall perpendicular to x with an x-component of velocity v_x leaves with an x-component of velocity $-v_x$ (see Fig. 2-1a). Its momentum has been changed by an amount $2Mv_x$. The wall has therefore exerted a force F for a very

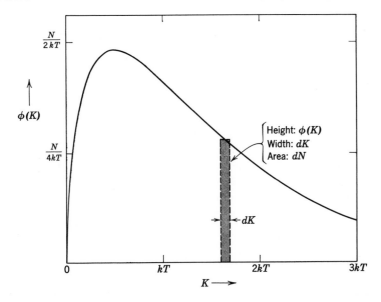

Fig. 2-2. Maxwell distribution of kinetic energies. The shaded area is the number of molecules with kinetic energies between K and $K + dK$.

Here $\phi(K)\,dK = dN$ is the number of molecules per unit volume with energies between K and $K + dK$, and N is the total number of molecules per unit volume. The function $\phi(K)$ is plotted in Fig. 2-2. [Note that, from its definition, $\phi(K) = dN/dK$, and distribution functions like eq. 2-9 are commonly written as an equation for the derivative (e.g., dN/dK); we prefer to define a new symbol $\phi(K)$ here in order that experience in working with distribution functions will be most directly applicable to the distributions to be encountered in Chapter 5 and later chapters, with a minimum change in notation.]

It is instructive to calculate the average kinetic energy $\overline{K}$ from eq. 2-9. The calculation is worth while because it illustrates the method of calculating an average when a distribution is given and because the result shows that eq. 2-9 and eq. 2-6 are consistent.

The method of calculating an average from a distribution will be introduced by the consideration of the distribution of heights of men in a given group. The measurement of the height H of every man in the group permits the computation of the number of men whose heights lie between H and $H + \delta H$, for each value of H; this number is called δN and is, of course, a function of H. The average height $\overline{H}$ can be calculated from δN in the same way that a weighted average is always computed: Each value H of the height is multiplied by the number of men $\delta N(H)$ whose height is between H and $H + \delta H$. The sum of all such products is then divided by the total number of men:

$$\bar{H} = \frac{\Sigma H \, \delta N(H)}{\Sigma \, \delta N(H)}$$

It should be noted that, although each δN would be larger if the interval δH were larger, the average would be unchanged by changes in δH since both the numerator and the denominator would be multiplied by the same factor. (Of course if δH were to become an appreciable fraction of the whole range of H there would be a dependence of the computed $\bar{H}$ upon δH, but δH should be very much smaller than the range of values of H.)

The distributions commonly encountered in physics are generally continuous distributions, like eq. 2-9, and the average value of some property of the distribution can be calculated by integration instead of summation. Thus, if the distribution of heights of men is available in the form of the function $\phi(H)$, where $\phi(H) \, dH = dN$ is the number with heights between H and $H + dH$, the average height $\bar{H}$ is calculated as follows:

$$\bar{H} = \frac{\int H \, dN}{\int dN} = \frac{\int H \, \phi(H) \, dH}{\int \phi(H) \, dH} \tag{2-10}$$

The limits on both integrals are the limits of the range of heights, and the denominator is still just the total number of men. The average of any other function of H (like H^2) could be obtained by substituting it for H in the integral in the numerator.

The average kinetic energy $\bar{K}$ of the molecules in a gas can be calculated from the Maxwell distribution function (eq. 2-9) by the general method of the preceding paragraph:

$$\bar{K} = \frac{\int K \, \phi(K) \, dK}{\int \phi(K) \, dK} = \frac{\int_0^\infty K 2N \left\{\frac{K}{\pi (kT)^3}\right\}^{\frac{1}{2}} e^{-K/kT} \, dK}{\int_0^N dN}$$

The integral in the denominator is the total number N of the particles in the gas. The numerator can be evaluated by the substitution of u

gas enclosed is $NMA\,dz$, where N is (as usual) the number of molecules per unit volume. The gas above the upper surface exerts a force $F_2 = p_2A$ downward on the gas in the little volume pictured; p_2 is the pressure on the upper surface and equals $p_1 + dp$. The force F_1 that the gas in the little volume exerts on the gas below it is the sum of this force and its own weight:

$$F_1 = F_2 + NMAg\,dz$$

$$p_2A = (p_1 + dp)A = p_1A - NMAg\,dz$$

$$dp = -NMg\,dz \tag{2-12}$$

Equation 2-12 relates the increment in pressure dp to the increment in height dz.

We can use the ideal gas law (eq. 2-1) and the fact that $V_m = N_0/N$ in order to eliminate either N or dp from eq. 2-12:

$$p(N_0/N) = \mathsf{R}T = N_0kT$$

$$p = NkT \tag{2-13}$$

$$dp = kT\,dN \tag{2-14}$$

(We could have combined eq. 2-3 and eq. 2-7 and differentiated to obtain the same result.) The combination of eq. 2-14 with eq. 2-12 gives

$$\frac{dN}{N} = -\frac{Mg}{kT}\,dz \tag{2-15}$$

Equation 2-15 can be integrated easily. It is convenient when doing so to let N_1 be the density of molecules at height z_1:

$$\int_{N_1}^{N} \frac{dN}{N} = -\frac{Mg}{kT}\int_{z_1}^{z} dz$$

$$\ln N - \ln N_1 = -(Mg/kT)(z - z_1)$$

$$N = N_1 e^{-(Mg/kT)(z-z_1)} \tag{2-16}$$

This equation states that the number of molecules per unit volume is an exponentially decreasing function of height.

The combination of eq. 2-13 with eq. 2-12 gives an equation similar to eq. 2-15 that can be integrated to give the companion equation to eq. 2-16:

$$p = p_1 e^{-(Mg/kT)(z-z_1)} \tag{2-17}$$

This equation is sometimes called the *law of atmospheres* because of its application to problems like probs. 2-13 and 2-14.

Since Mgz is the potential energy P of a gas molecule in the gravitational field of force, eq. 2-16 can be expressed in the form

$$N = N_1 e^{-(P-P_1)/kT} \qquad (2\text{-}18)$$

where P_1 is the potential energy at the point where the density is N_1 molecules per unit volume. This expression is plotted in Fig. 2-7.

The arguments of the last two paragraphs would apply equally well if, instead of the gravitational force, some other force that could be expressed as a function of position were acting on the molecules. Furthermore, we could have used eq. 2-3 and eq. 2-7 in place of the ideal gas law, and these expressions apply to *any* assembly of particles that interact only rarely. Thus eq. 2-18 is a *general* law and is applicable to many other systems as well as to the ideal gas. It is called the *Boltzmann distribution*.

In subsequent chapters of this book, the most common force will be the electrostatic force, and P will be the potential energy of a charged particle (usually an electron) in an electrostatic field. (In all problems in atomic, nuclear, and solid-state physics the gravitational forces are negligible compared to electrical forces.) The Boltzmann distribution will enable us to compute the fraction of electrons that are in a region where their potential energy is P; for an electron, $P = -eV$, where V is the electrostatic potential.

The Maxwell distribution shows that relatively few molecules have kinetic energies very much greater than kT. The Boltzmann distribu-

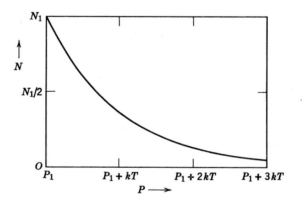

Fig. 2-7. Density of molecules as a function of *potential* energy. This plot of the Boltzmann distribution is the companion to Fig. 2-2, the Maxwell distribution of *kinetic* energies.

tion shows that relatively few molecules are in a region where the potential energy is more than a few kT larger than the minimum potential energy in the space in which the molecules are moving. Thus it is very unlikely that the total energy $E = K + P$ of a molecule is very much larger than kT, which is $\frac{1}{40}$ e.V. at room temperature. In studying nuclear and atomic structure we shall repeatedly seek out the lowest energy of the nuclei or atoms. This concentration of attention on the lowest states is justified by the fact that the energy difference between the lowest energy state of the system and any other possible state is almost always much greater than kT. Therefore only a very small fraction of the nuclei or atoms will be in states other than the state of lowest energy.

2-5 COLLISIONS, MEAN FREE PATHS, AND ATOMIC SIZES

A molecule in a gas at atmospheric pressure spends most of its time moving at a constant speed in a straight line, but it moves only a short distance, the *free path* length, in this line before it collides with another gas molecule. The process of collision occupies only a very short time and conserves energy and momentum. The two molecules start new free paths, to be terminated by second collisions, and so on. The *mean free path* $\bar{L}$ is the average distance between collisions, and many properties of gases depend on $\bar{L}$. Examples are the viscosity, thermal conductivity, and diffusion rate. By measurements of these properties and by the application of the theory of their dependence on $\bar{L}$, the values of $\bar{L}$ can be determined. For a given number of molecules per unit volume, $\bar{L}$ will be large if the molecules are small, since the chance of a collision is reduced as the molecules become smaller. Only one of the experiments that gives information about $\bar{L}$ and molecular sizes will be described. This experiment will first be interpreted in terms of molecular size and later in terms of the mean free path $\bar{L}$.

The size of molecules can be measured rather directly by an experiment employing the apparatus illustrated in Fig. 2-8. An oven S serves as a source of molecules that pass through several slits as a molecular beam. All the molecules of this beam would enter the detector D if it were not for collisions between beam molecules and the gas molecules in the chamber C. Those that do collide will be scattered out of the beam and will not reach the detector. (We assume here that the pressure is low enough so that the chance is slight that a beam molecule will suffer more than one collision; multiple collisions might scatter some molecules back into the beam.) The current to the detector is the number of beam molecules per second that have *not* collided with gas molecules.

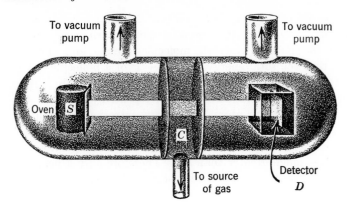

Fig. 2-8. Molecular or atomic beam apparatus for measuring collision cross sections and molecular or atomic sizes.

Figure 2-9 illustrates a beam molecule approaching a section of the gas a distance dx thick. Both beam molecules and gas molecules are considered here as spheres with definite radii r_1 and r_2, respectively. A collision takes place if the center of a beam molecule comes within a distance $(r_1 + r_2)$ of the center of a gas molecule. (The gas molecules are shown as if at rest, but their motion does not change the argument.) The number of gas molecules shown here is the product of N (the number

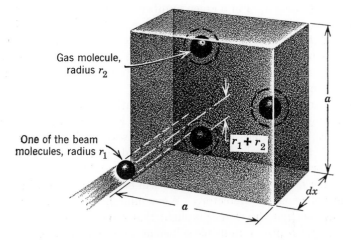

Fig. 2-9. A beam molecule passing through a thickness dx of a gas. It collides with a gas molecule if its center is within a distance $r_1 + r_2$ of the center of a gas molecule.

The mean free path $\bar{L}$ is the average value of L:

$$\bar{L} = \frac{L\phi(L)\,dL}{\phi(L)\,dL} = \frac{\displaystyle\int_0^\infty L e^{-N\pi(r_1+r_2)^2 L}\,dL}{\displaystyle\int_0^\infty e^{-N\pi(r_1+r_2)^2 L}\,dL}$$

If u is substituted for $N\pi(r_1 + r_2)^2 L$, we obtain

$$\bar{L} = \frac{\displaystyle\int_0^\infty u e^{-u}\,du}{N\pi(r_1 + r_2)^2 \displaystyle\int_0^\infty e^{-u}\,du}$$

The integral in the denominator equals unity, and so does the integral in the numerator (the latter integral can be integrated by parts or found in tables). Therefore the mean free path is

$$\bar{L} = \frac{1}{N\pi(r_1 + r_2)^2} \tag{2-22}$$

If the beam particles (radius r_1) are very small compared to the gas molecules (radius r_2), the term r_1 in this equation can be neglected.

It has already been noted that many properties of a gas depend on $\bar{L}$, and therefore the measurement of such properties determines $\bar{L}$ and the molecular size. The molecular beam experiment illustrated in Fig. 2-8 and analyzed in eq. 2-19 can readily be interpreted in terms of the mean

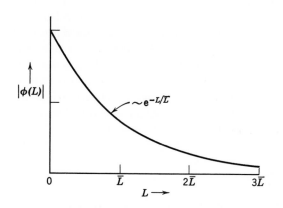

Fig. 2-10. Distribution of free path lengths from eq. 2-21.

free path. When eq. 2-22 is inserted into eq. 2-19, the latter becomes

$$J = J_0 e^{-x/\bar{L}}$$

The molecular beam is attenuated by a factor $1/e = 1/2.718$ in a distance of one mean free path.

The mean free path of a nitrogen or oxygen molecule in air at atmospheric pressure is very short (see prob. 2-17). After the collision that terminates each free path, the molecule has "forgotten" its original velocity (speed and direction). The path of a molecule is illustrated schematically in Fig. 2-11, although of course the real path does not lie in one plane. The distance of the end of the path from the starting point is very much less than the total path length. The rate at which a gas molecule can diffuse from one region to another is therefore very much less than the molecular speed.

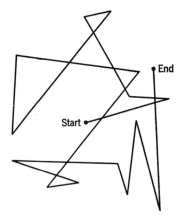

Fig. 2-11. The total path length is much greater than the distance between start and finish.

A numerical example will show how this interruption of the path by collisions is important for the properties of a gas. Suppose a small quantity of H_2S is released at the center of a room. The H_2S molecules move with an rms velocity of 460 m./sec at room temperature. But on the average a given H_2S molecule will have traveled only 0.04 m. from the point of release during the first 60 sec after release. Thus it will be some time before the odor of H_2S is apparent at the edges of the room, even though the speeds of the H_2S molecules are very high.

We shall encounter the concept of a mean free path again when we discuss the conduction of electricity by electrons in solids (Chapter 9) and in gases (Chapter 12). We shall encounter the concept of a collision cross section again in Chapters 12, 13, and 14.

REFERENCES

General

E. H. Kennard, *Kinetic Theory of Gases*, McGraw-Hill, New York, 1938.
R. D. Present, *Kinetic Theory of Gases*, McGraw-Hill, New York, 1958.

Thermal Noise

J. L. Lawson and G. E. Uhlenbeck, *Threshold Signals*, McGraw-Hill, New York, 1950, Chapter 4.

PROBLEMS

2-1.* The speed v_s of sound in oxygen gas (O_2) is 317 m./sec at 0°C. Compute the rms speed of the oxygen molecules from their average energy and compare with v_s. Could v_s be larger than the molecular speed?

2-2. A typical oil drop in the experiment sketched in Fig. 1-1 has a mass of 10^{-12} kg. What is its average kinetic energy at 290°K in electron volts? What is its rms speed? Could its thermal motion be observed with a microscope? Would its thermal motion seriously affect the oil-drop experiment with velocities of rise and fall like those in prob. 1-1?

2-3.* A television receiver at room temperature accepts and amplifies frequencies between 54 and 60 megacycles. Its input impedance is resistive and equals 150 ohms. What is the rms value $V_n = \{ \overline{V^2} \}^{1/2}$ of the thermal noise voltage at the input? A typical input signal from an antenna is $V_s = 100$ microvolts. With this input, what is the voltage signal-to-noise ratio V_s/V_n? (The *power* signal-to-noise ratio V_s^2/V_n^2 is more commonly used, and is usually expressed in decibels. "Channel 2" is 54 to 60 megacycles, and the *total* receiver noise in a typical receiver is about 8 decibels greater than thermal noise.)

2-4. It might be thought that the value of 150 ohms in the preceding problem is arbitrary and that the noise limitation is therefore not fundamental. Show that the noise *power* is independent of R; noise thus sets a limit to the smallest power that can be detected. [The power delivered from a source (such as an antenna) of impedance R is maximized if the load is also R; the value of 150 ohms in prob. 2-3 came from the parallel combination of a 300-ohm antenna and a 300-ohm receiver.]

2-5.* Calculate the fraction of molecules in a gas that have kinetic energies greater than $10kT$. *Hint:* In the integrand $K^{1/2}$ can be set equal to $(10kT)^{1/2}$ to a good approximation.

2-6. Show that the Maxwell distribution of the x components of velocities (eq. 2-11) gives $\frac{1}{2}kT$ for the average kinetic energy $\frac{1}{2}Mv_x^2$ associated with the motion in the x-direction.

2-7. Is $\overline{(1/v)}$, the average value of $1/v$, the same as $1/\bar{v}$ for a distribution of velocities? To answer this question, consider a group of objects half of which are 1 m. long and half 2 m. long. What is the average length? What is the average of the reciprocal of the length?

2-8.* Use tables of definite integrals to find the area under the curve of Fig. 2-2 from $K = 0$ to $K = \infty$. Give a rough check of your answer by graphically estimating the area under the curve in Fig. 2-2 and allowing for the exponential tail that is not shown.

2-9.* Let $\phi(v)\,dv = dN$ be the number of molecules with speeds (independent of direction) between v and $v + dv$. Obtain $\phi(v)$ from eq. 2-9 by noting that $\phi(v) = dN/dv$ and $\phi(K) = dN/dK$; since $K = \frac{1}{2}Mv^2$, dK/dv is easily obtained and thereby dN/dv.

2-10. Calculate the average value of the molecular speed v from the distribution function of prob. 2-9.

2-11.* In a molecular beam experiment, the number of molecules crossing a plane perpendicular to the direction of the beam per second with speeds between v and $v + dv$ is proportional to $v\phi(v)\,dv$, where $\phi(v)$ is the distribution found in prob. 2-9; this can be verified by an argument like that associated with Fig. 2-1. Calculate the average value $\overline{K} = \frac{1}{2}M\overline{v^2}$ of the molecules in the beam.

2-12. An experiment as illustrated in Fig. 2-5 is being performed to investigate the Maxwell distribution. The drum is 0.27 m. in diameter and rotates at 12,000 rpm. The source is an oven that contains zinc at 300°C. The point A (Fig. 2-5) is determined when the drum is stationary. How far from A will those zinc molecules with velocities such that $\frac{1}{2}Mv_x^2 = 2kT$ strike the plate?

2-13.* Compute the variation of pressure p with elevation h above the earth's surface. Assume an average molecular weight of 29 for air and assume a constant temperature $= 0°C$. What is p in atmospheres at 10,000 ft elevation? at 35,000 ft? (This assumption of an isothermal atmosphere is a poor one, and so we obtain only an approximation to the observed p vs. h.)

2-14. Use the assumptions of prob. 2-13 to estimate p in atmospheres at an elevation of 170 kilometers, which is about the minimum height at which an earth satellite can orbit without excessive air resistance. Estimate also the number of molecules per cubic meter at this height, using the fact that there are 2.7×10^{25} molecules/m.[3] at 0°C and $p = 1$ atmosphere. Discuss the meaning of "temperature" for densities as small or smaller than this.

2-15. Consider a mixture of two perfect gases. Show by analysis similar to Sec. 2-4 that in equilibrium in the earth's gravitational field an expression like eq. 2-17 gives the partial pressure of each gas.

2-16.* A balloon filled with hydrogen will, of course, rise to the ceiling of a room and remain there. If a small amount of hydrogen is released into the air of a room, will it rise to the ceiling and remain there? To answer this question, assume that equilibrium has been established (which may take several hours), calculate by the Boltzmann distribution the variation of pressure of hydrogen from the floor to a ceiling 2.5 m. higher, and compare with the similar calculation for nitrogen. Let $T = 290°K$. (This shows the relative importance of diffusion and of stratification by the gravitational force.)

2-17. Calculate the mean free path $\overline{L}$ for a nitrogen molecule in nitrogen gas at atmospheric pressure. Let $T = 290°K$.

2-18.* Calculate the mean free path $\overline{L}$ for a nitrogen molecule in nitrogen at a pressure of 10^{-6} mm of mercury (1 atmosphere equals 760 mm of mercury) and $T = 290°K$. If a vacuum tube of diameter 0.02 m. contains nitrogen at this pressure, is a free path usually terminated by a collision with the tube wall or with another molecule?

2-19.* A nitrogen molecule with energy $\frac{3}{2}kT$, $T = 290°$K, is initially directed horizontally and traverses 10^{-7} m. before its next collision with another gas molecule. What is the downward deflection produced by the gravitational field in this distance?

2-20.* A molecule in a good vacuum system, such as a molecular beam apparatus, frequently has a mean free path long compared to the dimensions of the system. Suppose that this is the case and that a horizontal beam of potassium atoms from an oven at $T = 500°$K traverses a tube 0.50 m. long. What is the downward deflection of a beam atom with $K = 2kT$ produced by the gravitational field in this distance?

2-21. One type of detector in a molecular beam experiment is a hot filament that is cooled by the arrival of the beam molecules; the resistance of the filament depends on its temperature and hence varies with the beam intensity. Explain *microscopically* (i.e., in terms of the motions of beam molecules and atoms in the solid filament) how this cooling occurs.

3
Nuclei and Atoms

3-1 INTRODUCTION

The chief conclusion of this chapter is that the problem of the interactions among electrons, protons, and neutrons in the atom can be divided into two separate problems: (1) The binding together of protons and neutrons into the tiny nucleus at the center of the atom. (2) The motions of the electrons around the nucleus. The division can be achieved because: (1) The energies of interaction between protons and neutrons in the nucleus are very much greater (millions of electron volts) than the energies of interaction between electrons and nuclei (a few electron volts to 10^5 electron volts). (2) The nucleus is much smaller than the atom. Therefore the size and energy scale of atomic experiments is such that the nuclei remain unchanged in an experiment concerned with atomic phenomena; we can consider the nucleus as a heavy particle and can ignore its size and internal structure. Nuclear experiments, on the other hand, involve such large energies and small distances that the presence and properties of the electrons surrounding the nuclei are of little consequence. Therefore we can deal separately with *atomic* physics (Chapters 4 to 7) and *nuclear* physics (Chapters 13 and 14).

The characteristic sizes of nuclei and energies of nuclear reactions are described in the present chapter. The characteristic sizes of atoms were discussed in Chapter 2. The characteristic energies of atomic reactions (such as ionization or excitation) will be described in the early sections of Chapter 4; meanwhile, the order of magnitude for the most useful atomic reactions can be inferred from the heats of chemical reactions, which are a few electron volts per atom (see probs. 3-1 and 3-2).

The first topic of the present chapter is the preliminary investigation of the structure of the atom. The conclusion is that an atom consists of a tiny core, the nucleus, surrounded by one or more electrons. Most of the remainder of this book is concerned with the spatial distribution, energies, and other properties of these electrons. The later parts of the present chapter discuss properties of the nucleus. Nuclear charge and size are considered in Sec. 3-2, nuclear masses in Sec. 3-3, and nuclear binding energies in Sec. 3-4. Only a brief introduction to nuclear physics is given here. We shall return to the study of nuclear physics in Chapter 13 after developing the laws of quantum physics in Chapters 4, 5, and 6. The physics presented in those three chapters enables us to provide a much more satisfactory explanation of nuclear phenomena than we could give in the present chapter.

3-2 THE NUCLEAR ATOM

The first fact of value in understanding the structure of atoms is that atoms are electrically *neutral*. They can be ionized by removing one or more electrons, but in their normal states they are neutral. If they were not, enormous fields would be exhibited by laboratory-size objects. For example, a sphere of iron weighing 1 gram would have a field of 4×10^{15} volts/m. at its surface if the number of electrons were only 0.01% more than the number of protons in it.

We shall assume that atoms are composed of at least electrons and protons, since these particles are the products of the ionization of the hydrogen atom. Later it will be demonstrated that neutrons are also present and are always found close to the protons, the protons and neutrons together constituting the nucleus.

We learned in Sec. 2-5 that a typical atom is about 3×10^{-10} m. in diameter. An atom contains at most a few hundred particles like protons and electrons. We learned in Chapter 1 that these particles are smaller than 10^{-14} m. in diameter. It is therefore apparent that an atom is mostly space and that the particles involved are not packed tightly together to form it.

Two conceivable structures of an atom are illustrated in Fig. 3-1. In (a) the positive charges are distributed uniformly, and in (b) they are concentrated into a core that is called the *nucleus*. A third conceivable structure with the electrons concentrated in the center is not consistent with experiment, since with that structure the products of ionization would be a proton and a negative ion, rather than an electron and a positive ion (as observed).

Experiments have decided in favor of the nuclear atom as sketched in Fig. 3-1b. The original experiment demonstrating this was the *Rutherford scattering* experiment. The basic idea of this experiment is to shoot energetic charged particles at a group of atoms and to examine the distribution in angle of the particles after passing through the atoms. From the deflections observed, inferences can be made about the structure of the atoms. The particles used in the original experiment were *α-particles* with kinetic energies of 4 to 8 M.e.V. α-particles are doubly charged helium ions, which have four times the mass and twice the charge of protons. They would be deflected through only very small angles by the kind of atom illustrated in Fig. 3-1a, since even a nearly head-on collision with a single proton would not deflect the α-particle appreciably. The proton would move rapidly away, like a light ball struck by a heavy ball. On the other hand, the atom of Fig. 3-1b makes large-angle deflections of the α-particles possible. Whenever one of the α-particles comes close to the nucleus, it is deflected by the electrostatic repulsion of *all* the protons of the nucleus, which is a much larger force than that of just one. Furthermore, the nucleus is heavier than the α-particle, and so the nucleus does not move rapidly away from the α-particle. This collision is like a light ball colliding with a heavy ball: The light ball suffers a large change in its direction after a collision.

The theory of the scattering of α-particles by a nucleus was developed by Rutherford. This theory assumes that the force between an α-particle and a nucleus is merely the electrostatic repulsion (Coulomb force) between two particles, one of charge $+2e$ (the α-particle) and the other

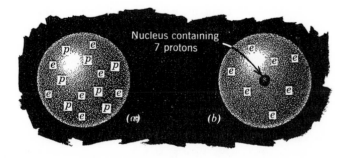

Fig. 3-1. Possible structures of an atom. (*a*) Uniform model. (*b*) Nuclear model.

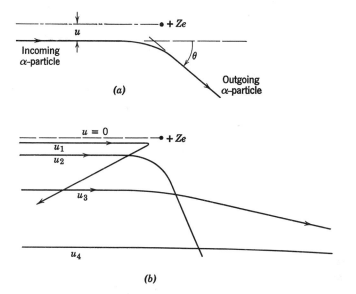

Fig. 3-2. (a) Geometry of the deflection of an α-particle by a nucleus. (b) Trajectories for various values of the impact parameter u.

of charge $+Ze$ (the nucleus, with Z protons). The force is therefore

$$F = \frac{2Ze^2}{4\pi\epsilon_0 r^2} \tag{3-1}$$

where r is the distance between the two particles.

The geometry of the collision is shown in Fig. 3-2a. The aim is such that the α-particle would pass a distance u away from the nucleus if there were no repulsion between it and the nucleus. u is called the *impact parameter*. Figure 3-2b illustrates how the angle θ of deflection depends on u. Small u (nearly a direct hit) means large θ; large u means small θ. The relation between u and θ can be obtained * from ordinary mechanics. It is

$$\operatorname{ctn}\frac{\theta}{2} = 4\pi\epsilon_0\,\frac{Ku}{Ze^2} \tag{3-2}$$

where K is the initial kinetic energy of the α-particle.

---

* See R. B. Lindsay, *Physical Mechanics*, Van Nostrand, New York, 2nd Ed., 1950, pp. 87–89; M. Born, *Atomic Physics*, Blackie, London, 6th Ed., 1956, Appendix IX; R. M. Eisberg, *Fundamentals of Modern Physics*, Wiley, New York, 1961, pp. 100–106. It is assumed that the target nucleus is so massive compared to the α-particle that it can be regarded as fixed.

We cannot choose a particular u and measure θ. If we could, we could test eq. 3-2 directly and determine whether the nuclear atom concept was correct. But u must be very small (of the order of 10^{-13} m.) in order to give an appreciable θ, and we cannot possibly obtain a beam this small nor can we fix the position of the target atom so precisely. The best we can hope to do is to compute the fraction of the α-particles which have an impact parameter between u and $u + du$, and to compute from this the fraction of deflected particles which have an angle of deflection between θ and $\theta + d\theta$. (This is the common situation in experimental nuclear physics; nuclear properties are usually inferred from the

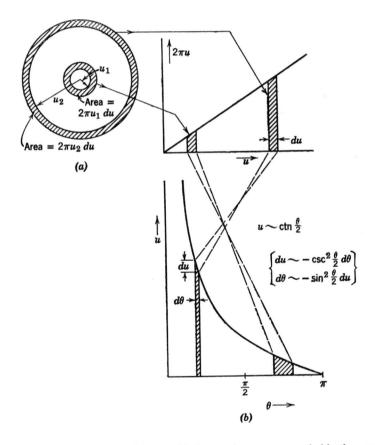

Fig. 3-3. Distribution of u values. (*a*) Large u's are more probable than small. (*b*) If a particle enters in a range du, it is deflected into a range $d\theta$. A certain range du becomes a smaller range $d\theta$ at small θ than at large θ.

comparison between an observed angular distribution and a distribution calculated from a tentative model of the nucleus.)

Large values of u are more probable than small values, as indicated in Fig. 3-3a. If there are n particles incident during the experiment and N nuclei per square meter in the foil of targets, the number of particles dn that will have an impact parameter between u and $u + du$ is

$$dn = (2\pi u\, du)(Nn) \qquad (3\text{-}3)$$

The problem here is just as if we had N holes per square meter in a screen, each hole of area $2\pi u\, du$ m.^{2} If n particles are fired at random at the screen, the number dn that go through holes is $(2\pi u\, du)Nn$. (This problem is very similar to the cross section problem discussed in conjunction with Fig. 2-9.)

In order to find the number dn that will be deflected into an angle between θ and $\theta + d\theta$, we must change eq. 3-3 into an expression in terms of θ, where θ and $d\theta$ come from eq. 3-2. First we obtain du from eq. 3-2:

$$du = \frac{Ze^2}{8\pi\epsilon_0 K}\left(-\csc^2\frac{\theta}{2}\right)d\theta$$

The minus sign appears because as u increases θ decreases; see Fig. 3-3b for a graphical representation of this step. dn can be found by inserting this expression for du and u from eq. 3-2 into eq. 3-3:

$$dn = \frac{-\pi Z^2 e^4 Nn}{(16\pi^2\epsilon_0{}^2)K^2}\,\text{ctn}\,\frac{\theta}{2}\csc^2\frac{\theta}{2}\,d\theta \qquad (3\text{-}4)$$

The ratio $|dn|/d\theta$ is plotted in Fig. 3-4. It rises very steeply as $\theta \to 0$, since there are many more wide misses (large values of u) than near hits and since u increases sharply as $\theta \to 0$, as shown by eq. 3-2.

The geometry of a typical experiment is shown in Fig. 3-5. The detector is on a platform that pivots about O as a center. All those particles that enter an area A are counted by the use of a photographic plate or a particle counter like those that will be described in Sec. 14-6. Equation 3-4 predicts the number of particles that emerge through the annular region in Fig. 3-5. The area of this region is

$$(2\pi R \sin\theta)(R\, d\theta)$$

The fraction f of these that enter the counter is therefore

$$f = \frac{A}{2\pi R^2 \sin\theta\, d\theta} \qquad (3\text{-}5)$$

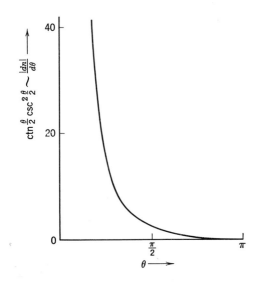

Fig. 3-4. Relative numbers of α-particles scattered at different angles.

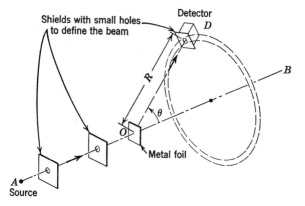

Fig. 3-5. Geometry of an α-particle-scattering experiment. Any α-particle scattered at an angle θ will go through the annular region between θ and $\theta + d\theta$ (dashed ring). Only a fraction f of these will go into the detector, which has a circular opening of area A.

If we combine eq. 3-4 and eq. 3-5, we find the number $\Delta n = |dn|f$ that enter the counter:

$$\Delta n = \frac{Z^2 e^4 N n A}{(32\pi^2 \epsilon_0{}^2) R^2 K^2} \operatorname{ctn} \frac{\theta}{2} \csc^2 \frac{\theta}{2} \csc \theta$$

This expression can be simplified by writing $\sin \theta = 2 \sin \frac{\theta}{2} \cos \frac{\theta}{2}$:

$$\Delta n = \frac{Z^2 e^4 N n A}{(64\pi^2 \epsilon_0{}^2) R^2 K^2 \sin^4 \dfrac{\theta}{2}} \tag{3-6}$$

Figure 3-6 compares data for silver with the theoretical expression, eq. 3-6. A logarithmic scale for the ordinate has been used because of

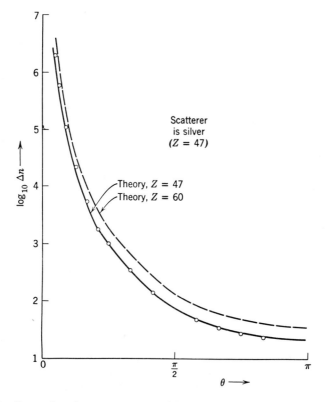

Fig. 3-6. Comparison between an α-particle-scattering experiment (data points) and theory. The solid line is eq. 3-6 with $Z = 47$. The dashed line is eq. 3-6 with $Z = 60$. (Data from Rutherford, Chadwick, and Ellis, *Radiations from Radioactive Substances*, Cambridge University Press, Cambridge, 1951.)

the enormous range of Δn values covered. All the quantities on the right side of eq. 3-6 can be observed directly except Z. Different Z's can be tried until agreement with experiment is secured. [Changing Z moves the curve up or down in Fig. 3-6 without changing its shape, because Z^2 is a factor multiplying the rest of the expression for Δn, and therefore $\log_{10} Z^2$ appears in $\log_{10}(\Delta n)$ as an additive constant.]

The agreement of the shape of the curve with the experiments is excellent over a range of Δn values such that the largest is $\sim 10^5$ times as large as the smallest, which constitutes proof that the nuclear model of the atom (Fig. 3-1b) is the correct model. Furthermore, the number of nuclear charges Z that gives the best fit to the data is 47. We therefore conclude that silver has a small, heavy nucleus with a charge $+47e$ and has 47 electrons surrounding the nucleus. The chemical properties of an atom depend on the number of electrons it possesses, as will be shown in Chapter 6. Therefore atoms of a chemical species (e.g., silver) are specified by giving the value of Z (e.g., 47), which is called the *atomic number*. Other, more accurate, methods of measuring the atomic number will be described in Chapter 6, and they agree with this Rutherford scattering method.

Disagreement between experiment and eq. 3-6 is expected if the incoming particles actually strike the nucleus. In deriving eq. 3-6 we have assumed that the nucleus acted just like a point charge, with zero dimensions. Scattering experiments show that this theory is in agreement with experiment for values of K and u such that the incoming particles approach to within 2×10^{-14} m. of the centers of silver nuclei. Thus a silver nucleus must be no larger in radius than 2×10^{-14} m., which is so small compared to the size of an atom (about 10^{-10} m.) that it can be considered as a *point* in our study of the physics of atomic structure.

In this discussion we have ignored the electrons. As explained before, the deflections produced by a diffuse distribution of small-mass particles would all be so small that they would not be observable. It might be thought that the nuclear charge $+Ze$ would be partially neutralized by electrons, and this neutralization becomes important when very low-energy α-particles are used with their correspondingly large distances of closest approach to the target nuclei. But for the usual α-particle energies, the deflection occurs so close to the nucleus that the chance of an electron's being inside the region where the deflection occurs is negligible. This is the common situation in the scattering experiments of nuclear physics.

This section can be summarized as follows: The scattering of high-energy particles demonstrates that the atom is composed of a heavy nucleus with charge $+Ze$ surrounded by Z electrons, where Z is the atomic number. The nucleus is so small that it can be considered as a point in studies of atomic phenomena.

3-3 MASS SPECTRA, ISOTOPES, AND ATOMIC MASSES

The mass of an atom is only slightly greater than the mass of its nucleus. The Z electrons, each with mass $1/1836$ of the proton mass, contribute only a small fraction to the total mass of the atom. Furthermore, since the mass of the electron is known, the nuclear mass can be calculated from the atomic mass, and vice versa.

The most convenient method of measuring an atomic mass is by measuring the q/M of the ionized atom, where q is its charge. If the ion is singly charged (i.e., it has lost *one* electron and $q = e$), we must then correct the measured M by adding one electron mass m. Similar corrections can be made for more highly ionized atoms. Since the instruments to be described measure q/M and not M, it might be thought that an ambiguity would arise as to whether $q = e$, or $2e$, or $3e$, etc. No such ambiguity arises because doubly and more highly charged ions are always accompanied by singly charged ions, and therefore the smallest q/M of a series e/M, $2e/M$, etc., corresponds to the singly charged ion. Furthermore, by changing the conditions in the ion source it is possible to get a beam composed exclusively of singly charged ions. For example, if the ion source consists of energetic electrons colliding with gas atoms, the electron energy can be decreased until only singly charged ions are formed. In the following discussion we shall assume that all ions are singly charged except in a few cases where the charge will be specified.

Instruments that permit measurement of q/M of ions, and hence of atomic masses, are called *mass spectrometers*. There is a large variety of these instruments, but all instruments fall into one or the other of two quite separate classes: (1) *Single-focusing* instruments. These instruments measure atomic or molecular masses with low precision (of the order of 0.1 to 1%) but measure accurately the relative amounts of the different atoms or molecules present in a sample. A typical instrument is the 180° mass spectrometer, which will be discussed below. A typical application is the analysis of gas mixtures such as those occurring in petroleum chemistry. (2) *Double-focusing* instruments. These instruments measure masses to a very high precision (of the order of 1 part per million). A typical instrument, developed by Bainbridge and Jordan, will be described later in this section. Such instruments are used to obtain precise data on the atomic masses. These data, which will be used in Sec. 3-4 and Chapters 13 and 14, are indispensable to an understanding of the nucleus.

(a) Single-focusing mass spectrometers

A typical single-focusing mass spectrometer is illustrated in Fig. 3-7. The glass vacuum tube is placed in a uniform magnetic field perpendicular to the plane of the tube. The three principal parts of the instrument are the *ion source*, the *sorting chamber*, and the *ion detector*. The effect of the magnetic field is slight in the source and the detector regions because the path lengths are short in these regions. The magnetic field in the sorting region forces an ion to move in a circular path, the radius of which can be obtained by combining eq. 1-12 and eq. 1-8:

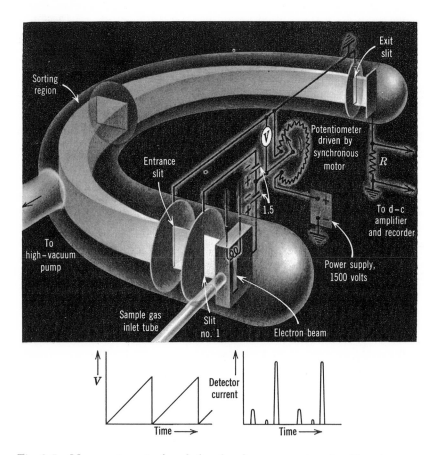

Fig. 3-7. Mass spectrometer for relative abundance measurements. There is a magnetic induction ℬ normal to the plane of the semicircular glass tube and directed upward. The plots at the bottom show how the accelerating voltage V is swept and how the plot of detector current vs. time presents the mass spectrum.

$$r = \frac{1}{\mathfrak{B}}\left(\frac{2MV}{e}\right)^{\frac{1}{2}} \tag{3-7}$$

Here $\mathfrak{B}$ is the magnetic induction, M the mass of the ion, and V the potential difference through which it has been accelerated before entering the sorting region. If the combination of $\mathfrak{B}$ and V is just right for a particular ion, r as computed above will be one-half the distance between entrance and exit slits, and the ion will go through the exit slit and will be detected. Since all the other quantities can be measured directly, M can be determined.

The *ion source* consists of an electron beam that ionizes some of the atoms of the gas in the small metal box, which is nearly closed at the left. Electrons are emitted from the coiled filament (Fig. 3-7) and formed into a beam by appropriate electrodes (not shown in the illustration). Ions formed have kinetic energies of 0 to 1 or 2 e.V. at the place of ionization, and they are accelerated toward slit 1 by a small electric field. Those ions that go through slit 1 are then accelerated in a much stronger field. The total acceleration is through a potential difference V of a few thousand volts, and hence the ion kinetic energy is nearly equal to V electron volts as it enters and traverses the sorting chamber. A motor-driven potentiometer provides the "sawtooth" $V(t)$ illustrated.

The *detector* begins with an electrode that collects all ions passing through the exit slit. The potential difference across the high resistance R (10^{10} or 10^{11} ohms) is amplified by a sensitive d-c amplifier and applied to a strip-chart recorder. Because V is proportional to time during any one cycle of the sawtooth, the time scale of the strip chart is also a voltage scale. Hence the trace on the chart gives ion current plotted vs. ion energy V.

The *sorting chamber* is a region of constant magnetic induction and zero electric field. The pressure is maintained very low in this region so that the probability that an ion will collide with a gas atom is very small (the mean free path is much longer than the path the ion follows). Only those ions whose circular paths have diameters equal to the separation of entrance and exit slits will pass through the exit slit, and therefore the instrument sorts ions according to their masses. All those ions that have a particular mass M will go through the exit slit when $\mathfrak{B}$, V, and M satisfy eq. 3-7. Since the recorder chart presents ion current vs. V, it also gives ion current (proportional to relative abundance of ions) as a function of ion mass M. Of course we must calibrate the mass scale either by knowing $V(t)$, $\mathfrak{B}$, and r accurately or else by using atoms of known mass intentionally mixed into the sample gas in order to give calibration points.

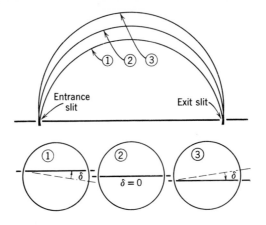

Fig. 3-8. Direction focusing of the 180° mass spectrometer.

Figure 3-8 shows the focusing property of this sorting chamber by exhibiting the paths of three ions of the same mass (and hence the same r) that entered the sorting chamber at slightly different angles. All go through the exit slit even though their paths differ considerably. The lower part of Fig. 3-8 shows how this focusing property arises. The circles represent the paths of each ion, and all the circles have the same radius. For path 2, the slits lie on a diameter of the circle. For paths 1 and 3, the slits lie on chords making a small angle with a diameter. Since the diameter of a circle is the chord of maximum length, chords making small angles with a diameter have lengths practically equal to a diameter.

The 180° deflection in this mass spectrometer was chosen to obtain this focusing property. Other types of mass spectrometers have other methods of focusing ions with different initial directions (but the same mass), but such focusing must always be provided. Without it we should be unable to obtain an appreciable current of ions through the instrument without an intolerable loss in resolution (the ability to distinguish ions of different mass). There are many possible arrangements of ion paths and magnetic fields that accomplish this focusing. We have described only a particular type, the 180° or *Dempster* mass spectrometer.

(b) Interpretation of mass spectra

A typical mass spectrum is illustrated in Fig. 3-9. Ion current is plotted as a function of ion accelerating voltage V at constant $\mathcal{B}$; each point on the abscissa scale corresponds to a particular mass M, according to eq. 3-7.

The only gas present in the ion source used to obtain Fig. 3-9 was neon ($Z = 10$), yet the mass spectrometer shows that there are three different masses present. Thus the mass spectrometer demonstrates the existence of *isotopes*, nuclei of different masses M but the same charge Ze. The masses of the three neon isotopes are approximately 20, 21, and 22 times the proton mass. The chemical properties of neon atoms with any one of these three nuclei are the same, since they have the same Z and hence the same number of electrons. Therefore isotopes cannot be separated chemically. There are methods of effecting partial separation by physical means, and these partial separations can be repeated over and over to provide nearly pure isotopes. The mass spectrometer can be used to separate very tiny quantities of isotopes (see prob. 3-16).

The relative abundance of the three isotopes in natural neon is also indicated in Fig. 3-9. The atomic weight of natural neon is evidently a weighted average of the three masses, each weighted according to its relative abundance (see probs. 3-13 and 3-14). Every element has several isotopes, but for some elements only one or even none is stable enough to occur in nature. (None is stable for $Z = 43$, $Z = 61$, and $Z > 83$.)

A specific kind of nucleus is referred to as a *nuclide*. A nuclide is thus a nuclear species with a particular mass and charge. The masses of nuclides are conveniently and conventionally measured in *atomic mass units* (amu). One amu is defined as exactly one-twelfth of the mass of

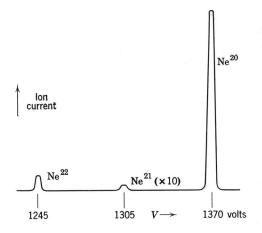

Fig. 3-9. Recorder trace of ion current as a function of accelerating voltage that shows the three isotopes of neon. (The amplifier gain was increased by a factor of 10 in order to present the middle peak.)

the most abundant isotope of carbon; this carbon nuclide thus has by definition a mass of precisely 12.000000 amu.*

Appendix C presents a partial but representative list of nuclides. Study of this list reveals the interesting feature that the masses of all nuclides are close to whole numbers of amu. This fact provides an especially convenient way of specifying a particular nuclide: We call the nearest whole number to the mass in atomic mass units the *mass number*, and we use the symbol A for it. Thus the neon isotopes (the nuclides with $Z = 10$) are $A = 20$, $A = 21$, and $A = 22$. The conventional way of specifying A for a nuclide is to write $_Z$(chemical symbol)^A; for example, the first neon isotope is designated $_{10}Ne^{20}$. These symbols are like the ordinary chemical symbols but give the additional information about mass. Giving both Z and the chemical symbol is convenient but unnecessary, and the subscript Z is frequently omitted. If the atoms are bound together into molecules, we write in the number of atoms per molecule in the usual way. For example, $_8O_2^{16}$ (or O_2^{16}) is the commonest molecule of oxygen, but $_8O^{16}{}_8O^{18}$ (or $O^{16}O^{18}$) molecules are also present in natural oxygen.

Some frequently used masses are

Electron	0.00055 amu	Neutron, $_0n^1$	1.00866 amu
Proton	1.00727 amu	C^{12} (by definition)	12.00000 amu
Hydrogen atom, $_1H^1$	1.00782 amu	1 amu $= 1.6605 \times 10^{-27}$ kg	

In our work, the principal contributions of the single-focusing type of mass spectrometer have been to demonstrate the existence of isotopes and to permit the measurement of their relative abundance. In chemical engineering, the mass spectrometer is a very valuable instrument for analyzing gas mixtures. It is frequently faster and more convenient to analyze gases with a mass spectrometer than by ordinary chemical analysis. The instrument is widely applied, especially in the petroleum-products industry.

* Until 1960–1961, the scale of atomic mass units was defined by setting the mass of the most abundant oxygen isotope equal to precisely 16.000000 amu. Also until 1960–1961, the chemical atomic weight scale was defined by setting the average mass of the naturally occurring distribution of the three oxygen isotopes equal to precisely 16.0000. Thus there was a small difference (a factor of about 1.0003) between the scales; this was a source of some confusion. In 1960–1961 the C^{12} scale was adopted for both atomic weights and atomic masses, and the atomic weights are now simply the weighted averages of the appropriate isotopic masses. The reader is warned that all publications prior to 1961, and some since, are in terms of the oxygen scales. Although the change is not important for chemical calculations based on atomic weights, it is quite important for nuclear masses.

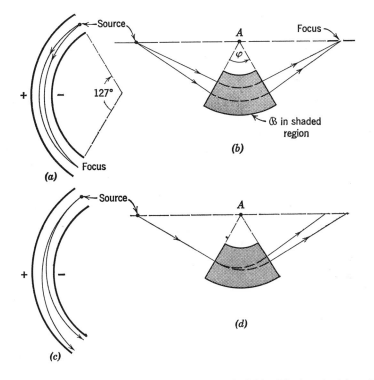

Fig. 3-10. A 127° electric field and sector magnetic field. The ions in (*a*) and (*b*) have the same initial velocities but different initial directions. The ions in (*c*) and (*d*) have the same initial directions but different initial velocities.

(c) Precision mass spectrometers and atomic masses

We shall next discuss the second type of mass spectrometer, which is capable of making precision mass measurements. Since the distribution of initial ion velocities (as well as the slit width) contributes to the widths of the lines in the mass spectrum in an instrument like that of Fig. 3-7, focusing in both direction and initial velocity (as well as reduction of slit width) is necessary to attain higher resolution. A typical instrument accomplishing this double focusing is illustrated in Fig. 3-11.

We shall first consider the components of this instrument that are illustrated in Fig. 3-10. Two concentric cylindrical electrodes with an electric field as shown provide direction focusing of positive ions at a point 127° (or $\pi/2^{1/2}$ radians) from the entrance slit (Fig. 3-10*a*). But if all the entering ions do not have the same velocity, their paths are different even if they have the same initial direction (see Fig. 3-10*c*).

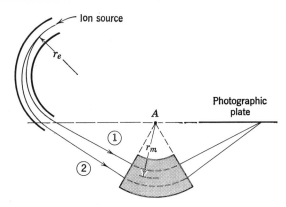

Fig. 3-11. Double-focusing mass spectrometer of Bainbridge and Jordan. The two ion paths shown have the same initial direction and the same mass but different initial velocities. If they had the same mass but different initial directions they would still be focused to the same point on the photographic plate.

The sector magnetic field of Figs. 3-10b and 3-10d also has direction focusing if the source, the apex A of the edges of the magnetic field, and the exit slit are all on the same straight line and if the ion paths are approximately perpendicular to the edge of the magnetic field. (The sector magnetic field by itself is frequently used as a direction-focusing mass spectrometer; the 180° instrument is a special case of the general focusing property stated here in which the angle θ of the sector is 180°.) Figure 3-10d illustrates the fact that the sector magnetic field gives different paths for ions of different initial velocities.

The double-focusing instrument of Fig. 3-11 combines the components of Fig. 3-10 in such a way that focusing in both direction and initial velocity occurs. Direction focusing is achieved by using the focus of the 127° electric field as the source for the sector magnetic field. Velocity focusing is achieved if the radius in the electric field r_e equals the radius in the magnetic field r_m. We shall not try to prove this focusing property, since it would take us far afield, but it is not difficult to show that the effects of the two fields on the final position of the ion are oppositely directed (prob. 3-17), that is, that focusing is possible by cancellation.*

* The double-focusing principle is quite similar to the principle of an achromatic lens in optics. Such a lens consists of two elements, each of which possesses chromatic aberration (that is, its focal length varies with wavelength). The combination is selected in such a way that the chromatic aberration of one lens just cancels that of the other.

An instrument incorporating this double-focusing principle is the mass spectrometer of Bainbridge and Jordan illustrated in Fig. 3-11. The ion source is similar to the source shown in Fig. 3-7. The detector is a photographic plate at a position and in an orientation such that ions are brought to a focus at the plate. After an exposure to the ion beam the plate is developed and the positions of lines on the plate are observed with a traveling microscope.

In order to use a double-focusing instrument to measure precisely the masses of nuclides it would ordinarily be necessary to make precision measurements of $\mathcal{B}$, V, and the positions of slits, electrodes, magnet pole pieces, and the photographic plate. All these precision measurements are difficult, but these difficulties can be avoided by the method of *doublets*. This method takes advantage of the fact that all nuclide masses are nearly integers. Suppose, for example, that we seek a precision determination of the mass of Li^6. We produce both singly charged Li^6 and doubly charged C^{12} ions in the same instrument. The resulting lines on the photographic plate lie very close together, since the ratio of charge to mass of $(C^{12})^{++}$ is nearly the same as that of $(Li^6)^+$. If these ratios were identical, the lines would coincide. We can determine by a separate experiment the approximate (perhaps to 0.1%) calibration of mass as a function of distance along the photographic plate; for example, nuclei whose masses were known to a precision of 0.1% could be used to calibrate the plate. Therefore we can compute from the measured separation of the two lines of the doublet the mass difference

$$Li^6 - \tfrac{1}{2}C^{12} = 0.01513 \text{ amu}$$

$$Li^6 = 6.00000 + 0.01513 = 6.01513 \text{ amu}$$

Once Li^6 has been determined, it can be used as a further "stepping stone" in order to measure other masses. Since carbon compounds are available with a wide variety of masses extending to hundreds of mass units, C^{12} is especially well suited to be the standard mass. The method of doublets provides a gain in precision of about a factor of 1000 over direct measurements.* A doublet that has been measured to a very high

* This method is an example of a widely applicable technique in physical measurements: An instrument is devised to measure *directly* the *difference* between an unknown and a standard. The scale reading of this instrument is subject to fluctuations and inaccuracies that may be, for example, 1% of the scale reading. If the unknown and the standard differ by only 0.1% in the property under investigation, the unknown can thus be measured to 0.001%. But if the unknown and standard were measured separately, a precision of only 1% would have been obtained in measuring the unknown.

He^4 H_2^2

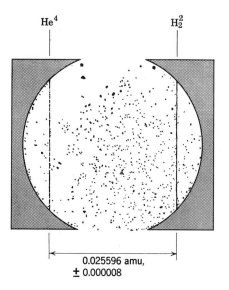

0.025596 amu,
± 0.000008

Fig. 3-12. Photograph of the $_2He^4$ and $_1H_2^2$ doublet made by H. Ewald with a double-focusing mass spectrometer. (From *Experimental Nuclear Physics*, Vol. 1, edited by E. Segrè, Wiley, New York, 1953.)

precision is reproduced in Fig. 3-12, which shows a small part of the photographic plate of a double-focusing mass spectrometer.

A table of nuclides is given in Appendix C. Many of the numbers in this table were obtained by precision mass spectrometry. It will become apparent in Sec. 3-4 why so much precision is desirable in measuring masses. Meanwhile, the example of the $Li^6 - \frac{1}{2}C^{12}$ doublet establishes an important conclusion: Although there are the same numbers of protons and of neutrons in two Li^6 nuclei as in one C^{12}, the mass of Li_2^6 differs from that of C^{12}. Thus the mass of a nucleus is *not* the sum of the masses of its protons and neutrons, a fact that can easily be verified for any of the masses tabulated in Appendix C.

3-4 NUCLEAR BINDING ENERGIES

A nucleus of atomic number Z and mass number A is composed of Z protons and $(A - Z)$ neutrons. It might be thought that A protons and $(A - Z)$ electrons would be the constituents. There are excellent reasons to favor the proton and neutron combination, but their explanation requires a knowledge of the content of Chapters 5 and 6. This explanation is therefore delayed until we return to nuclear physics in Chapter 13, where it will be shown that there is not room for electrons in the nucleus, and where additional arguments will be presented to show that there are no electrons in the nucleus. The constituents of the nucleus are called *nucleons;* a nucleon is either a proton or a neutron.

Two questions immediately arise about nuclear structure: What forces hold the protons and neutrons together? Why is the mass of a nucleus not exactly equal to the mass of Z protons and $(A - Z)$ neutrons? We shall discuss the answers to these questions, but not really answer them, by discussing the binding energies of nuclei.

Nuclear forces are imperfectly understood at the present stage of the development of physics. Several facts about these forces that have emerged from research are: (1) The nuclear forces are very *short range* attractive forces; they produce strong binding but decrease in strength rapidly to zero at distances greater than about 10^{-15} m. (2) The nuclear force between two neutrons is about the same as that between a neutron and a proton or between two protons; of course in the case of two protons there is the ordinary electrostatic repulsion in addition to the nuclear force. (3) The nuclear forces exhibit *saturation;* that is, each nucleon interacts with only adjacent nucleons, a behavior that should be contrasted with the electrostatic force by which a charged particle interacts with *all* other charged particles at the same time.

These three statements give the only elementary information known about nuclear forces. We cannot write an expression for the nuclear force as a function of distance. Such an expression has not been discovered, and its discovery is unlikely in view of the saturation property. We continue our study of nuclei without further information about the nuclear forces.

Fortunately the *binding energy* of a nucleus can be measured, and nuclear reactions can be investigated in terms of the energy instead of in terms of the actual force between particles. The concept of binding energy is also very useful and important for atomic structure. This concept can be explained in terms of the following example: A small steel ball is rolling without friction along a surface with a well in it. A cross-sectional view of this situation is given in Fig. 3-13a, which also gives a plot of the potential energy P as a function of position, since P equals Mgh in the gravitational field (h = height). We can conveniently measure h from the top of the well, and we set P equal to zero there. Figure 3-13b illustrates P, the total energy E, and the kinetic energy K as functions of position for a ball that is *bound* to the well. That is, the ball is oscillating back and forth with too little energy to permit it to escape. The total energy E is, of course, a constant, since no mechanical energy is being dissipated, and E is less than zero for a ball that is bound to the

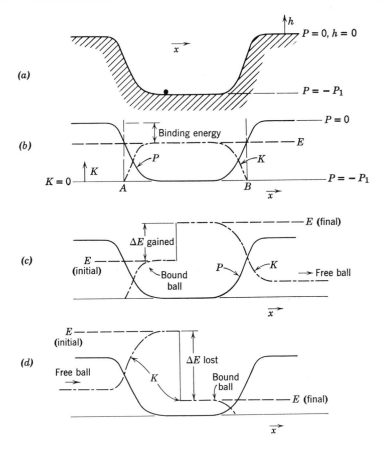

Fig. 3-13. Binding energy. (a) and (b) illustrate P, E, and K for a *bound* situation. (c) shows how the ball escapes if it absorbs an energy greater than the binding energy $|E|$. (d) shows how a *free* ball can be captured if it loses more energy than its initial K.

well. At the points A and B the value of K is zero, the velocity changes sign, and the ball is reflected.

The binding energy is $|E|$, the minimum energy that must be acquired by the ball if it is to escape from the well. Figure 3-13c illustrates the way K and E behave if an amount of energy ΔE greater than $|E|$ is absorbed (perhaps by collision with a more energetic ball). Since the increase in energy is greater than the binding energy, the ball can now escape. Figure 3-13d illustrates how a *free* ball can become bound if it loses energy while in the well. Although a gravitational force has been used in this example, the description of binding in terms of energy holds for any particle with any force acting on it.

We return now to the binding energies of nuclei and consider as an example the nucleus composed of a proton and a neutron. This is the *deuteron* or nucleus of deuterium, the heavy isotope $_1H^2$ of hydrogen. Usually when a proton and a neutron collide they merely change each other's energy and momentum, and no energy or momentum is lost. This collision process is called *elastic scattering*. During the collision the particles attract one another by the nuclear force and increase their kinetic energy (just like the steel ball in the gravitational field), but are then slowed down again as they move apart.

In an occasional collision, however, there will be an energy loss. This energy will be radiated away as a pulse of electromagnetic waves, called a *γ-ray*. If the energy loss is sufficient, the proton and neutron will be bound together. We write the reaction:

$$\text{Proton} + \text{Neutron} \rightarrow \text{Deuteron} + \gamma \qquad (3\text{-}8)$$

We shall later (in Chapters 13 and 14) be interested in such questions as: How probable is it that such a nuclear reaction occurs? But at present we are concerned simply with the energy balance and questions answerable by considering only this balance, such as: Is a spontaneous reaction possible? If so, how much energy is released? Conservation of energy, including of course the energy equivalent M_0c^2 of each rest mass M_0, applied to eq. 3-8 gives

$$(M_0 \text{ of proton})c^2 + (M_0 \text{ of neutron})c^2 = (M_0 \text{ of deuteron})c^2 + Q$$

$$(3\text{-}9)$$

where Q is the amount of energy released. The Q *values* of nuclear reactions can thus be computed from precise masses, or conversely mass differences can be computed from measurements of the energies released in nuclear reactions.

In the example under consideration the γ-ray carries away practically all the energy Q. Now the rest energy of 1 amu of mass is

$$M_0c^2 = (1.660 \times 10^{-27} \text{ kg}) \times (2.998 \times 10^8 \text{ m./sec})^2$$

$$1 \text{ amu} = 1.492 \times 10^{-10} \text{ joule} = 931 \text{ M.e.V.}$$

Therefore in our example

$$Q = (1.00727 + 1.00866 - 2.01355)931 = 2.22 \text{ M.e.V.}$$

(The deuteron mass was obtained by subtracting the mass of one electron from the deuterium atomic mass given in Appendix C.) This 2.22 M.e.V.

is the energy of the γ-ray emitted.   It is also the *binding energy* of the deuteron, since this is the energy that would have to be added to a deuteron to enable it to dissociate into a proton and a neutron.   (Measurement of the threshold amount of energy for disintegration permits the indirect measurement of the neutron's mass.)

It is much more convenient to make such computations in terms of *atomic* masses rather than in terms of proton, neutron, and deuteron masses.   The proton mass (1.00727 amu) plus the electron mass (0.00055 amu) equals the hydrogen *atomic* mass (1.00782 amu).   Tables of masses, like Appendix C, always tabulate atomic masses.   The electrons can be ignored in the nuclear reactions we shall consider here; there will always be the same number of electrons on the left side of the equation as on the right, and so it is immaterial whether we use nuclear masses or atomic masses if we use the same type of masses on each side.*   For example, the reaction of eq. 3-9 could equally well be written

$$_1\mathrm{H}^1 + {}_0n^1 \rightarrow {}_1\mathrm{H}^2 \tag{3-10}$$

Q would then be computed from the atomic masses as follows:

$$Q = 931 \times (1.00782 + 1.00866 - 2.01410) = 2.22 \text{ M.e.V.}$$

This is, of course, the same result as before; the effects of the electron mass included in 1.00782 and of the electron mass included in 2.01410 cancel.   Nuclear reactions are generally written in the form of eq. 3-10.

The binding energy of a more complicated nucleus is similarly defined as the energy required to break it up into its individual nucleons.   For example, the binding energy of $_9\mathrm{F}^{19}$, which has a mass of 18.99840 amu, is the energy required to make the following reaction go to the right:

$$_9\mathrm{F}^{19} \rightarrow 9_1\mathrm{H}^1 + 10_0n^1$$

$$18.99840 \times 931 + Q = 931(9 \times 1.00782 + 10 \times 1.00866)$$

$$Q = 0.1586 \times 931 = 147.7 \text{ M.e.V.}$$

Such binding energies are usually expressed by giving the average binding energy per nucleon.   In the example cited, the binding energy per nucleon is $147.7/19 = 7.8$ M.e.V. per nucleon.   A plot of binding energies per nucleon is given in Fig. 3-14.   The data of this figure will be fundamental to the discussions of nuclear properties in Chapter 13 and of atomic energy and the nuclear reactor in Chapter 14.

* There is one exception to this statement which will be considered in Sec. 13-14.

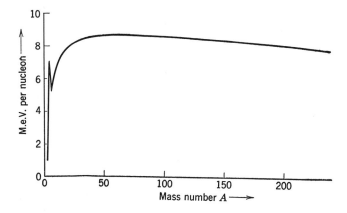

Fig. 3-14. Binding energy per nucleon plotted as a function of A. Most isotopes lie slightly above or below this curve. Note the relatively large binding energy of the α-particle ($_2$He^4), the sharp spike at 7.07 M.e.V. per nucleon. The nucleus $_4$Be^8 considered in the text lies to the right and very slightly *below* the α-particle.

We shall defer further discussion of nuclear reactions until Chapter 13, but it is advisable to consider briefly at this point the question of the *stability* of nuclei. It might be thought that any nucleus with a positive binding energy would be stable, that is, that it would not spontaneously disintegrate, but this is untrue. A positive binding energy means that energy would have to be supplied in order to break up the nucleus into protons and neutrons. Such a complete disintegration is only one kind, and a very rare kind, of disintegration process. If a nucleus is to be stable, it must be that there is *no* decomposition process that gives off energy. For example, $_4$Be^8 has a positive binding energy (56 M.e.V.) and therefore will not spontaneously disintegrate into protons and neutrons. But the reaction

$$_4\text{Be}^8 \rightarrow {_2\text{He}^4} + {_2\text{He}^4} \tag{3-11}$$

goes to the right, giving off a relatively small energy. Therefore $_4$Be^8 is unstable and is not found in nature.

The basic conclusion of this section is that nuclear binding energies are of the order of millions of electron volts, but we have also developed here some introductory nuclear physics that will be useful later. The binding energies of electrons to nuclei in atoms are much smaller than this (prob. 3-23). Thus our calculations in the preceding paragraphs were independent of whether the nuclei involved were bound into atoms,

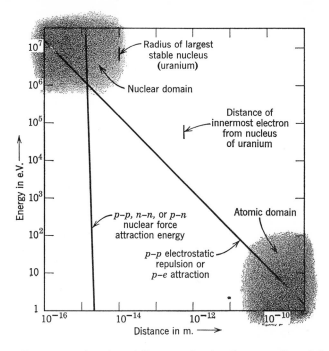

Fig. 3-15. Energy as a function of distance, showing the separation of the nuclear and atomic regimes. The electrostatic energy of two point charges, each of charge e, and a schematic diagram of the nuclear-force energy are plotted. The outermost nucleons in the largest stable nucleus (uranium) are bound with $\sim$6 M.e.V. and are only $\sim10^{-14}$ m. from the center of the nucleus. The innermost electrons of uranium are bound with 10^5 e.V. and are $\sim10^{-12}$ m. from the nucleus, but most atomic processes involve distances greater than 10^{-11} m. and energies less than 100 e.V.

molecules, or solids. Likewise in the discussion of atomic physics in the next nine chapters we can ignore the composition of the nucleus and the possibility of nuclear reactions; in all the experiments to be described there the energies involved are small enough to prevent the occurrence of nuclear changes.

The large binding energy, together with the small size demonstrated earlier in this chapter, of nuclei permits us to make separate studies of nuclear and atomic physics (see Fig. 3-15). In principle we could next concentrate on either study, but the forces in atoms are basically simpler than those in nuclei, which are still not well understood. Therefore we leave nuclear physics after this brief introduction and return to it in Chapter 13 after we have developed powerful tools that will originate from our understanding of atomic physics.

REFERENCES

The Nuclear Atom

F. K. Richtmyer, E. H. Kennard, and T. Lauritsen, *Introduction to Modern Physics*, McGraw-Hill, New York, 5th Ed., 1955.

D. Halliday, *Introductory Nuclear Physics*, Wiley, New York, 2nd Ed., 1955, Chapters 1 and 10.

H. Semat, *Introduction to Atomic and Nuclear Physics*, Rinehart, New York, 3rd Ed., 1954, Chapters 2 and 3.

Mass Spectrometers

M. G. Inghram in *Advances in Electronics*, Vol. I, edited by L. Marton, Academic Press, New York, 1948, pp. 219–268.

K. T. Bainbridge in *Experimental Nuclear Physics*, Vol. I, edited by E. Segrè, Wiley, New York, 1953, pp. 559–766.

H. E. Duckworth, *Mass Spectroscopy*, Cambridge University Press, Cambridge, 1958.

PROBLEMS

3-1.* The heat evolved when 1 kilomole of oxygen (O_2) combines with 1 kilomole of carbon (graphite) is 9.42×10^7 calories. What is this energy when expressed in electron volts per molecule of CO_2? (1 calorie equals 4.18 joules.)

3-2. A dilute solution of a strong base is neutralized by a strong acid in a calorimeter. By comparing the temperature rise with the temperature rise produced by an electric heater we find that the heat of neutralization is 5.80×10^7 joules per kilomole of H_2O formed in this reaction. What is this heat in electron volts per molecule of H_2O?

3-3. Use Gauss' law to verify the field calculated in the first paragraph of Sec. 3-2. (The density of iron is 7850 kg/m.^{3})

3-4.* In considering α-particle scattering our theory did not take into account the variation of mass with velocity. Is this an important error? To answer this question, compute M/M_0 for an α-particle with a kinetic energy of 7.7 M.e.V., which is the energy of the α-particles used in Rutherford's experiments.

3-5.* Suppose that an α-particle of 7.7-M.e.V. kinetic energy is directed precisely toward (i.e., $u = 0$) a nucleus of lead ($Z = 82$). Assume that the lead nucleus is so much heavier than the α-particle that it remains fixed. What is the distance of closest approach (the distance from the nucleus at which the α-particle turns around and goes backward over its initial path)?

3-6. A *collision radius* a is sometimes defined for a scattering process; it is the value of u that gives a 90° deflection. Calculate a for the case of Rutherford scattering and show that it coincides with the distance of closest approach in a collision for which $u = 0$.

3-7.* Compute the kinetic energy that an α-particle must have if it is to have a distance of closest approach of 1.3×10^{-14} m. from the center of a silver nucleus. Let $u = 0$ and assume that the silver nucleus is fixed. (For this and higher energies, the Rutherford scattering theory breaks down because the two particles interact by the nuclear force as well as by the Coulomb force.)

3-8. If the counting rate in a Rutherford scattering experiment is 20 α-particles per minute at $\theta = 150°$, what is the rate at $\theta = 10°$? How would you modify the experiment in order to avoid such a wide range of counting rates and still provide data for a plot like Fig. 3-6?

3-9.* In a Rutherford scattering experiment, 4.0 M.e.V. α-particles are incident on a silver foil. Calculate the impact parameter u corresponding to a deflection of 8° (the smallest scattering angle for which data appear in Fig. 3-6). Compare this value with the nearest neighbor separation (2.55 Å) of silver nuclei in metallic silver.

3-10.* Calculate the ratio of chord length to diameter length for $\delta = 2°$ and for $\delta = 5°$ (see Fig. 3-8). *Hint:* Since the angle is small, the calculation can be done quickly and accurately by using the series $\cos \delta = 1 - \delta^2/2 + \delta^4/24 - \cdots$, where δ is in radians.

3-11. The sorting chamber of a mass spectrometer is 0.25 m. long. Neon isotopes are being studied, and the gas in the sorting region is predominantly neon. What is the maximum pressure p (in millimeters of mercury) at $T = 300°K$ in the sorting region if fewer than 10% of the neon ions suffer collisions in traversing this region? Assume that the neon ion radius is the same as the neon atom; see Table 2-1.

3-12.* Calculate $\mathcal{B}$ for the 180° mass spectrometer that produced the data of Fig. 3-9. Assume that the radius of the circular path of the ions was 0.200 m.

3-13. Use the data of Appendix C to calculate the atomic weight of natural neon.

3-14. Use the relative abundance data for the calcium isotopes in Appendix C to calculate the atomic weight of natural calcium.

3-15. Mass spectrometers for measuring the relative abundance of isotopes are usually designed so that the peaks (Fig. 3-9) have flat tops. Why is this convenient? Show with a diagram why flat-top peaks are obtained if the exit slit is several times as wide as the entrance slit.

3-16.* A mass spectrometer is being used to produce a small quantity of pure Li^6. (Natural lithium contains 93% Li^7 and 7% Li^6.) The $\mathcal{B}$ and V of the spectrometer are set to let only Li^6 through the exit slit, and the current of Li^6 ions is 0.1 microampere. How long a time is required to deposit 1 milligram $(10^{-6}$ kg) of Li^6 on the collector? (This calculation shows that, in order to separate appreciable quantities of isotopes, much larger beam currents are required than are ordinarily used for mass spectrometry.)

3-17. Which of the two ion paths sketched in Fig. 3-11 corresponds to the higher velocity? Explain why the path that has the larger deflection in the electrostatic field has the smaller deflection in the magnetic field.

3-18. Sketch the paths of two ions in the apparatus illustrated in Fig. 3-11 if they have the same velocities but different initial directions.

3-19. Show with a sketch how the mass separation of the $Li^6 - \frac{1}{2}C^{12}$ doublet on a photographic plate can be determined from a knowledge of the positions of the lines on the plate corresponding to masses 5, 6, and 7. *Caution:* Mass as a function of position is nearly, but not quite, a linear function.

3-20.* The following doublets have been measured by Nier, Quisenberry, and Scolman with a double-focusing mass spectrometer:

$$C_4^{12} - S^{32}O^{16} = 0.033017 \text{ amu}$$

$$O_2^{16} - S^{32} = 0.017754$$

All the ions are singly charged. Find the masses of S^{32} and O^{16}.

3-21. The following doublets have been observed in a precision mass spectrometer by L. G. Smith:

$$O_2^{16} - S^{32} = 0.017756 \qquad B_2^{11}H_5^1 - C_2^{12}H_3^1 = 0.034257$$

$$B_5^{11}H_9^1 - S^{32}O_2^{16} = 0.15506 \qquad C_2^{12}H_4^1 - C^{12}O^{16} = 0.036383$$

All the ions are singly charged. Find the masses of S^{32}, H^1, B^{11}, and O^{16}. *Hint:* Solve the simultaneous equations for the *differences* between the mass of each isotope and its nominal mass.

3-22. Show that when the method of doublets is applied in order to determine atomic masses it is unnecessary to correct each ion mass by adding the appropriate number of electron masses to make the atom neutral. Show this by verifying that the corrections cancel out for the first doublet of prob. 3-20. Do the corrections cancel if one of the ions is doubly charged?

3-23.* The binding energy of an electron to a proton in a hydrogen atom is 13.6 e.V. Convert this to atomic mass units and compare with the present uncertainty in the measured mass of the proton, about 10^{-7} amu.

3-24. The binding energy of the KCl molecule is 4.40 e.V. When an atom of potassium and an atom of chlorine came together to form this molecule, 4.40 e.V. of energy was given off, and the mass of the molecule is less by an amount ΔM than the sum of the K and Cl masses. Calculate ΔM in amu. Does the magnitude of this ΔM, which is typical of molecules, justify neglecting it in problems like prob. 3-20?

3-25. If the curve for K in Fig. 3-13b were continued outside the points A and B it would have negative values. What is the significance of a negative value of K? Is there any particular significance to a negative value of E?

3-26.* Assume a ball of mass M to be at rest in the well of Fig. 3-13. A free ball of mass $2M$ has an initial kinetic energy outside the well equal to Mgh_1, where h_1 is the depth of the well. It enters the well and collides elastically with the lighter ball. Draw diagrams of E and K for each ball, like Fig. 3-13c and d. What is the kinetic energy of each ball immediately after the collision (in terms of Mgh_1)? Does either ball escape? (Both balls move only along the x-direction.)

3-27. Give arguments showing that the nuclear force cannot be the Coulomb's law electrostatic force. Give arguments showing that the nuclear force cannot be the gravitational force $F = -GM_1M_2/r^2$, where G = gravitation constant

$= 6.66 \times 10^{-11}$ joule m./kg^2, M_1 and M_2 are the masses of two particles, and r is the distance between the particles. (r is of the order of 10^{-15} m. for nucleons.)

3-28.* Compute Q in million electron volts for the disintegration of $_4\text{Be}^8$ by the process of eq. 3-11. Is any other spontaneous disintegration possible for $_4\text{Be}^8$?

3-29. Compute the Q value of the reaction $_1\text{H}^2 + _1\text{H}^2 \rightarrow _1\text{H}^3 + _1\text{H}^1$.

3-30.* Compute the Q value of the reaction $_3\text{Li}^6 + _1\text{H}^2 \rightarrow _2\text{He}^4 + _2\text{He}^4$.

3-31. Compute the Q value of the reaction $_3\text{Li}^6 + _0n^1 \rightarrow _2\text{He}^4 + _1\text{H}^3$.

3-32.* Compute the binding energy per nucleon of Be^9.

3-33. Compute the binding energy per nucleon in million electron volts of the three stable argon isotopes and of the two unstable isotopes Ar^{37} (36.96677 amu) and Ar^{39} (38.96432 amu).

4

Wave-Particle Experiments

4-1 INTRODUCTION

In this chapter we discuss in detail the principal experiments upon which the understanding of modern physics is based. The particles described in the earlier chapters are encountered in the present chapter in many different situations, and their behavior is analyzed. The analysis of these experiments produced a revolution in our thinking about atomic physics and made it apparent that the ordinary physics appropriate to laboratory-size objects could not explain the properties of atoms. Newton's laws and Maxwell's equations are the basic relations of ordinary mechanics and electromagnetism, and they are called the laws of *classical mechanics* and *classical electromagnetic theory*, respectively. In the description of each experiment we show how the predictions of classical physics disagreed with the experimental results. Such disagreements are not matters of quantitative detail but are fundamental differences between the nature of the phenomena observed and the nature of the phenomena predicted.

The present chapter establishes the *necessity* for a new physics to supplement classical physics. This new physics is the *quantum* physics or *wave mechanics* presented in Chapter 5. Wave mechanics will there be shown to be consistent with the experimental results described in the present chapter. It will be applied in the succeeding chapters to a wide variety of problems in atomic, molecular, solid-state, and nuclear physics. In every problem it meets the test of agreement with experiment.

The situation here is exactly like that with the relativity theory of Secs. 1-7 and 1-8. It will be recalled that experiments with high-velocity particles necessitated a new, more general relation for the kinetic energy of a particle as a function of its velocity. This relation was in agreement with experiments at high velocities and reduced to the familiar $K = \frac{1}{2}mv^2$ at low velocities. Thus it *extended*, rather than supplanted, Newton's laws. No experiments at low (laboratory-scale) velocities disagreed with the relativity expression, nor could such experiments have demonstrated the necessity for using the more general law at high velocities. The quantum-physics situation is precisely analogous. No experiments with laboratory-size objects can either prove or disprove quantum physics. But experiments where the results depend on what happens in lengths of the order of 10^{-10} m. (atomic sizes) show the necessity for the new laws. We shall from time to time show that the new laws do not contradict the old in the region where classical laws should apply, namely for laboratory-scale sizes and energies. In other words, we shall show that quantum physics *extends*, rather than supplants, classical physics; it is more *general* than classical physics.

The plan of this chapter is to present the fundamental experiments of atomic physics, to show how classical physics "does not work" in the analysis of each experiment, and to show what new concepts are required. In most of these experiments (Sec. 4-2 through Sec. 4-8) the new concept will be that electromagnetic waves exist only as discrete bursts of energy called *photons*. Each photon or *quantum* of the wave energy behaves very much like a particle. In Secs. 4-10 and 4-11 the motions of electrons and heavier particles will be shown to exhibit diffraction, which is a phenomenon associated with waves. Finally, in Sec. 4-12 the *Indeterminacy Principle*, which involves both particle and wave concepts, will be explained. This principle summarizes in a dramatic way the difficulties encountered when we attempt to apply classical physics to atomic phenomena.

4-2 THE PHOTOELECTRIC EFFECT

An experiment with an ordinary vacuum photocell is one of the most important experiments in modern physics. With such a tube, a voltmeter, a microammeter, a variable voltage supply, and a light source we can quickly demonstrate that classical electromagnetic theory is not adequate to deal with atomic physics.

When light of sufficiently short wavelength falls on the emitter electrode of the photocell, electrons are observed to come out of this electrode. (We can be sure that the charged particles emitted are electrons by measuring their e/m.) This process is called the *photoelectric effect*. We shall first consider what results of photoelectric experiments would be predicted if we tried to apply classical physics to this process. Then we shall describe the experiments and the experimental results and show that these results are completely inconsistent with classical theory. Next we shall apply quantum physics to these experiments and find satisfactory agreement between theory and experiment. In the analysis of the experiments we assume that the emitter is a metal, although commercial photocells with non-metallic emitters produce similar results.

Before discussing either theory, we must digress briefly to consider the concept of a *work function*, which concept is common to both theories. It is evident that there is a binding energy of electrons to a solid or, in other words, that energy must be given to an electron in order to remove it from the solid. If this were not true, we should expect the electrons to leave when even a very small electric field is applied. Electrons do not leave a solid spontaneously in large numbers; appreciable currents of electrons are emitted only when an additional source of energy is provided for the electrons. If a metal is heated to a high temperature in a vacuum, *thermionic emission* of electrons occurs; the additional energy in this case comes from the thermal energy ($\frac{1}{2}kT$ per degree of freedom) appropriate to the high temperature. In the photoelectric effect, the additional source of energy is the incident light.

We shall denote the binding energy of an electron to a solid by $e\phi$ e.V., where ϕ is measured in volts, and $e\phi$ is called the *work function*. More specifically, $e\phi$ is defined as the minimum energy that must be given to an electron in the solid in order to remove the electron from the solid. If there are electrons with various energies inside the solid (as in a metal, where energies from zero to several electron volts occur in the same metal), $e\phi$ represents the additional energy that must be given to the most energetic of these electrons in order to remove it. Figure 4-1 illustrates the surface-energy barrier and the work function.

The classical theory of photoelectric emission is: Electrons in the metal should be accelerated by the electric field of the electromagnetic wave, which increases as the light intensity increases. If this field is sufficiently strong, an electron could acquire the energy $e\phi$ and be emitted. At low intensities of incident light, no electrons would be emitted. As the intensity is increased above a threshold value, electrons ought to be emitted. The higher the intensity, the more kinetic energy the emitted electrons should have, since there would be more energy left over after surmounting the surface-energy barrier. Classical theory did

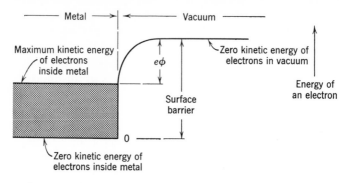

Fig. 4-1. Electron energy relations at the surface of a metal.

not predict any simple dependence of the kinetic energy of the emitted electrons on the frequency of the incident light (the frequency ν equals c/λ, where c is the velocity of light and λ is the wavelength).

The photoelectric experiment consists of measuring the number of electrons emitted and their energies as functions of the intensity and frequency of monochromatic incident light. The apparatus is illustrated schematically in Fig. 4-2. Either a retarding ($V < 0$) or accelerating ($V > 0$) electric field can be applied to the emitted electrons. The results of a typical experiment are presented in Fig. 4-3. The current is

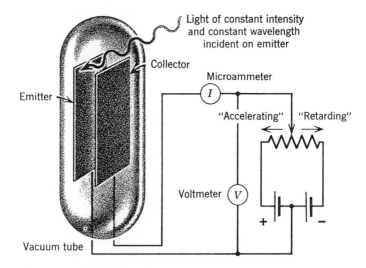

Fig. 4-2. Schematic diagram of the photoelectric experiment. The voltage V is considered positive ("accelerating") if the collector is positive.

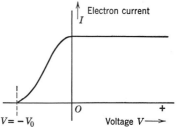
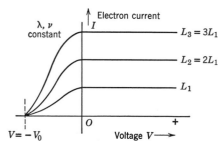

Fig. 4-3. Current as a function of voltage in the photoelectric experiment (constant intensity and λ).

Fig. 4-4. Photoelectric current for three values of light intensity L, with constant λ.

independent of voltage for $V > 0$. In this voltage region, all the emitted electrons are arriving at the collector. For retarding fields $(V < 0)$, the more energetic electrons are collected whereas those with small values of the kinetic energy K are turned back and made to re-enter the emitter. Evidently there is a maximum kinetic energy $K_{\max} = eV_0$ such that no electrons have more energy than this value of K. If the retarding voltage is made more negative than $-V_0$, no electrons reach the collector. The experimental results so far have yielded no surprises.

Now let us vary the intensity L of the light and keep ν constant. The resulting curves are shown in Fig. 4-4 for three different values of the intensity (watts/m.^{2}). The remarkable feature is that V_0 remains constant! The kinetic energies of the emitted electrons are unchanged, only their number is changed, when the light intensity and hence the electric field in the electromagnetic wave are varied. This result is in direct opposition to the classical prediction.

Furthermore, this experiment can be performed at very small intensities, and still the photoelectric current is strictly proportional to the light intensity. With modern techniques the individual electrons can be counted as they reach the collector. The emitted electrons are distributed randomly in time, and the average rate of emission is strictly proportional to the light intensity even with the weakest intensity that can be measured. We might try to salvage the classical theory by assuming that at low light intensities the electrons somehow accumulate energy from the light over a long enough time so that they acquire enough energy to leave the metal. But the computed storage time necessary would be measured in years in experiments at low light levels, whereas experiments indicate that the photoelectric current begins as soon as the light falls on the emitter.

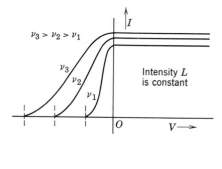

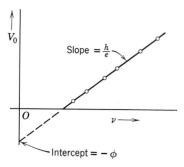

Fig. 4-5. Photoelectric current for three values of light frequency $\nu = c/\lambda$ with constant L.

Fig. 4-6. Maximum retarding voltage V_0 as a function of frequency ν.

Another surprising result occurs when we vary the frequency ν of the light and keep the intensity constant. The results of this experiment are illustrated in Figs. 4-5 and 4-6, and it is apparent that the maximum energy eV_0 of the photoelectrons changes with ν. Furthermore, V_0 is a linear function of ν. If the material of the emitter is changed, the data of Fig. 4-7 are obtained. Note that the slope of the lines is independent of the nature of the emitter, which forcefully suggests that this slope is a fundamental property of light. Classical theory is incapable of predicting results in agreement with these experiments.

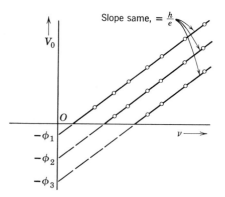

Fig. 4-7. V_0 as a function of ν for three different emitters, with different work functions.

We now postulate the *photon* theory of light and show how it succeeds where classical theory failed. We should be reluctant to accept this bold new theory on the basis of a single experiment (e.g., the photoelectric effect) even though the success of the theory is quite striking. But we study many other experiments in this chapter that will also demonstrate the agreement of the photon theory with experiment.

The photon theory of light is: (1) Light consists of pulses of electromagnetic waves called photons. (2) During the emission or absorption of light, photons are created or absorbed as indivisible units. Such processes are like a collision process between the photon, treated as a particle, and an electron in the atom that is creating or absorbing the light. The scattering of light is also a process of collision between a photon and the scatterer (e.g., an electron). (3) Each photon has the energy $h\nu$, where ν is the frequency (c/λ) of the light, and h is a universal constant called *Planck's constant*. (4) Photons travel through space and exhibit diffraction and polarization exactly like electromagnetic waves of frequency ν.

The photon theory thus *adds* properties to the common (diffraction and polarization) properties of light. It does not ask us to abandon the old concept of light; it asks us to superimpose the photon concept on the electromagnetic wave concept. Photons are also called *light quanta*. The photon theory is one aspect of the general *quantum theory* or *wave mechanics* underlying atomic physics.

An attempt is made in Fig. 4-8 to illustrate the properties of a photon. This drawing is a snapshot of the electric field $\mathcal{E}$ as a function of x at a particular time. We see a pulse or a wave train, which in the typical example of visible light emitted by an atom would have about 10^5 oscillations, moving with the velocity of light. As this photon moves it is guided in its motion by the properties of the wave (wavelength, polarization, velocity). When it was emitted, it was created as a unit.

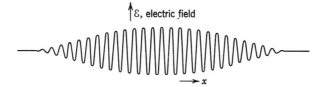

Fig. 4-8. Schematic representation of a photon as a packet of waves. The figure is a plot of $\mathcal{E}$ vs. x at a particular instant. The whole packet is moving with the velocity c of light. (There should be about 10^5 oscillations, instead of the 27 shown, for a typical spectral line.)

When it is absorbed, it will disappear as a unit. The energy carried by it is $h\nu$.

We shall now apply this photon theory to the photoelectric effect. A photon strikes an electron, and the photon's energy $h\nu$ appears as additional kinetic energy of the electron. Suppose that this electron is one of the group with the maximum kinetic energy. If this electron is moving in the proper direction (perpendicularly toward the surface of the emitter), its kinetic energy K after leaving the emitter is

$$K_{\max} = h\nu - e\phi \qquad (4\text{-}1)$$

It should be recalled that $e\phi$ is the energy that must be given to the most energetic electron in a metal to release it from the solid (see Fig. 4-9). If the electron involved had less than the maximum kinetic energy before absorbing the photon, or if it were not directed normal to the surface, its K after emission would be less than $K_{\max}$ from eq. 4-1 or it might not be emitted at all. (If the K calculated from eq. 4-1 is less than zero, the electron is not emitted.) Thus eq. 4-1 gives the maximum kinetic energy of emitted electrons, and this value is exactly what we measure as eV_0:

$$eV_0 = h\nu - e\phi \qquad (4\text{-}2)$$

This equation is the *Einstein photoelectric equation*. It clearly agrees well with the data of Figs. 4-6 and 4-7, since it predicts that V_0 should be a linear function of ν and that the slope $dV_0/d\nu$ should be independent of

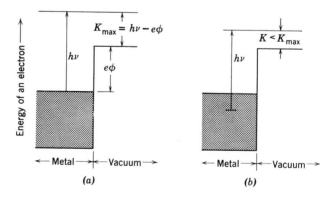

Fig. 4-9. In (a) a photon of energy $h\nu$ has collided with an electron with maximum initial kinetic energy. In (b) the photon has collided with a less energetic, and more typical, electron.

the material of the emitter. Planck's constant h is a universal constant and has been found from this and other experiments to be

$$h = 6.626 \times 10^{-34} \text{ joule sec}$$

The work function $e\phi$ is the product of e and the $\nu = 0$ intercept of the lines in Figs. 4-6 or 4-7 and is different for different emitter materials.

Equation 4-2 also agrees with the experimental fact that V_0 does not vary with the intensity of the incident light. As the light intensity increases, the number of photons striking the emitter per second increases, and the number of photoelectrons increases in proportion. But the kinetic energy of each emitted electron depends only on $h\nu$, ϕ, and the initial kinetic energy of that electron; it does not depend on how many photons are hitting the emitter per second. Another way of stating this result is to say that each photon absorption act is independent of all other such acts.

The quantum theory of light and the Einstein photoelectric equation, which is based on this theory, are thus in excellent agreement with the photoelectric experiments. Since experiments, rather than preconceived ideas, are the ultimate authority in science, we must use the quantum theory in preference to the classical theory when considering the interaction between light and electrons.

4-3 LINE SPECTRA

When an electrical discharge is produced in a low-pressure gas, the light emitted has a very striking nature. Only a series of discrete wavelengths is emitted, rather than the continuous spectrum emitted by a hot solid like a tungsten filament. These *spectral lines* can be observed with a prism or grating spectroscope and are very sharp and characteristic of the particular atom emitting. The width $\Delta\lambda$ of an individual line is of the order of 10^{-5} of the wavelength λ of the line. Some spectra are shown in Fig. 4-10.

The existence of sharp-line spectra cannot be explained satisfactorily by classical theory. We now attempt a classical explanation and show why it does not fit the facts. We assume that the electron moves in a circular orbit with the nucleus at the center, as shown in Fig. 4-11a. The force required to keep this electron in circular motion is the Coulomb attraction of the electron to the positive nuclear charge. If we were to observe this electron by looking in the direction of the arrow, we should see an electric charge $-e$ moving back and forth in simple harmonic motion. In other words, we should see the same motion of charge that exists on an ordinary dipole radio antenna. It can be shown by electro-

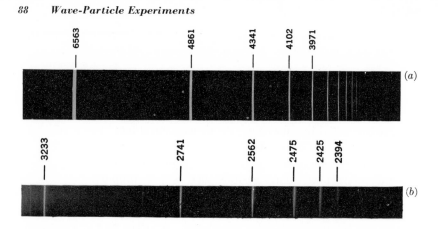

Fig. 4-10. Examples of line spectra. Wavelengths in angstroms are marked for the principal lines. Most spectra are much more complicated than the two examples shown here. (a) Balmer series of hydrogen (from G. Herzberg, *Atomic Spectra and Atomic Structure*, Dover, New York, 2nd Ed., 1944). (b) Principal series of lithium (photograph by P. L. Hartman, C. W. Gartlein, and J. N. Lloyd).

magnetic theory that any accelerated charge radiates electromagnetic waves. The frequency of the radiation is the frequency of the oscillations of the current (in the antenna problem) or of the revolution of the electron about the nucleus (in the atomic problem).

Classical theory can thus show why an atom radiates, but the trouble is that it predicts a *continuous* spectrum rather than a line spectrum. We can show this fact easily for the example of the hydrogen atom. An electron is attracted to a proton by the force

$$F = -\frac{e^2}{4\pi\epsilon_0 r^2} \tag{4-3}$$

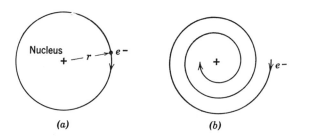

Fig. 4-11. Classical picture (wrong) of an electron in an atom. In (b) the loss of energy by radiation causes a spiral path and a continuous spectrum of radiated light.

where r is the radius of the electron's "orbit" about the proton (which is at $r = 0$). If the electron moves in a circular orbit, a force

$$F = -\frac{mv^2}{r}$$

is required to sustain this motion. By equating these two expressions for F, we can find the frequency ν of the circular motion as follows:

$$\nu = \frac{v}{2\pi r} = \frac{e}{4(\epsilon_0 \pi^3 r^3 m)^{1/2}} \tag{4-4}$$

The total energy E of the electron is the sum of its potential energy * $-e^2/4\pi\epsilon_0 r$ and its kinetic energy $\frac{1}{2}mv^2 = -Fr/2 = e^2/8\pi\epsilon_0 r$. This sum is $-e^2/8\pi\epsilon_0 r$, which varies with r. As the electron radiates energy, the law of conservation of energy requires that its total energy must decrease. Thus r continuously decreases (Fig. 4-11b) as the emission of light proceeds, and the ν predicted by eq. 4-4 changes continuously. In other words, a continuous spectrum, rather than a line spectrum, should be emitted, which clearly disagrees with the experimental facts. Although we have assumed circular orbits in this discussion, the nature of the result would be the same for any orbit.

The first step toward an explanation of sharp-line spectra was taken by Bohr. He based his work on the photon theory of light outlined in Sec. 4-2. He postulated that an electron in an atom could be in only one or another of a set of *discrete energy levels*. Bohr further postulated that only one photon was emitted at a time. Radiation of energy from the atom occurs, according to this view, when an electron with total energy E (kinetic plus potential) changes to total energy E', where E' is less than E. The energy lost by the atom appears as the energy of a single photon:

$$E - E' = h\nu \tag{4-5}$$

Thus only a discrete set of ν values (and hence λ values) occurs in the spectrum of an atom, since only a discrete set of E values exists for an electron in an atom. Thus the hypotheses stated lead to agreement with the experimental fact that all isolated atoms emit only line spectra.

Bohr also postulated a set of rules from which the E values for the hydrogen atom could be computed. He obtained these rules by rather arbitrarily requiring that the angular momentum of the electron be an

---

* The choice of the point where $P = 0$ is, as always, arbitrary; we set $P = 0$ at $r = \infty$ since this choice produces the simplest expression for $P(r)$.

integer multiplied by $h/2\pi$. He applied classical mechanics with this
assumption and found that

$$E_n = -\frac{13.60}{n^2} \text{ e.V.} \qquad n = 1, 2, 3, \cdots \qquad (4\text{-}6)$$

Here E_n is the energy of one of the states in electron volts, the factor
13.60 comes from a combination of the atomic constants e, m, and h,
and the *quantum number* n can be any positive integer. A plot of these
E values is given in Fig. 4-12. Arrows represent possible transitions.
All observed hydrogen spectral lines are in agreement with eq. 4-5 and
eq. 4-6. But Bohr's rules for calculating E values will not work for any

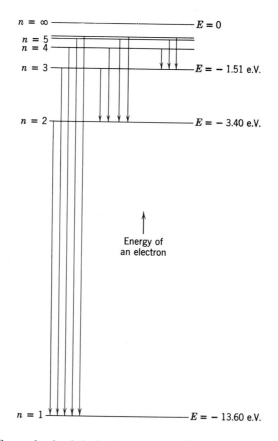

Fig. 4-12. Energy levels of the hydrogen atom. The vertical lines represent schematically transitions that radiate photons.

atom except hydrogen. We shall see in Chapter 6 how these values are calculated for other atoms as well as hydrogen; the calculation is based on the general theory of quantum mechanics.

The photon theory and the hypothesis of discrete energy levels thus give agreement with experiments on spectra. But this quantum explanation of line spectra leaves several points unexplained: What is the nature of the electron's motion in the atom? What is the nature of the transition from one energy state to another? Answers to these and similar questions will have to be delayed until further experiments are discussed and the general quantum mechanics theory of Chapter 5 is constructed on the basis of these experiments. Our chief conclusion from the present section is that there is evidence from spectra of a *discreteness* in nature unexplainable by classical theory.

4-4 X-RAY LINE SPECTRA

When we discussed line spectra in the preceding section we were considering visible light and other electromagnetic waves produced in the same way as visible light. These waves include the infrared (long-wavelength) and ultraviolet (short-wavelength) regions of the spectrum but nevertheless constitute only a tiny part of the total wavelength span that has been investigated; this span is illustrated in Fig. 4-13. The methods of producing radiation are different in different parts of the spectrum. Furthermore, at the time of discovery of X-rays and γ-rays, it was not immediately known that such rays were electromagnetic in nature. Hence we find different names for radiation that is fundamentally the same, differing only in the magnitude of the wavelength.

Our task in this section is to show that X-rays exhibit line spectra and hence to provide additional evidence for discrete energy levels in atoms. First, however, we shall have to digress to describe the production of X-rays and the measurement of X-ray wavelengths. These topics are

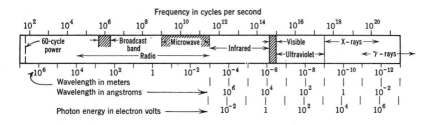

Fig. 4-13. The electromagnetic wave spectrum.

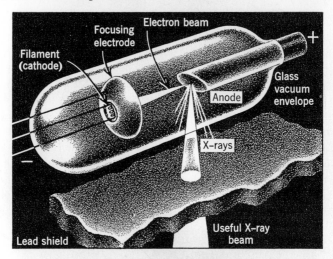

Fig. 4-14.   X-ray vacuum tube.

also of interest in themselves since X-rays are of considerable importance in chemistry, solid-state physics, and metallurgical engineering.

X-rays are produced when high-energy electrons strike atoms. A typical X-ray vacuum tube is illustrated in Fig. 4-14. A potential difference of the order of 50 kilovolts is maintained between the thermionic cathode (a hot tungsten filament) and the anode. The anode is usually a metal with large atomic number; it is usually water cooled, since even electron currents of only a few milliamperes dissipate hundreds of watts at the anode because of the high voltage. A photograph of an industrial X-ray tube is presented in Fig. 4-15 together with a sample radiograph. This radiograph was made by interposing a section of a weld in 1-inch steel between the X-ray tube and a photographic film and demonstrates that X-rays are very penetrating.

The index of refraction of X-rays in all materials is very nearly unity, and therefore it is impossible to make a prism spectrometer to measure X-ray wavelengths. A ruled diffraction grating can be used to measure wavelengths and, incidentally, to verify that X-rays are diffracted. This measurement is difficult because X-rays have such short wavelengths (about 0.1 to 1 Å) compared to the smallest grating spacing that can be ruled (about 10,000 Å). What we should like for these measurements would be a grating with a spacing between lines of the order of 1 Å. Since this is the same order of magnitude as the size of an atom, the atoms in a solid crystal are spaced from each other about the right

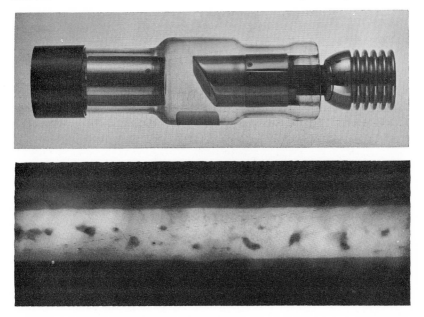

Fig. 4-15. The upper part is a photograph of an X-ray tube for industrial and medical radiography. The lower part is a radiograph of a weld and reveals gas holes (dark spots). (Courtesy of X-Ray Department, General Electric Company.)

distance to provide a diffraction grating for X-rays. Furthermore, these atoms form a regular array with constant spacing. The use of crystals is thus the most convenient method of measuring X-ray wavelengths.

The way a crystal serves as an X-ray diffraction grating is illustrated in Figs. 4-16 and 4-17. First consider a beam of monochromatic X-rays of wavelength λ incident on the first plane of atoms. This beam makes

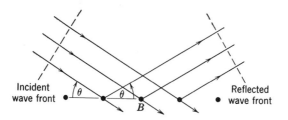

Fig. 4-16. Partial reflection of an X-ray beam by a plane of atoms. The dashed lines are lines of constant phase. Constructive interference of the scattered waves from the atoms in the plane occurs only if the two angles marked θ are equal.

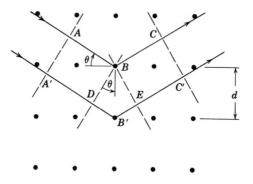

Fig. 4-17. Interference of partial reflections by two planes of atoms. The dashed lines are lines of constant phase. The path length $DB'E$ must be an integral number of wavelengths if the reflections are to add in phase.

an angle θ with the plane of atoms. Part of the beam will be reflected by this plane, and the angle that the reflected beam makes with the plane is also θ. It is easy to prove this fact by exactly the same reasoning (Huygens' principle) that shows that the angle of incidence equals the angle of reflection for the reflection of visible light from a mirror (see prob. 4-14).

It is clear from the great penetrating power of X-rays that no single plane of atoms can reflect a large fraction of the incident X-rays. Therefore we must add the partial reflections from many planes to get the resultant reflection. If these reflections add in phase (constructive interference), a strong reflection will result. The condition for constructive interference can be obtained by studying Fig. 4-17. If the wave reflected from B' is to add in phase to that reflected from B, the path length $A'B'C'$ must be an integral number of wavelengths longer than the length ABC. Since the path difference is $DB' + B'E$, and since $DB' = B'E = d \sin \theta$, we have

$$n\lambda = 2d \sin \theta \qquad (4\text{-}7)$$

where n is an integer, usually unity in our subsequent discussion.

Equation 4-7 is called *Bragg's law* of X-ray diffraction. If it is satisfied for one pair of planes, it will be satisfied for all planes with the same spacing d. If it is not exactly satisfied, there will be no reflected wave, since, as the contributions from more and more planes are considered, eventually the error in phase between the wave from the first plane and from the last will be $180°$. The strong reflection of X-rays when eq. 4-7 is satisfied is called *Bragg reflection*, and the integer n is called the *order* of reflection.

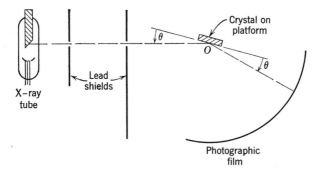

Fig. 4-18. Rotating crystal X-ray spectrometer. The crystal rotates about a vertical axis through O.

Bragg's law provides a powerful tool for the measurement of X-ray wavelengths. The experimental arrangement is shown in Fig. 4-18. A narrow beam of X-rays is provided by the small holes in the lead shields of the collimator. This beam strikes a crystal placed on a table, which is slowly rotated. When the crystal is in a position where eq. 4-7 holds, a strong reflected beam will occur. In this way the spectrum of λ's emitted by the tube can be explored.

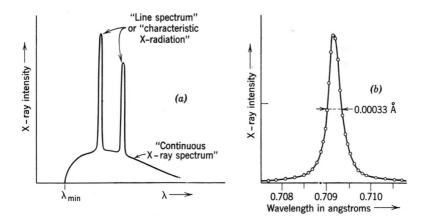

Fig. 4-19. (a) An X-ray spectrum of molybdenum taken with a low-resolution spectrometer. [From D. Ulrey, *Phys. Rev.*, **11**, 405 (1918).] (b) One of the X-ray lines of molybdenum observed with a high-resolution spectrometer. The continuous spectrum is only about 1/10,000 of the intensity of the line at the center of the line. [From C. H. Shaw and L. G. Parratt, *Phys. Rev.*, **50**, 1006 (1936).]

An X-ray emission spectrum studied in this way is illustrated in Fig. 4-19a. It is evident that there is some energy emitted at all wavelengths greater than λ_{min}; this is the *continuous X-radiation* to be investigated in Sec. 4-6. Superimposed on this continuous spectrum are sharp lines like the lines of visible-light spectra. Also like the visible-light phenomena is the fact that the wavelengths of these lines depend on the particular element used in the source (the target of the X-ray tube). We call these lines *characteristic X-radiation*. The characteristic lines are much sharper than Fig. 4-19a indicates; the spectrum presented in that figure was obtained from a low-resolution spectrometer, and the peaks were broadened and lowered by lack of resolution. The actual shape of an X-ray line is shown in Fig. 4-19b, which was obtained from a spectrometer with high resolution. The total intensity in the continuous spectrum is of the same order of magnitude as the total intensity of characteristic lines.

The existence of characteristic X-radiation verifies the existence of discrete energy levels in atoms in a quite different region of the spectrum from the optical spectra. We thus conclude that X-rays, like visible light, consist of photons. An X-ray photon is much more energetic than a visible-light photon, since typical X-ray wavelengths are of the order of 1/10,000 of the wavelengths of visible light. Atoms evidently contain some energy levels separated by a few electron volts (visible light) and others separated by many thousands of electron volts (X-rays). It will be our task in Chapter 6 to provide a quantum-physics understanding of atomic energy levels over this wide range of values.

We now return to the consideration of Bragg reflection in order to show how it can be used as a tool with which to investigate crystalline solids, and to show how the spacing d between planes can be measured. We consider as an example a crystal of potassium chloride. The geometrical arrangement of atoms is illustrated schematically in Fig. 4-20. If we ignore the distinction between potassium and chlorine atoms (that have about the same reflecting power for X-rays) this is a simple cubic crystal. Many planes could be drawn through sets of atoms, with different d values. We shall first be concerned with the cubic planes that are separated by the particular d value labeled d_0 in the figure. This d_0 is also called the *lattice constant*.

The lattice constant can be calculated from the density ρ and molecular weight W of KCl. We imagine the crystal to be broken up into cubes of edge length d_0 with an atom at the center of each. These cubes fill the entire volume, and each contains one atom; therefore there is one atom per d_0^3 m.^{3}, or d_0^{-3} atoms/m.^{3} Since there are two atoms per molecule, there are $1/2d_0^3$ molecules/m.^{3} or $1/2\rho d_0^3$ molecules/kg, where

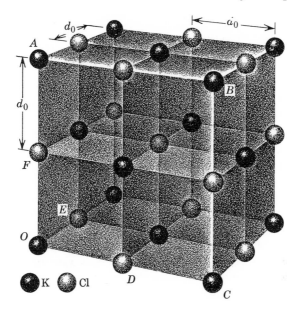

Fig. 4-20. The crystal structure of KCl. Many other diatomic crystals, such as NaCl, have this structure, which is named the *sodium chloride structure* for its most familiar example.

ρ is the density. We also know by the definition of Avogadro's number N_0 that there are N_0 molecules per kilogram-molecular weight. Therefore the number of molecules per kilogram can also be written as N_0/W:

$$N_0/W = 1/2\rho d_0{}^3$$

This expression can be solved for d_0:

$$d_0 = (W/2\rho N_0)^{\frac{1}{3}} \tag{4-8}$$

For KCl, for example, $\rho = 1990$ kg/m.^{3}, and

$$d_0 = \left(\frac{74.56}{2 \times 1990 \times 6.022 \times 10^{26}}\right)^{\frac{1}{3}} = 3.14 \times 10^{-10}\,\text{m.} = 3.14\,\overset{\circ}{\text{A}}$$

We have calculated d_0 from N_0 and used this value to measure X-ray wavelengths. This process can be reversed to give a good method of measuring Avogadro's number N_0. First we measure the wavelength λ of an X-ray line by an experiment employing a ruled grating. Then we reflect this X-radiation from the cubic planes of KCl (spaced a distance d_0 apart), and the application of Bragg's law (eq. 4-7) permits the computation of d_0. From the molecular weight (determined by chemical experiments) and density we can now compute N_0 from eq. 4-8.

Thus far we have treated a crystal as if there were a single set of planes a distance $d = d_0$ apart. There are actually many different planes with different d values in the same crystal. Figure 4-21 is an extension of the plane $OABC$ of Fig. 4-20; the distinction between K and Cl atoms has been dropped since they have about the same reflecting power for X-rays. We see that an X-ray with a given λ can be reflected from different planes, provided, of course, that $n\lambda = 2d \sin \theta$. Only a few planes at right angles to the plane of the figure are shown in Fig. 4-21, but many other planes occur; for example, another plane with a large density of atoms is the plane through EDF of Fig. 4-20. Investigation of reflections as a function of the angle at which they occur will therefore provide a series of d values, d_0, d_1, d_2, etc. For the cubic crystal and the planes illustrated, these ratios d_1/d_0, d_2/d_0, etc., are numbers characteristic of the crystal structure. If the crystal structure were hexagonal or another structure, different ratios would be obtained. In this way, X-ray reflection studies with monochromatic X-rays and a crystal on a rotating platform permit the determination of the crystal structure and the measurement of d_0.

Another convenient practical method of determining d_0 values and crystal structures with X-rays is the *powder method*. In this method a polycrystalline or powdered solid is placed in a beam of monochromatic X-rays as illustrated in Fig. 4-22. Since the crystals are oriented in all

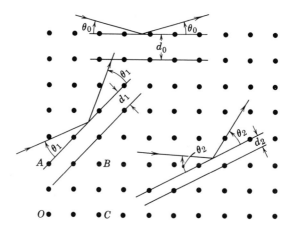

Fig. 4-21. Reflection of X-rays of the same wavelength from various planes. As d decreases, θ increases (eq. 4-7).

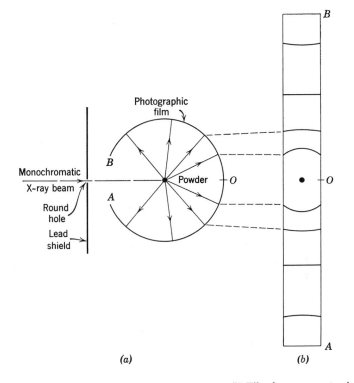

Fig. 4-22. (*a*) Powder X-ray diffraction apparatus. (*b*) Film from apparatus in (*a*), laid flat.

possible directions, there will be *some* crystals with the proper orientation of planes to satisfy Bragg's law for *any* plane spacing *d*. Thus lines on the film will be exposed corresponding to the various *d* values in the crystals. This method is frequently used as a method of analyzing mixtures of crystalline solids. Tables have been prepared of the principal *d* values for each of thousands of crystals. An unknown can be identified by comparing the observed set of *d* values with entries in the tables.

Since this section has been long and has contained digressions, it may be helpful to recall here the principal conclusions: (1) X-rays, although much different from visible light in wavelength, exhibit line spectra like visible light; the existence of X-ray line spectra increases our confidence in the photon theory and in the existence of discrete energy levels in atoms. (2) X-rays provide an important tool for measuring Avogadro's number, for studying the crystal structure of solids, and for inspecting opaque materials.

4-5 EXCITATION POTENTIALS

The existence of discrete energy levels in atoms can be demonstrated by a simple experiment using a gas-filled electron tube. This experiment, known as the Franck and Hertz experiment, is illustrated in Fig. 4-23. It can be performed with an ordinary mercury thyratron with a unipotential cathode. The tube is placed in an oil bath in order that the temperature of the walls of the tube can be controlled. Since there is liquid mercury in equilibrium with mercury gas in this tube, the pressure of the mercury gas is the vapor pressure of mercury at the wall temperature. Therefore the pressure can be controlled by controlling the temperature of the oil bath.

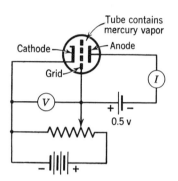

Fig. 4-23. Tube for excitation potential experiment.

Electrons emitted from the cathode are accelerated toward the grid. They also suffer ordinary *elastic* collisions with the mercury atoms; these are collisions in which the sum of the kinetic energies of the electron and of the gas atom is conserved. In such collisions, the electron loses a very small fraction of its energy. Therefore the electrons have a kinetic energy approximately equal to eV, where V is the grid-to cathode potential difference, as they pass between the grid wires. The small retarding voltage (0.5 volt) does not prevent them from being collected at the anode, provided that $V > 0.5$ volt. If, on the other hand, an electron should suffer an *inelastic* collision with a gas atom in which it lost nearly all its energy, it will be turned around by the retarding field between grid and anode and will not participate in the anode current I.

Some typical experimental results are shown in Fig. 4-24. The anode current increases until a critical voltage is reached, at which point it decreases sharply. We interpret this decrease to be the result of inelastic collisions that evidently occur as soon as the electron kinetic energy K reaches 4.86 e.V. An electron with this K loses all its energy to an atom, exciting an electron in the atom from one discrete energy level to a higher one. As the voltage is raised above 5.4 volts, an electron has sufficient energy to overcome the retarding voltage, even after making one inelastic collision. When V becomes 2×4.86, an electron can make *two* inelastic collisions, and another dip in the curve occurs. We conclude

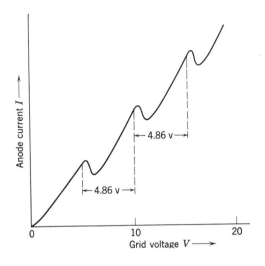

Fig. 4-24. Current as a function of voltage in the experiment illustrated in Fig. 4-23.

that the critical potential to excite an electron in the mercury atom is 4.86 volts. More complicated experiments have shown that there are other, higher excitation potentials in mercury. We do not see evidence for these in Fig. 4-24 since, almost as soon as an electron is accelerated to an energy of 4.86 e.V., its energy is reduced to zero by an inelastic collision. It is therefore very unlikely that an electron ever acquires enough energy to make one of the higher-energy excitations.

Another important result of this experiment is that ultraviolet light of $\lambda = 2536$ Å is observed to come from the tube as soon as V becomes greater than 4.86 volts. The light cannot be observed in an ordinary commercial thyratron tube since the glass is not transparent to ultra-violet light, but it can be observed with a tube made of quartz or ultra-violet-transmitting glass. Evidently, when an electron in a mercury atom is excited to an energy level E_2 that is 4.86 e.V. above its normal, ground state E_1, it returns to its normal state by radiating light. We can analyze the emission process by assuming only the conservation of energy and the emission of a single photon (as assumed in Sec. 4-3):

$$E_2 - E_1 = 4.86 \text{ e.V.} = 4.86e \text{ joules} = h\nu = hc/\lambda$$

$$\lambda = (hc/4.86e) = 2.551 \times 10^{-7} \text{ m.} = 2551 \text{ Å}$$

We therefore find excellent agreement between experiment and the photon theory. Furthermore, we can measure h/e by this and similar experiments. Numerical calculations like the foregoing one occur so frequently that it is convenient to compute and to remember the relation between the wavelength λ in angstroms and the energy E in electron volts of a photon:

$$\lambda(\text{Å}) = \frac{12,400}{E(\text{e.V.})} \qquad (4\text{-}9)$$

which comes from writing $E = h\nu = hc/\lambda$ and converting E to electron volts (1 e.V. $= 1.602 \times 10^{-19}$ joules) and λ to angstroms (1 Å $= 10^{-10}$ m.).

Excitation by collision is one of the ways in which atoms can be excited and can produce spectral lines. In gas-discharge light sources, the principal excitation of atoms occurs by this process of collision with high-energy electrons. The light-emission process after excitation is usually more complicated than the single-step process described in the preceding paragraph. The return of an electron in an atom to its ground state may occur by a succession of two or more steps: The atom emits first one photon as it goes to an intermediate level and later emits another.

Another way in which atoms can be excited is by *thermal excitation*. That is, the heating of the gas to a high temperature produces an appreciable number of atoms in an excited state E_2. These excited atoms can return to the ground state E_1 by the emission of light.

4-6   THE CONTINUOUS X-RAY SPECTRUM

In Sec. 4-4 the existence of a continuous spectrum of X-ray wavelengths was mentioned, but that section considered in detail only the sharp-line spectrum. We now examine this continuous spectrum more closely and show how one feature of the observed spectrum gives additional proof of the photon theory.

Experimental curves of X-ray intensity as a function of wavelength are given in Fig. 4-25. (The curve for the molybdenum target was also given in Fig. 4-19a; evidently tungsten and chromium do not happen to have lines in this wavelength region, and therefore only the continuous spectrum occurs for these targets and this voltage.) The striking fact is that there is no radiation with λ less than a critical value λ_{min}. Furthermore, the value of λ_{min} is the same for all three targets. This limiting value of λ is called the *Duane-Hunt limit*.

If the voltage V through which electrons in the X-ray tube are accelerated is increased, λ_{min} decreases. This result is best illustrated by plotting $\nu_{max} = c/\lambda_{min}$ as a function of V, with the result shown in Fig. 4-26. All the data lie on the line

$$h\nu_{max} = eV \qquad (4\text{-}10)$$

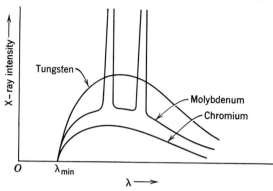

Fig. 4-25. Continuous X-ray spectra from three tubes, with target elements indicated, operated with the same accelerating voltage. [From D. Ulrey, *Phys. Rev.*, **11**, 405 (1918).]

The photon explanation of eq. 4-10 is: An electron with kinetic energy eV approaches the nucleus of an atom in the target. It is strongly accelerated by the Coulomb force between it and the nucleus, and therefore it radiates energy as mentioned in Sec. 4-3. If the electron radiates all its

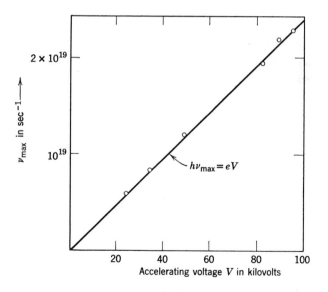

Fig. 4-26. Maximum frequency of the continuous X-ray spectrum ($\nu_{max} = c/\lambda_{min}$) as a function of the voltage applied to the X-ray tube. [From A. W. Hull, *Phys. Rev.*, 7, 157 (1916).]

kinetic energy and is brought to rest, a photon with energy $h\nu$ equal to the electron's original kinetic energy eV is emitted. Such an encounter between an electron and a nucleus produces a photon with the maximum frequency (or minimum λ). If the electron-nucleus collision is not a direct hit, the electron will not lose all its original energy and a photon with $\nu < \nu_{max}$ will be emitted. Therefore all frequencies of X-rays up to ν_{max} are emitted. Furthermore, this maximum value is determined by the electron energy eV and not by any property of the nuclei in the target, and thus ν_{max} should be the same for all target materials. The photon explanation therefore agrees perfectly with the data presented in Figs. 4-25 and 4-26. Classical theory is incapable of explaining satisfactorily why a minimum λ should occur.

The process described in the preceding paragraph is just the inverse process of the photoelectric effect, with two modifications. The first of them is that eq. 4-10 does not include the work function $e\phi$, and eq. 4-2 does. Strictly speaking, eq. 4-10 should have such a term, since an electron incident upon the anode of an X-ray tube acquires an additional energy $e\phi$ as it enters the anode. But, since V is of the order of tens of thousands of volts, a correction of a few volts would be negligible in data like those of Fig. 4-26. When we use the Duane-Hunt limit experiment for a *precise* determination of h/e, we must use eq. 4-2. The second modification is that in the photoelectric effect a photon is absorbed as a unit, and hence the kinetic energy it gives to an electron is either zero (photon passes by without absorption) or $h\nu$ (photon is absorbed). In the X-ray generation process, on the other hand, the electron can give up *part* of its kinetic energy eV; thus a succession of photons can be emitted with various energies. The details of this production depend on the target material, and its theory is complicated. The significant and simple fact remains, however, that the maximum-energy photon that can be produced is emitted when an electron loses *all* its energy in one collision, and such collisions are the reverse of the collisions producing the photoelectric effect.*

The dependence of the intensity of the continuous X-ray spectrum on target material and tube voltage is of interest from the practical

---

* The values of h/e that have been determined by the Duane-Hunt limit experiment agree with other determinations. If the charge of the electron e were a function of its velocity, the values of h/e from experiments with electrons in one velocity range (Duane-Hunt limit, $K \sim 50,000$ e.V.) would not agree with those from another range (photoelectric effect, $K \sim 3$ e.V.). The Duane-Hunt limit experiment involves electron kinetic energies which are large enough to produce a significant difference in e/m from the value for low-energy electrons. Therefore we have experimental proof that the variation of e/m with velocity must be attributed to a variation of m, rather than to a variation of e. See the beginning of Sec. 1-7.

point of view. The intensity is proportional to the atomic number of the target element. An illustration of this fact is seen in Fig. 4-25 for Cr ($Z = 24$), Mo ($Z = 42$), and W ($Z = 74$). The intensity is also approximately proportional to the square of the tube voltage.

4-7 THE COMPTON EFFECT

The experiments discussed in the preceding sections of this chapter have been concerned with the *energy* of a photon. We now consider an experiment that demonstrates that a photon has *momentum* and that gives additional evidence that the interaction between electromagnetic radiation and electrons must be considered as collisions between photons and electrons.

When waves are reflected or diffracted by obstacles we expect no change of wavelength or frequency to result. If monochromatic light of frequency ν_0 is incident on an arbitrarily complicated optical apparatus, changes in intensity and perhaps degree of polarization can be observed but no change in frequency. In the radio-frequency region of the electromagnetic wave spectrum, any appreciable change in ν when waves are reflected by hills or trees would be easily detectable and disastrous for radio communication. It was therefore a striking experiment when Compton showed that X-rays scattered by atoms exhibit a new frequency ν' as well as the incident frequency ν_0. This scattering with shift in frequency is called the *Compton effect.*

The apparatus for studying the Compton effect is illustrated schematically in Fig. 4-27. The X-ray tube is mounted on a rotatable platform, so that X-rays can strike the scattering block S at any selected

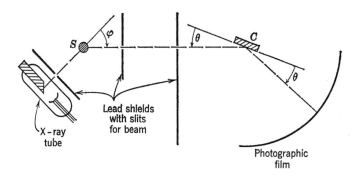

Fig. 4-27. Apparatus for studying the Compton effect. The X-ray tube is on a rotatable platform so that the angle φ can be varied.

angle φ with the line SC. The X-radiation studied is one of the characteristic X-ray lines. The data to be quoted were taken with a molybdenum target tube and with the 0.71-Å line of the molybdenum spectrum. First the tube is positioned so that $\varphi = 0$, the scattering block S is removed, and the crystal C is slowly rotated. For a value of θ satisfying Bragg's law for a principal plane of the crystal, an exposed line appears on the film as illustrated in Fig. 4-28a. The apparent width of this line is caused by the spread in angle of X-rays selected by the collimating slits.

The scattering block is next inserted, and φ is set at a succession of different values. For each φ value, the crystal is rotated and the film exposed in order to determine the wavelengths present in the scattered beam. The new and interesting experimental fact is that there is an *additional line* λ' in the X-ray spectrum after scattering that was not present in the output of the X-ray tube. The separation $\lambda' - \lambda_0$ depends on the scattering angle φ but does not depend on the wavelength λ_0 or upon the material used for the scatterer.

The photon theory of this process is simple in principle but complicated in detail. Therefore we shall describe the method and the results but not the detailed algebra. The basic assumption of the theory is that an X-ray photon collides with a relatively free electron and imparts some of its momentum and energy to the electron. In this process energy and momentum are conserved just as they would be in laboratory-scale experiments. The $h\nu'$ energy of the scattered photon is thus less than the $h\nu_0$ of the incident photon. The geometry of a collision is shown in Fig. 4-29, where vectors are drawn for the momentum p_0 of the incident photon, the momentum p of the scattered photon, and the momentum p_e of the *recoil* electron.

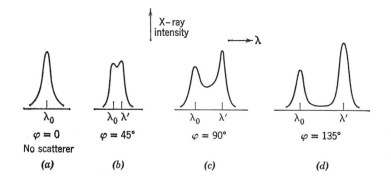

Fig. 4-28. Spectra of scattered X-rays at different scattering angles. The vertical scale is different for each φ. [From A. H. Compton, *Phys. Rev.*, **22**, 411 (1923).]

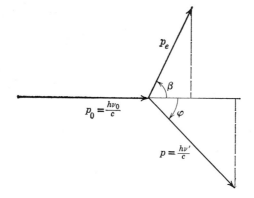

Fig. 4-29. Momentum vectors for a collision between a photon and an electron. The electron was approximately at rest before the collision.

The momentum of a photon can be calculated by using the relativistic expression for kinetic energy from eq. 1-17 ($K = Mc^2 - M_0c^2$). But M_0 (the rest mass) of a photon must be zero, since a photon travels with the velocity of light and a non-zero rest mass would give an infinite energy. Thus the only mass a photon has is the mass K/c^2 attributed to its motion. Its momentum p is its mass times its velocity c and can be calculated from its energy $h\nu = K$ as follows:

$$p = Mc = \left(\frac{K}{c^2}\right)c = \frac{h\nu}{c} \tag{4-11}$$

The momentum of the recoil electron is

$$p_e = \frac{m_0 v}{(1 - v^2/c^2)^{1/2}}$$

where v is its velocity and m_0 is as usual the rest mass of the electron. The energy of the recoil electron (eq. 1-16) is

$$\frac{m_0 c^2}{(1 - v^2/c^2)^{1/2}} - m_0 c^2$$

The photon theory simply equates the energy before the collision to the energy after the collision, and equates the sums of the x- and y-components of momentum before to the corresponding components after the collision. The principal result of this calculation is

$$\lambda' - \lambda_0 = \frac{h}{m_0 c} (1 - \cos \varphi) \tag{4-12}$$

The kinetic energy and angle of emission of the recoil electron can also be predicted. Equation 4-12 gives excellent agreement with experiment. Furthermore, the recoil electrons can be observed, and their energies and directions agree with the photon theory. The photon theory of the Compton effect therefore succeeds where classical theory failed.

It is noteworthy that the wavelength *shift* $\lambda' - \lambda_0$ from eq. 4-12 is independent of λ. If the λ's are expressed in angstroms, this equation becomes

$$\lambda' - \lambda_0 = 0.0243(1 - \cos \varphi) \overset{\circ}{\text{A}}$$

It should now be apparent why this effect gives an appreciable wavelength shift only for X-ray or γ-ray photons. With visible light or longer-λ radiation, the shift is a very small fraction of λ_0. Furthermore, the momentum of the incident visible-light photon is very small compared to the momentum of an electron in an atom, which can be a vector in any direction, and hence a spread in λ' values results. In the X-ray region, on the other hand, $\lambda' - \lambda_0$ is an appreciable fraction of λ_0, and our approximation that the atomic electrons are initially at rest is a better approximation because of the much larger momentum of the X-ray photon.

The fact that there is scattered radiation with $\lambda = \lambda_0$ means that some of the X-ray photons are scattered without the loss of energy or momentum. Evidently another type of collision occurs in which the binding of the electron to the atom as a whole cannot be neglected. If scattering occurred from an electron very tightly bound to the nucleus, rather than from a relatively free electron, the scattering would appear to be from a system (electron + nucleus) with mass $\gg m_0$. It is apparent from eq. 4-12 that a negligible change in λ would result. We shall see in Chapter 6 that atoms contain both loosely and tightly bound electrons. Atoms with large Z have more tightly bound electrons, and therefore the Compton effect is most easily observed with atoms of small Z. (A carbon block was used in the experiments described above.)

4-8 BLACK-BODY RADIATION

Any solid heated above $700°C$ emits visible light. At lower temperatures a solid still emits radiation, but the intensity of its spectrum in the visible light region ($4000 \overset{\circ}{\text{A}} < \lambda < 7000 \overset{\circ}{\text{A}}$) is too weak to be seen. The radiation of a solid is quite different from the radiation from a low-pressure atomic gas. In a gas, the radiation occurs only at certain λ values, the spectral lines. Each atom radiates almost independently of all the others. In a solid, the atoms are so close together that the radiation from each is strongly influenced by its neighbors. The radiation from a solid is a *continuous* spectrum, with some energy emitted at all λ's. [A high-pressure gas discharge (for example, a 1000-watt mercury lamp operating at a pressure of 80 atmospheres) gives a spectrum intermediate between a line and a continuous spectrum; it consists of very broad lines, almost a continuum, because the atoms collide so frequently.]

All solids exhibit nearly the same spectrum of continuous radiation when at the same temperature. The fundamental spectrum that all solids approximate is called the *black-body spectrum*. As the name implies, it is the radiation as a function of wavelength that is emitted by a perfectly *black* body, a solid with zero reflectivity at all wavelengths. The ratio of the energy radiated per second by a solid at temperature T to the energy radiated by a black body of equal area at temperature T is called the *total emissivity* ϵ. The similar ratio, but for only the energy in the narrow band between λ and $\lambda + d\lambda$, is called the *spectral emissivity* ϵ_λ and is a function of λ. The emissivity ϵ_λ can most easily be measured

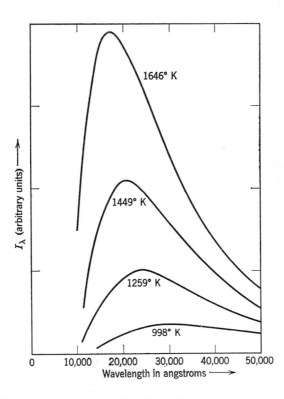

Fig. 4-30. Distribution of power radiated by a black body at various absolute temperatures. The curves are drawn through data by Lummer and Pringsheim.

by measuring the reflectivity r_λ in an obvious way; a simple thermodynamic argument shows that $\epsilon_\lambda + r_\lambda = 1$. Our chief concern here is with the radiation from a black body ($\epsilon_\lambda = 1$ for all λ). For experimental purposes a small area that emits black-body radiation can be obtained simply by drilling a hole in any solid; if the hole is several diameters deep, the radiation from the bottom of the hole is essentially black-body radiation.

The total power W radiated per unit area (in watts per square meter) of a black body can be measured by the use of a detector such as a thermocouple covered with carbon black in order to make it perfectly absorbing. The power radiated in any one wavelength range can be measured by interposing between the black body and the detector a spectrometer that passes only that wavelength range. We define $I_\lambda \, d\lambda = (dW/d\lambda) \, d\lambda$ as the power radiated by unit area of a black body in the wavelength range between λ and $\lambda + d\lambda$. Figure 4-30 shows experimental curves for I_λ as a function of λ for four different temperatures.

Classical theory is incapable of explaining the shape of the curves of emitted power as a function of wavelength. The analysis is rather abstract, and we shall only sketch it here. The classical approach treats the radiation inside an enclosure as if it were an assembly of oscillators, each at a different frequency. Each is assumed to have, on the average, $\frac{1}{2}kT$ kinetic energy, like the molecules of a gas. The abstract part of the argument is the calculation of the number of such oscillators in a wavelength range between λ and $\lambda + d\lambda$ as a function of λ. This calculation is not really difficult, especially for the student familiar with waveguides and resonant cavities, but it would take considerable time and space. The conclusion is that the number of oscillators increases as λ decreases and is proportional to $d\lambda/\lambda^4$. If each such oscillator is given $\frac{1}{2}kT$ kinetic energy, the power I_λ radiated is proportional to $kT\lambda^{-4}$. This theory works well at the extreme long-wavelength side of the black-body spectrum, but it obviously fails badly at short λ's, since it predicts a continual and rapid increase in I_λ as λ decreases (the dashed curve in Fig. 4-31).

Planck presented his theory of black-body radiation in 1900; it was the first of the quantum or photon theories of radiation. Planck saw that he could get agreement between theory and experiment if he *quantized* the energies of the oscillators. That is, he assumed that each oscillator could have only one or another of a set of *discrete* energy values, differing in energy by $h\nu$. In the region of the spectrum where $h\nu$ is very much less than kT (long λ), this assumption makes no appreciable change in the classical theory. But in the short-λ region, where $h\nu$ is

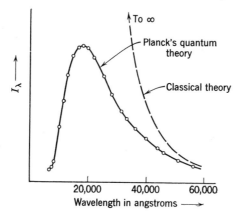

Fig. 4-31. Comparison of classical and quantum theories of black-body radiation with experiment, $T = 1600°K$. The data points (circles) are from Coblentz, *Natl. Bur. Standards Bull.*, **13**, 476 (1916).

very much greater than kT, Planck's hypothesis gives quite different results from classical theory, because almost all such oscillators are in their lowest energy states and therefore do not radiate (recall the argument at the end of Sec. 2-4). The number of excited oscillators with frequencies between ν and $\nu + d\nu$ turns out to be proportional to

$$\frac{\nu^2 \, d\nu}{e^{h\nu/kT} - 1} \tag{4-13}$$

Hence Planck's theory predicts that I_λ should drop rapidly to zero at large ν (small λ).

The actual expression from Planck's theory is

$$I_\lambda = \frac{2\pi c^2 h}{\lambda^5} \left\{ \frac{1}{e^{hc/\lambda kT} - 1} \right\} \tag{4-14}$$

This expression is plotted as the solid curve in Fig. 4-31 and is in excellent agreement with experiment over a wide range of λ and T. For practical purposes, we can evaluate the constant, express λ in angstroms, and obtain the following expression:

$$I_\lambda = \frac{3.74 \times 10^{34}}{\lambda^5} \left\{ \frac{1}{e^{(1.44 \times 10^8)/\lambda T} - 1} \right\} \text{ watts/m.}^2 \, \overset{\circ}{A} \tag{4-15}$$

Equation 4-15 gives the power radiated (watts) per unit area (m.^{2}) per unit wavelength range ($\overset{\circ}{A}$).

The total radiation emitted by a black body can be found by integrating $I_\lambda \, d\lambda$ over all λ's (the area under the curve in Fig. 4-31); in the integration, the value (from tables) of the integral

$$\int_0^\infty \frac{x^3 \, dx}{e^x - 1} = \frac{\pi^4}{15} \tag{4-16}$$

is needed. The result is

$$W = \int_0^\infty I_\lambda \, d\lambda = \frac{2\pi^5 k^4}{15c^2 h^3} T^4 = 5.67 \times 10^{-8} T^4 \text{ watts/m.}^2 \tag{4-17}$$

This equation, the *Stefan-Boltzmann law* of total radiation, agrees with experiment both in the T^4 dependence and in the value of the constant. If the body is not black, a factor ϵ (the total emissivity) appears on the right. ϵ is always less than 1; it usually lies between 0.2 and 0.9 but for highly polished metals may be as small as 0.01.

Of the many practical consequences of the black-body radiation theory we shall consider only one: The application of measurements of radiation to the determination of the temperature of a hot solid or liquid. The total radiation (eq. 4-17) from a small area of the emitter can be focused on a thermocouple, and the emf of the thermocouple is then proportional to T^4. In order to use this method, either the total emissivity ϵ of the emitter must be known or else the radiation from the bottom of a hole in the emitter must be observed in order to obtain black-body conditions.

Another method of temperature measurement is the method of *optical pyrometry*. In this method, an image of the emitter is viewed by eye. A tungsten filament is located in the plane of this image and also viewed. The radiation from both the emitter and the filament passes through a red filter, so that the eye sees only a narrow band of wavelengths near $\lambda = 6500$ Å. The observer adjusts the current through the filament until the image of the emitter and the filament have the same intensity (under these conditions the filament seems to disappear). The current through the filament is then measured; this current is a measure of the temperature of the filament and thus of the intensity of the radiation it emits. Therefore the filament current when the filament disappears is a measure of the radiation from the emitter in the small wavelength range near 6500 Å. By calibration, the filament current thus indicates the temperature of the emitter. If a hole cannot be drilled in the emitter in order to provide black-body conditions, an emissivity correction must be made. The correction procedure is to equate the intensity coming from the emitter at temperature T and with spectral emissivity ϵ_λ to the intensity that would have come from a black body at temperature T_b (and $\epsilon_b = 1$, of course). The instrument reads T_b. If ϵ_λ is known, T can be computed by using eq. 4-15 with a factor of ϵ_λ on the right.

4-9 THE STERN-GERLACH EXPERIMENT

In the experiments discussed thus far in this chapter, ample evidence has been gathered to demonstrate that energy is quantized. A different kind of quantization was revealed by an atomic beam experiment performed by Stern and Gerlach in 1922, in which it was demonstrated that the vector magnetic moment of an atom placed in an external magnetic field can be in only one or another of a set of *discrete directions*.

The experiment is illustrated in Fig. 4-32. Some of the silver atoms evaporated in the oven go through the narrow slits and pass along and very close to the knife-edge pole piece of the magnet. These atoms are electrically neutral and hence are not deflected by the force eq. 1-6. But silver (like most other atoms) has a magnetic moment, and in an inhomogeneous magnetic field a magnetic dipole of strength $\mathfrak{M}$ is deflected by a force

$$F_z = \mathfrak{M}_z \frac{\partial \mathcal{B}_z}{\partial z} \qquad (4\text{-}18)$$

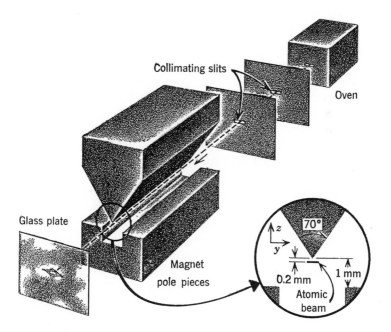

Fig. 4-32. The Stern-Gerlach atomic beam experiment. The inset shows the beam close to the knife-edge pole piece of the magnet.

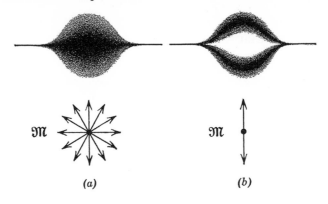

Fig. 4-33. (*a*) Pattern of silver deposit expected classically. (*b*) Pattern actually observed; only two *z*-components of the atomic magnetic moment $\mathfrak{M}$ are observed.

arising from the gradient $\partial \mathfrak{B}_z/\partial z$ of the magnetic induction. $\mathfrak{B}$ is strong and in the *z*-direction; the design of the magnet pole pieces produces a large $\partial \mathfrak{B}_z/\partial z$ near the knife-edge. Thus atoms are sorted according to their components $\mathfrak{M}_z$ of magnetic moment in the direction of the external field. The amount of the deflection produced by this force depends on the length of the atom's path and on its velocity. The Maxwell distribution of velocities thus spreads the deflected beam somewhat about an average deflection. Nevertheless $\partial \mathfrak{B}_z/\partial z$ can be made large enough to observe this average deflection, and measurement of the distribution of silver deposited on the glass plate permits calculation of the *z*-component of $\mathfrak{M}$.

Classically, the vector moment $\mathfrak{M}$ could be in any direction, and thus a continuous distribution of $\mathfrak{M}_z$ values would be expected (Fig. 4-33*a*). The observed silver deposit is shown in Fig. 4-33*b* and reveals *only two* components of $\mathfrak{M}_z$, one parallel to $\mathfrak{B}$ and one antiparallel. In other words, the experiment gives direct proof that the directions of $\mathfrak{M}$ with respect to a magnetic field are *quantized*.

The Bohr quantum theory had predicted values of $\mathfrak{M}$ for atoms, based on the angular momentum of the electrons in their "orbits" about the nucleus. The predicted value for silver did not agree with the experimental value, and furthermore an odd number of components was predicted (instead of two, as observed). The cause of the disagreement was later attributed to the intrinsic (spin) magnetic moment of the electron, which was discovered in 1925 by Uhlenbeck and Goudsmit in the course of line spectra investigations. The magnetic moment of silver arises exclusively from the electron spin, and thus the Stern-Gerlach experiment shows that the spin magnetic moment (and angular momentum) can have either of two discrete directions with respect to an external field. Later experiments with hydrogen atoms, and with alkali and other complex atoms, verified this conclusion.

4-10 ELECTRON DIFFRACTION

In this section we shall study the experiments that show that electrons have the properties of *waves*. Classical physics considers electrons as *particles;* the motion of an electron in electric and magnetic fields in the laboratory certainly agrees with the predictions of classical mechanics for the motion of a point charge. Only when experiments were performed that involved the interaction of electrons and atoms did it become apparent that classical mechanics was inadequate. In the crucial experiment performed by Davisson and Germer, a beam of monoenergetic electrons struck a nickel single crystal target, and the current of electrons reflected from the target was studied as a function of incident energy, emergent angle, and orientation of the crystal. This experiment showed that electrons are diffracted like waves when a suitable diffraction grating is provided.

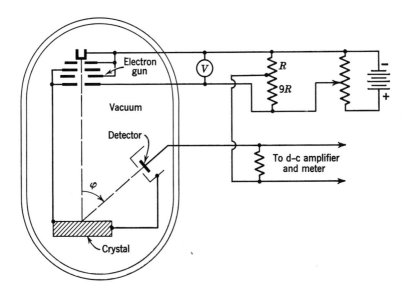

Fig. 4-34. The Davisson-Germer experiment. Electrons encounter a retarding field inside the detector so that the measured current consists of only those electrons that have been scattered with negligible loss of energy.

The Davisson-Germer experiment is illustrated in Fig. 4-34. The electron gun, target, and detector are all enclosed in an evacuated tube. The detector can be rotated so that φ takes on different values. Classically, we should expect the electrons to be scattered by the nickel crystal, and some of them to enter the detector. The detector current may vary with φ and the accelerating voltage V, but we do not expect a very sensitive or striking dependence on either parameter. Electrons of any kinetic energy K should be deflected into all angles. (Many of them will lose a large fraction of their energy in the scattering process, but they will not be collected since the detector is only a few volts positive with respect to the cathode, and any appreciable loss of energy will prevent an electron from arriving at the detector.)

Some of the results of this experiment are shown in Fig. 4-35. Superimposed on the (expected) background scattering is a peak in the current of reflected electrons centered on $K = 54$ e.V. and $\varphi = 50°$, which is quite unexpected on the basis of classical theory. This reflection is clearly selective in both angle and energy; we shall show that both forms of selectivity are exactly what would be expected if electrons have the properties of *waves*.

Before this experiment had been performed, de Broglie had proposed the hypothesis that electrons should be diffracted as if they had a wavelength:

$$\lambda = \frac{h}{\text{Momentum}} = \frac{h}{mv} = \frac{h}{(2mK)^{\frac{1}{2}}} \tag{4-19}$$

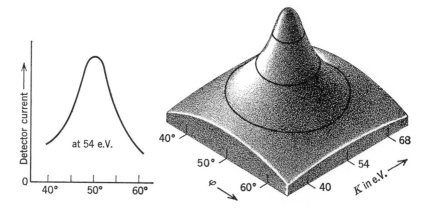

Fig. 4-35. Detector current as a function of angle and electron kinetic energy in the Davisson-Germer experiment.

For $K = 54$ e.V. we calculate from eq. 4-19 that $\lambda = 1.67 \times 10^{-10}$ m.
or 1.67 Å. Now this is the same order of magnitude as the spacing between atoms in the nickel crystal, and we therefore attempt to understand the selective reflection as the result of diffraction of waves of this
wavelength from the three-dimensional array of nickel atoms.

This diffraction is complicated by the fact that low-energy electrons
(unlike X-rays) are strongly scattered by atoms and hence do not penetrate more than a few atomic layers into the crystal. We begin by considering only the surface layer of atoms, which will be the major parti-
cipants in the reflection and are shown in Fig. 4-36. Rows of these
atoms separated by a distance d dif-
fract like the mirror diffraction gratings commonly used in practical
spectrographs for visible light (but
they are rarely described in textbooks,
most of which illustrate only *transmission* gratings). Constructive interference of the waves scattered by different rows of atoms occurs when
$n\lambda = d \sin \varphi$, where n is an integer.
Now d can be determined to be 2.15 Å
from X-ray measurements for this
particular orientation of the nickel
crystal, and therefore $\lambda = 2.15 \times$

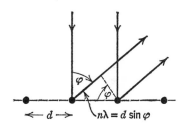

Fig. 4-36. Interference among
partial reflections from surface
atoms.

$\sin 50° = 1.65$ Å, in excellent agreement with de Broglie's hypothesis.
[We have let $n = 1$ since, if $n = 2$ or more, other reflections (not observed) would have appeared for $\varphi = 23°$ or less.]

Clearly this is not the complete analysis, however, since if only this
surface diffraction were occurring, many other combinations of λ and φ
would produce strong reflections; there would have been a *ridge* along
the curve $\lambda = d \sin \varphi$ (with λ related to K by eq. 4-19), rather than a
peak, in Fig. 4-35. The participation of atoms below the surface is considered in Fig. 4-37. Electrons entering normally to the surface are accelerated as they pass through the surface barrier (Fig. 4-1). When they
leave the crystal, their components of velocity normal to the surface are
decreased, but the components parallel to the surface are unchanged.
Thus *refraction* occurs for the exiting electrons; if the angle of reflection
is to equal the angle of incidence for the planes shown, the electrons must
be bent from 44° with respect to the normal inside the crystal to 50°
(observed) outside the crystal. Bragg reflection according to eq. 4-7
occurs with $n = 1$ from the planes spaced 0.81 Å apart when $\lambda = 1.62$

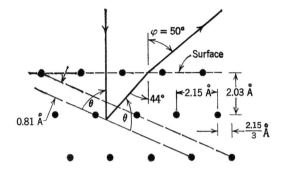

Fig. 4-37. Bragg interference among reflections from planes of atoms inside a nickel crystal.

$\sin \theta$. Since $\theta = 68°$, $\lambda = 1.49$ Å. From the refraction of the outgoing wave, we calculate that K of an electron inside the crystal is 13 e.V. more than outside. Thus the de Broglie λ for the electron inside the crystal should be $\lambda = (\frac{54}{67})^{\frac{1}{2}} 1.67 = 1.50$ Å, again in excellent agreement with experiment.

Agreement such as this was obtained by Davisson and Germer for widely different values of λ, n, d, and K. Even the refraction correction (frequently very small) did not have to be changed for different crystal orientations and different values of K but corresponded to the same change in K as an electron left the crystal. In every experiment, the electrons were found to behave like waves with a wavelength given by eq. 4-19. These experiments prove that electrons are diffracted by a grating of atomic-size spacing just as waves are diffracted.

The selectivity in angle and in K exhibited in Fig. 4-35 resembles a diffraction pattern from a grating of only a few lines, rather than from thousands of lines or atom planes as is common in visible light or X-ray diffraction experiments. Surface imperfections restrict the interference diagrammed in Fig. 4-36 to a small number of rows of atoms. The small penetrating power of low-energy electrons restricts the Bragg interference diagrammed in Fig. 4-37 to a small number of atomic planes. Thus the breadth of the peak in Fig. 4-35 can be well understood and reveals details of surface structure and electron penetration.

Electron diffraction experiments and apparatus are now common. The usual technique, quite similar to the powder diffraction method with X-rays, was discovered independently by G. P. Thomson almost simultaneously with the Davisson-Germer experiment. Electrons are accelerated through a potential difference of the order of 50,000 volts and strike a powder sample. The diffraction pattern is observed on a photographic film and exhibits much sharper lines than the diffraction shown in Fig. 4-35, since hundreds of atomic planes participate in the interference with these much more penetrating electrons.

Electron diffraction is frequently used to study the structure of surfaces, for example, in studies of corrosion or catalysis, since the observed diffraction pattern reveals the structure of the last few or last few hundred atoms at the surface of a solid. Electron diffraction is also used to study molecular gases. Suppose, for example, that the molecules of a gas are each composed of two atoms with a distance R_0 between their centers. These molecules will be randomly oriented, like the little crystals in a powder diffraction experiment. Some will be oriented just right to give constructive interference between the partial scatterings at the two atoms. A ring pattern like the powder pictures will result. From such measurements the interatomic spacing and structure of molecules can be measured, and such measurements are very useful in studying molecular structure (Sec. 7-2).

These experiments indicate that an electron moves as if guided by a wave motion. Yet we know that on a laboratory scale of sizes electrons behave like particles. Our picture of the electron motion is thus like that of a photon with three exceptions: (1) The electron has a charge. (2) It has a rest mass. (3) Its velocity is dependent on its energy. The electron is very small, but the wave packet that guides it can extend over many angstroms. If the electron were large enough to extend from one atom to the next, electrons would not penetrate solids at all. On the other hand, the guiding wave *must* extend over distances of at least several interatomic spacings; otherwise, we could not get the constructive interference among partial reflections that is responsible for Bragg reflection.

Our picture of this wave packet is just the same as Fig. 4-8 for a photon except that the ordinate is no longer the electric field. The electron diffraction experiments do not tell us directly what the ordinate should be. The experiments indicate only that a wave motion is involved. We shall learn in the next chapter more about the nature of the wave. For the time being we need only the fact that the electron behaves as if it is carried along by the wave packet; when the wave packets are strongly reflected in a certain direction, a large fraction of the electrons go in that direction.

We should show that such a wave packet moves in electric and magnetic fields of laboratory size as if it were a point charge moving according to Newton's laws. We shall not do this, but it can be done by using the results of Chapter 5.

4-11   NEUTRON DIFFRACTION

Wave properties are not confined to the electron but are a universal attribute of matter.   Any mass M exhibits the diffraction associated with a de Broglie wavelength

$$\lambda = \frac{h}{\text{Momentum}} = \frac{h}{Mv} \tag{4-20}$$

Experiments have been performed using atoms or molecules as the particles and a crystal as the grating.   These experiments are difficult because of the low beam strengths available in atomic beams, but the results give evidence for eq. 4-20.   Much better evidence comes from diffraction experiments with neutrons.

Nuclear reactors provide a copious supply of neutrons (such reactors are described in Sec. 14-3).   The apparatus for demonstrating diffraction with these neutrons is illustrated in Fig. 4-38.   Neutrons from the reactor are slowed down in the graphite thermal column.   Neutrons interact only very weakly with carbon, and therefore very few neutrons are lost in the column.   They collide elastically many times with carbon nuclei, however, and keep losing energy as long as their energies are greater than the thermal energy ($\frac{1}{2}kT$ per degree of freedom) of the carbon atoms. When their energies have been reduced to this point, collisions no longer change the distribution of neutron energies.   The neutrons are in thermal equilibrium and have an energy spectrum just like the particles in a gas at temperature T, namely a Maxwellian distribution.   From this energy

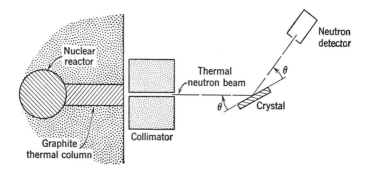

Fig. 4-38.   Apparatus for demonstrating the wave properties of neutrons.

distribution and the de Broglie expression, the wavelength distribution of the beam can be computed.

The wavelength distribution of the beam can be *measured* by introducing a crystal and a neutron detector as shown in Fig. 4-38 (detectors suitable for this experiment are described in Sec. 14-6). The crystal and detector are turned so that the two angles marked θ are kept equal. The arrangement is thus like the X-ray diffraction apparatus of Fig. 4-18 except that a different detector is used. Reflection from the crystal occurs only when λ and θ satisfy the Bragg relation, and therefore the fraction of neutrons reflected as a function of θ gives an experimental determination of the wavelength distribution of the incident neutrons. This measured distribution is found to be just what the Maxwellian distribution and the de Broglie relation predict. Of course, the fact that the Bragg relation works at all is evidence that neutrons have wave properties.

Neutron diffraction has proved to be a powerful tool for studying the structure of crystals, especially organic crystals containing hydrogen. X-ray and electron diffraction are not very useful for such studies, since hydrogen contains only one electron and one nuclear charge. X-rays interact primarily with the electrons of atoms, and electrons interact with the atomic electrons and the nuclear charge. Therefore, the scattering of X-rays or electrons by hydrogen is very weak compared to scattering by heavier atoms. Thus the presence of hydrogen cannot usually be detected, and the position of hydrogen atoms in a crystal structure cannot be measured by X-ray or electron diffraction. Neutrons, on the other hand, interact with nuclei through the nuclear force and also by the interaction between the magnetic moments of the neutron and the atom. This interaction is different from nucleus to nucleus but is quite strong for hydrogen. Neutron diffraction is also frequently applied to the study of the magnetic properties of solids.

The arrangement for applying neutron diffraction to the study of crystal structure is illustrated in Fig. 4-39. The first crystal is the same as the crystal of Fig. 4-38, and a beam of thermal neutrons is incident upon it as in that figure. The neutrons reflected from this crystal, which is now kept fixed, consist of only those with a single wavelength, and this monochromatic beam is incident on the crystal being studied. This crystal and the detector are rotated in synchronism so that the two angles marked φ are kept equal. Detector current is observed only when λ and φ satisfy the Bragg relation for a set of planes in the crystal. In this way the spacing d between planes can be investigated, as in X-ray diffraction. The powder method of studying crystals is also used with a

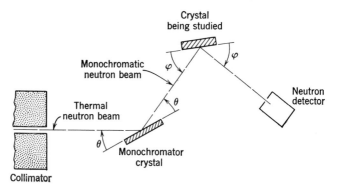

Fig. 4-39. Apparatus for using neutron diffraction to study crystal structure.

monochromatic beam of neutrons provided by a crystal monochromator.

These diffraction experiments with neutrons have thus verified the de Broglie relation and have provided an important tool for studying crystal structure.

4-12   THE INDETERMINACY PRINCIPLE

The experiments described in the preceding sections prove that classical mechanics is inadequate to predict phenomena on an atomic scale. It was shown that electromagnetic radiation was emitted and absorbed as if it consisted of photons, and that electrons, neutrons, and other masses moved as if guided by waves. In both kinds of experiments we were compelled to look upon the photon or the electron as if it were a *wave packet*, as illustrated in Fig. 4-8. These packets are emitted or absorbed as units, and their behavior in diffraction experiments is determined by their wavelengths.

Many individual inadequacies or false predictions of classical mechanics have been stated in the preceding sections. These difficulties can be summarized by discussing the *Indeterminacy Principle* of Heisenberg (also called the Uncertainty Principle). This principle can be shown to follow from the experiments already discussed, or it can be shown to follow from the wave-packet concept that in turn follows from these experiments. We shall first state this principle and then show how it is related to experiments and wave packets.

The Indeterminacy Principle is concerned with the simultaneous measurement of certain pairs of variables. It is possible to state it very generally, but such a statement would necessarily be rather abstract. Instead, we shall state two aspects of the principle that involve by far the most important pairs of variables and that illuminate the physical meaning of the principle.

The first aspect involves the simultaneous measurement of the momentum and position of a particle (e.g., an electron or a photon). The principle states that experiment cannot fix these to an unlimited precision, but that the momentum p_x is determinable only to within a range Δp_x, and the position within a range Δx, where

$$\Delta p_x \, \Delta x \geqq h \qquad (4\text{-}21)$$

(An expression of this type also holds for other components of linear momentum and for angular momentum and angular position.) We could devise an experiment that would give much *poorer* determinations of p_x and x than $\Delta p_x \, \Delta x = h$. The principle states that we cannot do *better* than this. Note that there is no restriction on Δx or on Δp_x, but only on their product. Therefore we could, for example, devise an experiment to measure x very accurately, but only at the sacrifice of the knowledge of p_x. Note also that there is no restriction on products like $\Delta p_x \, \Delta y$.

The second aspect is a similar limitation on the simultaneous measurement of the energy E and time t. For example, E might be the energy of a photon and t the time it was emitted. The principle states that

$$\Delta E \, \Delta t \geqq h \qquad (4\text{-}22)$$

Here again there are no restrictions on the accuracy with which E or t can be measured, but only on the product $\Delta E \, \Delta t$.

The Indeterminacy Principle is quite foreign to classical mechanics, which recognizes no fundamental limitations on measurements of any kind. Note that, in common with all quantum theories and phenomena, Planck's constant appears here.

An example of the way these limitations follow from the experiments already discussed is illustrated in the experiment illustrated in Fig. 4-40. In this experiment we are attempting to find the position of an electron along the axis of x and the x component p_x of its momentum. We naturally use a microscope of some kind, and we immediately inquire: What is the fundamental limitation on the precision with which x can be measured? The answer to this question is provided by physical optics, which tells us that diffraction makes a broadened image of a point object. The position x of the object can be determined with a precision limited by the wavelength λ of the light being used and the numerical

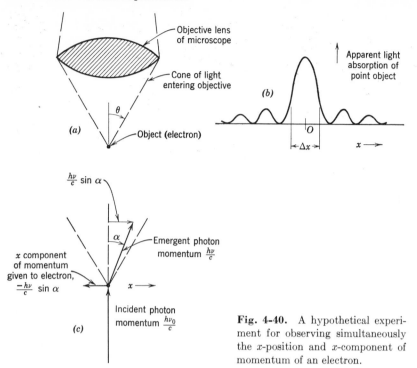

(a)

Objective lens
of microscope

Cone of light
entering objective

θ

Object (electron)

(b)

Apparent light
absorption of
point object

O

Δx

$x \longrightarrow$

$\frac{h\nu}{c} \sin \alpha$

x component
of momentum
given to electron,
$\frac{-h\nu}{c} \sin \alpha$

α

Emergent photon
momentum $\frac{h\nu}{c}$

$x \longrightarrow$

Incident photon
momentum $\frac{h\nu_0}{c}$

(c)

Fig. 4-40. A hypothetical experiment for observing simultaneously the x-position and x-component of momentum of an electron.

aperture (a function of the apex angle 2θ of the cone of light entering the optical system). The expression from physical optics is

$$\Delta x = \frac{\lambda}{2 \sin \theta} \qquad (4\text{-}23)$$

The image of the point object is "fuzzy" by the amount Δx as illustrated in Fig. 4-40b. Δx could be defined in several different ways, and for each definition the factor 2 in eq. 4-23 would be slightly different, but there would be no change in the physics.

Clearly we can make λ very small and therefore Δx very small. There is no *fundamental* limitation on Δx. There may be practical limitations on making microscopes using very short wavelengths, but these do not concern us here. If we had poor lenses, we might do very much worse than eq. 4-23 indicates, but we cannot do *better*, because of the wave nature of the light used and the diffraction accompanying waves.

What is the precision with which p_x can be measured? We might at first think that there is no fundamental limitation, since we could measure two positions of the electron at different times and compute the velocity and hence the momentum. The trouble with this approach is that we have ignored the possibility that the momentum of the object may be changed by the observation process. Suppose that we illuminate the object from below with radiation of wavelength λ. We know from the Compton effect experiments that this radiation consists of photons with momentum $h\nu/c = h/\lambda$. Unless one of these is scattered by the electron, we will not have any evidence in the microscope of the position of the object. If one of these *is* scattered, it gives some momentum to the electron.

We now see that the electron's momentum will be different after the scattering process. If we knew in which direction the scattered photon went, we could compute the change in the electron's momentum by applying the conservation of linear momentum. In other words, we could correct for the disturbance produced by the measuring process. But we do not know the direction of the emergent photon; all we know is that it entered the microscope lens system. In other words, it emerged within the cone of apex angle 2θ. The x component of momentum of the emergent photon can therefore be anything from $-(h/\lambda)\sin\theta$ to $+(h/\lambda)$ $\sin\theta$ (see Fig. 4-40c). Before striking the electron the photon had zero x momentum. Hence the x component of momentum given the electron can be anything from $+(h/\lambda)\sin\theta$ to $-(h/\lambda)\sin\theta$:

$$\Delta p_x = 2\,\frac{h\sin\theta}{\lambda}$$

Again, there is no fundamental limitation on Δp_x. We need only make λ very large to attain a very small Δp_x.

It is now clear, however, that we cannot *simultaneously* make Δx and Δp_x arbitrarily small, since the operation that makes one small makes the other large. In fact,

$$\Delta p_x\,\Delta x = \left(2\,\frac{h\sin\theta}{\lambda}\right)\left(\frac{\lambda}{2\sin\theta}\right) = h$$

Thus by assuming ideal conditions we have just reached the limiting value of this product given by the Indeterminacy Principle expressed in eq. 4-21. Note that the only parameters that we can vary (namely, λ and θ) do not enter into this product, and therefore there is nothing we can do to improve the precision of the simultaneous measurement of x and p_x. (If we had used a different definition of Δx or Δp_x we might have a numerical factor like $1/2\pi$ multiplying h in this equation, but this fact does not alter the basic result.)

This experiment has not been performed and perhaps never will be performed. But we have used the results of experiments that *have* been performed (resolving power of a lens and Compton effect) to show that simultaneous measurement of x and p_x is possible only to the precision indicated.

The second aspect (eq. 4-22) of the Indeterminacy Principle can be shown to follow from the properties of waves and from the experimental fact that $E = h\nu$. We shall first show that

$$\Delta\nu \, \Delta t \geqq 1 \qquad (4\text{-}24)$$

for any packet of waves. This demonstration can be accomplished rigorously on the basis of the Fourier transform theory, which is widely used in electrical engineering. We shall use only elementary ideas here to show that a relation like eq. 4-24 must hold. The Fourier theory is presented in Appendix E.

A train of waves with a single frequency ν_0 is illustrated in Fig. 4-41a, which is a plot of the function $u = \cos 2\pi\nu_0 t$, representing the oscillations of a wave function u at a fixed position. For example, u might be the pressure of a sound wave or the electric field of an electromagnetic wave. This variation continues from $t = -\infty$ to $t = +\infty$. The frequency ν is precisely determined ($\Delta\nu = 0$), and the frequency spectrum is simply the single line illustrated in Fig. 4-42a. But the time of arrival of the wave at the position considered is not known at all ($\Delta t = \infty$).

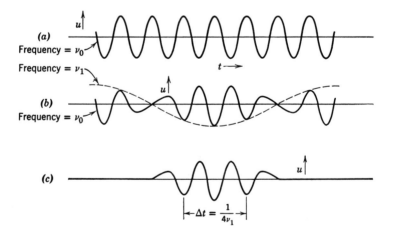

Fig. 4-41. The wave packet illustrated in (c) is obtained from (a) by modulating the continuous wave.

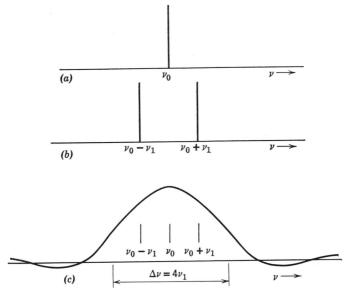

Fig. 4-42. Spectra of the waves (a), (b), and (c), respectively, illustrated in Fig. 4-41. The magnitude of the ordinate at any value of the abscissa ν is the relative amount of that ν in the complex wave.

A wave packet must have a beginning and an end, unlike the infinite wave train of Fig. 4-41a. A pulse of electromagnetic waves emitted by a radar antenna, a wave packet of light (a photon), or a wave packet guiding an electron appears at a given position for only a finite (and short) time. In order to obtain a function with these properties we must *modulate* the wave of Fig. 4-41a to give it a beginning and an end. We do this in two steps. First we multiply u by another, more slowly varying, sinusoidal function of time. u then becomes

$$u = \cos 2\pi\nu_1 t \cos 2\pi\nu_0 t \qquad (4\text{-}25)$$

and is illustrated in Fig. 4-41b. This wave function now has a frequency spectrum containing more than just the original frequency. In order to determine this spectrum we need to change eq. 4-25 into a form in which u is the *sum* of terms with various frequencies. This can be accomplished by the application of the following trigonometric identities:

$$\cos (y + x) = \cos x \cos y - \sin x \sin y$$
$$\cos (y - x) = \cos x \cos y + \sin x \sin y$$

From these expressions it follows that:

$$\cos x \cos y = \tfrac{1}{2} \cos (y - x) + \tfrac{1}{2} \cos (y + x)$$

When this result is inserted into eq. 4-25, it becomes apparent that the frequencies $\nu_0 - \nu_1$ and $\nu_0 + \nu_1$ are present in the modulated wave. The superposition of a simple wave at each of these frequencies gives exactly the modulated wave. This spectrum is illustrated in Fig. 4-42b; it is still a line spectrum, but it now contains a spread of frequencies because of the modulation.*

We still do not have a good wave packet, however, since eq. 4-25 gives u values for all t. An acceptable packet can be obtained by "chopping off" all except one-half cycle of the ν_1 oscillation, as shown in Fig. 4-41c. The spectrum of this cannot be found so easily as the result above, but the spectrum can be found by the methods of Appendix E and is illustrated in Fig. 4-42c. The spectrum is concentrated in a frequency range of the same order of magnitude as the spectrum of Fig. 4-42b, but now there is a continuous distribution of frequencies.

The time at which the wave packet of Fig. 4-41c arrives at a particular position can be determined with only a finite precision. A good estimate of this precision is the time interval $\Delta t = 1/4\nu_1$ identified on the figure. The spectrum of this packet (Fig. 4-42c) occupies a frequency range such that the frequency of the wave packet can be determined with only a finite precision. A good estimate of this precision is the frequency interval $\Delta \nu = 4\nu_1$ identified on the figure. The product of these precisions is $\Delta \nu \, \Delta t = 4\nu_1(1/4\nu_1) = 1$, which is just the result (eq. 4-24) we set out to prove. We might have estimated $\Delta \nu$ and Δt somewhat differently, in which case a number like $\frac{1}{2}$, or 2, or $1/\pi$ would have been obtained instead of 1. The important point is that this product is of the order of magnitude of unity, regardless of how long or short the wave packet is. Of course if we had an unsuitable shape of wave packet or a poor experimental arrangement we might have far *less* precision than indicated by eq. 4-24. We cannot hope to attain higher precision. Thus eq. 4-24 has been shown to be a property of wave packets.

The result $\Delta E \, \Delta t \geqq h$ follows from eq. 4-24 immediately, since the experimental evidence presented earlier in this chapter requires that $E = h\nu$. Therefore $\Delta E = h\Delta \nu$, and eq. 4-22 follows.

Both of the two parts of the Indeterminacy Principle impose definite restrictions on the information that can be obtained from experiments. These restrictions are quantitatively unimportant for the relatively large distances, momenta, energies, and times appropriate to laboratory-

---

* This is a familiar result to those who have studied radio communication: The audio frequency ν_1 modulates the carrier frequency ν_0, producing the side bands $\nu_0 - \nu_1$ and $\nu_0 + \nu_1$. In the present case of 100% modulation the carrier is suppressed and only the side bands remain; if the modulation were less than 100%, a frequency ν_0 would also be present in the spectrum.

scale experiments, but they are crucial for the tiny magnitudes of these quantities encountered in atomic, molecular, solid-state, and nuclear physics. The principle shows that we cannot expect answers to such questions as: Where, precisely, is a certain electron? When, precisely, was a certain photon emitted by an atom? In order to answer such questions it would be necessary to forego all knowledge of the electron's momentum (first question) or its energy (second question). Thus we cannot expect a theory to give orbits or trajectories, such as the radius r of an electron as a function of the polar angles θ and φ in the hydrogen atom; even if we had such a theory, we could not compare it with experiment. We must find and follow a quite different theoretical approach that *can* be compared with experiment.

4-13 SUMMARY

This chapter has presented a considerable number of experiments that showed that classical theory cannot be simply extrapolated to atomic problems. The theory failed for fundamental, qualitative reasons and predicted phenomena quite different from those observed.

The concepts of photons and of waves guiding the motion of particles were introduced. The introduction of these concepts led to the consideration of both beams of light or X-rays and beams of material particles as groups of wave packets. Because such a wave packet is extended in both space and time, we could show that experiments cannot simultaneously measure p_x and x (or E and t) with infinite precision. This Indeterminacy Principle imposes a definite restriction on the questions that we can hope an experiment will answer and denies the possibility of a comfortable "orbit" approach to the analysis of atomic structure.

If we remember all these experiments, these concepts, and the Indeterminacy Principle, we shall not make any erroneous predictions about atomic phenomena. But if we did not develop quantum theory beyond the stage reached in the present chapter we should be left in a very weak position for three reasons: (1) There is a wide variety of experiments and concepts to keep in mind. (2) Many points are left vague. For example, what is oscillating in an electron wave? In the photoelectric experiment, why are electrons emitted at random intervals instead of regularly spaced in time? The experiments do not provide answers to these questions or to many other questions. (3) We make no errors, but we make very few predictions. In short, our understanding is sterile and unproductive of predictions beyond the experiments on which it is based.

What is required is a *theory*, based on these experiments, but more general than any one experiment. Such a theory would first be tested by comparing its predictions with the experiments of this chapter. If successful, it would unify and explain the diverse experiments. Thus the objections of points 1 and 2 of the preceding paragraph would be met. Then it would be applied to wholly new problems (like molecular structure or the thermionic emission of electrons by metals). It might there give us an understanding of problems where experiments were lacking. It might suggest new experiments and ultimately new phenomena and practical devices. Thus a successful theory would meet the objection of point 3 of the preceding paragraph.

The next chapter will present the desired theory. This theory is the *wave mechanics* or *quantum mechanics*, which is the cornerstone of modern physics.

REFERENCES

General

F. K. Richtmyer, E. H. Kennard, and T. Lauritsen, *Introduction to Modern Physics*, McGraw-Hill, New York, 5th Ed., 1955.

P. L. Copeland and W. E. Bennett, *Elements of Modern Physics*, Oxford, New York, 1961.

J. C. Slater, *Modern Physics*, McGraw-Hill, New York, 1955, Chapters 1–6.

Photoelectric Effect

A. L. Hughes and L. A. DuBridge, *Photoelectric Phenomena*, McGraw-Hill, New York, 1932.

Line Spectra

W. Finkelnburg, *Atomic Physics*, McGraw-Hill, New York, 1950, Chapter 3.

X-Rays and the Compton Effect

A. H. Compton and S. K. Allison, *X-Rays in Theory and Experiment*, Van Nostrand, New York, 1935.

G. L. Clark, *Applied X-Rays*, McGraw-Hill, New York, 4th Ed., 1955.

H. A. Liebhafsky, H. G. Pfeiffer, E. H. Winslow, and P. D. Zemany, *X-ray Absorption and Emission in Analytical Chemistry*, Wiley, New York, 1960.

B. D. Cullity, *Elements of X-ray Diffraction*, Addison-Wesley, Reading, 1956.

Excitation Potentials

G. P. Harnwell and J. J. Livingood, *Experimental Atomic Physics*, McGraw-Hill, New York, 1933, pp. 314–323.

Black-Body Radiation

R. M. Eisberg, *Fundamentals of Modern Physics*, Wiley, New York, 1961, Chapter 2.

R. L. Weber, *Heat and Temperature Measurement*, Prentice-Hall, New York, 1950, Chapters 6 and 7.

Electron Diffraction

G. P. Thomson and W. Cochrane, *Theory and Practice of Electron Diffraction*, Macmillan, London, 1939.

Neutron Diffraction

D. J. Hughes, *Pile Neutron Research*, Addison-Wesley, Cambridge, Mass., 1953, Chapter 10.

PROBLEMS

4-1. What is the *smallest* experiment (e.g., smallest amplitude of oscillation of a pendulum or smallest collision experiment) in which you have personally shown that Newton's laws are approximately correct? What is the smallest experiment that you can conceive for verifying Newton's laws by direct observation with ordinary laboratory apparatus?

4-2.* What is the energy in electron volts of a photon with wavelength $\lambda = 5000$ Å? with $\lambda = 0.5$ Å?

4-3.* The work function of tungsten is 4.52 e.V., and that of barium is 2.50 e.V. What is the maximum wavelength of light that will give photoemission of electrons from tungsten? from barium? Would either of these metals be useful in a photocell for use with visible light?

4-4. Photoemission of electrons from calcium is being studied. The following threshold voltages (from retarding-field plots like Fig. 4-5) are found: $\lambda = 2536$ Å, $V_0 = 1.95$ volts; $\lambda = 3132$ Å, $V_0 = 0.98$ volt; $\lambda = 3650$ Å, $V_0 = 0.50$ volt; $\lambda = 4047$ Å, $V_0 = 0.14$ volt. Plot these data, and find Planck's constant.

4-5.* The work function of tantalum is 4.19 e.V. If light of wavelength 2536 Å is incident on a tantalum emitter in a phototube, what value of V_0 will be measured (the collector is also of tantalum)? What value of V_0 do you compute for $\lambda = 3650$ Å and what is the physical interpretation of this value?

4-6. A phototube is constructed with identical emitter and collector mounted symmetrically as shown in Fig. 4-2. Because of reflection, only 80% of the incoming light is absorbed by the emitter, the remaining 20% being absorbed by the collector. Sketch the replacement for Fig. 4-3 appropriate to these conditions.

4-7. Why does not the electron current I rise abruptly (vertically in Fig. 4-3) to the maximum value when V is increased beyond $-V_0$ in the photoelectric effect experiment?

4-8. The photoelectric process in the human eye is so sensitive that the brain learns of a light signal even if only ~100 photons per second enter the dark-adapted eye. In order to estimate the *range* of intensities over which the eye can function, compare this intensity with the light reflected from a square millimeter of white paper illuminated by direct sunlight. This reflected intensity is about 1 milliwatt per square millimeter, with a typical wavelength $\lambda = 5500$ Å.

4-9.* What is the energy of a quantum of radiation with a frequency of 1 megacycle? Suppose that the LC resonant circuit of a 1-megacycle oscillator has a stored energy of 10^{-5} joule (a value that might arise if the oscillator output is about 1 watt). How many quanta of energy is 10^{-5} joule? Energy changes by only an integral number of quanta are permitted. Does this fact cause any observable effects in this case?

4-10. Derive from Bohr's formula a general expression for the wavelength of hydrogen spectral lines that result from transitions from a general state n to the state with quantum number $n = 2$. Calculate from this expression the four lines with longest λ's. These are the first four lines of the *Balmer series*, Fig. 4-10a.

4-11.* The Balmer series hydrogen line $\lambda = 6563$ Å has a width of about 0.05 Å in a typical discharge tube, and the light energy emitted in this one line is 0.5 watt. What is the spectral power density (watts per Å)? How does this compare (roughly) with the spectral power density of a 100-watt incandescent lamp?

4-12. A crude way of determining whether relativistic effects are important in the hydrogen atom is to calculate the classical velocity v of an electron in this atom. Make this calculation, starting from eq. 4-4 with r set equal to 0.53 Å (this value comes from setting $-e/8\pi\epsilon_0 r$ equal to -13.6 e.V.). Estimate m/m_0 from this v, and comment on the necessity of a relativistic treatment of the hydrogen atom.

4-13.* Calculate the energy difference between the first excited state E_2 of hydrogen ($n = 2$ in eq. 4-6) and the ground state E_1 ($n = 1$), in electron volts. Suppose that a sample of hydrogen in a flame contains 10^{20} atoms in equilibrium at a temperature of 3300°K. About how many (n_2) are in the state E_2? The number of photons emitted per second will be about $10^8 n_2$, since the mean lifetime of the excited state is about 10^{-8} sec. How many watts of light are emitted? Assume that $n_2/n_1 \cong e^{-(E_2-E_1)/kT}$, as suggested in Sec. 2-4.

4-14. Show from Huygens' principle that, in the partial reflection of X-rays by a plane of atoms, the angle of incidence equals the angle of reflection. Note that your argument is valid even if the atoms in the plane are not evenly spaced.

4-15.* Calculate the lattice constant d_0 of NaCl. The density of NaCl is 2165 kg/m.³, and NaCl has the same crystal structure as KCl.

4-16. Construct a model of the crystal structure of KCl from toothpicks and gumdrops.

4-17. Sodium iodide has the sodium chloride crystal structure (Fig. 4-20), but the X-ray scattering power of the iodine ions is very large compared to that of the sodium ions. The sodium ions are therefore "invisible" in an X-ray crystal structure experiment. Draw the replacement for Fig. 4-21 appropriate for NaI, and sketch the two reflections with largest d values.

4-18.* The wavelength of a particular X-ray line from a molybdenum target is 0.709 Å, as determined by the use of a ruled grating spectrometer. This line is incident on an NaCl crystal (density = 2165 kg/m.³), and the first-order reflection from the cubic planes (spacing = d_0) is found at $\theta = 7.27°$. Calculate Avogadro's number from these facts.

4-19.* Compute the spacing d_1 in Fig. 4-21, in terms of d_0. Compute the spacing d between the plane DEF in Fig. 4-20 and the nearest parallel plane, in terms of d_0.

4-20. In the powder method of X-ray diffraction, the particle size must be neither too small nor too large. Describe the change in the appearance of the rings of Fig. 4-22b to be expected as (a) the particles become so small that a typical particle diameter is only a few X-ray wavelengths, and (b) the particles become so large that the sample contains only a few hundred crystals.

4-21. Why are the first, second, and fourth lines on either side of the center of Fig. 4-22b curved? Why is the third line nearly straight? Draw a perspective sketch of Fig. 4-22a to illustrate your answer.

4-22.* Consider an elastic collision between an electron and a mercury atom. Suppose that the aim is perfect, and so the electron is reversed in direction. Use the conservation of energy and momentum to compute the fraction $\Delta K/K$ of the electron's kinetic energy K that it loses at each collision. (The gas atoms are moving so slowly compared to the electrons that the mercury atoms can be considered to be at rest before the collisions.)

4-23.* Make a plot of electrostatic potential vs. distance from the cathode for the tube of Fig. 4-23; assume plane-parallel symmetry with the grid midway between anode and cathode, no space charge, a grid of very fine mesh, and a grid-cathode potential difference of 5 volts. Superimpose on this a plot of an electron's kinetic energy as a function of distance on the assumption that the electron makes an inelastic collision near the grid and loses 4.86 e.V. Where does the electron reverse its direction?

4-24.* If a Franck and Hertz experiment could be performed with atomic hydrogen gas, how much energy would an electron lose in an inelastic collision?

4-25. In the Franck and Hertz experiment, assume that the cross section σ_e for excitation is 2×10^{-20} m.² whenever the electrons have $K > 4.86$ e.V. Estimate [from eq. 2-20 with $\pi(r_1 + r_2)^2 = \sigma_e$] the pressure required in order that 10% of the electrons make excitation collisions in each 1 cm of path once they have sufficient energy. Use the vapor pressure of mercury tables in handbooks to estimate the temperature at which a Franck and Hertz tube should be operated.

4-26. Sketch the replacement for the tungsten or chromium curve of Fig. 4-25 if intensity is plotted vs. *frequency* ν, instead of vs. λ. *Caution:* The ordinate in Fig. 4-25 is X-ray intensity I per unit wavelength, or $dI/d\lambda$; the new ordinate will be $dI/d\nu$.

4-27. In the *transmission Laue* method of crystal structure determination, a beam of X-rays with a continuous spectrum of wavelengths is collimated by passing through a pin hole in a lead shield. This beam is incident on one side of a single crystal specimen, and a plane photographic film is placed perpendicular to the beam a few centimeters from the other side of the crystal. Explain (with a diagram) why only *spots* (rather than lines, as in the powder method) appear on the film.

4-28.* In the Laue method (see the preceding problem), what is the minimum accelerating voltage in the X-ray tube that will permit spots to occur from reflections by planes separated by 2.20 Å?

4-29.* What is the minimum wavelength emitted by an X-ray tube with an accelerating voltage of 45,000 volts?

4-30. A particular tungsten target X-ray tube can be operated with an anode power dissipation of not more than 200 watts. If we wish the maximum output of the continuous X-ray spectrum, would it be better to operate at 0.010-amp electron current and 20,000 volts, or 0.005-amp and 40,000 volts?

4-31. Show that the three vectors of Fig. 4-29 must lie in one plane.

4-32. Consider the special case of the Compton effect in which $\varphi = \pi/2$, and make the simplifying assumptions appropriate to low-energy photons, namely, that $p_e = m_0 v$, $K_e = \frac{1}{2} m_0 v^2$, and $\lambda' - \lambda_0 \ll \lambda_0$. Prove from the conservation of energy and momentum that eq. 4-12 gives the correct wavelength shift for this case. *Hints:* Eliminate β from the two momentum equations by squaring and adding; eliminate v from the resulting equation and the energy equation; let $\lambda' = \lambda_0 + \delta\lambda_0$, where δ is a number $\ll 1$; use the binomial expansion eq. 1-19 and neglect δ with respect to unity in order to simplify terms like $1/\lambda'^2$.

4-33.* The momentum diagram of Fig. 4-29 is drawn for $\lambda_0 = 0.71$ Å and $\varphi = 45°$. Use eq. 4-12 and the conservation of momentum to calculate the angle β of the recoil electron.

4-34.* A 255,000-e.V. X-ray photon strikes an electron (initially at rest) head on, and the photon is deflected through an angle $\varphi = \pi$. Use conservation of momentum and energy to find the ratio of the electron's velocity after the collision to the velocity of light, and verify eq. 4-12 for this particular collision.

4-35. Calculate the Compton wavelength shift $\lambda' = \lambda_0$ for $\varphi = \pi/2$ if the mass of a whole carbon atom is substituted for m_0 in eq. 4-12. Compare with the argument in the last paragraph of Sec. 4-7.

4-36. What is the wavelength of the Compton-scattered photon at $\varphi = \pi/2$ for $\lambda_0 = 0.71$ Å? What is the angle β of the recoil electron, and what is its energy? Is a relativistic treatment of the electron's momentum and energy necessary in this case?

4-37.* A KCl crystal is used in the X-ray spectrometer of a Compton effect experiment. Reflection from the principal planes ($d_0 = 3.14$ Å) is used, and the first order ($n = 1$) reflection is observed. The X-ray line is the $\lambda = 0.71$ Å line of molybdenum. What value of $\Delta\theta = \theta' - \theta_0$ will be measured for a scattering angle $\varphi = 90°$? *Hint:* Calculate θ_0, and use the fact that $\Delta\theta \ll \theta_0$ to write

$$\Delta\theta \cong \frac{d\theta}{d\lambda} \Delta\lambda = \frac{d\theta}{d\lambda} (\lambda' - \lambda_0)$$

4-38. Show that $\epsilon_\lambda + r_\lambda = 1$ by considering a black body and a non-black solid in equilibrium at temperature T. Let these solids be two parallel plates separated by a distance much less than their lateral extents, isolated from all other bodies, and interacting by radiation only.

4-39. Show that the emissivity ϵ_λ of the bottom of a hole that is several diameters deep is very close to 1. In order to do this, consider the amount of radiation reflected from the bottom of the hole and assume that the reflections at the sides and bottom are diffuse. Then use the result of problem 4-38.

4-40.* Find an expression for the value of λ at which the maximum of I_λ occurs at any T. *Hint:* The solution of the equation $(5 - x)e^x = 5$ is $x = 4.965$.

4-41. At what wavelength does the maximum of I_λ occur for $T = 2900°K$? Would a tungsten-filament lamp, which ordinarily operates at this temperature, be a more efficient producer of visible light if it could be operated at a higher temperature? (See the dashed line in Fig. 10-23 for the sensitivity of the human eye as a function of λ.)

4-42. The maximum I_λ for the radiation from the sun occurs at $\lambda = 4700$ Å. Assume that the sun is a black body, and use the result of prob. 4-40 to estimate the surface temperature of the sun.

4-43.* The total emissivity of tungsten at 2000°K is 0.26. How much power is required to maintain the temperature of a radio transmitting tube filament at this value if the area of the filament is 0.001 m.^{2} and if there are no power losses other than radiation?

4-44.* An optical pyrometer is focused on the outside of a platinum crucible, and a reading of 1200°C is found. If platinum had an emissivity at $\lambda = 6500$ Å equal to unity, this would be the temperature of the crucible, but the actual spectral emissivity of platinum is 0.18. What is the temperature of the crucible?

4-45. Sketch eq. 4-13 for values of ν from $kT/2h$ to $2kT/h$.

4-46. Verify the λ dependence of I_λ expressed in eq. 4-14 by starting from eq. 4-13 for the *number* of oscillators at each ν interval; multiply by $h\nu$ to get the *energy* of each group of oscillators; and convert this expression for $dW_\nu/d\nu$ into an expression for $I_\lambda = dW/d\lambda$ by using $\lambda = c/\nu$. Do not attempt to verify the factor $2\pi c^2 h$ in eq. 4-14, since eq. 4-13 is only a proportionality.

4-47. Explain qualitatively the shapes of the distributions sketched in Figs. 4-33a and 4-33b by sketching magnetic field lines or magnetic equipotentials for the pole pieces sketched in Fig. 4-32 and by referring to eq. 4-18.

4-48.* In the original Stern-Gerlach experiment, silver atoms, each with a magnetic moment of 0.93×10^{-23} joule m.^{2}/weber, were emitted from a source at 1320°K. The most abundant atoms had velocities such that their energies were $1.8kT$. The magnet pole pieces were 0.03 m. long, and the maximum field gradient at the path of the atoms was 2300 webers/m.^{3} What was the maximum separation of the two components at the detector plate when that plate was placed at the end of the pole pieces?

4-49.* In order to obtain the high field gradient in the Stern-Gerlach experiment it is necessary to have values of the field $\mathcal{B}$ in excess of 1 weber/m.^{2} Suppose the silver atoms in prob. 4-48 had been singly charged. Compare the force exerted by the field gradient on the magnetic moment with the force eq. 1-6 exerted by a field of 1.5 webers/m.^{2} on the charged atom. Comment on the possibility of performing a Stern-Gerlach experiment on free electrons.

4-50. Calculate the 0.81 Å spacing indicated in Fig. 4-37 from the other spacings given in that figure.

4-51.* Calculate the gain in energy of an electron entering a nickel crystal from the refraction of the exiting electron shown in Fig. 4-37. *Hint:* Resolve the momentum of the 54-e.V. electron outside the crystal into components parallel to and normal to the surface; only the latter is different inside the crystal.

4-52. An electron diffraction experiment on KCl is being performed with 50,000-e.V. electrons. What is the angle θ for first-order Bragg reflection from the cubic planes $(d = d_0)$? Compare this θ with the comparable angle for photons with $h\nu = 50,000$ e.V.

4-53.* What is the de Broglie wavelength of a laboratory-scale particle (for example, a mass of 1 gm) moving at a laboratory-scale velocity (for example, 10 m./sec)? Is it necessary to consider the wave properties of matter in this case?

4-54. What is the de Broglie wavelength of an electron with a kinetic energy of 24.6 volts (the ionization energy of helium)? How does this λ compare with the estimate of the diameter of the helium atom from the radius tabulated in Table 2-1? Is it necessary to consider the wave properties of matter when studying the motion of an electron in the helium atom?

4-55.* What is the de Broglie wavelength of an α-particle (He nucleus) with a kinetic energy of 7.7 M.e.V.? In the Rutherford scattering experiments, distances of the order of 10^{-13} m. were involved, yet the analysis of the experiment did not include the wave properties of the α-particle. Was this justified?

4-56.* What is the velocity of a neutron with a kinetic energy equal to $\frac{3}{2}kT$ at room temperature (300°K)? What is its de Broglie wavelength? (This is a typical thermal neutron.)

4-57.* A pair of discs is attached at right angles to a rapidly rotating shaft. A small sector in each is left open for neutrons to pass through. The open sector in the second disc lags 20° behind the open sector in the first. If the discs are 1 m. apart, what must the shaft rotational speed (in rpm) be in order that neutrons with the velocity of prob. 4-56 go through both open sectors? (This *mechanical velocity selector* has been used to make a beam of monoenergetic neutrons.)

4-58. What is the de Broglie wavelength of a He atom with an energy $2kT$, where $T = 290°K$? (In 1929 Estermann and Stern conducted diffraction experiments with such atoms incident upon NaCl and LiF crystals.)

4-59.* Consider a pendulum bob of mass 0.10 kg moving at 3 m./sec. Suppose that the momentum p_x need not be known more accurately than $\Delta p_x = 10^{-6} p_x$. What limitation does the Indeterminacy Principle impose on the simultaneous measurement of x?

4-60. Estimate crudely the size of the hydrogen atom in the following way: If the electron is known to be bound to the proton, its momentum is determined at least as precisely as can be calculated by setting Δp_x equal to the momentum of an electron with $K = 13.6$ e.V. The resulting indeterminacy in position Δx is a crude measure of the diameter of the H atom.

4-61. Compare the *fundamental indeterminacy* $\Delta p_x \, \Delta x \geq h$ with the *actual* experimental *uncertainties* in the Millikan oil-drop experiment. In order to do this, use the data of prob. 1-1, assume that "21.2 sec" means "21.20 ± 0.05 sec" and similarly for the other data given, and compute the actual δx, δp_x, and $\delta x \delta p_x$ product, which then can be compared with h. Also, compute the minimum uncertainty δx that could have been observed using light of $\lambda = 5000$ Å and a microscope with an aperture angle $\theta = 20°$ if all experimental uncertainties other than diffraction could have been eliminated.

4-62.* The *lifetime* of an excited state of an atom is about 10^{-8} sec; an atom can radiate at any time from $t = 0$ to $t = \infty$ after it is excited, but the average time is $\sim 10^{-8}$ sec. Using this as the Δt for the emission of a photon, compute the minimum $\Delta \nu$ permitted by the Indeterminacy Principle. What fraction of ν is this if the wavelength of the spectral line involved is 5000 Å? (This calculation gives the limiting sharpness of a spectral line if no other processes, such as Doppler effect, broaden the line.)

4-63.* A radar transmitter sends out pulses of radio-frequency waves like the wave packet of Fig. 4-41c. The delay between the time of radiation and of reception of the pulse reflected from a distant object permits the measurement of the distance R to the object. The range R is to be measured with a precision of $\frac{1}{2}$ mile (805 m.). What should be the width Δt of the pulse? How wide a band of frequencies must be passed by the amplifier in the receiver?

5

Introductory Quantum Mechanics

5-1 INTRODUCTION

The experiments described in the preceding chapter tell us that classical physics is inadequate for atomic-scale phenomena and for the interaction between electrons and electromagnetic waves. We could proceed directly from these experiments to an explanation of the structure of atoms, molecules, and solids, but such a procedure would have to be very vague. The ideas expressed in $E = h\nu$ and $\lambda = h/mv$ are basic to all these problems, but it is not possible to build a complete understanding of atomic phenomena *directly* upon these ideas.

Our position at this point is similar to the situation in classical mechanics before Newton's laws were developed. There existed a variety of experimental information, such as the motion of pendulums, falling bodies, and bodies on inclined planes; there was also some sort of explanation of each experiment. But there was no unifying theory and no way of predicting the results of problems in mechanics, such as astronomical problems, where experiments could not be performed.

What we require is a *theory* that is based on the experiments of Chapter 4 and that enables us to predict and explain more complicated atomic phenomena. The theory that accomplishes this is the *wave mechanics* or *quantum mechanics*, which is the subject of the present chapter. The heart of this theory is the Schrödinger equation, which

138

will be stated in Sec. 5-2. The test of its validity (like the test of New-ton's laws) is that predictions from it must agree with experiment. Ac-cordingly solutions of this equation for several simple examples will be worked out in the remainder of the chapter and will be compared with the conclusions from the experiments of Chapter 4.

In this chapter, the attitude will be taken that quantum mechanics is on trial and must prove itself. Although in subsequent chapters we consider that the validity of quantum mechanics has been demonstrated by agreement of its predictions with experiment, the experiments dis-cussed there strongly support the theory and so could be used as addi-tional proof if the reader feels that it is needed.

In Sec. 5-5 the quantum-mechanical theory of harmonic oscillations is discussed, since it is important for such applications as the thermal properties of solids and is useful in introducing the Correspondence Principle in Sec. 5-6. This principle shows how quantum mechanics transforms into ordinary classical mechanics for sufficiently large-scale phenomena. In other words, it shows that quantum mechanics gives the same answers as ordinary mechanics in the region of laboratory sizes where the latter theory is known to be valid.

This chapter is by necessity somewhat mathematical, but most of the mathematics is just the same as the mathematics of simple electrical circuits. As we begin the more complicated problems (like the harmonic oscillator and the hydrogen atom) we shall not present the rather com-plicated mathematics, since we are not attempting in this book to teach theoretical physics. We are attempting to develop an understanding of the *nature* of nuclei, atoms, molecules, and solids. With this object in mind, we shall present the problems and their solutions, but not the detailed mathematics of their solution. The physical ideas and processes are the important thing; we shall try to learn these by applying quantum mechanics to the simple problems of Secs. 5-3 and 5-4. These problems are artificial, but the mathematics is relatively easy, and the physical concepts are therefore best developed by studying such problems.

Two notes should be added here: (1) In all the experiments of modern physics there appears to be no reason to question the validity of the laws of conservation of energy and of momentum; all our theoretical work will be consistent with these laws. (2) We can always tell whether a particular theory is a classical or a quantum theory by inspecting its results to see whether Planck's constant h enters. If it contains h, either explicitly or hidden in a numerical factor with other constants, the theory is a quantum theory; if it does not, the theory is classical.

5-2   THE SCHRÖDINGER WAVE EQUATION

In this section the Schrödinger equation will be stated and the technique of using it will be explained.   This equation replaces $\mathbf{F} = m\mathbf{a}$ for the motion of particles on an atomic scale of sizes.   We shall usually apply the equation to the motion of an electron, and so unless otherwise stated we shall be considering an electron in this chapter.   But the equation applies as well to any other particle if the appropriate charge and mass are substituted for the e and m of the electron.

Before we state the Schrödinger equation it is worth while to consider what the nature of this equation must be.   It must be a *wave equation*, like the equations for electromagnetic or acoustic waves.   We know this fact because otherwise the wave solutions for electron diffraction experiments could not be obtained.   A wave equation is a partial differential equation with second derivatives, and the independent variables are space and time.   But what is the dependent variable to be?   In other words, our equation will give us the space and time coordinates of something, but of what?   For sound waves, this variable is the pressure; for electromagnetic waves, it is the electric field (or the magnetic field).   In our problem we must introduce a different kind of variable, which at first sight appears to be more abstract than quantities like the electric field.

Our variable will be called the *wave function*, and we shall use the symbol Ψ for it.   It will be defined in the following way: $|\Psi|^2 \, \Delta v$ is proportional to the *probability* that (if an experiment is performed) the electron will be found in the volume Δv.   We attach no physical significance to Ψ itself, only to the square of its absolute magnitude $|\Psi|^2$.   Thus it does not matter if Ψ is a complex (instead of a real) variable.   This may appear to be a poor substitute for "knowing where the electron is."   Actually, the electron diffraction and Compton effect experiments and the Indeterminacy Principle have demonstrated that there is no meaning to the question: "Where, precisely, is the electron?" except in the uninteresting case where we were willing to tolerate an infinite uncertainty in the electron's momentum.   Also, in the photoelectric effect the emitted electrons were observed to be randomly distributed in time, which suggests that there is not an exact equation predicting the number emitted per unit time but only a certain probability of emission.   Thus a determination of the *probability* of finding the electron in a certain region of space is the most we can expect to accomplish with our wave equation, and therefore Ψ is an acceptable dependent variable.

The predictions of our equation can be compared with experiment in the following way: We calculate Ψ and then calculate from Ψ the interesting properties of the electron's motion like the energy, the momentum, or the probability that the electron will arrive at a certain position. Ψ itself is just a *construct*, a means to an end. Its role is quite similar to the role of the field vectors $\boldsymbol{\mathcal{E}}$ and $\boldsymbol{\mathcal{B}}$ in radio waves or microwaves: $\boldsymbol{\mathcal{E}}$ and $\boldsymbol{\mathcal{B}}$ are not measured directly in any experiments at high frequencies, but a theoretical treatment of an antenna or transmission line can be carried out in terms of $\boldsymbol{\mathcal{E}}$ and $\boldsymbol{\mathcal{B}}$. The theory predicts results like antenna patterns or power flows that can be compared with experiment. For example, we could calculate $|\Psi|^2$ as a function of position on the photographic plate for an electron in the diffraction experiments of Sec. 4-10. We perform the experiment with a large number of electrons and determine (in the example, from the relative exposure of the plate) the fraction of the electrons that are found in a volume element Δv as a function of the location in space of this element. This fraction should be the same as the computed probability if the theory is correct, the experiment well performed, and a sufficient number of electrons observed so that the *statistical fluctuations* in the position of each electron do not create an intolerable uncertainty in the result.

The question of statistical fluctuations may require some explanation. Suppose that we perform an electron diffraction experiment with only three or four electrons. All these *might* follow such paths that each appeared near a diffraction maximum, but in another exposure with the same experiment some of them *might* be found at locations where the probability of finding them was very small (but not zero); poor agreement between theory and experiment would result. On the other hand, if we use a large number of electrons, we should obtain a good picture of the relative probability of finding electrons at various points on the photographic plate. Since the charge of the electron is only 1.6×10^{-19} coulomb, 10^{12} electrons pass during a 1-minute exposure at a current of less than $1/100$ of a microampere. Therefore ordinary experiments with electrons usually involve a sufficient number of particles so that the measured quantities should compare closely with the theoretical predictions.

The Schrödinger wave equation for an electron is

$$\frac{h^2}{8\pi^2 m}\left(\frac{\partial^2 \Psi}{\partial x^2} + \frac{\partial^2 \Psi}{\partial y^2} + \frac{\partial^2 \Psi}{\partial z^2}\right) - P\Psi = \frac{h}{2\pi i}\frac{\partial \Psi}{\partial t} \tag{5-1}$$

Here h is Planck's constant, P is the potential energy of the particle, and $i = \sqrt{-1}$. Although in most physical problems this three-dimensional form of the equation must be used, we shall work exclusively with the one-dimensional form in which neither P nor Ψ is a function of y or z:

$$\frac{h^2}{8\pi^2 m}\left(\frac{\partial^2 \Psi}{\partial x^2}\right) - P\Psi = \frac{h}{2\pi i}\frac{\partial \Psi}{\partial t} \tag{5-2}$$

This restriction to one dimension is introduced here to avoid the mathematical complexity of the three-dimensional problems. We are seeking to understand the *nature* of atomic phenomena, and one-dimensional problems illustrate the principal concepts and basic physics of quantum theory. As mentioned above, this equation is valid for *any* particle if the appropriate mass is substituted for m and if the appropriate potential energy, which usually depends on the charge of the particle, is inserted.

Equation 5-1 is *not* valid in the *relativistic* region of velocities where m is a function of velocity; the equation that replaces it there (the Dirac equation) is part of an intrinsically more complicated theory and is not simply obtained by substituting the velocity-dependent mass (eq. 1-13) for m in eq. 5-1.

For a large class of problems, the Schrödinger equation can be simplified by dealing separately with the time dependence. These are problems in which the potential energy depends only on position, not upon time. When P is a function of x alone, the simpler form of the Schrödinger equation can be obtained by *separating the variables x and t*. In order to do this, we try a solution of the form

$$\Psi = \psi(x)\phi(t) \tag{5-3}$$

where ψ is a function of x alone and ϕ is a function of t alone. If we can find a number of such product functions we can write the general solution of eq. 5-2 as the superposition of such solutions (just as in electric circuit theory we can superimpose individual solutions of a network problem for, say, different frequencies). If we insert eq. 5-3 into eq. 5-2 and perform the indicated differentiations we obtain

$$\frac{h^2}{8\pi^2 m}\left(\frac{d^2\psi}{dx^2}\right)\phi(t) - P\psi(x)\phi(t) = \frac{h}{2\pi i}\frac{d\phi}{dt}\psi(x)$$

When this equation is divided by $\psi\phi$ it becomes

$$\frac{h^2}{8\pi^2 m}\frac{1}{\psi}\frac{d^2\psi}{dx^2} - P = \frac{h}{2\pi i}\frac{1}{\phi}\frac{d\phi}{dt}$$

None of the quantities on the left is a function of t; hence the left side can be only a function of x or a constant. None of the quantities on the right is a function of x; hence the right side can be only a function of t or a constant. Since the two sides are equal, the only possibility is that each is equal to the same constant. We shall call this constant $-E$ for reasons that will become apparent in the discussion below eq. 5-21. We now have two equations:

$$\frac{h^2}{8\pi^2 m}\frac{d^2\psi}{dx^2} + (E - P)\psi = 0 \tag{5-4}$$

and

$$\frac{d\phi}{dt} = \frac{-2\pi i}{h}E\phi$$

The second equation can be integrated at once by multiplying both sides by dt/ϕ:

$$\int \frac{d\phi}{\phi} = -\int \frac{2\pi i}{h}E\,dt$$

$$\ln\phi = -\frac{2\pi i}{h}Et + \ln\phi_0$$

$$\phi = e^{(-2\pi iE/h)t} \tag{5-5}$$

The constant of integration $\ln\phi_0$ is unnecessary and has been set equal to zero (i.e., $\phi_0 = 1$). No loss in generality occurs, because ϕ will always be used in conjunction with ψ to give $\Psi = \psi\phi$; since ψ will contain an arbitrary constant as a multiplying factor it is unnecessary to include such an arbitrary constant in eq. 5-5.

What we have accomplished is the reduction of a *partial* differential equation to two *ordinary* differential equations, and this reduction greatly simplifies application. In typical problems, the force on the electron as a function of position will be known. This force will usually be the electrostatic (Coulomb) force. From the force, the potential energy $P(x)$ can be calculated. Equation 5-4 can then be solved, giving $\psi(x)$, which also gives the way Ψ depends on x. The properties of interest can usually be calculated from $\psi(x)$, but if we wish the complete wave function $\Psi(x, t)$, we can multiply our solution $\psi(x)$ by the $\phi(t)$ from eq. 5-5.

There are three conditions on ψ or Ψ that are as important as the Schrödinger equation itself. They are:

$$\int_{-\infty}^{\infty} |\Psi|^2\,dx \quad \text{must be finite} \tag{5-6}$$

$$\Psi \quad \text{must be continuous and single-valued} \tag{5-7}$$

$$\partial\Psi/\partial x \quad \text{must be continuous} \tag{5-8}$$

These statements have been made for Ψ but apply as well to ψ. If Ψ is a function of three space variables, the integral of eq. 5-6 is a triple integral over all space of $|\Psi|^2 \, dv$, and all three partial derivatives $\partial\Psi/\partial x$, $\partial\Psi/\partial y$, and $\partial\Psi/\partial z$ must be continuous. These conditions can be looked upon as statements, like the Schrödinger equation itself, to be tested by experience. But we shall see as we proceed that these conditions are necessary. For example, the necessity for eq. 5-7 follows from the fact that a discontinuity in Ψ would produce a discontinuity in $|\Psi|^2 \, \Delta v$ and hence in the probability of finding the electron in Δv; this probability should vary continuously from point to point if no electrons are created or destroyed. Similarly, $\partial\Psi/\partial x$ will be related to the electron's momentum, which must be a continuous function of x (except at a point where $P = \infty$, and at any such point eq. 5-8 need not be satisfied). Ψ must be single-valued in order that there be no ambiguity in the predictions of the theory. These conditions frequently enable us to select the *one* actual solution to a physical problem from a number of possible solutions to the Schrödinger equation.

We have already indicated the physical significance of $|\Psi|^2$, namely, that $|\Psi|^2 \, \Delta v$ is proportional to the probability that the electron will be found in Δv. The probability of an occurrence is a number $\mathcal{P}$ between 0 and 1 such that, in many experiments with identical starting conditions, the fraction of the experiments in which this occurrence happens equals $\mathcal{P}$. Thus, if the probability of finding an electron in a certain region of space is 0.1 in a particular experiment, and if we perform this experiment many times, we should find that the electron actually was found in that region in 10% of the total number of experiments. $\mathcal{P} = 1$ corresponds, of course, to "certainly" and $\mathcal{P} = 0$ to "certainly not." Our physical interpretation of $|\Psi|^2$ can hence be written in quantitative, probability form by the following expression for the probability $\mathcal{P}$ of finding the electron in the space between x and $x + \Delta x$:

$$\mathcal{P} = \frac{|\Psi|^2 \, \Delta x}{\displaystyle\int_{-\infty}^{\infty} |\Psi|^2 \, dx} \tag{5-9}$$

Since Ψ in general is a function of x, $\mathcal{P}$ has different values at different values of x. Equation 5-9 has been constructed to satisfy the two requirements: (1) $\mathcal{P}$ is proportional to $|\Psi|^2 \, dx$, since the denominator is not a function of x. (2) $\mathcal{P}$ takes on only values between 0 (numerator equals 0, electron certainly *is not* in Δx) and 1 (numerator equals denominator by summing $|\Psi|^2 \, \Delta x$ over all space, electron certainly *is* in Δx). We can now see the necessity for the condition expressed in eq. 5-6: If this condition were not obeyed, the probability would be zero of finding the electron in any (finite) interval Δx, and such an answer could not apply to any physical problem.

All problems could be worked by using eq. 5-9 to determine the probability of finding an electron in Δx, but it is much more convenient to apply the process of *normalization* to the solution Ψ of each problem. Suppose that we have found a solution (call it ψ_1) of eq. 5-4. It follows that $\psi_2 = b\psi_1$ (where b is a constant) is also a solution. (This can easily be verified by substituting $b\psi_1$ into eq. 5-4 and by using the fact that ψ_1 is a solution.) Let us choose b by setting

$$\int_{-\infty}^{\infty} |\psi_2|^2 \, dx = 1 \tag{5-10}$$

which implies

$$b^2 \int_{-\infty}^{\infty} |\psi_1|^2 \, dx = 1 \quad \text{or} \quad b = \left[\int_{-\infty}^{\infty} |\psi_1|^2 \, dx \right]^{-\frac{1}{2}} \tag{5-11}$$

When ψ_2 is computed in this way it is said to be a *normalized* wave function.

In order to obtain a normalized $\Psi = \psi\phi$, all we need to do is to normalize ψ and multiply by ϕ from eq. 5-5, since

$$\Psi = e^{-(2\pi iE/h)t} \psi$$

and

$$|\Psi|^2 = |e^{-(2\pi iE/h)t}|^2 |\psi|^2 = |\psi|^2$$

(The absolute magnitude of an exponential with an imaginary exponent is unity.)

When ψ or Ψ has been normalized in this way, the integral in the denominator of eq. 5-9 is unity and

$$\mathcal{P} = |\Psi|^2 \, \Delta x$$

or, in terms of ψ,

$$\mathcal{P} = |\psi|^2 \, \Delta x \tag{5-12}$$

In other words, we have been able to replace our statement "$|\Psi|^2 \, \Delta x$ is proportional to the probability that the electron will be found in Δx" by the statement "$|\Psi|^2 \, \Delta x$ is *equal* to the probability that the electron will be found in Δx." Note that the condition expressed in eq. 5-10 means that "the electron is certainly somewhere." We shall apply this process of normalization to all the wave functions we compute.

Up to this point we have expressed interest only in the relative probability of finding the electron in different regions of space, but its energy and momentum are usually of more interest than its position. The wave mechanics must give ways of computing these and other observable quantities from Ψ. The general procedure will be outlined in Sec. 5-7 and Appendix F. This procedure is useful and necessary in problems that are more complicated than the ones we consider in this chapter. For our present purposes, however, we shall learn by a more restricted approach in the next section how values of the energy are predicted.

5-3   ELECTRON IN FIELD-FREE SPACE

We learn how the energy is computed from the Schrödinger equation, and incidentally find other interesting properties of the solution of this equation, by considering here a simple special case in which the answer is already known. This example is an electron traveling in a region of constant potential energy P and with a constant momentum p. By comparing the solution of the Schrödinger equation for this problem with the known form of a traveling wave we shall learn that the constant E introduced in the process of obtaining eqs. 5-4 and 5-5 is actually the total energy.

Let the constant potential energy be P_0. Then eq. 5-4 becomes

$$\frac{d^2\psi}{dx^2} + \frac{8\pi^2 m(E - P_0)}{h^2}\,\psi = 0 \qquad (5\text{-}13)$$

The coefficient of ψ is a constant. The form of eq. 5-13 is identical with the pendulum equation, the equation of simple harmonic motion or of an electrical circuit consisting of an inductance and a capacitance.

The general solution of this equation is

$$\psi = A'e^{\frac{2\pi i}{h}\sqrt{2m(E - P_0)}\,x} + B'e^{\frac{-2\pi i}{h}\sqrt{2m(E - P_0)}\,x} \qquad (5\text{-}14)$$

where A' and B' are arbitrary constants to be evaluated for the specific conditions of the problem (e.g., electron moving toward $+x$ or toward $-x$). Since eq. 5-13 is a linear differential equation with constant coefficients, systematic methods are available for finding the solution eq. 5-14. But it is more characteristic of the usual process of solution of differential equations to write down an assumed, trial solution (usually with coefficients like the coefficients of x in eq. 5-14 left to be evaluated) and to insert it into the differential equation. This operation in this problem verifies that eq. 5-14 is a solution; furthermore, eq. 5-14 contains two independent arbitrary constants and therefore it is the *general* solution of the second-order eq. 5-13. An alternate, equivalent form of eq. 5-14 is

$$\psi = A \sin\frac{2\pi}{h}\sqrt{2m(E - P_0)}\,x + B \cos\frac{2\pi}{h}\sqrt{2m(E - P_0)}\,x$$

$$(5\text{-}15)$$

in which new (real) arbitrary constants A and B appear. This form, which contains no complex or imaginary quantities, will be more useful for later work, but we prefer the form eq. 5-14 here because it simplifies the next step.

The next step is to obtain $\Psi = \psi\phi$, which we do by multiplying eqs. 5-14 and 5-5:

$$\Psi = A'e^{-2\pi i\left(\frac{E}{h}t - \frac{\sqrt{2m(E-P_0)}}{h}x\right)} + B'e^{-2\pi i\left(\frac{E}{h}t + \frac{\sqrt{2m(E-P_0)}}{h}x\right)}$$

(5-16)

This rather complicated expression must contain the properties of an electron traveling in a region of space without an electric field ($P_0 = -eV = $ constant). The experiments of Sec. 4-10 tell us that such electrons exhibit the interference effects characteristic of a wavelength

$$\lambda = \frac{h}{p} = \frac{h}{mv} = \frac{h}{\sqrt{2mK}}$$

(5-17)

The equation for *any* one-dimensional wave (e.g., a sound or electromagnetic wave) can be written in terms of the frequency ν and wavelength λ as

$$u = C_1 \cos 2\pi[-\nu t + (x/\lambda)]$$

(5-18)

if it is traveling toward $+x$, or

$$u = C_2 \cos 2\pi[-\nu t - (x/\lambda)]$$

if it is traveling toward $-x$. C_1 and C_2 are constants, and u is the wave function; for a sound wave u is the pressure, and for an electromagnetic wave u is the electric or magnetic field. In general there might be a wave of one amplitude going toward $+x$ and a wave of another amplitude going toward $-x$. The superposition of these two waves (to give the total pressure for a sound wave or electric field for an electromagnetic wave) is in general

$$u = C_1 \cos 2\pi[-\nu t + (x/\lambda)] + C_2 \cos 2\pi[-\nu t - (x/\lambda)]$$

This expression can be written in the exponential form, in which u becomes the real part of

$$C_1e^{-2\pi i\left(\nu t - \frac{x}{\lambda}\right)} + C_2e^{-2\pi i\left(\nu t + \frac{x}{\lambda}\right)}$$

(5-19)

Comparison of eq. 5-19 with eq. 5-16 shows that these are the same in *form* and are equivalent if we let

$$\lambda = \frac{h}{\sqrt{2m(E - P_0)}} \tag{5-20}$$

and

$$E = h\nu \tag{5-21}$$

We now see by comparing eqs. 5-17 and 5-20 that our Schrödinger equation theory agrees with the known properties of an electron wave if we identify $(E - P_0)$ with the kinetic energy K. In other words, we have identified E with the *total energy*. This is, of course, the reason we used the symbol "E" for this quantity, but up to this point we had known only that E was a constant with the dimensions of an energy.

The potential energy P for a given problem is arbitrary in that we can choose the *origin* for P as we please. For example, one person could set $P = 0$ for this problem (eqs. 5-13 to 5-21) and another could consider the same problem, with the same kinetic energy of the electron, and set $P = P_0$. They must get the same answer for any quantity that can be compared with an actual observation. (This situation is more familiar in the theory of electric circuits, where we can set $V = 0$ at any point we please, and the predicted currents and potential differences do not depend on the choice of reference potential or potential energy.) We can see how this requirement is satisfied in the foregoing theory. If we had used $P = 0$ instead of $P = P_0$, the total energy $E = K + P$ would have been different for the same physical conditions (same kinetic energy). But eq. 5-20 contains only $E - P_0 = K$, and hence the wavelength would have been unchanged.

On the other hand, ν in eq. 5-21 would definitely have been changed. It is thus apparent that this ν can have no physical significance. If we exercise our freedom of choice of reference for the potential energy we find different ν values for the same problem. What *is* significant is a *change* in the value of ν. Suppose that the total energy of the electron changes from E_1 to E_2 by emitting radiation. The photon emitted will have a frequency equal in magnitude to the *difference* in ν values before and after the change in E. That is,

$$E_1 - E_2 = h\nu_1 - h\nu_2 = h(\nu_1 - \nu_2)$$

This difference, which is the only observable quantity related to ν, is independent of the choice of reference potential energy.

The expressions of eq. 5-16 and eq. 5-19 are for infinitely long trains of waves with exactly constant λ and p. If we were to describe the motion of a single electron, we should have to use a *wave packet* as in Sec. 4-12 and Fig. 4-8 or Fig. 4-41c. We should then have a wave train of finite length and a finite spread in λ and p. Suppose this electron to be subject to electric and magnetic fields in a vacuum tube of ordinary, laboratory-size dimensions. Such fields have a negligible variation over distances of the order of magnitude of the size of an electron wave packet. It can be shown that for these conditions the Schrödinger equation predicts the same motion of this wave packet that Newton's laws predict for a particle; the packet travels with the velocity $(2K/m)^{1/2}$ that would be computed by classical physics. Thus the wave mechanics predicts the electron diffraction effects without sacrificing agreement with laboratory-scale experiments.

5-4 ELECTRON IN A "SQUARE-WELL" POTENTIAL

In this section the Schrödinger equation will be solved for a group of rather artificial problems. These problems have been chosen to have the same nature as problems of electrons in atoms but with very much simpler mathematics. Our aim here is to learn the qualitative features of the solutions of the Schrödinger equation and how they differ from the predictions of classical mechanics.

(a) Square well with infinitely high sides

The first problem is illustrated in Fig. 5-1, where the potential energy $P = -eV$ is plotted as a function of x for an electron. This is an extremely crude approximation to the problem of the electron in an atom, but it contains the essential feature for the present application, namely that the electron is bound to a small region of x. The "walls" of this "box" are infinitely steep and infinitely high. Once the electron is in such a well, it cannot escape: The probability of finding the electron in the regions $x < -x_0$ and $x > x_0$ is zero, and $\psi = 0$ in those regions. Thus the *boundary conditions* on our solution of the Schrödinger equation are that $\psi = 0$ at $x = \pm x_0$. (These conditions will be established more rigorously in connection with the next example of this section.)

For $-x_0 < x < x_0$, the potential energy is

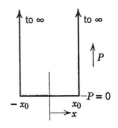

Fig. 5-1. Potential energy as a function of distance for the square well with infinite sides.

constant, and we may as well set $P = 0$, since we can define P by setting its zero at any place we wish. Equation 5-4 then becomes

$$\frac{d^2\psi}{dx^2} + \frac{8\pi^2 mE}{h^2}\psi = 0 \tag{5-22}$$

The general solution of this equation is (as explained in connection with eq. 5-15):

$$\psi = A \sin \frac{2\pi}{h}\sqrt{2mE}\, x + B \cos \frac{2\pi}{h}\sqrt{2mE}\, x \tag{5-23}$$

We now choose A and B in such a way that $\psi = 0$ at $x = \pm x_0$. Let us first define a constant $\beta = (2\pi/h)\sqrt{2mE}$ to save frequent repetition of this expression. At $x = x_0$,

$$\psi = A \sin \beta x_0 + B \cos \beta x_0 = 0$$

and at $x = -x_0$

$$\psi = -A \sin \beta x_0 + B \cos \beta x_0 = 0$$

By adding and subtracting these two equations, we learn that

$$A \sin \beta x_0 = 0 \quad \text{and} \quad B \cos \beta x_0 = 0$$

But *both* the sine and the cosine of the same argument (βx_0) cannot be zero. Hence there are only two possible ways of satisfying these relations:

$$(a) \qquad A = 0, \qquad \cos \beta x_0 = 0$$

$$(b) \qquad B = 0, \qquad \sin \beta x_0 = 0$$

If $\beta x_0 = n\pi/2$, where n is an *odd integer*, (a) will be satisfied. If $\beta x_0 = n\pi/2$, where n is an *even integer*, (b) will be satisfied. Thus the final solutions are of two classes, and either

$$\psi = x_0^{-\frac{1}{2}} \cos \beta x \qquad (n = 1, 3, 5, \cdots)$$

or

$$\psi = x_0^{-\frac{1}{2}} \sin \beta x \qquad (n = 2, 4, 6, \cdots) \tag{5-24}$$

where

$$\beta = n\pi/2x_0 = (2\pi/h)\sqrt{2mE} \tag{5-25}$$

(We have normalized the solutions, and thus $x_0^{-\frac{1}{2}}$ replaces A and B in eq. 5-24.) $\psi(x)$ for the three lowest values of n is plotted in Fig. 5-2.

The most important feature of this result is that we have shown that solutions are possible *only if the energy E takes on one or another of a set of discrete values.* These values are from eq. 5-25:

$$E_n = n^2 \frac{h^2}{32mx_0^2} \text{ joules} \tag{5-26}$$

where n is an integer.

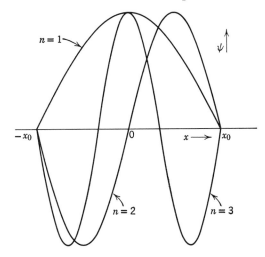

Fig. 5-2. Wave functions for the three quantum states with lowest energies for the square well with infinite sides.

The n's introduced above are called *quantum numbers*. The E values are called *energy levels*. An electron that is described by the wave function with a certain n value is said to be in the *quantum state n*. The quantum state with lowest energy ($n = 1$ in the present example) is called the *ground state*. It should be noted from eq. 5-26 that the kinetic energy is *not equal to zero* in the ground state, which is a general result of quantum mechanics applicable to all problems and is completely foreign to classical mechanics.

The existence of quantum numbers and discrete energy levels was proved here only for a very special potential energy as a function of distance, but the discrete levels and quantum numbers are characteristic of *all* problems where a particle is bound to a small region of space. We shall find additional verification of this general statement in the examples presented later in this chapter.

Classical mechanics, of course, has no such requirement of discrete energy levels for bound systems. We thus see that we are making some progress toward understanding the discrete levels observed in atoms. Furthermore, our quantum-mechanics theory does not disagree with classical mechanics in the region of sizes (namely, laboratory-scale sizes) where classical theory is known to apply. Problem 5-16 shows that for laboratory sizes the energy levels are so closely spaced as to be experimentally indistinguishable from a continuous set.

It should be noted that we could have obtained our quantum condition eq. 5-26 by considering the interference of de Broglie waves reflected back and forth between the walls. For any E not satisfying eq. 5-26, this interference is destructive, and for the E_n's it is constructive. Satisfying the Schrödinger equation is evidently equivalent to demanding constructive interference of de Broglie waves, at least in the present example. But for any problem more complicated than this one, the direct de Broglie wave approach is not powerful enough to produce results.

This square-well problem is mathematically very similar to the vibrations of a violin string; the displacement y of a point on the string takes the place of ψ. The string is fixed at each end, and therefore y equals zero at these ends (call them $x = \pm x_0$). The solutions of the acoustic wave equation are then just like eqs. 5-24 since the general solution of the wave equation gives sinusoidal oscillations, and the boundary conditions at $x = \pm x_0$ are the same as for the electron in the square well. Such a string has, of course, a *fundamental* mode of vibration ($n = 1$) and a set of *overtones* ($n = 2, 3, \cdots$), each with a characteristic frequency.

Another close analogy is a section of coaxial transmission line or waveguide, shorted at both ends, and here the voltage V takes the place of ψ. V equals zero at each end ($x = \pm x_0$) because the termination is a short (impedance $Z = 0$) at each end. The electromagnetic wave equation is of the same form as the Schrödinger equation. Since the differential equation and the boundary conditions are the same, the solutions for $V(x)$ are again just like eqs. 5-24; an appreciable voltage can appear on the line section only for a discrete set of frequencies. (Such a shorted transmission line or waveguide is a special case of the *resonant cavities* that are indispensable in microwave engineering. If the reader is familiar with resonant cavities he will be able to find ψ for a *three*-dimensional square well without any additional mathematics.)

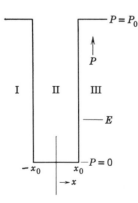

Fig. 5-3. Potential energy as a function of distance for the square well with finite sides.

(b) Square well with finite sides; bound states

We next turn to a more realistic model of an atom, even though it is still one-dimensional and artificial. The potential as a function of distance is illustrated in Fig. 5-3. We consider here the *bound* states, for which $E < P_0$; classically, an electron in such a state would be confined to the well (compare Fig. 3-13). In each of the three regions identified in Fig. 5-3, P is a constant, but $P > E$ in regions I and III and $P < E$ in region II. The technique of attacking such a problem is first to solve for ψ in each region and then to evaluate the arbitrary constants of each of the three solutions in such a way that the ψ's fit together at $x = \pm x_0$ with no discontinuities in ψ or in $d\psi/dx$.

We have already found the general solution in region II, namely eq. 5-23. In region III, eq. 5-4 becomes

$$\frac{d^2\psi}{dx^2} - \frac{8\pi^2 m(P_0 - E)}{h^2}\psi = 0 \tag{5-27}$$

with $P_0 > E$. The general solution of this equation is

$$\psi = Ce^{-\frac{2\pi}{h}\sqrt{2m(P_0-E)}\,x} + De^{\frac{2\pi}{h}\sqrt{2m(P_0-E)}\,x} \tag{5-28}$$

which can be verified by substituting eq. 5-28 into eq. 5-27. Note that the coefficients of x in the exponents are now *real*. Since ψ cannot increase without limit (in order to satisfy the condition eq. 5-6), D must be zero in region III. If $D \neq 0$, ψ rapidly becomes infinite as x increases without limit. Similarly, in region I, where the general solution is identical with eq. 5-28, the constant C must be zero or ψ would approach ∞ as x approaches $-\infty$ and eq. 5-6 would not hold. Thus the solutions are

$$\psi = Ce^{\frac{-2\pi}{h}\sqrt{2m(P_0-E)}\,x} \quad \text{in region III} \tag{5-29}$$

and

$$\psi = De^{\frac{2\pi}{h}\sqrt{2m(P_0-E)}\,x} \quad \text{in region I} \tag{5-30}$$

These solutions must join eq. 5-23 with continuous ψ and $d\psi/dx$ at the points $x = \pm x_0$. The joining at $x = +x_0$ is illustrated in Fig. 5-4 for two values of P_0. It is instructive to examine the value of ψ at $x = \pm x_0$ as P_0 is changed, and in particular to study $\psi_{x=x_0}$ as P_0 approaches ∞. It appears from Fig. 5-4 that $\psi_{x=x_0}$ approaches zero as P_0 approaches ∞. This result can be verified analytically as follows: From eq. 5-23 (region II),

$$\left(\frac{d\psi}{dx}\right)_{x=x_0} = \frac{2\pi}{h}\sqrt{2mE}\left(A\cos\frac{2\pi}{h}\sqrt{2mE}\,x_0 - B\sin\frac{2\pi}{h}\sqrt{2mE}\,x_0\right)$$

$$\tag{5-31}$$

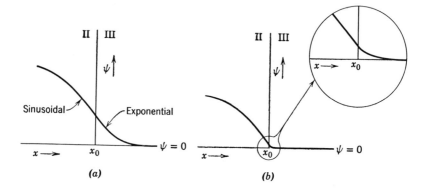

Fig. 5-4. Plots of $\psi(x)$ near boundary $x = x_0$ of the well. (*a*) is for a small value of P_0. (*b*) is for a very large value of P_0.

From eq. 5-29 (region III),

$$\left(\frac{d\psi}{dx}\right)_{x=x_0} = \frac{-2\pi}{h} \sqrt{2m(P_0 - E)}\; Ce^{-\frac{2\pi}{h}\sqrt{2m(P_0-E)}\; x_0}$$

$$= \left(-\frac{2\pi}{h} \sqrt{2m(P_0 - E)}\right)\psi_{x=x_0} \qquad (5\text{-}32)$$

Note that $(d\psi/dx)_{x=x_0}$ from eq. 5-31 does not change appreciably as P_0 approaches ∞; it changes only because of a slight change in E. But $(d\psi/dx)_{x=x_0}$ from eq. 5-32 approaches $-\infty$ as P_0 approaches ∞ unless $\psi_{x=x_0} = 0$, since the coefficient of ψ approaches $-\infty$. Therefore, since these two expressions for $(d\psi/dx)_{x=x_0}$ must be equal, as P_0 approaches ∞, ψ at x_0 must approach zero. We can make a similar argument for ψ at $x = -x_0$. We have thus established the boundary conditions used in example (*a*) above, namely, $\psi_{x=\pm x_0} = 0$.

The joining of solutions at $x = \pm x_0$ for any value of P_0 other than ∞ involves a rather tedious calculation of the constants A, B, C, and D, and therefore we are content with illustrating the result for a particular P_0 in Fig. 5-5. The general shape of ψ within the well is the same as in Fig. 5-2. But note that there are now *exponential tails* on the wave functions in regions I and III. Note further that, for larger E (E_2, for example), and therefore smaller $(P_0 - E)$, the tails have larger amplitudes and fall off less rapidly with distance away from the well. The larger amplitude can be inferred from the argument of the preceding paragraph, and the slower decrease in magnitude follows from the smaller coefficient of x in eqs. 5-29 and 5-30.

The fact that ψ (and therefore $|\psi|^2$) is not zero in regions I and III is a new result that is not expected on the basis of classical theory. In these regions the kinetic energy is *negative!* As we saw in Sec. 3-4, a negative kinetic energy means that classically the electron should have been turned around at the edge of the well and never appear in the negative K regions. The quantum mechanics therefore predicts a probability of penetrating some distance into a *classically forbidden* region of nega-

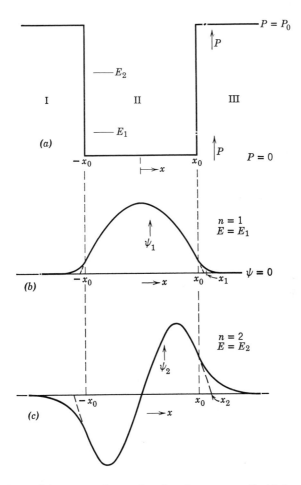

Fig. 5-5. Potential energy and wave functions for square well with low sides. The dashed lines in (b) and (c) are the extrapolations of the sinusoidal functions into the classically forbidden regions. The position x_1, which is to be used in prob. 5-26, is defined by writing $\psi(x_1) = e^{-1}\psi(x_0)$.

tive kinetic energy. As explained in the preceding paragraph, ψ will remain appreciable in size for greater distances beyond the barrier if (P_0-E) is small. Thus the degree of penetration is a rapidly varying function of the negative kinetic energy.

In addition to the new phenomenon of penetration into a classically forbidden region, the lower sides of the well (compared to the preceding example) have introduced a modification of the energy levels. Since ψ is no longer reduced to zero at the edges of the well, the wavelength of the oscillations within the well is somewhat longer, and therefore the energy levels are somewhat lower, than in the preceding example. In effect, the walls have been separated by a distance $2x_2$ (see Fig. 5-5c) somewhat greater than $2x_0$. The wave function need not complete $\frac{1}{2}$, 1, $\frac{3}{2}$, $\cdots$, oscillations within the well but has a "little left over" to join smoothly with the exponential tails. Nevertheless, eq. 5-26 is still a good approximation, especially for large P_0-E.

It should be noted how the requirement arose that the energy could take on only one or another of a set of discrete energy levels. The necessity for $\displaystyle\int_{-\infty}^{\infty} |\psi|^2\, dx$ to remain finite and the continuity conditions forced ψ to equal nearly zero at $x = \pm x_0$. This requirement then forced ψ to have an integral number of half-cycles of oscillation in the distance $2x_2$ (see Fig. 5-5c). Since the distance for a half-cycle depended on E, this compelled us to have only certain discrete values of E. For any value of E *not* in the set, ψ would approach infinity outside the box, and hence, after normalizing, ψ would equal zero inside the box. In other words, the probability of finding the electron in the box with an energy other than one of these E_n's is zero.

In drawing Fig. 5-5 we have tacitly assumed that there were only two energy levels such that E was less than P_0, which occurred because of a particular choice of the product $P_0x_0{}^2$. If the well had been wider or deep r, more levels would have been obtained. For any well size, however, eventually an energy level would be reached such that there were no more E_n's less than P_0. If E is greater than P_0, our solutions in region I and III are no longer correct. We shall consider this situation in example (e).

(c) Transmission through a barrier; tunnel effect

The penetration of an electron into a region of negative kinetic energy cannot be verified experimentally for a $P(x)$ as in Fig. 5-5. The details are left for a problem (prob. 5-26), but the argument is briefly as follows. If we are to observe this effect, we must have simultaneous information demonstrating that the electron is outside the well ($x > x_0$) and that its

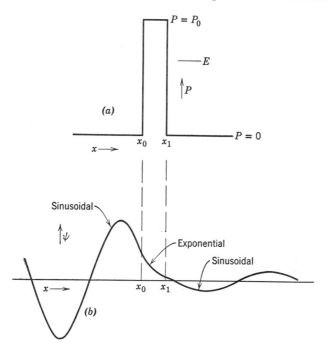

Fig. 5-6. Penetration of an electron wave through a barrier. The potential energy as a function of x is shown in (a), and $\psi(x)$ for an electron incident from the left is shown in (b).

kinetic energy is negative $(E < P_0)$. We could in principle focus a microscope on, say, the point x_1 of Fig. 5-5b with sufficient resolution (sufficiently small wavelength of radiation) that if we "saw" the electron we could be confident it was outside the well. But we should then find, as in the argument associated with Fig. 4-40, that we could have no confidence that the electron had a $K < 0$, since the radiation would have to be energetic enough to increase E to a value greater than P_0.

It is quite possible, however, to devise a $P(x)$ such that this penetration *can* be observed by avoiding the necessity for an accurate position determination. An electron coming from the left is incident on the *barrier* of Fig. 5-6, in which the height (P_0-E) and width (x_1-x_0) of the barrier are small enough that the exponential tail has not been reduced to zero at x_1. At x_1 the kinetic energy becomes positive again and ψ becomes sinusoidal. The wave for $x > x_1$ does not have so large an amplitude as the wave for $x < x_0$, and therefore the probability of penetration is considerably less than 1, but it is not zero. This transmission through a classically forbidden region is called the *tunnel effect*. In order to verify this striking prediction of quantum mechanics we do not need an accurate position determination; if we know that an electron with energy E was initially anywhere in the region $x < x_0$ and later is found anywhere in $x > x_1$, we know it has penetrated the barrier. Of course sufficiently low and thin barriers occur only with atomic and nuclear magnitudes, not with macroscopic sizes, and the tunneling probability is a very sensitive function of the barrier height and width (prob. 5-28). We shall study many illustrations of the tunnel effect in the motions of electrons in molecules and solids, in the field emission of electrons from solids (Sec. 12-5), and in the phenomena of radioactivity (Sec. 13-2).

(d) Partial reflection of electrons

As an introduction to example (e) and as an interesting problem in its own right, we next consider electrons incident from the left upon the abrupt energy step illustrated in Fig. 5-7. The solutions are now traveling waves in both regions I and II, but with different wavelengths in these two regions. The complete wave function (including the time) for the incident wave is merely the first term of eq. 5-16, which is

$$\Psi_{\text{in}} = A'e^{-2\pi i\left(\frac{Et}{h}-\frac{x}{\lambda_{\text{I}}}\right)} \tag{5-33}$$

where $\lambda_{\text{I}} = h/\sqrt{2m(E - P_0)}$. If there were no reflection at $x = 0$, the only other wave present would be the *transmitted* wave in region II, with

$$\Psi_{\text{tr}} = C'e^{-2\pi i\left(\frac{Et}{h}-\frac{x}{\lambda_{\text{II}}}\right)} \tag{5-34}$$

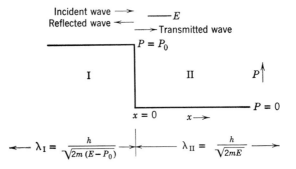

Fig. 5-7. Partial reflection of an electron wave at a step in the potential energy. The height of the step is P_0.

where $\lambda_{II} = h/\sqrt{2mE}$. It is easy to see that both Ψ and $\partial\Psi/\partial x$ (or ψ and $d\psi/dx$) cannot be continuous at $x = 0$ without another wave in addition to this transmitted wave. In order to make Ψ continuous, A' would have to equal C'. But then $\partial\Psi/\partial x$, which is proportional to the product of $1/\lambda$ and Ψ, would be discontinuous. Thus an additional wave is necessary in order to satisfy the boundary conditions at $x = 0$. The only possibility is that there be a *reflected* wave

$$\Psi_{\text{ref}} = B'e^{-2\pi i\left(\frac{Et}{h}+\frac{x}{\lambda_I}\right)} \tag{5-35}$$

traveling toward the left in region I (a wave traveling toward the left in region II is not a possibility in this problem, since here there is only a single source of electrons at the left of $x = 0$). The sum of eqs. 5-33 and 5-35 is the general solution for Ψ in region I.

The fraction of the electrons that are reflected, the *reflection coefficient* r, is easily obtained from Ψ, since the probability of finding an electron is proportional to Ψ^2. Therefore r equals $|\Psi_{\text{ref}}|^2/|\Psi_{\text{in}}|^2$, which is just B'^2/A'^2. When the calculation outlined above is performed (prob. 5-30) it develops that

$$r = \left(\frac{\sqrt{1 - P_0/E} - 1}{\sqrt{1 - P_0/E} + 1}\right)^2 \tag{5-36}$$

Thus an electron beam encountering an abrupt change in potential will always experience some reflection, and the amount of reflection is sensitive to the change in kinetic energy $(E - P_0)$. Since r is neither 0 nor 1, it may seem that we are requiring the electron to be divided, part transmitted and part reflected. But r is, of course, only the *probability* of reflection. For example, an r value of 0.1 means that of 10^{13} electrons incident (about 1 microampere for 1 second), about 10^{12} electrons will be reflected and about 9×10^{12} will be transmitted.

It is interesting to note that if the step illustrated in Fig. 5-7 is reversed ($P = 0$ in region I and $P = P_0$ in region II), the reflection coefficient is unchanged (prob. 5-31). But the phase of the reflection (the phase difference if any between the incident and the reflected waves) is different. The reflection is *180° out of phase* ($B'/A' < 0$) for an electron incident from the left on the potential step of Fig. 5-7 and is *in phase* ($B'/A' > 0$) for the reversed step (prob. 5-32).

The student might well now study Sec. 4-10 again with the aim of appreciating how the interference of partial reflections produces the phenomena described there.

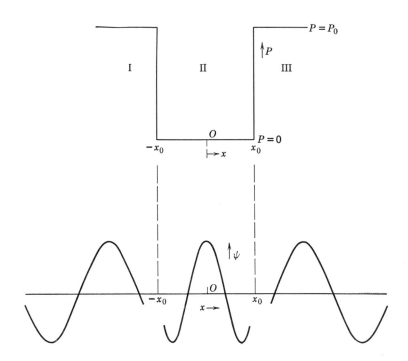

Fig. 5-8. Free states with square-well potential. The wave function in II has a smaller λ than the others. The amplitudes of the three solutions have not been adjusted to make ψ and $d\psi/dx$ continuous, since these amplitudes depend on the particular initial conditions (e.g., electron initially in one of the three regions).

(e) Square well with finite sides; free states

We now return to the square-well potential, but we consider not an electron that is bound ($E < P_0$) but one that is *free* ($E > P_0$), as diagrammed in Fig. 5-8. The solutions in all three regions (I, II, and III) are now sinusoidal in form. The only difference between the solutions for I and III and the solution for II is that the wavelength of the oscillations is different. The solutions for the different regions must join at $x = \pm x_0$ with no discontinuities in ψ or $d\psi/dx$. None of these solutions becomes large without limit, and therefore there are no arguments like those of example (*b*) that forced E to have only one of a set of discrete values. For free states, then, E has a *continuous* distribution of allowed values. This contrasts with bound states, for which a *discrete* distribution of allowed values occurred.

An electron can therefore enter this region of space with any E greater than P_0. It will have a smaller λ in the region of the well, since its kinetic energy is increased. Upon leaving the well it will regain its original K and λ. Partial reflections occur at the two steps in the potential at $x = \pm x_0$. Thus, for an incident wave from the left, there are waves going both to the right and to the left in regions I and II and a transmitted wave in III. The reflection from the change in P at $x = x_0$ interferes with that from $x = -x_0$, and the algebra of the general solution is rather involved.

There is an interesting special case, however, for which the interference is simple. Since the reflections at $x = \pm x_0$ are equal in magnitude and 180° different in phase, if $2x_0 = \lambda_{II}/2$ the two reflections exactly cancel. The time of travel from $-x_0$ to x_0 (where reflection occurs) and back to $-x_0$ is just one cycle, and therefore this reflection adds algebraically to the reflection from $-x_0$. Because the two reflections are opposite in phase, cancellation occurs. If the kinetic energy of the electron is just right for the particular depth and width of the well, there is thus no reflected wave in region I. In other words, electrons are 100% transmitted, just as if the well were not present.

The simple theory of the preceding paragraph explains the *Ramsauer effect*, one of the famous experiments of atomic physics, which might have been described in Chapter 4 but can better be considered at this point together with the quantum mechanical understanding of the phenomenon. This effect is the almost complete transparency of the noble gases argon, krypton, and xenon for electrons with a critical kinetic energy. The experimental arrangement to study this effect is similar to Fig. 2-7, but employs a beam of electrons instead of a beam of molecules. Measurement of the loss of electrons from the beam enables us to compute the collision cross section for electron scattering by the gas molecules.

The way this cross section σ depends on the electron velocity is shown in Fig. 5-9. For most atoms and molecules, σ monotonically increases as the velocity decreases as illustrated for zinc in that figure. This increase is caused by the fact that slow electrons are near a gas atom for a longer time during an electron-atom collision, and therefore the Coulomb forces are more effective in deflecting the electron out of the beam. For the noble gases like krypton, however, a sharp dip is superimposed on this variation. At just the right electron velocity v_0 (and hence λ),

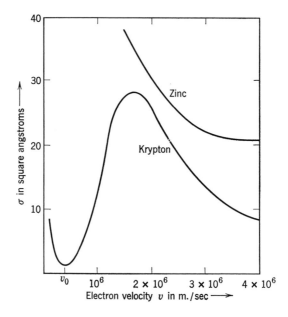

Fig. 5-9. Scattering cross section as a function of electron velocity. Electrons with $v_0 = 0.4 \times 10^6$ m./sec pass through krypton atoms with almost no reflection. (From H. S. W. Massey and E. H. S. Burhop, *Electronic and Ionic Impact Phenomena*, Clarendon Press, Oxford, 1952.)

the cross section is nearly zero, and this is the Ramsauer effect. Its explanation is that, at $v = v_0$, the electron wavelength is such that the partial reflections at the beginning and the end of the atom cancel, and 100% transmission occurs. The Ramsauer effect does not occur for helium and neon because the *strength* of the potential well (product of the width and the square root of the depth) is insufficient in these atoms to produce cancellation ($2x_0 < \lambda_{II}/2$ in Fig. 5-8). It does not occur for gases other than the noble gases because only the noble gases are spherically symmetrical with reasonably sharp outer boundaries, and they approach the square-well model more closely than any other atoms.

The mathematics of the multiple reflections in the square-well problem is identical with that for partial reflections of radio waves in transmission lines or light waves in thin films. Some special cases in which the partial reflections cancel to give no reflection are illustrated in Fig. 5-10. Practically useful transmission-line arrangements are shown in *b* and *c* of the figure, and the arrangement shown in *d* is the basis of the modern low-reflecting coatings for optical lenses.

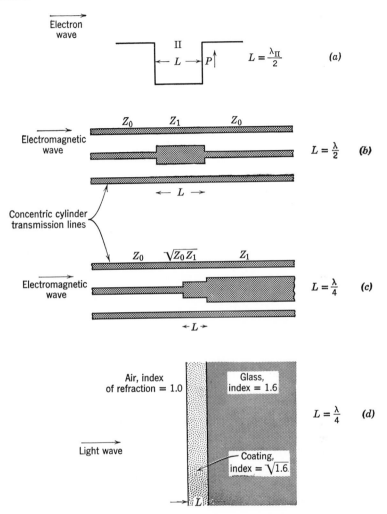

Fig. 5-10. Examples of the prevention of reflection: (a) electron incident on a potential well; (b) the characteristic impedance Z of the line changes because the diameter of the central wire changes; (c) like (b), but second reflection is *in phase* with first; (d) a non-reflecting coating for lenses is made by a fluoride coating $\lambda/4$ thick with an index of refraction intermediate between glass and air.

5-5 THE HARMONIC OSCILLATOR

In classical mechanics the problem of a particle moving in one dimension and attracted to a fixed point $x = 0$ by a force

$$F = -C^2 x \tag{5-37}$$

where C is a constant, is a very important problem called the *harmonic oscillator* problem. The change in the particle's position, x, as a function of t is called *simple harmonic motion*. Examples are the small-amplitude oscillation of a pendulum and the vertical oscillation of a mass supported by an ideal spring. The importance of this particular force law extends far beyond such simple cases. The reason for its importance is that a force like eq. 5-37 occurs in *all* cases of vibrations about a position of stable equilibrium if the amplitude is small enough.

It is easy to see why this is true. At equilibrium F equals zero; let this be the point $x = 0$. Then *any* force that is a function of x can be expressed by Maclaurin's series as

$$F = F_{x=0} + x \left(\frac{dF}{dx}\right)_{x=0} + \frac{x^2}{2}\left(\frac{d^2F}{dx^2}\right)_{x=0} + \cdots \qquad (5\text{-}38)$$

$F_{x=0}$ equals zero since the origin of x was chosen at the equilibrium position. Furthermore, for oscillations with sufficiently small amplitudes, x^2 is very much less than x, and all the terms but the second are negligible. Hence

$$F = x\left(\frac{dF}{dx}\right)_{x=0} = x \times \text{(a constant)} \qquad (5\text{-}39)$$

This constant must be negative if the equilibrium is stable. That is, the force when x is displaced from zero must be a *restoring* force. Thus, for any force, small-amplitude oscillations about a position of stable equilibrium are described by eq. 5-37. (This proof would break down if $(dF/dx)_{x=0}$ happened to equal zero, but it never does in physically interesting problems.) Problems in atomic physics like the vibrations of a diatomic molecule or the vibrations of the atoms in a crystal thus involve a force law like eq. 5-37.

We now seek the quantum theory of such harmonic oscillators. Because our applications will usually involve atoms, rather than electrons, as the oscillating particles, we shall use the symbol M (instead of m) for mass in the following theory. In order to be able to compare classical and quantum results, we first recall the classical theory. The classical solution is, of course, found by substituting eq. 5-37 into Newton's second law, $F = M(d^2x/dt^2)$:

$$M\frac{d^2x}{dt^2} + C^2x = 0$$

The solution of this equation is

$$x = A\cos\left[(C/\sqrt{M})t + \phi\right] = A\cos\left(2\pi\nu_0 t + \phi\right)$$

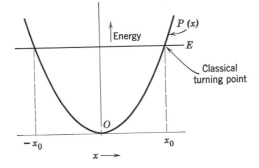

Fig. 5-11. Potential energy for the harmonic oscillator; the potential energy as a function of x is a parabola. At the points where $P = E$, the particle turns around.

where A and ϕ are arbitrary constants (cf. eq. 5-23). We have written this expression in terms of the classical frequency of oscillation ν_0, which is related to the mass and force constant by $\nu_0 = C/(2\pi\sqrt{M})$.

The potential energy as a function of x can easily be computed from eq. 5-37 and is

$$P = C^2 x^2/2 \tag{5-40}$$

in which P has been set equal to zero at $x = 0$. The amplitude of oscillation is determined by the total energy E, as shown in Fig. 5-11. The particle reverses its direction (its velocity goes to zero and changes sign) at the points $x = \pm x_0$ where $E = P$, since $K = 0$ at these points, which are therefore called the *classical turning points*. Thus

$$E = C^2 x_0^2/2 \quad \text{or} \quad x_0 = \sqrt{2E}/C \tag{5-41}$$

We now turn from the classical solution to the quantum-mechanical solution. The Schrödinger equation with the potential energy eq. 5-40 inserted is

$$\frac{h^2}{8\pi^2 M}\frac{d^2\psi}{dx^2} + \left(E - \frac{C^2 x^2}{2}\right)\psi = 0 \tag{5-42}$$

This may not seem to be a difficult differential equation, but the x^2 term complicates it substantially. We do not attempt the solution here, but we shall give some of the results. The *nature* of the solution is the same as for bound states in the square-well problem. As in that problem, the conditions of finiteness and continuity of ψ require a discrete set of energy levels. The solution ψ for any E not a member of this set

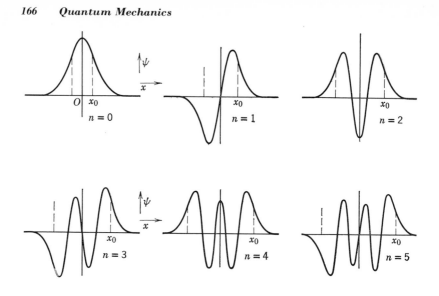

Fig. 5-12. Wave functions for the harmonic oscillator. The dashed lines are the limits between which a classical oscillator with the same energy would oscillate. The full width of the scale in each case is $5\sqrt{2}/a$. (From L. Pauling and E. B. Wilson, *Introduction to Quantum Mechanics*, McGraw-Hill, New York, 1935.)

would become infinite for either large x, small x, or both. The allowed energy values are

$$E_n = (n + \tfrac{1}{2})h\nu_0 \qquad (5\text{-}43)$$

where n takes on the values 0, 1, 2, $\cdots$.*

The three wave functions with lowest energies are:

$$n = 0 \qquad \psi_0 = 2^{1/4}\pi^{-1/4}a^{1/2}e^{-a^2x^2} \qquad (5\text{-}44)$$

$$n = 1 \qquad \psi_1 = 2^{5/4}\pi^{-1/4}a^{3/2}xe^{-a^2x^2} \qquad (5\text{-}45)$$

$$n = 2 \qquad \psi_2 = 2^{-1/4}\pi^{-1/4}a^{1/2}(4a^2x^2 - 1)e^{-a^2x} \qquad (5\text{-}46)$$

which can be verified by substituting these functions into eq. 5-42. Here we have written $a^2 = 2\pi^2 M\nu_0/h$.

These and other wave functions are plotted in Fig. 5-12. In each part of Fig. 5-12 the classical turning points $\pm x_0$ for a classical oscillator with equal energy are indicated. The values of x_0 are different for each value of n, because E depends on n (eq. 5-43) and x_0 depends on E

---

* It is general practice to write E_n as in eq. 5-43 and therefore to begin the set of E_n's with $n = 0$, but we could equally well have written $E_n = (n - \tfrac{1}{2})h\nu_0$ and have begun with $n = 1$.

(eq. 5-41). The wave functions are, of course, different in detail from those for the square well, but the following *similarities* are significant: (1) There is a set of discrete energy levels for bound states. (2) The particle has substantial kinetic energy in the ground state. (3) $|\psi|$ is largest inside the classical turning points but has exponential-like tails outside these points. (4) ψ makes an additional crossing of the x-axis for each increase in quantum number from n to $n + 1$ (zero crossings in ground state, one in first excited state, etc.). These features are in fact shared by *all* bound-state wave functions for *any* potential energy as a function of distance.

The wave functions for large n are also interesting. In Fig. 5-13 the quantity $|\psi_{10}|^2$ is plotted, and this curve gives the relative probability of finding the particle at various positions, computed by wave mechanics. The dashed line is the similar probability but computed classically; since the particle is moving faster near $x = 0$, the chance of finding it in a region Δx near $x = 0$ is smaller than for a region Δx near the turning points. The similarity of the wave-mechanical and classical results will be considered in the following section.

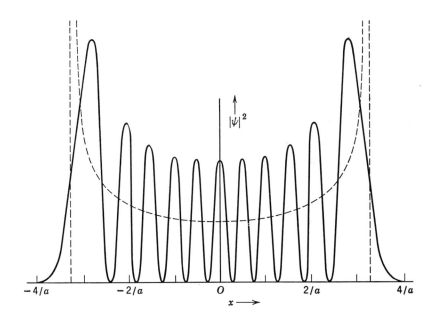

Fig. 5-13. Position probability density for the state $n = 10$ of the harmonic oscillator (solid curve) and for a classical oscillator with the same energy (dashed curve). (From L. Pauling and E. B. Wilson, *Introduction to Quantum Mechanics*, McGraw-Hill, New York, 1935.)

5-6 THE CORRESPONDENCE PRINCIPLE

The *Correspondence Principle* states that the predictions of quantum mechanics and classical physics agree in the limit of large sizes and energies, as the quantum number n approaches infinity. The quantum-mechanical solutions of problems approach classical-mechanical solutions as the lengths involved become much larger than the de Broglie wavelength of the particle involved. This principle merely states a result that could be proved for each separate problem directly from the Schrödinger equation; it is *not* an additional assumption of quantum mechanics. The Correspondence Principle implies that quantum mechanics may be successfully extrapolated to the laboratory scale of sizes. The converse (extrapolation of classical mechanics to atomic sizes) is certainly *not* successful.

Examples of classical results occurring as limiting cases of quantum-mechanical results have already been discussed (for example, the last paragraph of Sec. 5-3). Others will be explored in problems at the end of this chapter and at the end of Chapter 6.

We now study the way this classical limit is approached in the example of the harmonic oscillator. We start by investigating the $n = 0$ state of this oscillator. If $|\psi_0|^2$ is plotted as a function of x it would not be much different in shape from ψ_0 itself, which is plotted in Fig. 5-12. Thus there is a larger probability of finding the particle near $x = 0$ than near the classical turning points. This is just the reverse of the classical result, which gives a higher probability of finding the particle near the classical turning points where the velocity is least. But we do not expect any correspondence between quantum and classical theories for this lowest state of the oscillator, since this is the limiting case at *small* sizes. Also, the solutions for $n = 1$ to $n = 6$ do not show much correspondence.

We now consider the plot of $|\psi_{10}|^2$ in Fig. 5-13, and here the correspondence is becoming quite striking. The probability rises toward the classical turning points and then sharply decreases to zero outside these points, just as expected classically. Yet the rapid oscillations in $|\psi|^2$ as a function of x remain even as n becomes large enough to describe an oscillator of ordinary laboratory size (e.g., a pendulum). Thus it may at first appear that there is an important disagreement between the quantum mechanical and the classical description of the motion even in the limit $n \rightarrow \infty$. But if we design an experiment in an attempt to observe this fine structure, we must use a short enough wavelength in our microscope that the maxima can be resolved. To provide such resolution, the radiation will necessarily be sufficiently energetic (short wavelength) that it can readily change the quantum state n (see prob. 5-52). Even a change of unity in n alters the fine structure of $|\psi|^2$ completely, changing $|\psi(0)|^2$ from a maximum to zero, for example. Thus as we repeat the experiment in order to obtain a good statistical average, we should in fact be measuring not $|\psi|^2$ as a function of x for a given fixed n, but $|\psi|^2$ as a function of x for a mixture of states with different values of n. Inspection of Figs. 5-12 and 5-13 shows that in such a mixture the rapid oscillations of $|\psi|^2$ are suppressed. A mixture of n's from 100 to 110, for example, gives a curve for $|\psi|^2$ as a function of x that closely resembles the dashed curve in Fig. 5-13. Thus the predictions of quantum theory for actual experiments are the same in the limit as $n \rightarrow \infty$ as the predictions of classical mechanics.

The Correspondence Principle is of most interest in the problem of the *absorption* or *radiation* of energy. If an electron is in harmonic motion with frequency ν_0, we know that classically it should absorb or radiate energy at the frequency ν_0. The oscillating motion of the charge, like the oscillating current in a dipole radio transmitting antenna, radiates an electromagnetic wave of frequency ν_0. This same motion of charge, like the induced currents in a dipole radio receiving antenna, can absorb strongly waves of frequency ν_0. When this classical result is compared with the energy levels for the oscillator given by eq. 5-43, it is apparent that a transition $\Delta n = \pm 1$ gives the radiation or absorption of the same frequency as that predicted classically. But two points about this comparison should be noted: (1) For the simple-harmonic oscillator the agreement is independent of n, and correspondence occurs for large or small n. This result is a peculiarity of the harmonic oscillator related to the classical peculiarity that the frequency is independent of amplitude; for all other problems, correspondence occurs only for large n. (2) The quantum theory seems to predict frequencies $2\nu_0$, $3\nu_0$, $\cdots$ (corresponding to $\Delta n = 2, 3, \cdots$) in addition to the frequency ν_0, which is the only frequency predicted by classical theory. In order to obtain correspondence between the classical and quantum results, it must be true that the *only* changes in n that can occur are $\Delta n = \pm 1$. This statement is called a *selection rule*, and the quantum-mechanical theory should produce this selection rule for the harmonic oscillator. That it does will be demonstrated in Sec. 5-8.

5-7   AVERAGE ENERGIES AND GROUND-STATE WAVE FUNCTIONS

It has already been noted that once the wave function has been found for a particular problem, predictions of properties and phenomena in that problem can be made by calculations based on this wave function.

It will be useful for our later work with molecules and solids to sketch here the method of determining from the wave function such quantities as the average kinetic energy (the average value of the K's observed in a large number of observations on identical systems). This method will be described more fully in Appendix F.

Equation 5-12 provides information about what may crudely be called the "probable position" of an electron. To be more precise, if we make repeated observations on a system with wave function $\psi(x)$, the fraction of these observations that will show that the electron is between x and $x + dx$ will be $\mathcal{P}(x)\,dx = |\psi(x)|^2\,dx$. Or, if we make observations on a large number of identical systems, the electrons will be between x and $x + dx$ in a fraction $|\psi(x)|^2\,dx$ of them. Thus $|\psi|^2$ plays the role of a distribution function, like the distribution functions ϕ that we encountered in Secs. 2-3 and 2-4.

If we wish to calculate an average quantity, such as the average position $\bar{x}$, from such a distribution, we apply the method described in connection with eq. 2-10. Thus

$$\bar{x} = \int_{-\infty}^{\infty} x\,|\psi|^2\,dx \qquad (5\text{-}47)$$

in which we have been able to omit the integral from the denominator of eq. 2-10 because $\int_{-\infty}^{\infty} |\psi|^2\,dx = 1$ by normalization. Similarly the average potential energy is

$$\bar{P} = \int_{-\infty}^{\infty} P(x)\,|\psi|^2\,dx \qquad (5\text{-}48)$$

The physical interpretation of such equations is simple: If the electron is at a particular position x, its potential energy is $P(x)$. To get the average P, we multiply the probability that the electron is at each position (more accurately, within dx of that position) by the $P(x)$ it has when there, and then sum over all possible positions.

The calculation of the average kinetic energy takes a slightly different form, since we have no function $K(x)$. But the basic procedure, which is described in Appendix F, is actually quite similar. The result is

$$\bar{K} = \int_{-\infty}^{\infty} \frac{h^2}{8\pi^2 m} \left| \frac{d\psi}{dx} \right|^2 dx \qquad (5\text{-}49)$$

We can verify this equation easily for the special case of $n = 1$ of the square well with infinite sides; $\bar{P} = 0$ for such a well, $\bar{K}$ must therefore equal E_1 from eq. 5-26, and a short calculation shows that eq. 5-49 is obeyed. This equation can likewise be verified for the higher states, in which $\bar{P}$ still equals zero and the increased energy of these states arises from the sharper rise and fall (larger $|d\psi/dx|$) of the wave functions appropriate to these states.

The calculations of $\bar{P}$ and $\bar{K}$ can be profitably applied to finding the ground state wave function for any potential $P(x)$, since it can be shown (Appendix F) that the actual ground state $\psi(x)$ is that function of x that makes the total energy $\bar{E} = \bar{P} + \bar{K}$ a *minimum*. We can illustrate this method by finding the shape of the ground state ψ_0 of the harmonic oscillator, as illustrated in Fig. 5-14. If we try a ψ (Fig. 5-14b) that is

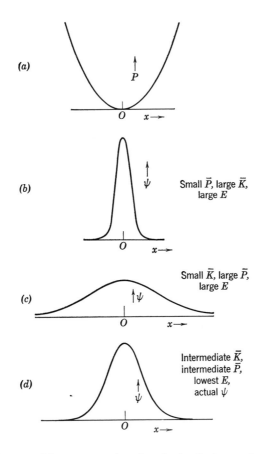

Fig. 5-14. The potential energy as a function of x for the harmonic oscillator problem is shown in (*a*). Trial wave functions are shown in (*b*) and (*c*). The actual ground state ($n = 0$) wave function is shown in (*d*).

concentrated near $x = 0$, $\bar{P}$ is very small since the electron is nearly always in the region of lowest potential energy. But concentrating ψ in this way necessitates large values of $|d\psi/dx|$. Therefore $\bar{K}$ is large and $\bar{E}$ is large. On the other hand, we might try a ψ with very small $|d\psi/dx|$ and therefore small $\bar{K}$ as in Fig. 5-14c. But this ψ necessitates a large $\bar{P}$, since $|\psi|^2$ is appreciable where P is large. The lowest $\bar{E}$ is obtained by a compromise between these extremes. The wave function of Fig. 5-14d, which is also the $n = 0$ wave function in Fig. 5-12, is the actual wave function and represents the result of this compromise.

This method of estimating the shape of the ground-state wave function of a system has many applications, such as in determining the nature of the binding energy in molecules and solids. It is instructive to apply these energy concepts to every ground state ψ that is encountered. The method has been explained here for one-dimensional problems, but it can easily be extended to three dimensions.

5-8 RADIATION AND ABSORPTION

The observed facts of the radiation and absorption of light by atoms can be adequately explained by wave mechanics. We shall only sketch the theory, since the complete presentation is extensive and complicated. We shall study the frequencies of light emitted or absorbed, without attempting to determine the rate of emission or absorption. We shall see how the Schrödinger equation predicts sharp-line spectra and Bohr's frequency equation (eq. 4-5). We shall also see how selection rules arise and shall compare the result for the harmonic oscillator with the Correspondence Principle. The selection rules tell which transitions from one quantum state to another are *allowed* and which are *forbidden*. It will develop that the allowed transitions are those in which the electrical charge oscillates as in a dipole antenna.

In order to determine whether radiation occurs from some system containing a moving electron, we must find the average position $\bar{x}$ of the electron. If this $\bar{x}$ oscillates at a frequency ν, we expect radiation of frequency ν to be emitted or absorbed, because such an oscillation of the probable position of the electron means an oscillation of electrical charge. Radiation from such an oscillating charge is a consequence of ordinary electromagnetic theory.

We have seen in eq. 5-47 how to calculate $\bar{x}$ from ψ, but here we must work in terms of Ψ since we seek the possibility of an $\bar{x}$ oscillating in time. Also, it will be convenient to apply the method of calculating the square of the absolute magnitude of a complex quantity Ψ that is common in electric circuit theory, namely, to multiply Ψ by its *complex* *conjugate* Ψ^*. Ψ^* is simply Ψ but with i changed to $-i$; of course if Ψ is real (does not contain i), $\Psi^* = \Psi$. Thus eq. 5-47 becomes

$$\bar{x} = \int_{-\infty}^{\infty} x\Psi\Psi^* \, dx \tag{5-50}$$

Let us first consider that the electron is in a single quantum state with quantum number n and energy E_n. Its wave function is

$$\Psi_n = e^{-(2\pi i E_n/h)t} \, \psi_n \tag{5-51}$$

When this Ψ_n is substituted into eq. 5-50, we obtain:

$$\bar{x} = \int_{-\infty}^{\infty} x e^{-(2\pi i E_n/h)t} e^{(2\pi i E_n/h)t} |\psi_n|^2 \, dx$$

$$= \int_{-\infty}^{\infty} x |\psi_n|^2 \, dx \tag{5-52}$$

$\bar{x}$ from eq. 5-52 is not a function of time; there is no oscillating charge, and hence no radiation will occur, even though the electron is in motion with appreciable kinetic energy. (Compare this result with the example illustrated in Fig. 4-11, in which classical theory predicted radiation at all times.)

Let us now suppose that the electron is changing from one quantum state n to another m, as in absorption $(m > n)$ or emission $(n > m)$. We suppose further that n is the ground state; if no radiation has been incident on the system (harmonic oscillator, atom, or other system) it must be in this state. At time $t = 0$, light is turned on and radiation is incident on the system. The system may henceforth be in the state n or in the state m (or higher states, if there are any, but we assume that there are only two states for the time being). We describe this situation by writing

$$\Psi = a\Psi_n + b\Psi_m$$

where a and b are changing with time (at $t = 0$, $a = 1$, $b = 0$). The Indeterminacy Principle specifically denies any possibility of learning the energy of the system, and therefore the state the system is in, except to a time precision Δt given by $\Delta t > h/(E_m - E_n)$. Therefore we cannot plot a and b as functions of time for any single system (we could, however, plot the average values of a and b for a number of systems with identical starting conditions). We can infer that the system was in the state m (that is, $b = 1$, $a = 0$) at *some* time if we observe the emission of radiation as the system returns to the ground state. Now $\bar{x}$ can be calculated for this wave function as follows:

$$\bar{x} = \int_{-\infty}^{\infty} x(a\Psi_n + b\Psi_m)(a\Psi_n{}^* + b\Psi_m{}^*)\, dx$$

In the product in the integrand, the terms $\Psi_n\Psi_n{}^*$ and $\Psi_m\Psi_m{}^*$ will lead to stationary (non-oscillating) charge distributions as in eq. 5-52 and therefore will not give rise to radiation or absorption. We call the sum of the remaining cross-product terms $\overline{x'}$ and insert eq. 5-51: *

$$\overline{x'} = \int_{-\infty}^{\infty} xab(\Psi_n\Psi_m{}^* + \Psi_m\Psi_n{}^*)\, dx$$

$$= ab\int_{-\infty}^{\infty} x\{ e^{-\left(\frac{2\pi i E_n}{h}\right)t} e^{\left(\frac{2\pi i E_m}{h}\right)t} + e^{-\left(\frac{2\pi i E_m}{h}\right)t} e^{\left(\frac{2\pi i E_n}{h}\right)t} \}\psi_n\psi_m\, dx$$

In order to calculate the rate of radiation we should have to calculate the average value of ab for a large number of oscillators. But in order to determine the frequencies of allowed transitions it is necessary to investigate only the integral that multiplies ab; this integral can be simplified as follows:

$$\int_{-\infty}^{\infty} x\{ e^{\frac{2\pi i}{h}(E_m-E_n)t} + e^{-\frac{2\pi i}{h}(E_m-E_n)t} \}\psi_n\psi_m\, dx$$

$$= 2\cos\left\{\frac{2\pi}{h}(E_m - E_n)t\right\}\int_{-\infty}^{\infty} x\psi_n\psi_m\, dx \quad (5\text{-}53)$$

The average position of the electron according to eq. 5-53 is a cosine function of time multiplied by some number (the definite integral). There is therefore an oscillating charge, and hence radiation, at the frequency

$$\nu = (E_m - E_n)/h$$

which is just Bohr's postulate (eq. 4-5). The wave-mechanical theory has thus led to line spectra and an explanation of Bohr's postulate: The only photons emitted or absorbed have frequencies such that $h\nu$ equals the difference $E_m - E_n$ between two energy levels. Note that, although

---

* The terms ψ_n and ψ_m are real in all the problems we have discussed thus far; therefore in the following equations we shall let $\psi_n = \psi_n{}^*$ and $\psi_m = \psi_m{}^*$. In three-dimensional problems the ψ's are usually not real, but in such problems the distinction between allowed and forbidden transitions arises in the same way as in the present analysis.

complex quantities were used, the final result is real, as are all the results of wave mechanics.

No oscillating charge will occur if the integral in eq. 5-53 happens to equal zero:

$$\int_{-\infty}^{\infty} x\psi_n\psi_m \, dx = 0 \tag{5-54}$$

Such a case constitutes a forbidden transition, and therefore selection rules arise. Only those absorption or emission transitions are allowed that give a non-zero value to this integral. It might seem at first sight that zero values would occur rarely; this is not true, however, because of the symmetry of the wave functions of most problems.

The comparison of the above derivation of selection rules with the results of the Correspondence Principle for the harmonic oscillator can now be made. If we let $n = 0$ and $m = 1$, eqs. 5-44 and 5-45 give for the *form* (dropping constants) of the integral of eq. 5-54:

$$\int_{-\infty}^{\infty} x^2 e^{-2a^2x^2} \, dx$$

This integral cannot equal zero, since the integrand is always positive. Similarly, if we let $n = 1$ and $m = 2$, the form is

$$\int_{-\infty}^{\infty} x^2(4a^2x^2 - 1)e^{-2a^2x^2} \, dx \tag{5-55}$$

This integral is not zero, but a graphical analysis or evaluation of it by integral tables is required to demonstrate this fact. Therefore the transitions from 1 to 0 and from 2 to 1 are allowed transitions.

We now consider the transition $n = 0$ and $m = 2$. The integral is

$$\int_{-\infty}^{\infty} x(4a^2x^2 - 1)e^{-2a^2x^2} \, dx$$

It is easy to see that this integral equals zero, since for every contribution to the integral from a region at $+x_1$ there is an equal-in-magnitude but opposite-in-sign contribution from the similar region near $-x_1$. This cancellation occurs because x multiplies a *symmetric* (same value for $+x$ as for $-x$) function of x. The transition from 2 to 0 is forbidden.

Thus far we have agreement with the selection rule that Δn must equal $+1$ or -1. We could continue this examination of integrals for ψ_3, ψ_4, etc., but it is unnecessary since the general result can be proved from the nature of the wave functions for the harmonic oscillator. We shall not, however, take the space to prove this statement.

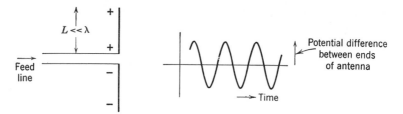

(a) Antenna driven "out of phase," radiation

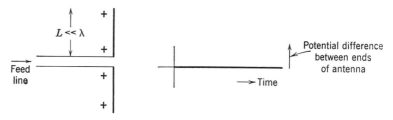

(b) Antenna driven "in phase," no radiation

Fig. 5-15. Antenna analogues of a radiating harmonic oscillator or an atom. *(a)* is similar to an allowed transition. *(b)* is similar to a forbidden transition, and the symmetry prevents radiation. It is assumed in both cases that the antenna is very far removed from other conductors and from the ground.

Wave mechanics has thus explained the principal features of the radiation and absorption of light. The general approach is valid for all emitters and absorbers, but the particular selection rule described is valid only for the harmonic oscillator. Wave mechanics predicts a different set of selection rules for the hydrogen atom, for example, but these rules arise in the same way.

The physical requirement for emission or absorption is that a charge distribution must oscillate in space at the desired frequency. The distributions of charge in the two states involved must be such that this back-and-forth motion of the electronic charge is produced. The problem of the *forbidden* transition is similar to the problem of a short dipole radio antenna in which the two lead wires are driven *in phase* (instead of 180° out of phase as usual). The antenna problem is illustrated in Fig. 5-15, in which both dipoles are much shorter than a wavelength of the radiation, just as an atom is much smaller than the λ of the light it radiates. In Fig. 5-15*b* a symmetrical motion of charge occurs: Charge flows outward on both limbs at the same time and inward at the same later time. No radiation or absorption occurs since the effects on the two limbs just cancel; for example, in the absorption case, the electric field outside the antenna could only make one limb positive with respect to the other and could not make both positive at the same time. It is the similar cancellation produced by the symmetry of wave functions that results in selection rules in radiation and absorption by atoms.

5-9 SUMMARY

The basic approach of quantum mechanics has been developed in this chapter through study of a number of artificial examples. In each one, the potential energy of an electron was specified as a function of position, the Schrödinger equation was solved subject to the important conditions on ψ (finite integral of $|\psi|^2$, continuous ψ and $d\psi/dx$, single-valued ψ), and the solution was interpreted to reveal the phenomena predicted. An outline of these examples is provided in Fig. 5-16. The first column in Fig. 5-16 is a sketch of each potential energy function, an idealized function chosen to make the mathematics easy and yet to provide the essential feature of a physical situation (e.g., a potential well as a crude model of an atom). The third column summarizes the new, non-classical phenomena predicted.

It should be noted how the wave and particle concepts are joined in the quantum theory. At the end of Chapter 4 it seemed as if we might need a whole catalogue of rules that would specify when to treat the electron or photon as a particle and when as a wave; either decision committed us to a mode of analysis with restricted applicability. But the quantum mechanics, a synthesis of these two approaches, follows basically the same approach for all problems. For example, we start with the same assumptions and mode of analysis for the problem of emission of light by atoms (photons, particles) as for the problem of interference resulting in the Ramsauer effect or electron diffraction (waves).

Although we have not actually carried out the quantum theory of any of the experiments of Chapter 4, it should be apparent from the summary in Fig. 5-16 that quantum mechanics is capable of predicting the phenomena observed in the wave-particle experiments, such as discrete energy levels for bound states or the interference among partial reflections of electron waves. Detailed, three-dimensional quantum calculations have been made for *all* the Chapter 4 problems, and they are in excellent agreement with experiment. Rather than stopping here to present these details, we prefer to move on to the application of the principles and techniques learned in the present chapter to the understanding of phenomena in atoms, molecules, solids, and nuclei.

$P(x)$	Studied in Section	New Features of the Solution
	5-3	E = total energy $\Delta E = h\Delta\nu$ Solutions have wave properties with $\lambda = h/mv$
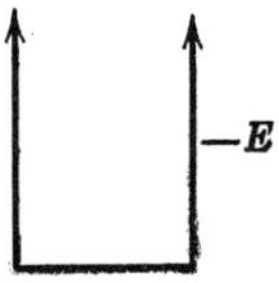 $-E$	5-4a	Discrete energy levels for bound states Quantum numbers Kinetic energy in the ground state One additional zero in $\psi(x)$ for each increase of unity in the quantum number
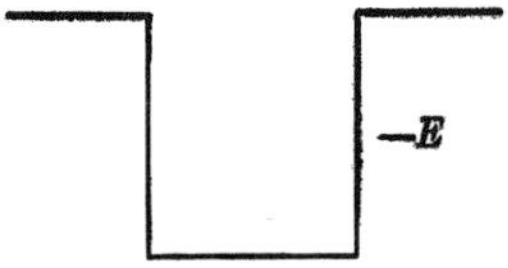 $-E$	5-4b	Penetration into classically forbidden regions Lowering of E_n's as walls become less high
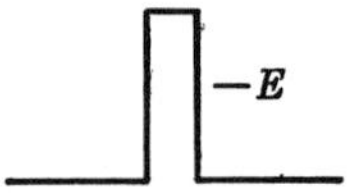 $-E$	5-4c	Penetration through a barrier ("tunneling")
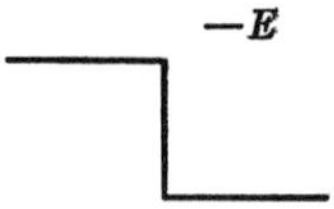 $-E$	5-4d	Partial reflection at a step in potential energy
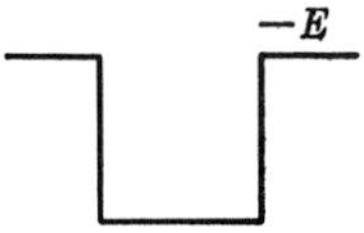 $-E$	5-4e	Continuous distribution of energy levels for free states Interference of partial reflections
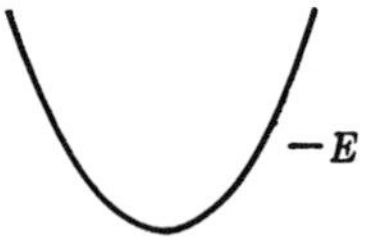 $-E$	5-5 5-6	Oscillatory $\psi(x)$ inside classical turning points, damped exponential-like $\psi(x)$ outside Correspondence with classical theory in limit $n\to\infty$
	5-7	Calculation of observables from ψ Ground-state $\psi(x)$ to minimize E
	5-8	Radiation when system is changing from one quantum state to another Selection rules

Fig. 5-16. Summary of problems discussed in this chapter and the principal new features of quantum mechanical solutions revealed in each.

REFERENCES

L. Pauling and E. B. Wilson, *Introduction to Quantum Mechanics*, McGraw-Hill, New York, 1935, Chapter 3.

J. C. Slater, *Quantum Theory of Matter*, McGraw-Hill, New York, 1951, Chapters 2–4.

C. W. Sherwin, *Introduction to Quantum Mechanics*, Holt, New York, 1959, Chapters 2 and 3.

PROBLEMS

5-1. Prove the superposition property asserted immediately below eq. 5-3. To do this, let Ψ_1 and Ψ_2 each be solutions of eq. 5-2; show that $\Psi_1 + \Psi_2$ is also a solution.

5-2. Show that if ψ_1 is a solution of eq. 5-4, $b\psi_1$ is also a solution, where b is a constant.

5-3.* If ψ_1 is a function of three coordinates, what is the replacement for eq. 5-11? Write the new equation first in rectangular coordinates, then in spherical polar coordinates, and finally in spherical polar coordinates for the special case where ψ_1 is a function only of the radius r (not a function of the angles).

5-4. When the Schrödinger equation is written for an electron, the electron mass m appears explicitly but the charge e does not, yet surely the charge is important in determining the motion of the electron. Explain.

5-5. Show that eqs. 5-14 and 5-15 are both solutions of eq. 5-13.

5-6. Show that eq. 5-16 is a solution of eq. 5-2 with constant $P = P_0$.

5-7. Show that eq. 5-18 represents a wave of frequency ν and wavelength λ traveling toward $+x$. To do this, first show that, at constant x, u varies with time with frequency ν. Then show that, at constant time, u is a sinusoidal function of x with wavelength λ. Finally, show the direction of propagation by considering some point on the wave (say $u = 0$) at time t_1, position x_1. Show that, at a short time Δt later, the x position of this point on the wave is at a point $x_2 > x_1$.

5-8. Write down the differential equation for the motion of a simple, small-amplitude pendulum in terms of the mass M, length l of the string, and angle θ that the string makes with the vertical. Compare the form of this equation with that of eq. 5-13.

5-9. Write down the differential equation for the voltage V across the capacitor in a simple series circuit consisting of an inductance L and a capacitance C. Compare the form of this equation with that of eq. 5-13.

5-10. Schrödinger tried $h^2/8\pi^2m$ as the coefficient of $\partial^2\Psi/\partial x^2$ because of his knowledge of how this coefficient would enter into the solutions of his equation. Discuss how we know now that this is the "correct" coefficient and what we mean by "correct."

5-11. Outline the solution of the square well with infinitely high sides (Sec. 5-4a) if we let $P = P_1$ for $-x_0 < x < x_0$, where P_1 is a constant. From the formula you find for the energy levels (similar to eq. 5-26), calculate the kinetic energy in the ground state ($n = 1$) and the difference in total energy between the states $n = 1$ and $n = 2$. Compare these two results with the comparable predictions from the development in the text (in which P_1 was set equal to zero).

5-12.* We let $x = 0$ at the center of the square well for ease of comparison with later examples. Find the replacement for eqs. 5-24 if we let $x = 0$ at the left edge of the well and $x = 2x_0$ at the right edge.

5-13. Show that the wave functions of eqs. 5-24 have been normalized.

5-14. Sketch $\psi(x)$ for $n = 3$ and $n = 4$ for the electron in the square well with infinitely high sides. Check to make sure that these ψ functions satisfy the conditions at $x = \pm x_0$.

5-15.* Compute from eq. 5-26 the lowest three energy levels for an electron in a square well of width 3 Å. Express your answers in electron volts.

5-16.* Compute from eq. 5-26 the energy levels for a 1 gm mass particle in a square well of width 1 cm. What must n be in order that the kinetic energy be 1 joule? What is the separation in joules between E for this n and E for $n + 1$? Will the discreteness of energy states be apparent in laboratory-size experiments?

5-17.* Compute from eq. 5-26 or from prob. 5-15 the kinetic energy of an electron in the ground state of a square well of width 3 Å. What fraction is this of the rest energy m_0c^2? Is relativity important in calculations with this order of magnitude of well width, which is typical of the outer electrons in atoms and molecules?

5-18.* Compute from eq. 5-26 the kinetic energy of a *proton* (substitute M for m) in the ground state of a square well of width 6×10^{-15} m. What fraction is this of the rest energy M_0c^2? Is relativity important in calculations involving protons or neutrons with this order of magnitude of well width, which is typical of the nucleons involved in nuclear reactions?

5-19. Find the wave functions for the *three-dimensional* square well with infinitely high sides. *Hints:* Write the analogue of eq. 5-4 but in three dimensions (like eq. 5-1). Try as a solution the product of three terms like eqs. 5-24, one in each of the space variables x, y, and z.

5-20. What is the *classical* probability of finding the electron as a function of x in the square well with infinite sides? With finite sides?

5-21. In Fig. 5-2 we chose to plot $\psi \sim +\cos \beta x$ for $n = 1$ rather than $\psi \sim -\cos \beta x$. What is the difference, if any, between the Ψ's derived from these two ψ's? Is there any difference in energy, momentum, or probability of finding the electron as a function of x?

5-22. Estimate by measurements on the plots of ψ's in Fig. 5-5 how good an approximation eq. 5-26 is to the actual E_n's for $n = 1$ and $n = 2$ for the particular well that is sketched there.

5-23.* Compute the ratio of the probability of finding the particle in a small range δx of x at a distance Δx outside the edge of the well (Fig. 5-5) to the probability of finding it in the same range δx at $x = x_0$. That is, compute $|\psi(x_0 + \Delta x)|^2 / |\psi(x_0)|^2$. Use values of Δx and $P_0 - E$ typical of atomic problems, namely, $\Delta x = 1$ Å and $P_0 - E = 1$ e.V. Solve the problem for both electrons and protons.

5-24. Sketch carefully the ground state ψ for a square well with finite sides of height $P_0 > E$. Sketch to the same scale the ground state ψ for a square well with the same width but higher sides. Compare the ψ's and from this comparison show why one of the E's is larger than the other.

5-25. Consider the square well Fig. 5-3 with high enough sides that eqs. 5-24 are a good approximation, but try a value $n = \frac{1}{2}$, which of course does *not* satisfy all the arguments leading up to these equations. Let $\psi \cong 0$ at $x = -x_0$. Plot $\psi(x)$ within the well. What is the slope $d\psi/dx$ at $x = x_0$? What are the relative values of C and D in eq. 5-28 for region III if $d\psi/dx$ is to be continuous at $x = x_0$? Sketch $\psi(x)$ in region III.

5-26.* Discuss the connection between the Indeterminacy Principle and the penetration of an electron into a classically forbidden region like region III of Fig. 5-5a. In order to do this, first estimate the uncertainty Δx that can be tolerated in an experiment if the experiment is to convince us that the electron is probably outside the well; this Δx is illustrated on Fig. 5-5b and is defined as $x_1 - x_0$ such that $\psi(x_1) = e^{-1}\psi(x_0)$. From the Indeterminacy Principle calculate the minimum uncertainty in momentum Δp_x and the uncertainty in kinetic energy corresponding to this Δp_x. Could the experiment verify that the kinetic energy was negative?

5-27. The wave functions of Figs. 5-5 and 5-12 are concave toward the OX axis in regions of positive kinetic energy and concave away from OX in regions of negative kinetic energy. Show that this general property of wave functions follows from the Schrödinger equation by comparing the signs of $d^2\psi/dx^2$ and ψ in the two kinds of regions. Do not assume P is constant.

5-28.* The ratio of $|\psi|^2$ evaluated at x_1 to $|\psi|^2$ evaluated at x_0 is called the *transmission* or *tunneling probability* of the barrier of Fig. 5-6. Calculate this probability for an electron if (a) $P_0 - E = 1$ e.V. and $x_1 - x_0 = 1$ Å; (b) $P_0 - E = 10$ e.V. and $x_1 - x_0 = 10$ Å.

5-29. It is frequently of interest in understanding phenomena in molecules or solids to learn whether electrons tunnel through a barrier or whether they go over it by *thermal activation* (i.e., of a large group of electrons with average energy E at temperature T, a few will have enough energy to go over the barrier). What is the difference in temperature dependence of the probability of getting past the barrier by the two processes? Let $P_0 - E \gg kT$.

5-30. Verify eq. 5-36. Check the reasonableness of this equation by considering the limiting situations $E = P_0$ and $P_0 = 0$.

5-31. Show that if the step in potential energy in Fig. 5-7 is reversed (i.e., $P = 0$ in I and $P = P_0$ in II), r for an electron coming from the left is still given by eq. 5-36.

5-32. The *phase* of the reflection at $x = 0$, Fig. 5-7, is determined by the phase difference (if any) between the incident wave (eq. 5-33) and the reflected wave (eq. 5-35). If B'/A' is a positive real number, the reflection is *in phase;* if it is a negative real number, the reflection is *180° out of phase*. Show that for an electron coming from the left the reflection is 180° out of phase. Next find the phase of the reflection of an electron incident on this same potential step *from the right*.

5-33.* Find approximate forms of eq. 5-36 that are valid (*a*) when $P_0/E \ll 1$; (*b*) when $(1 - P_0/E) \ll 1$. (In the latter part, let $P_0 = E(1 - \delta)$, where $\delta \ll 1$.)

5-34. In the Davisson-Germer experiment the atomic planes used were *not* parallel to the surface. Why was this a better experimental arrangement than an arrangement with the planes parallel to the surface? *Hint:* As mentioned in Sec. 4-2, the potential energy at the surface of the solid looks much like Fig. 5-7.

5-35. The reflection coefficient r in eq. 5-36 does not contain h and yet is a quantum phenomenon. Explain. Why does not eq. 5-36 apply to a macroscopic object like a baseball? *Hint:* If the index of refraction of a transparent medium varies slowly with distance (less than a few percent per wavelength), there is no reflection of light; if the variation is abrupt (in a distance $\ll \lambda$), the reflection is like eq. 5-36.

5-36. Consider the total internal reflection of light in a block of glass when light is incident from the inside upon a surface at a sufficiently large angle of incidence. If another block of glass is placed within a distance $\ll \lambda$ from the reflecting surface, the light is no longer reflected but is largely transmitted into the second block. Compare (with diagrams) this situation with the electron behavior in region III of Fig. 5-5 and in Fig. 5-6.

5-37.* Calculate λ for electrons with velocity v_0, Fig. 5-9. This is, of course, the wavelength *outside* the well. Is the wavelength inside larger or smaller? Can you establish an upper limit to the diameter of the krypton atom?

5-38.* The diameter of the krypton atom is about 4.1 Å. Suppose that an electron with 0.7 e.V. kinetic energy encounters a one-dimensional square well with a width of 4.1 Å (this is a crude model of the krypton atom). What must be the depth of the well for 100% transmission of the electron wave?

5-39. Verify eq. 5-40.

5-40. Show that ψ_0 of eq. 5-44 satisfies the Schrödinger equation with the appropriate E.

5-41. Show that ψ_0 of eq. 5-44 has been normalized.

5-42.* Calculate the *classical* probability $\mathcal{P}(x) \, dx$ of finding the particle in dx as a function of x in the harmonic oscillator, which is proportional to the reciprocal of the classical velocity. Evaluate $\mathcal{P}(x)$ at $x = 0$, at the classical turning points $x = \pm x_0$, and at points a fraction 0.8 of the distance from $x = 0$ to the classical turning points. Sketch $\mathcal{P}(x)$.

5-43.* A typical value of C^2 of eq. 5-40 for a problem in the oscillations of a diatomic molecule is 1.3×10^3 joules/m.[2] Show that this value leads to a potential energy of about 10 e.V. when $x = 0.5$ Å. Find the frequency ν_0 and the first two energy levels, E_0 and E_1, if C^2 has this value and if the mass of the oscillating particle is the mass of an oxygen atom. Light of this frequency ν_0 is in what region of the spectrum?

5-44.* A typical value of C^2 of eq. 5-40 for a problem in the laboratory (say the vibration of a simple pendulum) is 1 joule/m.[2] Find the frequency ν_0 of this oscillation if the mass of the oscillating particle is 1 kg. What must the quantum number n be (in the wave-mechanical description of this experiment) if the total energy E is 0.1 joule? What is the separation in joules between E_n and $E_n + 1$? Will the discreteness of energy levels be apparent in laboratory-size experiments?

5-45. Make a sketch of $|\psi_{15}|^2$ as a function of x for the harmonic oscillator, with the correct number of oscillations and roughly the correct shape. Do this by inspecting Figs. 5-12 and 5-13 and by studying Sec. 5-5 and the first part of Sec. 5-6.

5-46. Why did we not consider free states for the harmonic oscillator?

5-47. Discuss the connection between the Indeterminacy Principle and the fact that there is always kinetic energy in the ground state, using the harmonic oscillator as an example.

5-48.* Find a formula for the number N of zeros (as a function of n) of $\psi(x)$ for the square well, that is, the number of times $\psi(x)$ crosses the OX axis; do not include the zeros at $\pm x_0$. Do the same for the harmonic oscillator. Make a general statement about N valid for any one-dimensional problem.

5-49. Sketch $|\psi(x)|^2$ for the state $n = 11$ of the square well with finite sides high enough that $E_{11} < P_0$. Sketch the classical probability of finding the particle as a function of x on the same plot. Compare the quantum result with the harmonic oscillator $|\psi(x)|^2$ for $n = 10$ in Fig. 5-13.

5-50. Let the potential energy as a function of distance be $P = -ax$ for $x < 0$ and $P = ax$ for $x > 0$, where a is a constant. Sketch to the same horizontal scale $P(x)$ and $\psi(x)$ for the state of lowest energy E (put a horizontal line on the $P(x)$ sketch to indicate the value of E). *Hints:* Use the result of prob. 5-27, the relative magnitude of $d^2\psi/dx^2$ as $(E - P)$ changes in eq. 5-4, and the conditions on ψ provided by eqs. 5-6, 5-7, and 5-8.

5-51. Sketch $\psi(x)$ for the state $n = 11$ (where the ground state is called $n = 1$) of the triangular well described in the preceding problem.

5-52.* Consider an experiment in which we are attempting to measure the position x of a local maximum of the ψ^2 curve of the harmonic oscillator with $n = 10$ without changing E_n by more than $h\nu_0$ (i.e., without changing n to 9 or 11). The allowed maximum change in momentum Δp is about $p_{10} - p_9$, evaluated at $x = 0$. Estimate Δp from $\Delta p \cong (dp/dE)\Delta E$ and use the Indeterminacy Principle to calculate Δx. For the experiment to succeed, Δx must be $\leqq \sim \frac{1}{2}a$ (see Fig. 5-13). What is the value of $\Delta x \div (\frac{1}{2}a)$? Will the experiment succeed?

5-53. Work the preceding problem without the specializing assumption that $n = 10$. Let n be large, and calculate Δp by setting $\Delta E_n = 1$ in $2\Delta p/p \cong \Delta E_n/E_n$, which comes from $p^2/2m = K$ with K proportional to E. Calculate Δp and use the Indeterminacy Principle to calculate Δx. For the experiment to succeed, Δx must be less than the distance between classical turning points $(2x_0)$ divided by the number of maxima $(n + 1 \cong n)$. What is the value of $\Delta x \div (2x_0/n)$? Will the experiment succeed?

5-54. Show by sketching the factors in the integrand that $\bar{x} = 0$ for the states $n = 1$ and $n = 2$ of the square well with infinite sides.

5-55.* Calculate the *root-mean-square position* $(\overline{x^2})^{1/2}$ for the state $n = 1$ of the square well with infinite sides. (The integral can be found in most tables of indefinite integrals and in some tables of definite integrals.)

5-56.* What is $(\overline{x^2})^{1/2}$ (as in prob. 5-55) for the square well with infinite sides when n is very large? *Hint:* Sketch $|\psi(x)|^2$ and then replace $|\psi(x)|^2$ in the integrand by a suitable approximation.

5-57.* Calculate $(\overline{x^2})^{1/2}$ (as in prob. 5-55) for the state $n = 0$ of the harmonic oscillator in terms of a and compare with the distance to the classical turning point x_0, also expressed in terms of a.

5-58.* Calculate $\overline{P}$ for $n = 0$ and for $n = 1$ of the harmonic oscillator in terms of $h\nu_0$.

5-59. Calculate $\overline{K}$ for $n = 0$ and for $n = 1$ of the harmonic oscillator in terms of $h\nu_0$. Compare with $\overline{P}$ from prob. 5-58 and with eq. 5-43.

5-60.* Calculate $\overline{P}$ and $\overline{K}$ for the state $n = 1$ of the square well with infinite sides and compare with eq. 5-26.

5-61. Compare qualitatively the total energies E for the square well with infinite sides and with finite sides by applying the concepts of eqs. 5-48 and 5-49. In going from infinite to finite sides, what effect lowers E? What effect raises E? (The net effect lowers E.)

5-62. Show by the method described at the end of Sec. 5-7 that the ground state $\psi(x)$ of any bound system will have only a *single* maximum (i.e., $\psi(x)$ monotonically decreases as x increases or decreases from $x = 0$; no crossing of OX axis).

5-63. Draw a figure like Fig. 5-14 but for the potential energy illustrated in Fig. 5-5a. Explain how the ψ_1 sketched in Fig. 5-5b arises from the compromise between kinetic and potential energy.

5-64. Compute $\overline{E}$ for the wave function $\psi = 2^{-1/2}(\psi_1 + \psi_2)$, using ψ_1 and ψ_2 from eqs. 5-24.

5-65. Sketch ψ_0, ψ_1, x, and the product of these three for the harmonic oscillator. Show from the last sketch that the transition from $n = 0$ to $n = 1$ is allowed.

5-66. Sketch ψ_0, ψ_2, x, and the product of these three for the harmonic oscillator. Show from the last sketch that the transition from $n = 0$ to $n = 2$ is forbidden.

5-67. Evaluate the integral eq. 5-55.

5-68. Show graphically or analytically that the selection rule for the square well is that the quantum number must change from odd to even or from even to odd.

5-69. The probability of an allowed transition is proportional to the integral on the right side of eq. 5-53. Compare qualitatively (by sketching the integrands) the probability of the transition $n = 1$ to $n = 2$ with the transition $n = 1$ to $n = 10$ in the square well with infinite sides.

5-70. Before the times of Galileo and Newton, motion dominated by friction was considered to be the basic or fundamental motion; for example, many scholars thought that force was proportional to velocity. Newton showed that such motion was only a special case of a more general and fundamental law. Discuss the comparison between this situation and the quantum mechanics, using the motion of an electron as an example.

6
Atomic Structure and
Spectra

6-1 INTRODUCTION

Quantum mechanics will be applied in this chapter to the problem of the structure of atoms. The hydrogen atom is of course the simplest atom, since it contains only one electron, and therefore it will be treated first. Because the hydrogen problem is more difficult than the artificial problems of Chapter 5, we shall be unable to find solutions of the Schrödinger equation for the hydrogen atom in terms of simple mathematics. The mathematical development will therefore not be given, but the solutions will be indicated. For atoms with more than one electron, the Schrödinger equation cannot be solved in terms of known functions. This fact does not mean, however, that wave mechanics cannot deal with these situations; it means only that the solutions must be carried out by numerical or approximation techniques.

In the preceding chapter the simple artificial examples introduced the concepts of discrete energy levels for bound states, of penetration into classically forbidden regions, of kinetic energy in the ground state, and of correspondence with classical theory at large quantum numbers. It should be noted in this chapter that these same concepts will also appear in the more complicated problems that are of physical interest, and this was, of course, the reason for discussing the artificial problems. The new concepts were shown to be a natural consequence of the basic theory in

185

situations simple enough for the mathematical treatment to be straight-forward.

The principal new physics of this chapter is the Exclusion Principle, which is important in all atoms with more than two electrons. It is the physical principle underlying the size of such atoms, the size of mole-cules, and the density of solids. Its implications are nearly as extensive as those of the Schrödinger equation itself.

The Exclusion Principle will be stated in Sec. 6-3, and the demonstra-tion of its validity and applicability will be found in the study of the electronic structure of atoms, in Sec. 6-4. The quantum-physics explan-ation of the periodic system of chemical elements will be provided in that section also. Additional comparisons of theory and experiment will be made in Sec. 6-5 for optical spectra and in Sec. 6-6 for X-ray line spectra.

With the statement of the Exclusion Principle we shall complete the exposition of quantum theory. The remainder of this book will be de-voted to the study of problems in which quantum theory is indispensable.

6-2 THE HYDROGEN ATOM

The hydrogen atom consists of one proton and one electron. The proton is so much heavier than the electron that it can be considered fixed. (If the hydrogen problem is solved without making this approxi-mation, the result is not different qualitatively and only slightly different quantitatively.) The potential energy P of the electron is that P pro-duced by a point charge $+e$. We define r as the distance of the electron from the proton, which is situated at $r = 0$. From Coulomb's law,

$$P = \frac{-e^2}{4\pi\epsilon_0 r} \tag{6-1}$$

is the potential energy of the electron. This function $P(r)$ is plotted in Fig. 6-1. Here we have exercised our freedom of choice of the zero for P by setting $P = 0$ at $r = \infty$.

The variation of P with r and (especially) the three-dimensional na-ture of the problem make the mathematics of the solution of the Schrö-dinger equation here much more difficult than it was in the problems of Chapter 5. This equation can be solved by separating the three varia-bles r, θ, and φ of spherical polar coordinates * by a technique like that

---

* Let OR be the line from the origin O to any point R, a distance r from O. Then θ is the angle OR makes with the OZ axis, and φ is the angle the plane ZOR makes with the OX axis.

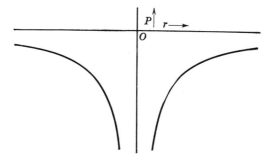

Fig. 6-1. Potential energy of an electron as a function of its distance r from the proton. [The section to the left of $r = 0$ should not appear in a plot of $P(r)$, since r in polar coordinates is never < 0, but this section helps us to visualize the potential energy. The plot gives, in fact, the way P varies along a line through the proton.]

used in Sec. 5-2. We shall not carry out the solution but merely describe properties of the solutions and give some examples.

Quantum numbers arise here just as in Secs. 5-4 and 5-5, but now we have *three* quantum numbers, represented by the symbols n, l, and m_l (three quantum numbers would arise in any three-dimensional problem). Furthermore, although n can be any positive integer, l can have only one of the values $0, 1, \cdots, (n - 1)$; and m_l can have only one of the values $-l, -l + 1, \cdots, 0, \cdots, l - 1, l$. Thus, for $n = 1$, only $l = 0$ and $m_l = 0$ are permitted; for $n = 2$, we may have $l = 0$ (in which case $m_l = 0$) or $l = 1$ (in which case m_l can be either -1, 0, or 1). These rather complicated rules are a consequence of the Schrödinger equation and the conditions on ψ (eqs. 5-6, 5-7, and 5-8). They are *not* special assumptions for the problem.

The number n is called the *principal* or *radial* quantum number, l the *azimuthal* quantum number, and m_l the *magnetic* quantum number. In order to specify a particular quantum state, we must specify the values of the three quantum numbers. A state in which $l = 0$ is called an *s* state, $l = 1$ a *p state*, and $l = 2$ a *d state*. The three quantum numbers n, l, and m_l completely specify the way ψ varies from point to point in space.

The energy levels corresponding to the various quantum states depend only on n:

$$E_n = -\frac{e^4 m}{n^2 h^2 8 \epsilon_0{}^2} \text{joules} \qquad (6\text{-}2)$$

If we divide by 1.602×10^{-19} joule/e.V. and evaluate the constants,

$$E_n = -\frac{13.60}{n^2} \text{ e.V.} \tag{4-6}$$

These energies are illustrated in Fig. 4-12.

If the effects of electron spin, relativity, and the magnetic moments of the electron and the proton are included in the theory, very minor additions to the energy expressed in eq. 6-2 must be included. These additions depend on l, m_l, and the orientation of the spin of the electron relative to the angular momentum of its motion about the proton. This orientation is specified by giving the value of the *spin* quantum number m_s, which takes on only the values $+\frac{1}{2}$ or $-\frac{1}{2}$. All these effects together make corrections of only about 1 part in 10^5 in eq. 6-2.

The two simplest wave functions for hydrogen are

$$n = 1 \quad (l = 0, \, m_l = 0): \quad \psi = \pi^{-1/2} \rho^{-3/2} e^{-r/\rho} \tag{6-3}$$

$$n = 2 \quad l = 0 \quad m_l = 0: \quad \psi = \pi^{-1/2} 2^{-5/2} \rho^{-3/2} \left(2 - \frac{r}{\rho} \right) e^{-r/2\rho} \tag{6-4}$$

Here the constant ρ has been defined as

$$\rho = \frac{h^2 \epsilon_0}{\pi m e^2} = 5.3 \times 10^{-11} \text{ m.} = 0.53 \text{ Å} \tag{6-5}$$

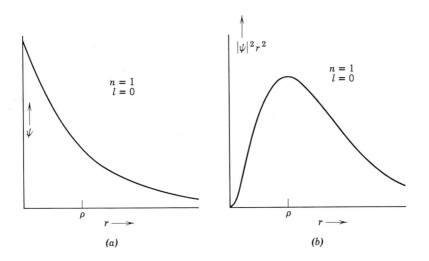

Fig. 6-2. (*a*) Wave function of the hydrogen atom 1*s* state. (*b*) Probability density of the 1*s* state.

All the wave functions with $l \neq 0$ contain functions of θ, and all those with $m_l \neq 0$ contain functions of φ. Thus only for the states with $l = 0 = m_l$ (s states) can ψ be specified in terms of r alone. ψ is spherically symmetrical for such states. For other states, ψ depends on angles and in a very complicated way for large l values.

The ground-state wave function (eq. 6-3) is plotted in Fig. 6-2a. We call this the "$1s$" wave function, the "1" indicating that $n = 1$ and the "s" indicating that $l = 0$ (as it must if $n = 1$). Note that ψ has the same sign at all values of r, like all other ground-state ψ's.

The probability that the electron is between r and $r + dr$ is of considerable interest. This probability is

$$|\psi|^2 \, dv = |\psi|^2 4\pi r^2 \, dr \qquad (6\text{-}6)$$

which comes from integrating $dv = r^2 \sin \theta \, dr \, d\theta \, d\varphi$ over the range $0-\pi$ in θ and $0-2\pi$ in φ. (If ψ is a function of θ and φ, the $|\psi|^2$ in eq. 6-6 is the average $|\psi|^2$ at the radius r in question.) Note that the volume element dv is the volume of the spherical shell of radius r and thickness dr, and this volume increases rapidly with r. Thus a plot of $|\psi|^2$ does not give a good picture of the probability of finding the electron as a function of r, since the larger values of r should be weighted heavily. Figure 6-2b presents a plot of $|\psi|^2 r^2$ as a function of r, which *does* give a good picture of this probability. The average distance of the electron from the proton appears from the figure to be about $\frac{3}{4}$ Å, which is as good a definition of the atomic radius as we can obtain. There is no sharp outer boundary to the atom, but the probability is quite small that the electron will be farther from the proton than about twice this average value. Although experiments that give the size of the hydrogen atom directly have not been performed, the predicted size is in accord with the magnitudes of Table 2-1 and with many indirect experimental data.

The $2s$ wave function ($n = 2$, $l = 0$) is plotted in Fig. 6-3a, and the corresponding $|\psi|^2 r^2$ in Fig. 6-3b. The electron in this state is on the average about four times as far from the nucleus as it would be in the $1s$ state. The average potential energy is thus nearer to zero by a factor of 4 (see Fig. 6-1 or eq. 6-1). The total energy E is also nearer to zero by a factor of 4, as shown by eq. 6-2.

The $2p$ wave functions ($n = 2$, $l = 1$, $m_l = 0$ or ± 1) cannot be clearly drawn in two dimensions, since they vary with θ and φ. An illustration of one of these ($m_l = 0$, cylindrically symmetrical about OZ) is given in Fig. 6-4. ψ is > 0 for $z > 0$ and < 0 for $z < 0$. In Fig. 6-3a, ψ is plotted as a function of r, where r is measured along the $\theta = 0$ line (OZ axis). ψ in the XOY plane is zero. The $2p$ wave functions with

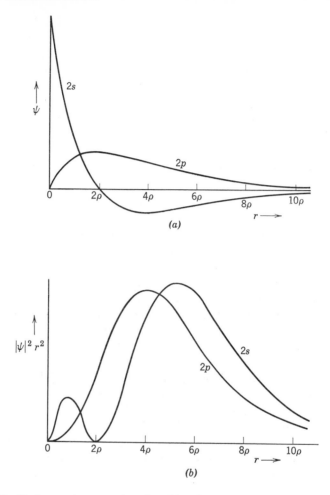

Fig. 6-3. Hydrogen atom wave functions (a) and probability densities (b) for $n = 2$, $l = 0$ ($2s$) and $n = 2$, $l = 1$ ($2p$).

$m_l = \pm 1$ are identical in shape, but each is cylindrically symmetrical about one of the other Cartesian coordinates.

Wave functions with $n = 3$ have about nine times the radial extension of the $1s$ wave function. The dependence of these ψ's on angles for $l = 1$ or 2 is quite complicated. As before, the ψ for $l = 0$ is spherically symmetrical.

The wave functions for hydrogen exhibit considerable penetration of the electron into classically forbidden regions. Problem 6-23 shows that the kinetic energy is negative at a distance from the nucleus where there is still an appreciable probability of finding the electron.

Another important property of the hydrogen atom is the angular momentum, which can be determined by a mathematical operation on ψ like the operations described in Appendix F. The result is that the magnitude of the angular momentum equals $(h/2\pi)\sqrt{l(l+1)}$. Thus, for example, the angular momentum of an s state equals zero. The angular momentum is said to be *quantized* since it, like the energy, takes on only one or another of a set of discrete values. Since l is a constant for any one quantum state, the angular momentum is a constant. This result should have been expected since we have asserted that the conservation of energy and momentum apply in quantum mechanics as well as in classical mechanics. The force on the electron is in the $-r$ direction (there is no torque acting on the electron), and therefore the angular momentum is conserved. If the atom has an angular momentum, it will have a magnetic moment proportional to the angular momentum. This magnetic moment arises because the electron is charged, and therefore its orbital motion produces a magnetic field like that of a small magnetic dipole.

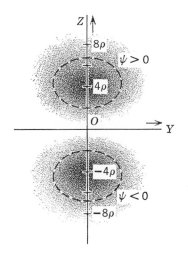

Fig. 6-4. The hydrogen $2p$ wave function with $m_l = 0$. The intensity of the shading is proportional to the magnitude of $|\psi|^2 r^2$ in the YOZ plane. The maximum of this probability density is at $y = 0$, $z = 4\rho$, as shown in Fig. 6-3b, and $|\psi|^2 r^2$ on the dashed curve above is equal to one-half this maximum value. The probability distribution is cylindrically symmetrical about the OZ axis and thus has the shape of a dumbbell with a fuzzy outer boundary.

The component of the angular momentum in any one direction is also quantized and has the value $(h/2\pi)m_l$ in the z-direction. This result may seem strange, since the atom "cannot know which is the z-direction" in space unless there is some external influence such as a magnetic field, which is fixed in that direction. But there is no real anomaly here. In the absence of such an influence, there is no experimental way of distinguishing the various m_l states. When a magnetic field is present, however, the different m_l values give different components of the magnetic moment of the atom in the direction of the field. These m_l values can now be distinguished since the energy of an atom is slightly different for different values of the component of the magnetic moment in the direction of the magnetic field. Thus the emission spectrum of an atom in a magnetic field will have several lines with nearly the same wavelength where only one existed in the absence of the magnetic field. This phenomenon, called the *Zeeman effect*, will be considered further in Sec. 6-5.

Selection rules for the absorption or emission of radiation by the hydrogen atom can be determined as explained in Sec. 5-8. Transitions from one state to another are allowed only if the integral

$$\int_{\text{All space}} u\psi^*_{n,l,m_l}\psi_{n',l',m_{l'}}\,dv \qquad (6\text{-}7)$$

does not equal zero. Here u equals either x, y, or z, and the resulting radiation will be polarized like the radiation of a dipole antenna oriented along OX, OY, or OZ, respectively.

We can see from the wave functions of eqs. 6-3 and 6-4 that the transition $2s{\to}1s$ is forbidden. These wave functions are spherically symmetrical, and their product is the same at some positive value of the coordinate u as at a negative u with the same $|u|$. Therefore contributions to the integral cancel in pairs. The oscillation of charge is spherically symmetrical, like the oscillations of a spherical rubber balloon alternately filled and emptied. This symmetrical change in the distribution of charge, like the symmetrical antenna of Fig. 5-15b, does not radiate. The transition $2p{\to}1s$ *is* allowed, since it gives an oscillation of charge; comparison of Fig. 6-4 with Fig. 6-2 may make this seem reasonable. This result can be proved by verifying that the integral of eq. 6-7 does not vanish when the two ψ's are the $2p$ and $1s$ wave functions.

We have noted in the preceding paragraph two special cases of the general selection rules:

$$\Delta l = \pm 1 \qquad \Delta m_l = \pm 1 \quad \text{or} \quad 0 \qquad (6\text{-}8)$$

There are no selection rules for n. Any change or zero change in n is permitted. We shall not attempt to prove these rules here, but they can be proved by applying the test described in conjunction with eq. 6-7 to the general wave functions.

The observed spectrum of hydrogen demonstrates the validity of the results for energies and wave functions and helps to establish the whole wave-mechanical theory. It has already been noted in Sec. 4-3 that excellent agreement is obtained between the observed wavelengths of spectral lines and Bohr's formulas given in eqs. 4-5 and 4-6. The wave-mechanics theory predicts both these expressions. The former (eq. 4-5), which is the general expression found for the radiation or absorption process, can now be expressed in modified form in order to include the specification of a state by three quantum numbers:

$$\nu = \frac{1}{h}\left(E_{n,l,m_l} - E_{n',l',m_{l'}}\right) \tag{6-9}$$

The latter (eq. 4-6), which is the calculation of the energy levels for hydrogen, was given in eq. 6-2. Since there are only fundamental atomic constants in both expressions, the agreement is quite striking. Observations of the Zeeman effect also confirm the prediction of the quantization of the angular momentum.

Agreement between theory and experiment for the selection rules is also excellent. Because of the small (but measurable) dependence of E on l and m_l, the transitions between states with various values of n, l, and m_l can be identified. The selection rules are obeyed for all observed transitions, and every transition allowed by the selection rules is actually observed as a spectral line.

It is interesting to examine the behavior of the wave functions for large quantum numbers. The simplest example is a wave function with $l = n - 1$, $m_l = l$. The $|\psi|^2 r^2$ vs. r plot for this example looks just like the $2p$ wave function shown in Fig. 6-3b, except that the maximum of this function occurs at $n^2\rho$ (instead of the special value $2^2\rho$ shown in that figure). The azimuthal dependence of ψ is illustrated in Fig. 6-5 for $n = 9$. The sectors are labeled according to the sign of ψ. As we make a complete circle about the origin, there are 16 points at which ψ goes through zero when $l = 8$; in general there are $2l$ such points. The connection between such wave functions and de Broglie waves is sketched at the bottom of Fig. 6-5 and is investigated in prob. 6-13. It is shown in the problem that, at large n, the wave-mechanics solutions are similar to a set of de Broglie standing waves. The "orbits" of the electrons for this relation between the quantum numbers $(l = n - 1)$ approach circles.

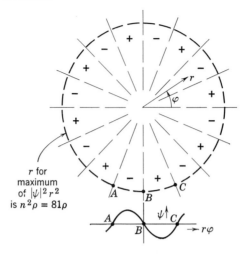

Fig. 6-5. The $n = 9$, $l = 8$, $m_l = 8$ wave function changes sign sixteen times as φ goes from 0 to 2π. The curve at the bottom shows the way ψ varies along the circle of radius 81ρ; distance along this circle is $r\varphi$, where $r = 81\rho$.

For such wave functions ($l = n - 1$), a transition from n to $n - 1$ gives a $\Delta l = -1$ and hence is an allowed transition. The frequency of the radiation emitted in such a transition is computed in prob. 6-16 *for large n*. This frequency turns out to be exactly the classical frequency of rotation of an electron in a circular orbit of radius equal to the average radius $\bar{r}$ of the wave-mechanical treatment. Thus at large n there is correspondence between the quantum-mechanical and classical results for the frequency of the energy radiated.

Wave functions with $l < n - 1$ approach classical *elliptic* orbits as n becomes very large. Wave functions with the special value $l = 0$ approach an ellipse with zero minor axis, that is, a straight-line oscillation. At large n, the electron in a state with $l = 0$ is therefore oscillating in and out along a line through the nucleus. Note that this picture agrees with the fact that, when $l = 0$, the angular momentum equals zero. The spherical symmetry of the wave functions with $l = 0$ comes about because all orientations in space of this line are equally probable.

Practically all hydrogen atoms at ordinary temperatures are in their ground states ($n = 1$, $l = 0$, $m_l = 0$, m_s either $+\frac{1}{2}$ or $-\frac{1}{2}$) because the first excited state in hydrogen is 10.2 e.V. (which is $\gg kT$) above the ground state (see the argument at the end of Sec. 2-4). We can easily calculate the relative numbers of atoms in a sample of hydrogen in various quantum states. We apply an expression from quantum statistical mechanics * that we shall not prove, but its similarity to the classical distributions of Secs. 2-3 and 2-4 should be noted. Let N_{1i} be the number of atoms in quantum state i with energy E_1, and N_{2j} the number in quantum state j with energy E_2. Then in equilibrium at temperature $T°K$

$$\frac{N_{2i}}{N_{1j}} = e^{-(E_2-E_1)/kT}$$

It usually happens that there are several quantum states with the same energy. If there are w_2 states with energy E_2, and w_1 with E_1, the likelihood of finding atoms in states with energy E_2 is increased by the ratio w_2/w_1, just as the likelihood of throwing a particular sum of numbers of two dice is proportional to the number of different ways this sum can be obtained. Thus

$$\frac{N_2}{N_1} = \frac{w_2}{w_1} e^{-(E_2-E_1)/kT} \tag{6-10}$$

The w's are called *statistical weights*. For hydrogen, w_1 equals 2 for the ground state, since there are two wave functions with $n = 1$, $l = 0$, $m_l = 0$, and with $m_s = +\frac{1}{2}$ or $m_s = -\frac{1}{2}$. The statistical weight for the first excited state is $w_2 = 8$. There are two wave functions with $n = 2$, $l = 0$, $m_l = 0$; two with $n = 2$, $l = 1$, $m_l = 1$; two with $n = 2$, $l = 1$, $m_l = 0$; and two with $n = 2$, $l = 1$, $m_l = -1$. Therefore there is a total of eight.

The most important part of eq. 6-10 is the exponential factor, which is called the *Boltzmann factor*. This factor will be encountered again in the study of molecules and solids. If $E_2 - E_1$ is greater than a few kT, the Boltzmann factor is very small, and almost all the atoms are in the ground state.

6-3 THE EXCLUSION PRINCIPLE

Only a single electron has been involved in each of the applications of wave mechanics that have been considered thus far. As we begin consideration of atoms with two or more electrons, we must examine the interactions among electrons. The most obvious interaction is the electrostatic repulsion; the potential energy of one electron depends not only on its distance from the nucleus but also on its distance from each of the other electrons. This interaction is familiar physics but leads to very difficult mathematics. Even the helium atom problem, which involves only two electrons and a nucleus, can be solved only approxi-

* See, for example, R. W. Gurney, *Introduction to Statistical Mechanics*, McGraw-Hill, New York, 1949, Chapter 1.

mately. (The classical-mechanical treatment of three particles is equally complicated, and the problem of the motions of the sun and two planets cannot be solved exactly.) But the wave functions for helium can be determined by approximation methods to any desired degree of precision, and the energy levels have been calculated to an accuracy which gives agreement with experiment to 0.01%.

There is another subtler, but vital, interaction among electrons that is described by the *Exclusion Principle* discovered by Pauli. This principle can be stated: *There can be at most one electron in each quantum state in an atom.* Since each quantum state is specified by a particular set of the quantum numbers n, l, m_l, and m_s, the Exclusion Principle implies that at most one electron in an atom can have any particular combination of these numbers. If one electron is in a particular quantum state in an atom, a second electron added to the atom must be in a different state. The motion of the second electron is therefore affected by the presence of the first; this is what we call the *Exclusion Principle interaction*.

This principle cannot be derived from any theory, nor is it based directly on a single experiment, but abundant proof of its correctness will be presented in Sec. 6-4. If nature were constructed without such a principle, we should not find the great variety of properties of the chemical elements. All matter would be nearly alike. Furthermore, all matter except hydrogen and helium (but including other atoms, molecules, and solids) would be much more dense than it is observed to be.

The foregoing statement of the Exclusion Principle is clearest and easiest for use in problems of atomic structure, but it is not the only or the most general statement. We shall not be concerned with the most general statement, which involves considerable mathematics. It will be valuable, however, to state an alternative form of the principle, since it is necessary to use this form in problems (such as those in solid-state physics) involving many atoms. This statement is: *There can be at most two electrons in an interval of momentum and position given by*

$$\Delta p_x \, \Delta p_y \, \Delta p_z \, \Delta x \, \Delta y \, \Delta z = (h/2)^3 \qquad (6\text{-}11)$$

One of these electrons has a spin quantum number $m_s = +\frac{1}{2}$, and one has $m_s = -\frac{1}{2}$. By an "interval" we mean that the three components of momentum agree to within Δp_x, Δp_y, Δp_z, and the three coordinates agree to within Δx, Δy, Δz. In other words, at most two electrons can have such agreement of momentum and position, and they must disagree in the spin quantum number. If a third electron has momentum components in this interval, it must be in a different region of space (outside the interval Δx, Δy, Δz). If a third electron is in this interval of space, it must have different components of momentum.

The connection between this statement and the earlier statement is not easy to prove, but it is worth stating the way the connection occurs. It can be proved that the quantum states of any system are packed together just closely enough to allow one n, l, m_l state for each interval expressed in eq. 6-11. That is, the separation in either momentum or position or both of electrons in the various quantum states of a system is just enough to give one n, l, m_l state for each such interval. In the hydrogen atom, the lower energy states are close together in position ($\Delta x\ \Delta y\ \Delta z$ is small) but far apart in energy and momentum ($\Delta p_x\ \Delta p_y\ \Delta p_z$ is large). The higher energy states are extended in space but close together in energy and momentum. This subject will be explored further in Appendix G; see eq. G-12 in particular.

The Exclusion Principle interaction between electrons cannot be simply described as a force between electrons. (The *electrostatic* interaction can, of course, be described in terms of a repulsive force $+e^2/4\pi\epsilon_0 r^2$ between electrons a distance r apart, or equally well in terms of an energy $+e^2/4\pi\epsilon_0 r$.) The only feasible way of describing the Exclusion Principle interaction is in terms of the energy, and even this procedure is not so simple as the electrostatic interaction. We shall illustrate the interaction by discussing the example of an atom that has five electrons. When the atom is in its ground state (the observed, lowest energy state), these five electrons occupy the five quantum states with lowest energies. A sixth electron is added to this system. If it were not for the Exclusion Principle, this electron could be put into one of the five states. Because of this principle, it must be put into another quantum state, which necessarily has a higher energy than the states already filled. Thus the electron must have a higher energy in order to enter the atom, which is the effect that would be produced by a repulsive force. The amount of this energy increase depends on the energy levels of the atom, not just on the distance r between electrons, and therefore cannot be expressed in as simple a form as the electrostatic repulsion. Because of the repulsion energy that develops in this way electrons are not packed closer together in atoms, molecules, or solids.

6-4 ELECTRONIC STRUCTURE OF ATOMS

To decide how the electrons of an atom are distributed among the available quantum states we use only two principles: (1) There can be at most one electron with a given combination of n, l, m_l, and m_s. (2) Subject only to that restriction, in the normal state of an atom each electron occupies the quantum state with the lowest energy possible. Statement 2 is a consequence of eq. 6-10 and the fact that the lower quantum states are many kT apart at room temperature. It should be recalled from Sec. 6-2 that there are 2 "*s*" states ($l = 0$) for any value of n (one with $+\frac{1}{2}$ spin, one with $-\frac{1}{2}$ spin). There are 6 "*p*" states ($l = 1$) for any value of n ($m_l = -1$, $m_s = \pm\frac{1}{2}$; $m_l = 0$, $m_s = \pm\frac{1}{2}$; $m_l = +1$, $m_s = \pm\frac{1}{2}$). There are 10 "*d*" states ($l = 2$) for any value of n and 14 "*f*" states ($l = 3$).

We begin now a study of the electronic structure of atoms. The results are summarized in Table 6-1, which should be studied in conjunction with the following paragraphs.

$Z = 1$ (HYDROGEN). We have already discussed this atom. There is one electron in $n = 1$, $l = 0$, $m_l = 0$, and $m_s = \pm\frac{1}{2}$. The energy difference between $m_s = +\frac{1}{2}$ and $m_s = -\frac{1}{2}$ is so small that it is of no importance for this study. The *ionization energy* E_i (the energy required to remove the electron) is 13.6 e.V. The *ionization potential* V_i is 13.6 volts. The electron's mean distance from the nucleus is about $\frac{3}{4}$ Å.

$Z = 2$ (HELIUM). There is one electron in $n = 1$, $l = 0$, $m_l = 0$, $m_s = -\frac{1}{2}$, and one in $n = 1$, $l = 0$, $m_l = 0$, $m_s = +\frac{1}{2}$. In other words, there are two 1s electrons. The wave functions are somewhat different in shape from the wave functions of hydrogen, because of the electrostatic repulsion of the electrons, but the chief difference is a difference in scale. Because the nuclear charge is $+2e$, the energies E_n are considerably larger in magnitude.

In eq. 6-2 the factor e^4 in the numerator came from the square of the product of the nuclear charge ($+e$) and the electronic charge ($-e$). Thus, for an atom like hydrogen but with $+Ze$ nuclear charge,

$$E_n = -\frac{Z^2 e^4 m}{n^2 h^2 8\epsilon_0^2} \text{ joules} \tag{6-12}$$

Similarly the characteristic length ρ_Z is

$$\rho_Z = \frac{h^2 \epsilon_0}{\pi m e^2 Z} \text{ m.} = \frac{0.53}{Z} \text{ Å} \tag{6-13}$$

in place of the quantity $\rho = 0.53$ Å for hydrogen ($Z = 1$). These equations apply accurately only for an atom with a *single* electron, such as the He$^+$ *ion* ($Z = 2$), which is just like the hydrogen atom except for the increased nuclear charge. In the helium *atom*, on the other hand, eqs. 6-12 and 6-13 are only very crude approximations, since they ignore the electron-electron repulsion. Thus the actual energies E_n for the helium atom are intermediate between the values computed from eq. 6-12 with $Z = 1$ and the values computed with $Z = 2$. The ionization potential

TABLE 6-1 *

Electronic Structure of Atoms

Principal Quantum Number n			1	2		3			4	
Azimuthal Quantum Number l			0	0	1	0	1	2	0	1
Letter Designation of State			$1s$	$2s$	$2p$	$3s$	$3p$	$3d$	$4s$	$4p$
Z		Element	V_i volts							
1	H	Hydrogen	13.60	1							
2	He	Helium	24.58	2							
3	Li	Lithium	5.39	Helium core	1						
4	Be	Beryllium	9.32		2						
5	B	Boron	8.30		2	1					
6	C	Carbon	11.26		2	2					
7	N	Nitrogen	14.54		2	3					
8	O	Oxygen	13.61		2	4					
9	F	Fluorine	17.42		2	5					
10	Ne	Neon	21.56		2	6					
11	Na	Sodium	5.14		Neon core		1				
12	Mg	Magnesium	7.64				2				
13	Al	Aluminum	5.98				2	1			
14	Si	Silicon	8.15				2	2			
15	P	Phosphorus	10.55				2	3			
16	S	Sulfur	10.36				2	4			
17	Cl	Chlorine	13.01				2	5			
18	Ar	Argon	15.76				2	6			
19	K	Potassium	4.34		Argon core						1
20	Ca	Calcium	6.11								2
21	Sc	Scandium	6.56						1	2	
22	Ti	Titanium	6.83						2	2	
23	V	Vanadium	6.74						3	2	
24	Cr	Chromium	6.76						5	1	
25	Mn	Manganese	7.43						5	2	
26	Fe	Iron	7.90						6	2	
27	Co	Cobalt	7.86						7	2	
28	Ni	Nickel	7.63						8	2	
29	Cu	Copper	7.72						10	1	
30	Zn	Zinc	9.39						10	2	
31	Ga	Gallium	6.00						10	2	1
32	Ge	Germanium	7.88						10	2	2
33	As	Arsenic	9.81						10	2	3
34	Se	Selenium	9.75						10	2	4
35	Br	Bromine	11.84						10	2	5
36	Kr	Krypton	14.00						10	2	6

* From Charlotte E. Moore, *Atomic Energy Levels*, Vol. II, National Bureau of Standards Circular 467, Washington, 1952.

V_i is 24.58 volts (it would be 13.60 volts if $Z = 1$ and $13.60 \times 4 = 54.4$ volts if $Z = 2$, and if eq. 6-12 were accurately applicable). This is the highest ionization potential of any element. Helium is very inert chemically because of its large ionization potential and because there are no vacant electron states in the $n = 1$ group. The helium atom can therefore neither give up nor take on an electron without the expenditure of a prohibitive amount of energy. It does not form molecules with any element.

$Z = 3$ (LITHIUM). Here the Exclusion Principle becomes vital in determining the electronic structure. The first two electrons can go into $n = 1$, $l = 0$, $m_l = 0$, $m_s = \pm\frac{1}{2}$. They have wave functions almost exactly like the helium wave functions except that they are closer to the nucleus and more tightly bound (because $Z = 3$ instead of 2). The third electron cannot go into either of these states because of the Exclusion Principle. Furthermore, there are no more states with $n = 1$, and therefore this electron must go into a state with $n = 2$. This state has a much higher energy and extends to much larger r than the $n = 1$ states. Its average r is so large that for a first approximation we can treat the nucleus and the two 1s electrons as a point *core* with net charge $+e$ (see Fig. 6-6). In this approximation, the third electron can be treated like the electron in the hydrogen atom. For $n = 2$, the average $\bar{r}$ is then $\bar{r} = n^2(\frac{3}{2}\rho) \cong 3$ Å. The ionization energy is only the energy required to remove an electron from the $n = 2$ state of hydrogen, which is 13.6/4 e.V., or $V_i = 3.4$ volts. This is, of course, only an approximation, since the core is *not* a point charge; that it is a good approxi-

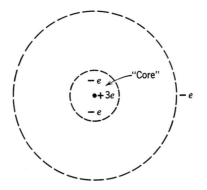

Fig. 6-6. The lithium nucleus and the two $n = 1$ electrons constitute a core that is much smaller than the average radius of the $n = 2$ electron. The circles represent the average radial positions of the electrons.

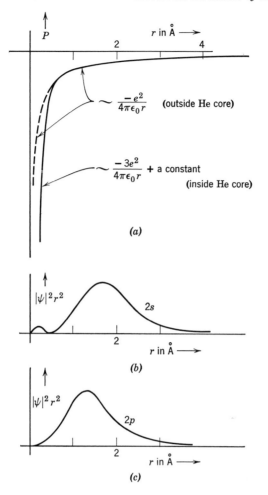

Fig. 6-7. Potential energy and $n = 2$ probability densities for the lithium atom. The $2s$ wave function gives a higher probability that the outer electron will be in the especially low potential energy region inside the core.

mation is demonstrated by the fact that the observed ionization potential *is* very small, being only 5.39 volts.

It is easy to see that, unlike the situation in hydrogen, the $n = 2$, $l = 0$ ($2s$) energy is somewhat lower than the $n = 2$, $l = 1$ ($2p$). This can be seen by first noting the way the potential energy P varies with r, as shown in Fig. 6-7. Note that near the nucleus there is an especially low potential energy. An electron that has an appreciable probability of being in this region will thus have a lower energy (other factors being equal) than an electron that is unlikely to be in this region. Comparison of the $2p$ and $2s$ wave functions in Fig. 6-3 shows that the $2s$ has a larger probability of being near $r = 0$. (It should be recalled that the higher l values mean higher angular momentum, which keeps the electron away from the region near $r = 0$, and that the $l = 0$ wave functions have zero angular momentum and correspond at high n's to oscillations along a line through the nucleus.) The $2s$ and $2p$ wave functions have the same energy for hydrogen. In lithium, where P decreases sharply at small r below the values it has for hydrogen, the $2s$ wave function thus has the lower energy.

The ground state of lithium therefore consists of two $1s$ electrons and one $2s$ electron (with spin either $+\frac{1}{2}$ or $-\frac{1}{2}$). The low ionization potential means that positive ions are formed with little expenditure of energy; this explains the extremely high reactivity of lithium and its *electropositive* nature in chemical compounds. Note that the low ionization potential of lithium is excellent proof of the Exclusion Principle. If no such law were at work, all three electrons in lithium would be in the $n = 1$ state, all would be tightly bound, and the ionization potential would be much greater than the 24.58 volts of helium. The size of the lithium atom also helps prove the Exclusion Principle, since in the absence of this principle it would be smaller than helium instead of much larger, as observed.

The energy required to remove a *second* electron from lithium is very large, since this electron must come from the $n = 1$ pair. The *second ionization potential* (75.6 volts) is even more than the first ionization potential (24.58 volts) of helium because of the larger Z. Thus lithium always appears in compounds with a valence of $+1$ (giving up one electron), never with $+2$ (giving up two electrons).

$Z = 4$ (BERYLLIUM). The arguments used for lithium show that the lowest energy state for the fourth electron is $n = 2$ and $l = 0$, with spin opposite to the third electron. Thus the ground state for beryllium consists of two $1s$ electrons and two $2s$ electrons. In other words, it consists of a *helium core* plus two $2s$ electrons. The first ionization potential is 9.32 volts, somewhat more than V_i for lithium because of the increased Z. The second ionization potential is not much larger, since this electron also comes from an $n = 2$ state. Thus beryllium has a valence of $+2$ in compounds.

$Z = 5$ (BORON). The electronic structure is the helium core, two $2s$ electrons and the fifth electron in $n = 2$, $l = 1$. (The m_l and m_s values of $2p$ electrons like this one will not be followed, since they are not the dominating influences on the energy values of light elements.) The ionization potential V_i equals 8.30 volts, which is somewhat less than V_i for beryllium because the $2p$ states are of higher energy than the $2s$ states, and this effect outweighs the increase in Z. All three of the $n = 2$ electrons are usually removed in chemical compounds, giving a valence of $+3$.

$Z = 6$ (CARBON). The structure is the helium core, two $2s$ electrons, and two $2p$ electrons. The value of V_i has increased to 11.26 because of the increase in Z. The usual valence is $+4$.

$Z = 7$ (NITROGEN), $Z = 8$ (OXYGEN), AND $Z = 9$ (FLUORINE). Additional $2p$ electrons are added in these elements. Although V_i is generally increasing here, V_i for oxygen is actually a little less than for nitrogen. This is a situation where the detailed effects of spin and the geometrical arrangement of wave functions dominate over the general trend.

All three of these atoms are chemically *electronegative*. That is, they commonly form negative ions in compounds. Fluorine forms a stable *negative* ion F^-. The additional electron is bound to the fluorine atom with a binding energy of 4.2 e.V. The usual way of expressing this is to say that fluorine has an *electron affinity* of 4.2 volts. The concept of electron affinity is thus like that of the ionization potential except that the former applies to an *additional* electron, not present in the neutral atom. This additional electron has no electrostatic attraction to or repulsion from the neutral atom at large distances. Its binding energy arises because its wave function *extends into* the charge distribution of the remaining electrons, as illustrated in Fig. 6-8. Thus part of the time this additional electron is attracted by a net positive charge ($+9e$ of the nucleus minus somewhat fewer than 9 negative electronic charges). The additional electron experiences an electrostatic repulsion from the other valence electrons, but the nuclear attraction dominates for fluorine and other atoms with an electron affinity. A second additional electron is not bound at all since it now experiences an electrostatic repulsion from the F^- ion as a whole, and since the Exclusion Principle compels it to go into an $n = 3$ state. Fluorine thus has a valence of -1.

The same argument for the existence of an electron affinity is valid for oxygen, which has an electron affinity of 2.2 volts. In the oxygen atom, a second additional electron, although still repelled from the O^- ion, is not repelled so strongly as in fluorine because it need not go into an $n = 3$ state. The ion O^{--} is thus not stable, but only a moderate energy (about 9 e.V.) is required in order to form this ion. In chemical compounds oxygen commonly appears with a valence of -2. Similarly, the most common valence for nitrogen is -3.

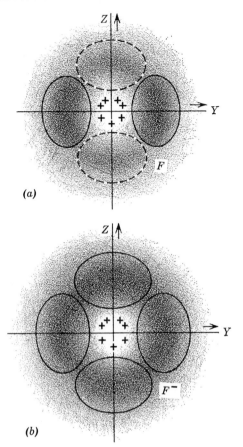

Fig. 6-8. The origin of the electron affinity $e\chi$ in fluorine. The intensity of the shading is proportional to the magnitude of $|\psi|^2 r^2$ in the YOZ plane. (a) Neutral fluorine atom. There is *one* electron with a $2p$ wave function like that shown in Fig. 6-4, there are *two* electrons with similar wave functions oriented along OY, and there are *two* electrons (not shown in this two-dimensional representation) with similar wave functions oriented along OX. (b) The F⁻ ion. An additional $2p$ electron has filled the vacant state (oriented along OZ). All of the $2p$ electrons are now a little less tightly bound and a little farther from the helium core because of the increased electron-electron repulsion, but the additional electron is as strongly bound as any of the others. The net effect is that the energy in (b) is $e\chi = 4.2$ e.V. *less* than in (a). (The $2s$ electrons are not shown; the helium core is indicated by the seven plus signs.)

$Z = 10$ (NEON). At this point, all ten possible states with $n = 1$ and $n = 2$ have been *filled* with electrons ($2 - 1s$, $2 - 2s$, $6 - 2p$). We say that the $n = 1$ and $n = 2$ *shells* have been completely filled. The value of V_i, which has been increasing almost steadily from the value for lithium, is now 21.56 volts. It is therefore very difficult to remove an electron from neon in order to make a positive ion. It is also difficult to add one to make a negative ion, since it would have to be in the $n = 3$ state, which would have a much higher energy than the $n = 2$ states. Neon is thus a noble gas like helium and is quite inert chemically.

Although each of the $2p$ wave functions varies with the polar angles θ and φ, the sum of the probability densities of all six of them is spherically symmetrical (prob. 6-6). Furthermore, the electron density for all six electrons falls to zero fairly sharply at the edge of the atom. Atoms with a completely filled shell are thus as close to the hard sphere concept of an atom as any real atoms are. It is for this reason that the Ramsauer effect (Sec. 5-4e) can be observed with noble-gas atoms. The compactness and the stability of this filled-shell structure make the existence of an electron affinity for fluorine more plausible, since the F^- ion has this noble-gas structure (so does O^{--} and N^{---}).

$Z = 11$ (SODIUM) TO $Z = 18$ (ARGON). In this series the $n = 3$ shell is filling in exactly the same way that the $n = 2$ shell filled from lithium to neon. The chemical properties of a pair of elements in these two series with the same outer electronic structure (e.g., sodium and lithium) are very nearly alike. This is, of course, an example of the general similarity of chemical properties that is expressed in the periodic system of chemistry. The periodic table (Appendix B) should be compared with Table 6-1.

$Z = 19$ (POTASSIUM) AND $Z = 20$ (CALCIUM). We might expect that these elements would start filling in the $3d$ shell ($l = 2$), but the lowest energy states are now the two $4s$ states. Here we see a more extreme illustration of the effect that was described in connection with the lithium structure. The $4s$ ($l = 0$) wave functions penetrate into the region of low potential energy near the nucleus. This gives them a lower energy than the $3d$, which do not penetrate much into this region, even though their value of n is greater.

$Z = 21$ (SCANDIUM) TO $Z = 30$ (ZINC). The energies of the $4p$ states are higher than those of the $4s$ state, and the difference in energy is enough to cause the $3d$ states to lie between the $4s$ and the $4p$. Hence the $3d$ shell fills for this group of elements. The closeness of the competition for lower energy between $4s$ and $3d$ states manifests itself in two elements where an extra electron goes into the $3d$ shell at the expense of removing one from the $4s$ shell. At $Z = 29$ (copper) the $3d$ shell is filled since it has ten electrons.

The elements with an incomplete $3d$ shell are called *transition elements*. (We shall see in Chapter 9 that ferromagnetism occurs with some of these elements.) These elements are very similar in chemical properties, since the filling of the $3d$ shell has little effect on the ionization potential and other properties of the outer ($4s$) electrons.

$Z = 31$ (GALLIUM) TO $Z = 36$ (KRYPTON). The $4p$ shell fills as expected and becomes filled for the noble-gas element krypton.

$Z > 36$. We shall not continue this study through the rest of the elements. The electronic structures of the remaining elements present no new physics and are tabulated in the references.

SUMMARY

The chemical properties to be expected from the various electronic structures have been indicated in the foregoing survey. The Exclusion Principle is effectively proved by the agreement between these predicted properties and experiment. The regularities of chemical properties summarized in the periodic table are precisely the regularities predicted by quantum physics and the Exclusion Principle. Without that principle there would be a general uniformity of chemical properties of all elements. As Z increases, the energies of the $1s$ states would change, but all elements would have only $1s$ electrons. The variety of chemical properties of the elements is thus in itself good evidence for this principle.

6-5 OPTICAL SPECTRA

The hydrogen spectrum was described in Sec. 4-3, and the wave-mechanics explanation of this spectrum was presented in Sec. 6-2.

As we proceed to the spectra of atoms with more than one electron the problem becomes very complicated even for the elements with small values of Z. The complications are caused by the dependence of the energy upon n, l, m_l, and m_s (instead of upon n alone as in hydrogen) and by the fact that more than one electron can be in an excited state. If one electron is in an excited state the charge distribution is different from the distribution in the ground state, and hence the energy levels of all the other electrons are changed. We therefore have to consider the change in energy of the atom *as a whole* when it absorbs or radiates light, which is obviously a complicated problem since many electrons may be involved.

There is one kind of atom in which the problem of the frequencies emitted or absorbed is much clearer. These atoms are the alkali atoms,

Li, Na, K, Rb, and Cs (see Fig. 4-10*b* for a section of the lithium spectrum). In each of these atoms there is a single electron in a shell by itself. In sodium, for example, the $n = 1$ and $n = 2$ shells are completely filled, and there is a single 3*s* electron. All the other electrons are much more tightly bound to the nucleus than this electron is. If a sodium atom is excited in an arc or flame, the excited state can be simply described: The 3*s* electron is excited to a 3*p*, a 3*d*, a 4*s*, or a higher state. The electrons in the filled inner shells are nearly unaffected, and only slight modifications of their energies occur. Thus the general problem of changes in energy for the atom as a whole reduces in the alkali atoms (to a good approximation) to the changes in energy of a single electron.

The possible excited states of the sodium atom are illustrated in Fig. 6-9. It should be noted that the *p* states lie above the corresponding *s* states. This is the effect of the relatively greater penetration of *s* electrons into the low potential energy of the neon core. (This effect was

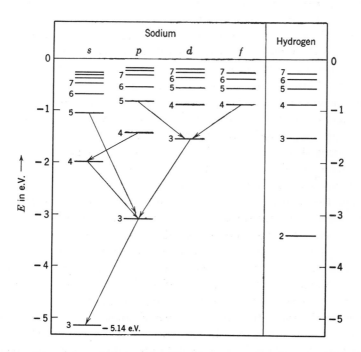

Fig. 6-9. Energy level diagram for sodium compared with hydrogen. The energies are labeled with the values of *n*. The spin splitting of the *p*, *d*, and *f* levels is too small to be revealed on this vertical scale. The arrows show only a few of the allowed transitions, which follow selection rules like those for hydrogen.

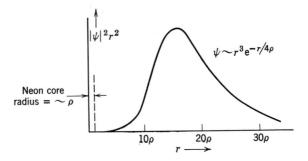

Fig. 6-10. The probability density for the 4f wave function of hydrogen compared to the size of the neon core.

explained in connection with the lithium 2s electron in Sec. 6-4.) The hydrogen atom states are plotted for comparison. Note the great similarity between the energy levels of f states for sodium and the energy levels for hydrogen. The 4f, 5f, etc., wave functions have very small values in the region of the neon core. For example, the 4f wave function (for which the radial probability density is plotted in Fig. 6-10) is proportional to $r^3 e^{-r/4\rho}$, and its value is almost zero inside the neon core. Thus the electron is almost never in that region, and it does not matter what happens to the potential energy P there. In the region outside the neon core, P is just $-e^2/4\pi\epsilon_0 r$, the potential energy of a point charge $+e$. This is precisely the situation in the hydrogen atom, and therefore it is no accident that the sodium f levels resemble hydrogen levels. Even the sodium d levels closely resemble hydrogen levels, but the correspondence is not so close as in the f levels.

From the observed sodium spectrum we can verify the electronic structure we have asserted for sodium, namely, the neon core plus one 3s electron. If the structure were different, we should obtain quite different series of lines. The analysis of spectra is, in fact, the method by which all the electronic structures of Sec. 6-4 were obtained. Theoretical estimates of the relative energies of various possible structures were made in Sec. 6-4, and careful computations have been made to determine the theoretical ground states of atoms. But computational difficulties of deciding, for example, whether the nineteenth electron in potassium goes into a 3d or a 4s state are serious. The experimental evidence is that the potassium spectrum is qualitatively identical with the sodium spectrum, and thus potassium must have the same outer electron structure as sodium. Of course for this example the chemical evidence also strongly supports the $4s$ structure, but for elements like chromium and manganese only the spectra can be used.

Even the simplest, one-electron energy diagrams are more complicated than Fig. 6-9 when plotted on an expanded ordinate scale. The principal complicating factor is the electron spin. For example, the p, d, and f states are shown as single lines in Fig. 6-9, but they actually consist of pairs of lines too close together to be distinguished in the illustration. This splitting is caused by the different orientations of the electron spin angular momentum relative to the electron's orbital angular momentum. The possible components of the spin angular momentum in any one direction are $(h/2\pi)m_s$ with $m_s = \pm\frac{1}{2}$, whereas (Sec. 6-2) the components of orbital angular momentum are quantized in units twice as large, namely, $(h/2\pi)m_l$ with m_l taking on the values $0, \pm1, \cdots \pm l$. The vector sum j of these two momenta, which has many possible values when l is large, is quantized, and thus there are many more energy levels than there would have been without spin. For example, the $3p$ level in Fig. 6-9 is actually two closely spaced levels separated by 0.00214 e.V.; the upper corresponds to the orbital and spin momentum in the same direction ($j = \frac{3}{2}$) and the lower to opposite directions ($j = \frac{1}{2}$). The most familiar lines of the sodium spectrum are the two yellow lines $\lambda = 5890$ Å and $\lambda = 5896$ Å, which result from the transition from the two closely spaced $3p$ states to the $3s$ state.

Accompanying, and proportional to, both the orbital and spin angular momenta are magnetic moments $\mathfrak{M}$, which can be conveniently expressed in terms of the *Bohr magneton:*

$$\mathfrak{M}_B = \frac{eh}{4\pi m} = 0.927 \times 10^{-23} \text{ joule m.}^2/\text{weber} \qquad (6\text{-}14)$$

For the spin momentum, $\mathfrak{M}$ has the two possible values $\pm\mathfrak{M}_B$ corresponding to the two values $\pm\frac{1}{2}$ of the spin quantum number m_s.*

A test of all these assertions, and much additional information, can be obtained by observations on atoms placed in a magnetic field. We shall consider only the simplest example, namely, an atom with only a single electron outside a closed state and that electron in a state without orbital angular momentum, such as the $3s$ level of sodium or the $5s$ level of silver. The two possible values $m_s = \pm\frac{1}{2}$ of spin correspond to aligning this atom with its spin moment $\mathfrak{M}$ either parallel or anti-parallel

---

* It is common practice to introduce the *spectroscopic splitting factor* g and to write $\mathfrak{M} = gm_i\mathfrak{M}_B$, where m_i is either the spin (m_s) or orbital (m_l) quantum number or in general a combination of the two. For electron spin alone, $g = 2$, but if orbital momentum is also present g takes on other values calculated by the vector addition of momenta.

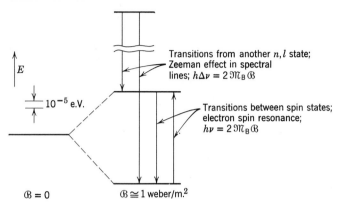

Transitions from another n, l state; Zeeman effect in spectral lines; $h\Delta\nu = 2\,\mathfrak{M}_B\,\mathfrak{B}$

10^{-5} e.V.

Transitions between spin states; electron spin resonance; $h\nu = 2\,\mathfrak{M}_B\,\mathfrak{B}$

$\mathfrak{B} = 0$ $\mathfrak{B} \cong 1$ weber/m.^{2}

Fig. 6-11. Splitting of energy levels in a magnetic field.

to the magnetic induction $\mathfrak{B}$, as sketched in Fig. 6-11. Since the potential energy of a dipole of moment $\mathfrak{M}_B$ in a magnetic induction $\mathfrak{B}$ is $\mathfrak{M}_B \cdot \mathfrak{B} = \mathfrak{M}_B \mathfrak{B} \cos \theta$, where θ is the angle between $\mathfrak{M}_B$ and $\mathfrak{B}$, the difference in energy between the two levels is $2\mathfrak{M}_B\mathfrak{B}$. Thus a spectral line originating from a transition to or from such a split level will be split into two components with nearly the same, but measurably different, frequencies; this is the Zeeman effect.

A more direct observation of this splitting is obtained in the *electron spin resonance* experiment. If atoms with moments $\mathfrak{M}_B$ are placed in a strong magnetic induction $\mathfrak{B}$, radiation incident on them with frequency ν given by

$$h\nu = 2\mathfrak{M}_B\mathfrak{B} \tag{6-15}$$

will be absorbed since it can produce the transition between the two levels indicated in Fig. 6-11. Since this frequency is in the microwave region of the electromagnetic spectrum for typical values of $\mathfrak{B}$ (see prob. 6-39), the experiment is actually performed by placing the specimen in a resonant cavity excited by a microwave oscillator. As $\mathfrak{B}$ is varied slowly through the resonance condition, pronounced absorption (reduction in the "Q" of the cavity) occurs when ν and $\mathfrak{B}$ are related by eq. 6-15.

Atomic beam experiments like the Stern-Gerlach experiment, line spectra investigations, the Zeeman effect, and electron spin resonance experiments are capable of providing extremely detailed information about the electronic structure of atoms. The examples we have discussed briefly above have been only the simplest applications (one-electron spectra, only spin moments) of these techniques and thus do not reveal their full power. The complications produced by the simultaneous transitions of more than one electron and by combining spin with orbital momenta can be successfully predicted by quantum theory, and these predictions can be tested by this array of experiments.

For many years, applications of line spectra were confined to light sources and to spectrochemical analysis, the identification of elements by their spectral lines (which are quantitatively different for each element). More recently, the development of *masers* and *lasers* from the atomic beam and spin resonance techniques has provided remarkably stable oscillators and sensitive amplifiers of electromagnetic waves; some of these devices will be discussed in Secs. 10-6 and 12-8d.

6-6 X-RAY LINE SPECTRA

As we considered atoms with increasing atomic number Z in Sec. 6-4 we concentrated on what was happening to the *outer* (valence) electrons in each atom. We now return to the problem of atomic structure as a function of Z and consider the *inner* electrons. We shall thereby obtain a quantum-mechanical explanation of the X-ray line spectra described in Sec. 4-4.

Every atom except hydrogen has two 1s electrons. Each of these electrons moves in a region of potential energy $-Ze^2/4\pi\epsilon_0 r$ as modified by the presence of the other electrons. As Z increases, the energy levels of these 1s electrons decrease rapidly. We saw in eqs. 6-12 and 6-13 that the energies of these electrons became more negative roughly as Z^2, and their mean distances from the nucleus were about proportional to $1/Z$. Thus the energy necessary to remove such an electron increases about as Z^2.

In Fig. 6-12 a plot is given of the energies of the 1s, 2s, 2p, and 3s electrons in sodium. The energy necessary to remove an electron to infinity from a particular state in the sodium atom is the absolute magnitude of the energy listed for that state. Note that we are considering here a process different from the process of second ionization. The second ionization energy of sodium (47.2 e.V.) is the energy required to remove one of the 2p electrons *after* the 3s electron has already been removed. The energy 30.7 e.V. is the energy required to remove a 2p electron from the neutral atom.

Figure 6-13 presents the conventional way of diagramming energies for X-ray absorption and emission spectra. The ordinate is the energy of the *entire atom*, and zero corresponds to the neutral atom in its ground state. Thus excitation of the atom is portrayed by an upward directed arrow. For example, one of the 1s electrons can be removed from a

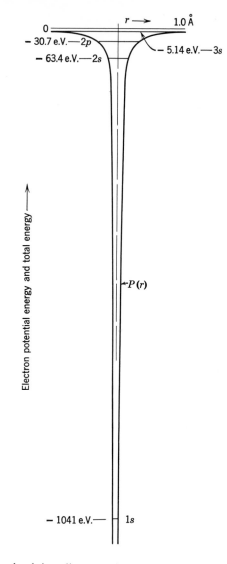

Fig. 6-12. Energy levels in sodium superimposed on a sketch of the potential energy of an electron.

sodium atom by bombardment with electrons having kinetic energies greater than 1041 e.V. This excitation process leaves a *hole*, a vacant quantum state, in the $n = 1$ shell, which is called the *K-shell* in X-ray terminology. The atom is now in an excited state with an energy 1041

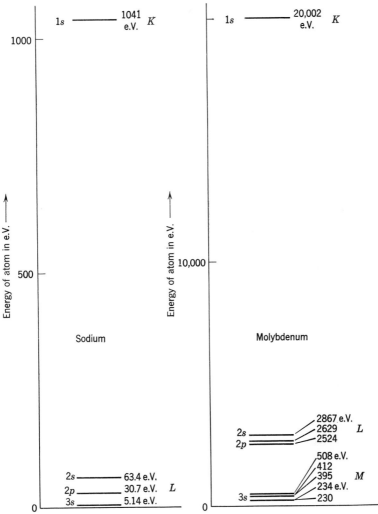

Fig. 6-13. Conventional X-ray energy level diagrams for sodium and molybdenum. The energy marked, for example, "1s" is the energy of the atom when there is a 1s hole, or, in other words, when one of the 1s electrons has been removed from the atom.

e.V. above the ground state. A 2p electron (*L-shell*) can make a transition to 1s, filling the *K*-shell and emitting an X-ray photon; such photons constitute the *K*α spectral line. [Another transition allowed by the selection rules would be 3p→1s (*K*β-line), but this does not occur in sodium because there are no 3p electrons.]   In turn, the L-shell can be filled by a transition 3s→2p, giving one of the L-emission lines.*

The minimum energy of the bombarding electron required in order to produce a hole in the K-shell of sodium is actually a little less than 1041 volts because it is not necessary to remove a 1s electron from the atom but only to excite it to an unoccupied state.   The lowest-lying unoccupied state of sodium is the second 3s state.   The energy of this state is so close to zero on the scale of Fig. 6-13 that we shall make only a very small error if we assume that the threshold energy of the bombarding electron is the same in magnitude as the energy of the 1s states. Furthermore, this error is of even less consequence for atoms with higher Z.

X-ray lines for practical applications have wavelengths of the order of 0.1 to 2.5 Å, and therefore atoms with much higher atomic numbers than sodium must be used.   An energy-level diagram for molybdenum is also presented in Fig. 6-13.   This diagram was constructed from the observed values of the wavelengths of emission lines; two of these lines, the molybdenum Kα- and Kβ-lines, were illustrated in Fig. 4-19.   In order to produce any K-line of molybdenum, an electron must be removed from the K-shell, which requires at least 20,000 e.V., the energy required to excite a 1s electron to the lowest unoccupied quantum state. The lowest voltage at which an X-ray tube can operate and still produce K-lines from a molybdenum target is therefore 20,000 volts.   This threshold voltage is the same for the various lines of the K-series of a given element, even though the energies of these various photons are different.

Another way of removing an electron from a K- or L-shell is by the X-ray photoelectric effect.   If X-ray photons with sufficiently high energy are incident upon an atom, an electron can be ejected from the K- or L-shell.   The threshold photon energy for this process is, of course, the same as the threshold energy for ejection by a bombarding electron. We expect and find that the *absorption* of X-rays increases sharply as the incident photon energy increases beyond this threshold, as shown in Fig. 6-14, which is a plot of the absorption coefficient for X-rays in solid molybdenum as a function of the X-ray photon energy.

---

* There is another process, the *Auger process*, by which the atom can return to the ground state after a hole is created in an inner shell.   An electron from an outer shell goes to the inner shell and simultaneously another outer-shell electron is ejected from the atom.   The energy therefore appears largely as kinetic energy of this ejected Auger electron rather than as the hν of an X-ray photon.   The Auger process actually occurs with a higher probability than the X-ray emission process, but the Auger electrons quickly lose energy and are rarely emitted from a solid X-ray target, whereas X-ray photons (which are very penetrating) *are* emitted.

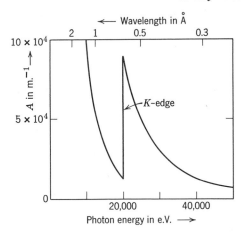

Fig. 6-14. X-ray absorption coefficient A of molybdenum as a function of photon energy or wavelength.

The attenuation of a beam of X-rays upon passing through matter occurs primarily by this photoelectric absorption. Each absorption act removes a photon from the beam, and an analysis similar to that leading to eq. 2-19 gives an expression for the intensity I of X-rays transmitted through a slab of matter of thickness x

$$I = I_0 e^{-Ax} \tag{6-16}$$

where I_0 is the intensity incident on the slab. This expression defines the absorption coefficient A. (The *mass absorption coefficient* is A/ρ, where ρ is the density of the slab; this is the quantity that is usually tabulated.)

The absorption coefficient generally decreases with increasing photon energy, but Fig. 6-14 shows that in molybdenum it increases sharply at an energy of 20,000 e.V., the minimum energy for K-excitation, which is therefore called the *K-absorption edge*. Three *L-edges* occur at much lower energies (2867, 2629, and 2524 e.V.), and similar K- and L-absorption edges occur, of course, in all other elements.

The dependence upon atomic number of the $K\alpha$ photon energy is illustrated in Fig. 6-15, in which the square root of this energy is plotted as a function of Z. A nearly straight line with intercept $Z = 1$ fits the data. Thus it must be that

$$E_{2p} - E_{1s} \cong (\text{constant})(Z - 1)^2 \tag{6-17}$$

The quantum theory gives a very satisfactory explanation of this equation. We saw in eq. 6-12 that the energy of the $n = 1$ state for an atom

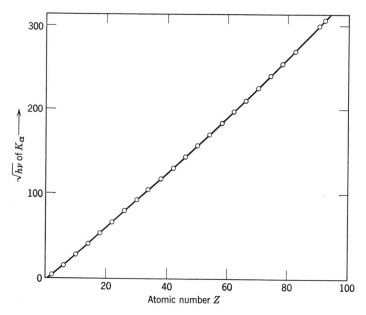

Fig. 6-15. The square root of the $K\alpha$ photon energy as a function of atomic number. Only a few of the available data are plotted.

like hydrogen, but with a nuclear charge of $+Ze$ in place of $+e$, was proportional to Z^2. In an atom with large Z all the electrons except the $1s$ electrons are usually much farther from the nucleus than the $1s$ electrons. Thus the $1s$ electrons have energy levels almost as if the other electrons were not present. Each $1s$ electron moves in the attractive potential energy of the nucleus $+Ze$ reduced by the repulsion of the other electron. To a crude approximation, the negative charge of the other electron effectively reduces the nuclear charge from $+Ze$ to $+(Z-1)e$. In this approximation, then, the energy of the $1s$ electron is $\sim -13.60\,(Z-1)^2$ e.V. Similarly, the energy of the $2p$ state is $\sim -(13.60/2^2)(Z-1)^2$ e.V. Although both these expressions are in error because of the neglect of the outer electrons, their *difference* should be a good approximation to the $K\alpha$ transition energy. Thus eq. 6-17 is predicted.

Plots such as Fig. 6-15, called *Moseley diagrams*, can be drawn for the various X-ray lines and absorption edges. The atomic number Z of an element can be determined from its X-ray spectrum by such a diagram far more accurately and conveniently than the method described in Sec. 3-2.

In practical applications of X-rays the K-emission lines are generally used in experiments for determining crystal structure and for identifying unknown compounds, such as the experiments that were described in Sec. 4-4. Continuous X-rays of 200,000- to 2,000,000-e.V. energy are used for the inspection of thick plates and machinery. These high energies are used to take advantage of the decrease in absorption constant with increasing photon energy. Since the highest photon energies of any X-ray lines are obtained in the K-lines of uranium (about 110,000 e.V.), line spectra are not present in the energy range above 110,000 e.V.

In this section we have discussed X-ray emission and absorption spectra as they would be observed for atoms in the gaseous state, but almost all X-ray experiments are actually performed on *solids*. We shall learn in Chapter 8 that the lower-lying, inner-shell levels are practically the same in gases as in solids. Our discussion here is hence an excellent approximation to the actual situation, and this discussion will be refined when the energy levels in solids are explored in Chapter 8.

REFERENCES

General

J. C. Slater, *Quantum Theory of Matter*, McGraw-Hill, New York, 1951, Chapters 5 and 6.

F. O. Rice and E. Teller, *The Structure of Matter*, Science Editions, New York, 1961.

L. Pauling and E. B. Wilson, *Introduction to Quantum Mechanics*, McGraw-Hill, New York, 1935, Chapters 5, 8, and 9.

F. K. Richtmyer, E. H. Kennard, and T. Lauritsen, *Introduction to Modern Physics*, McGraw-Hill, New York, 5th Ed., 1955.

R. M. Eisberg, *Fundamentals of Modern Physics*, Wiley, New York, 1961.

Optical Spectra

G. Herzberg, *Atomic Spectra and Atomic Structure*, Dover, New York, 2nd Ed., 1944.

X-Ray Spectra

A. H. Compton and S. K. Allison, *X-Rays in Theory and Experiment*, Van Nostrand, New York, 2nd Ed., 1935.

A. E. Sandström in *Encyclopedia of Physics*, edited by S. Flügge, Springer, Berlin, 1957, Volume 30.

PROBLEMS

6-1. Show that the wave function of eq. 6-3 has been normalized.

6-2. Figure 6-2a exhibits a discontinuity in $d\psi/dr$ at $r = 0$ when ψ is plotted along a line through the nucleus. What is the potential energy P at $r = 0$? Compare with the behavior of $d\psi/dx$ at $x = x_0$ in Fig. 5-4b, where $P \to \infty$ at $x = \pm x_0$. Can you state an exception to eq. 5-8? What is the physical interpretation of this exception in terms of momentum? Compare also the behavior at the wall in Fig. 2-1.

6-3.* Calculate the energy of the photon radiated in the transition $2p \rightarrow 1s$ for hydrogen. Calculate the wavelength of this photon.

6-4.* Calculate the value of r that gives a maximum of $|\psi|^2 4\pi r^2$ for the $1s$ wave function of hydrogen.

6-5. Show that the average value of r for the $1s$ wave function of hydrogen is $\frac{3}{2}\rho$.

6-6. The wave functions for the six $2p$ states of hydrogen are given by the following expressions:

$$m_l = -1, \quad m_s = \pm\tfrac{1}{2}: \qquad \psi = A(r/\rho)e^{-r/2\rho} \sin\theta \cos\varphi$$

$$m_l = 0, \qquad m_s = \pm\tfrac{1}{2}: \qquad \psi = A(r/\rho)e^{-r/2\rho} \cos\theta$$

$$m_l = +1, \quad m_s = \pm\tfrac{1}{2}: \qquad \psi = A(r/\rho)e^{-r/2\rho} \sin\theta \sin\varphi$$

Here A is a constant and r, θ, and φ are the usual spherical polar coordinates. Show that the sum of the $|\psi|^2$ for all six is a function only of r.

6-7.* In prob. 6-6, evaluate A when $m_l = 0$ in order that ψ should be normalized.

6-8. Show that the two $2p$ wave functions with $m_l = \pm 1$ are identical in shape to that with $m_l = 0$, but that they are each cylindrically symmetical about one of the other Cartesian coordinates. *Hint:* Write x, y, and z in terms of r, θ, and φ; express each of the ψ's in prob. 6-6 in terms of r and one of these Cartesian coordinates.

6-9.* Calculate the average value $\bar{r}$ of r for the hydrogen $2p$ state with $m_l = 0$; use the wave function from prob. 6-6 and the value of A from prob. 6-7.

6-10. Consider a hydrogen atom in a $2s$ state and one in a $2p$ state. Compare the wave functions. (What features do they have in common? What features are different?) Discuss the difference in properties of the two atoms.

6-11.* Calculate the average value $\bar{P}$ of the potential energy P for the hydrogen atom in the $1s$ state. Compare with E.

6-12.* Calculate the average value $\bar{K}$ of the kinetic energy K for the hydrogen atom in the $1s$ state. Compare $\bar{P} + \bar{K}$ with E. When ψ is a function of r alone, eq. 5-49 becomes

$$\bar{K} = (h^2/8\pi^2 m) \int_0^\infty 4\pi r^2 |d\psi/dr|^2 \, dr$$

6-13.* Consider an electron in a state with $l = n - 1$ in the hydrogen atom. The average value of r for such states is $\bar{r} = n^2\rho[1 + (1/2n)]$, which equals approximately $n^2\rho$ for large n. Calculate the kinetic energy K of the electron by subtracting P (from eq. 6-1 with $n^2\rho$ in place of r) from E (from eq. 6-2). Calculate the momentum mv from K, and the de Broglie λ from mv. If the electron were moving in a circle of radius $n^2\rho$, how many wavelengths would there be in the circumference of this circle? Compare with Fig. 6-5, using the fact that, at large n and $l = n - 1$, $n \cong l$.

6-14.* The theory of the hydrogen atom presented in Sec. 6-2 did not include the variation of the mass of the electron with its velocity. Estimate the velocity of the electron in the state $n = 4$, $l = 3$ by the method of the preceding problem. What is the ratio of the actual mass of the electron to the rest mass m_0 at this velocity?

6-15.* Calculate the frequency ν_c of rotation of a classical electron in a circular orbit about a proton, as a function of the radius r_c of this orbit. (Set the force $-e^2/4\pi\epsilon_0 r_c{}^2$ equal to $-mv^2/r_c$, and calculate $\nu_c = v/2\pi r_c$.)

6-16.* For large n, calculate the frequency ν_w radiated by the hydrogen atom according to wave mechanics for the transition $n \rightarrow n - 1$. Express this in terms of the mean radius $\bar{r}$ (prob. 6-13). Compare ν_w with ν_c from the preceding problem. (*Hint:* Since the change $\Delta n = -1$ is very much less than n, write $\Delta E \cong (dE/dn)\Delta n$.)

6-17. Sketch the energy level diagram for the hydrogen atom up to and including $n = 4$, label the energy scale in electron volts, and designate the states by the quantum numbers n and l. Show on the diagram all the possible transitions that produce photons with $\lambda = 6570$ Å.

6-18. In the integral eq. 6-7, let $u = z$ and consider the hydrogen $1s$ and $2p$ (with $m_l = 0$; see prob. 6-6) states. Show that the integral does not vanish and therefore that the transition $2p \rightarrow 1s$ is allowed. *Hint:* Express z in terms of r and θ.

6-19. Prove the result of prob. 6-18 by actually evaluating the integral.

6-20. The probability of an allowed transition is proportional to the integral on the right side of eq. 5-53. Compare qualitatively (by rough sketches of the integrands) the probabilities of the transitions $np \rightarrow 1s$ and $2p \rightarrow 1s$ in the hydrogen atom, where $n \gg 2$. Consider only the radial extent of ψ; neglect the changes in shape (additional zeros of ψ) of the wave function as n increases from 2.

6-21.* Calculate the relative number of hydrogen atoms in the ground state and in the first excited state for a gas in thermal equilibrium at a temperature of 3000°K.

6-22. Explain how the ratio of the intensity of radiation from the $3p \rightarrow 2s$ transition to the intensity from the $2p \rightarrow 1s$ transition can be used as an indication of the temperature in an atomic hydrogen flame. (Assume equilibrium.)

6-23.* Demonstrate the penetration of an electron into a classically forbidden region by calculating the kinetic energy $K = E - P$ of an electron in the ground state of the hydrogen atom when it is at a distance 3ρ from the nucleus. At what value of r is $K = 0$?

6-24. Estimate roughly the ionization potential that the lithium atom would have if the electron-electron interaction described by the Exclusion Principle were absent, and compare with the observed value.

6-25. Discuss the connection between the Indeterminacy Principle and the statement of the Exclusion Principle given with eq. 6-11. What paradoxical situation would arise if the right side of eq. 6-11 were $\ll (h/2)^3$?

6-26.* Calculate the second ionization potential of helium (this is numerically equal to the energy in electron volts necessary to remove the remaining electron from the He^+ ion). Calculate the third ionization potential of lithium. (The experimental values are 54.4 volts and 122.4 volts, respectively.)

6-27. * The size of the uranium nucleus corresponds to a sphere of radius $R_0 \cong 9 \times 10^{-15}$ m. Assume a hydrogen-like ψ, and hence use eq. 6-13 in eq. 6-3 to find an approximate ψ for the $1s$ state in uranium. What fraction of the time is the electron inside the nucleus? (Although this number is small, it is not zero; since the theory of the present chapter has successfully neglected interactions other than electrostatic, the electron must interact only very weakly or not at all with the nucleus.)

6-28. * The $2p{\rightarrow}2s$ transition in lithium gives a spectral line with $\lambda = 6708$ Å. What is the energy of the $2p$ state? Why is it closer to the hydrogen $2p$ state energy than the lithium $2s$ state is to the hydrogen $2s$?

6-29. The second ionization potential of helium could be accurately calculated from the first ionization potential of hydrogen (prob. 6-26). Why cannot the second ionization potential of lithium (75.6 volts) be similarly calculated from the first ionization potential of helium?

6-30. Apply qualitatively to the lithium $2s$ wave function the method of finding wave functions that was outlined at the end of Sec. 5-7. How should the hydrogen $2s$ function be modified to lower the energy? Why does the zero of $|\psi|^2 r^2$ in Fig. 6-7b occur at a smaller r (i.e., more rapid variation of ψ near $r = 0$) than the zero of the $2s$ function in Fig. 6-3b?

6-31. At the beginning of the discussion of lithium in Sec. 6-4, a crude theory gave $V_i = 3.40$ volts. Why was this value *lower* than the observed value of 5.39 volts?

6-32. From Table 6-1 and the periodic table in Appendix B, determine the electronic structures of bromine and barium.

6-33. The ionization potential of barium is 5.19 volts. Plot V_i for Be, Mg, Ca, and Ba vs. the number of the row in the periodic table where each occurs. Estimate V_i for Sr.

6-34. Show by using Gauss' law of electrostatics and letting $P = 0$ at $r = \infty$ that the potential energy P outside of the neon core of the sodium ion Na^+ is $-e^2/4\pi\epsilon_0 r$. This is the potential energy which governs the motion of the $4f$ and similar electrons.

6-35. * What excitation potential would be found for atomic sodium in an experiment like those of Sec. 4-5?

6-36. * Use the wavelengths given for the yellow sodium lines to calculate the difference in energy (in electron volts) between the two $3p$ states of sodium. *Hint:* Use the method of the calculus: $\Delta E \cong (dE/d\lambda) \, \Delta\lambda$.

6-37. What are the total angular momentum and the total magnetic moment of a closed shell of electrons?

6-38. * What is the difference in energy (in electron volts) between the two spin states of the $3s$ level in sodium in a magnetic induction of 0.8 weber/m.²? To find the order of magnitude of the Zeeman splitting of spectral lines, assume that transitions occur to these two states from the same $3p$ state. What is the difference in λ of the two lines?

6-39.* Calculate the frequency ν at which electron spin resonance occurs in a magnetic induction $\mathcal{B}$ of 0.4 weber/m.^{2}

6-40.* For spin resonance absorption to occur, the number N_2 of atoms in the upper energy state must be fewer than the number N_1 in the lower (if these numbers were equal, there would be as much emission as absorption). Find an expression in terms of $\mathcal{B}$ and T for the difference $N_2 - N_1$ between these populations in thermal equilibrium by using the Boltzmann factor from eq. 6-10, by assuming equal statistical weights for the two states, and by making the appropriate approximation when $(E_2 - E_1) \ll kT$.

6-41. Use the result of prob. 6-40 to discuss the following questions: (*a*) It becomes more difficult to conduct experiments as ν and $\mathcal{B}$ increase, yet a great deal of effort is made to use higher ν's and $\mathcal{B}$'s. Why? (*b*) If too much microwave power is incident upon a sample, the absorption decreases. Why? *Hint:* Thermal *equilibrium* was assumed in prob. 6-40.

6-42.* Assume that the kinetic energy of a K-shell electron equals the magnitude $|E|$ of the total energy in uranium. What is the ratio of the actual mass m to the rest mass m_0 for this kinetic energy? (The K-absorption edge in U has a photon energy of 115,600 e.V.)

6-43. What is the minimum X-ray tube voltage required to generate the $K\alpha$-line with a molybdenum anode? Each electron that excites a Mo atom loses this amount of energy, but the $K\alpha$-line has a lower energy. What happens to the energy difference?

6-44. Calculate from Fig. 6-15 the experimental value of the "constant" in eq. 6-17, and compare it with the theoretical value from the hydrogen-like model described just below eq. 6-17.

6-45. Find the energy of the K-absorption edge in tungsten by the use of tables of absorption coefficient or mass absorption coefficient as a function of wavelength.

6-46. Suppose that it is desired to detect small inclusions of iron particles in a graphite block by the use of X-rays. A photographic plate is placed behind the block and an X-ray-line source some distance in front of the block. What X-ray wavelength should be used to provide maximum contrast (i.e., to make the absorption by iron as strong as possible compared to that by carbon)? Suggest an anode material for the X-ray tube to generate this wavelength.

7

Molecules

7-1 INTRODUCTION

Quantum physics will be applied in this chapter to the study of the structure of molecules. The first question to which we seek an answer is: What holds two atoms together in a diàtomic molecule? We shall find two rather different answers to this question for different combinations of atoms. The *ionic* form of binding will be discussed in Sec. 7-2; this type of binding occurs in most inorganic molecules. The *covalent* form of binding will be discussed in Secs. 7-3 and 7-4; this type occurs to some extent in inorganic molecules and in all organic molecules. Molecular binding is studied for its own interest and also because this study provides an introduction to binding in solids.

The energy levels of electrons in molecules will be discussed briefly in Sec. 7-5. One purpose of this discussion is to serve as an introduction to the study of energy bands in solids in Chapter 8. Molecular spectra and the process of dissociation will be discussed in Sec. 7-6. Molecular spectra provide valuable information about the structure of molecules, especially the geometrical arrangement of the constituent atoms.

Throughout this chapter we shall concentrate on diatomic molecules, but the principles of binding and energy levels apply also to polyatomic molecules.

7-2 THE KCl MOLECULE AND IONIC BINDING

Our task in this section is to develop an understanding of ionic molecules, which we shall do by studying the KCl molecule and the nature of the forces holding potassium and chlorine together. The binding of K and Cl will be discussed in terms of the *binding energy* or *dissociation energy* $|E_d|$, instead of in terms of binding forces. (The concept of binding energy was explained in Sec. 3-4 and Fig. 3-13.) The binding energy of KCl is the energy that must be supplied in order to remove the K from the Cl. This energy is also called the dissociation energy because the process of separating the constituents of a molecule is called dissociation. To prove that K and Cl will be found together in a stable molecule, we must prove that the energy of the KCl molecule is lower than the energy of a K atom and a Cl atom with a large distance between these atoms. The difference in these energies is the dissociation energy $|E_d|$; E_d is the energy of the bound KCl molecule and is a negative number.

One possible structure of the KCl molecule is a K^+ ion attached closely to a Cl^- ion. We shall now show that this structure is stable and calculate the energy E_d for it. Let E be the total energy of the system of one K and one Cl atom or ion in any stage of the development of the molecule. We define the zero of E by setting E equal to zero for the two atoms, each in its ground state, when they are separated by a large distance.* Thus E equals the change in total energy of the system as the atoms are brought together. Figure 7-1 illustrates the first steps in the computation of E_d. Starting at $E = 0$ with the two atoms a large distance apart, we first ionize K. This "costs" 4.34 e.V., the ionization energy of potassium. Next we put the electron obtained from this ionization process onto the Cl atom and thereby produce a Cl^- ion. This step gives us back 3.82 e.V., the electron affinity energy E_a of chlorine. Now we have a K^+ and a Cl^- ion with a net energy expenditure of 0.52 e.V. This is not a stable situation, since 0.52 e.V. would be given off if the ions reverted to the neutral atoms.

These two oppositely charged ions will attract each other strongly, however, and this electrostatic attraction is responsible for the stability of the molecule. Both K^+ and Cl^- are spherically symmetrical ions, since they both have the noble-gas, closed-shell electronic structure. Furthermore, the radial extent of the charge distribution is fairly

---

* This is not the only zero of E in common use. Other choices are that $E = 0$ for the separated *ions* or for the separated constituents in their standard states, the naturally occurring states at the temperature in question and 1 atmosphere pressure (for chlorine, a gas of Cl_2 molecules; for potassium, solid potassium metal).

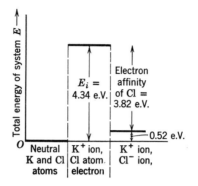

Fig. 7-1. The "net cost" of a pair of K^+ and Cl^- ions is 0.52 e.V.

sharply defined for these ions.   There is a large concentration of charge up to a radius called the *ionic radius* and nearly zero charge outside this radius.   Coulomb's law for the force between spherical distributions of

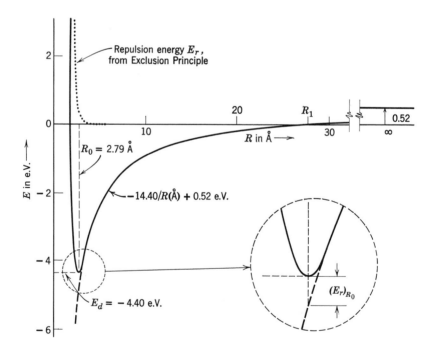

Fig. 7-2.   Total energy of K^+ and Cl^- as a function of the internuclear separation R. The dashed curve coincides with the solid curve for $R > {\sim}4$ Å.

charge with their centers a distance R apart is the same as Coulomb's law for point charges a distance R apart if the charge distributions do not overlap appreciably. Hence if we set $P = 0$ at $R = \infty$, the potential energy of the two ions is $-e^2/4\pi\epsilon_0 R$ joules; or, if R is expressed in angstroms, $P = -14.40/R$ e.V. If we add 0.52 e.V. to this P we retain the definition of $E = 0$ used in Fig. 7-1. The dashed curve of Fig. 7-2 is therefore drawn as the algebraic sum of the net energy required to form the ions and the energy of the electrostatic attraction (Coulomb energy).

Another energy term becomes important as soon as the closed shells of K^+ and Cl^- begin to overlap appreciably, and this is a repulsive interaction energy E_r. This energy arises primarily from the Exclusion Principle. Since all available $3s$ and $3p$ states are filled in both ions, electrons are packed as closely together as the Exclusion Principle permits. If these two ions approach so closely that the $n = 3$ shells begin to overlap, some of the electrons must go to higher quantum states such as the $3d$ or $4s$. If they did not, we should have more than one electron per quantum state. Raising one or more electrons to these higher unoccupied states raises the energy of the system as a whole. The Exclusion Principle interaction thus produces a repulsive energy, which rises very rapidly as the ions begin to interpenetrate. This energy E_r is plotted as the dotted line on Fig. 7-2.

The argument of the preceding paragraph might make it appear that the dotted line in Fig. 7-2 should be a discontinuous function, corresponding to first one, then two, etc., electrons raised to higher states. But the wave functions of the individual ions are strongly distorted as the ions approach, and the quantum states no longer have the pure form of the states of the individual ions. A $3p$ quantum state in K^+, for instance, has some of the properties (extension in space, angular dependence, energy) of the $4s$ state and other higher states mixed into the actual wave function when the ions start to interact. As the interpenetration becomes greater, the state becomes more like the higher states, and the energy continuously increases as R decreases.

An alternative explanation of the repulsive energy from the Exclusion Principle can be based on the alternative form of this principle given in eq. 6-11. Interpenetration of the ions requires more electrons to be put into the same region $\Delta x \, \Delta y \, \Delta z$ of space (the region of overlap). Therefore some of these electrons must go into other regions $\Delta p_x \, \Delta p_y \, \Delta p_z$ of momentum. Since all the lowest-momentum regions are filled, this necessitates raising the momentum and therefore the energy of some of the electrons. The energy E of the system thus increases rapidly as the region of overlap becomes larger. This region grows rapidly in volume as R decreases after the ions have touched (see Fig. 7-3).

The repulsion arising from the Exclusion Principle is the principal repulsion in all molecules except the lightest (e.g., H_2). Repulsive energy also arises from the electrostatic repulsion of the nuclei, but this is a small contribution to the total energy except in very light molecules.

It is difficult to calculate the precise shape of the repulsive energy curve. It is clear, however, from the fact that E_r rises *rapidly* as R decreases that the sum of this energy and the attractive energy described above (dashed line in Fig. 7-2) must have a *minimum*. Thus there is a minimum value E_d for the total energy E of the system at the value of R labeled R_0. (At this value of R very little interpenetration of the ions has occurred, and therefore our calculation of the electrostatic attraction energy is not seriously in error.) The stable position of this system is therefore a K^+ and a Cl^- ion bound together with their nuclei separated by a distance R_0. To dissociate a KCl molecule into K and Cl atoms, energy $|E_d|$, the dissociation energy, must be supplied.

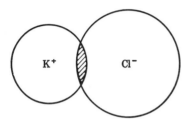

Fig. 7-3. The region of overlap of K^+ and Cl^- ions, approximated by sharply defined radii.

The arguments presented above do not prove that the KCl molecule actually is an *ionic* molecule, composed of K^+ and Cl^- ions. They show only that such a molecule is possible. To prove that KCl is ionic, we should have to investigate the binding energy on the assumption of the *covalent* form of binding to be discussed in the next two sections, and to show that the minimum energy is smaller ($|E_d|$ larger) for ionic than for covalent binding. We would then know that the actual KCl molecule would be ionic, since the system found in nature will always be in the lowest possible energy state (provided, of course, that the next higher state is many kT above the lowest state).

There is adequate experimental proof that molecules like KCl are ionic. The most direct evidence comes from the *dielectric constant* or *relative electric permittivity* $\kappa_e = \epsilon/\epsilon_0$ of gases of such molecules. In an electric field, the electric dipole (a $+e$ and a $-e$ charge separated by a distance R_0) of the molecule experiences a torque tending to align it with the field. The alignment is far from complete at ordinary fields and temperatures because of the disorienting effects of collisions between the molecules, which have an average thermal energy equal to $\frac{3}{2}kT$. Although the dielectric constants of all gases are very nearly unity, the differences $\kappa_e - 1$ are much greater for ionic molecules than for others.

We can learn a little more about the repulsive energy E_r by using observed values of $|E_d|$ and R_0. The dissociation energy can be determined by physical-chemistry experiments in which the fraction of KCl ions that dissociate is measured as a function of temperature. This fraction is proportional to

$$e^{-|E_d|/kT} \tag{7-1}$$

from the Boltzmann factor (eq. 6-10). E_d can also be determined from molecular-spectra experiments to be described in Sec. 7-6. R_0 can be determined from these experiments and from electron diffraction experiments in gases (the latter experiments were described in Sec. 4-10).

For KCl, R_0 is 2.79 Å (from electron diffraction) and E_d is -4.40 e.V. (from molecular spectroscopy). From the definitions given in Fig. 7-2 we can write (with R in angstroms):

$$E_d = (E_r)_{R_0} - 14.40/R_0 + 0.52 \text{ e.V.} \tag{7-2}$$

It is possible to estimate E_r quantum mechanically, but it is more accurate and more revealing here to use the measured values of E_d and R_0 to determine $(E_r)_{R_0}$, the value of E_r at $R = R_0$:

$$(E_r)_{R_0} = -4.40 + 5.16 - 0.52 = 0.24 \text{ e.V.}$$

The repulsive energy is thus a small part of the total energy at R_0, and we could have obtained a good approximation to E_d by neglecting it. This does not mean, of course, that E_r is not important, since in the absence of this part of our theory there would be no explanation of why K^+ and Cl^- do not go closer together than the observed separation. In fact, it can be proved that if Newton's laws and the laws of classical electrostatics are obeyed, *no* array of charges can be in stable static equilibrium (*Earnshaw's theorem*). Thus there is no completely classical theory of ionic molecules.

Since E has a minimum at $R = R_0$, dE/dR evaluated at $R = R_0$ must be zero. Therefore

$$\left(\frac{dE_r}{dR}\right)_{R=R_0} = \frac{-14.40}{R_0{}^2} \text{ e.V. per Å} \tag{7-3}$$

Thus the magnitude of the rate of change of E_r is equal to that of the rate of change of the electrostatic energy. But the magnitude of E_r (0.24 e.V.) is much less than the magnitude of the electrostatic energy (5.16 e.V.). Therefore the *fractional* change dE_r/E_r in E_r is much greater than the fractional change in the electrostatic energy. This confirms our prediction that the repulsion by the Exclusion Principle should give a rapid variation of E_r with R, a prediction that is quantitatively confirmed in probs. 7-4 and 7-5.

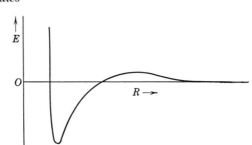

Fig. 7-4. Schematic plot of the energy of two atoms as a function of the internuclear separation.

There are two minor contributions to $|E_d|$ that have been neglected. The first is the *zero-point energy*, which will be discussed in Sec. 7-6. It decreases $|E_d|$ by a few tenths of an electron volt. The second is the *van der Waals attractive energy*, which increases $|E_d|$ by a few tenths of an electron volt. The van der Waals attraction originates from the distortion of the charge distributions of the ions when each is subject to the field of the other. Each ion develops a small induced dipole moment, and the van der Waals attraction is the attraction between these two dipoles. Although it is a minor effect in the present problem, it is the only type of binding energy possible in such substances as the solid noble gases. It is a kind of "last resort" binding that holds atoms together weakly if no other binding is possible.

All alkali halide molecules (e.g., NaCl, LiF, LiBr) are ionic. Alkaline-earth oxides (e.g., MgO, BaO), sulfides, and other compounds between elements of group II and elements of group VI of the periodic table are also ionic; in such molecules, the net "cost" of producing ions is much more than the 0.52 e.V. of the KCl example, but the electrostatic attraction energy is also greater because the ions are doubly charged. Most other inorganic molecules have partially ionic, partially covalent binding.

The calculation of the binding energy of KCl was carried out in four steps, but it must not be thought that a KCl molecule is formed from K and Cl atoms by a process consisting of these steps. Molecules are generally formed between atoms that are present on the surfaces of solids or between ions in solutions, and in either event the situation is complicated. In most reactions there is an energy "hump" that must be surmounted before the atoms can form a molecule, as illustrated in Fig. 7-4. The reaction rate is then the rate at which the constituent atoms acquire enough energy to pass over this hump to the low-energy state at smaller R. In KCl, for example, some energy is required to transfer an electron from K to Cl. Once the atoms have been brought closer together than the distance R_1 (Fig. 7-2), the stable form is the ions, but at larger distances the stable form is the neutral atoms. The energy to get the atoms "over the hump" may be supplied thermally; therefore the rate of most chemical reactions increases rapidly as the temperature is raised.

7-3 THE HYDROGEN MOLECULE ION AND RESONANCE

The *nature* of the binding in the hydrogen molecule ion H_2^+ is the same as in the hydrogen molecule, H_2, and therefore the consideration of H_2^+ makes a convenient transition to the more interesting problem of H_2. As in the discussion of the KCl molecule, we seek the binding energy or dissociation energy $|E_d|$. We shall show that the energy of the system is lower for the bound molecule H_2^+ than the sum of the energies of the constituents (an H^+ ion and an H atom).

The potential energy as a function of distance along the line joining two protons is shown in Fig. 7-5. P is the sum of the energies produced by two point charges situated at $x = \pm R/2$. The curve presented in Fig. 7-5 therefore has the form

$$P(x) = \frac{-e^2}{4\pi\epsilon_0}\left(\frac{1}{|x - \frac{1}{2}R|} + \frac{1}{|x + \frac{1}{2}R|}\right) \tag{7-4}$$

Of course this is a three-dimensional problem, and P depends on all three coordinates.

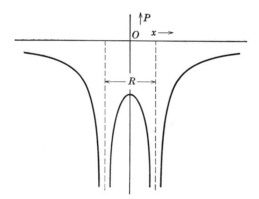

Fig. 7-5. Potential energy of an electron as a function of x along a line between two protons a distance R apart.

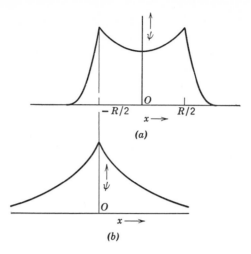

Fig. 7-6.   (*a*) Wave function of H_2^+ as a function of x along a line between the two nuclei.   (*b*) Wave function of H atom plotted to same scale as (*a*).

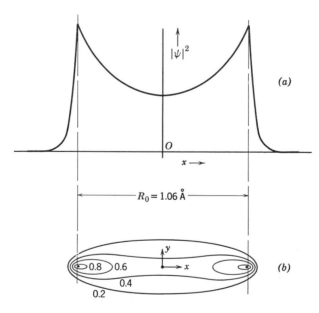

Fig. 7-7.   Electron density distribution for H_2^+ from calculations by O. Burrau. (*a*) $|\psi|^2$ along OX axis.   (*b*) Contour map of $|\psi|^2$ in x,y plane; the numbers give the relative values of $|\psi|^2$ on the various contours.

The Schrödinger equation for the motion of an electron in this potential energy $P(x, y, z)$ can be solved without the use of approximations other than the approximation that the proton motion is so sluggish compared to the electron motion that we can regard the protons as fixed. The proton-proton spacing R is thus treated as a parameter, as in the KCl example.

A plot of the x dependence of the wave function with lowest energy is given in Fig. 7-6a. A similar plot for the hydrogen *atom* is given in Fig. 7-6b. Note that ψ for H_2^+ is similar to the sum of two hydrogen atom ψ's. But ψ for H_2^+ has a larger magnitude near $x = 0$ and therefore, of course, a smaller magnitude for large $|x|$ values. This change is even more striking in Fig. 7-7, where $|\psi|^2$ for H_2^+ is plotted as a function of x and where a contour map of $|\psi|^2$ is plotted in the x,y plane. There is clearly a much greater probability of finding the electron in the region between the two nuclei than at comparable distances from a nucleus but "outside the nuclei" (i.e., where $|x| > \frac{1}{2}R$). The region between the two nuclei is a region of especially low P (see Fig. 7-5), and therefore the electron in H_2^+ on the average has a considerably lower potential energy than in H. It is this *decrease in potential energy* that is primarily responsible for the binding in H_2^+, H_2, and other molecules with covalent binding.

In addition to the reduction in potential energy the electron's kinetic energy is also changed from its value in the H atom. Inspection of Fig. 7-6 shows, however, that this change is not very great. We first recall that the kinetic energy is proportional to $|d\psi/dx|^2$ for one-dimensional problems (eq. 5-49) and note that a similar proportionality holds in three dimensions. Now in Fig. 7-6a the average value of $|d\psi/dx|^2$ is not much different from its average for the H atom of Fig. 7-6b. There is, therefore, little change in the kinetic energy K when an H^+ ion and an H atom combine to give H_2^+. Actually K is somewhat lower in the bound state, and this fact contributes to the binding energy.

Another contribution to the energy of the H_2^+ ion is the repulsive energy of the two protons, which is, of course, simply

$$E_r = \frac{e^2}{4\pi\epsilon_0 R} \text{ joules} = \frac{14.40}{R(\text{Å})} \text{ e.V.} \tag{7-5}$$

In Fig. 7-8 the two contributions to the energy have been plotted as functions of R. The dashed line is the decrease in E of the electron that originates largely from the lower potential energy. The dotted line is the repulsion energy from eq. 7-5. A stable molecule is predicted since the sum of the two energies has a minimum. The predicted values

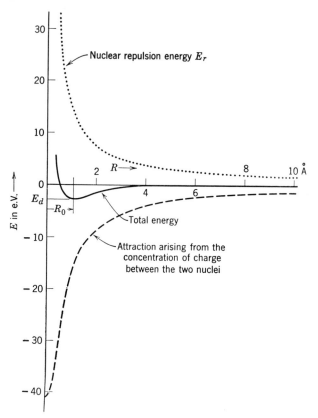

Fig. 7-8. Contributions to the energy of H_2^+ as functions of R. The zero of E is defined by setting $E = 0$ for an H^+ ion and an H atom (in its ground state) at infinite separation ($R = \infty$).

of R_0 (1.06 Å) and E_d (-2.65 e.V.) are in good agreement with values from molecular spectroscopy.

We have shown that the ψ calculated from the Schrödinger equation predicts binding for H_2^+. But why is this particular ψ the solution? Or, to phrase the question a little differently, what is the reason that $|\psi|^2$ is concentrated between the atoms? The most accurate answer to the question is the answer already given: This ψ is the solution of the Schrödinger equation, and adequate proof of the validity of this equation has already been supplied. We shall describe briefly, however, two less abstract answers.

The first is a semiclassical approach in terms of *resonance*. We could show that the electron in the H_2^+ ion *resonates* back and forth between the state with a high probability of finding the electron near nucleus A and the state with a high probability of finding it near nucleus B. We could actually develop a wave-packet description of the electron and show that (on the average) the wave packet moves back and forth between the two nuclei. There must be a large probability, therefore, of finding the electron between the two nuclei, and the concentration of charge in this region leads to binding.

The second is an accurate quantum approach. It was pointed out in Sec. 5-7 that the ground-state wave function for any system can be calculated by adjusting ψ as a function of the coordinates in order *to minimize the total energy.* Thus any trial ψ becomes closer to the actual ground-state ψ if it is modified in such a way that the E calculated from it decreases. In the present problem, one trial description of the ground state is the electron in a hydrogen atom $1s$ wave function ψ_A centered on nucleus A. Another is the electron in a hydrogen atom $1s$ wave function ψ_B centered on nucleus B. Since the two nuclei are identical, the energy of the system in the state with ψ_A is the same as in that with ψ_B. These solutions for hydrogen atoms are, of course, the ψ's that make E a minimum for isolated atoms. We should not expect that they would make E a minimum for the $H_2{}^+$ ion. If we can distort ψ_A or ψ_B or some combination of them in such a way as to decrease E, the distorted wave function will be closer to the correct ground-state wave function for $H_2{}^+$. We have already seen that the sum of the average potential energy and the average kinetic energy can be reduced if ψ is concentrated in the region between the two nuclei. The correct wave function plotted in Fig. 7-6a is precisely the distortion of $\psi_A + \psi_B$ that produces this result. The molecule is bound because the energy E corresponding to this wave function is less than the energy of the isolated H atom and H^+ ion, and because this binding contribution to the energy is greater in magnitude than the nuclear repulsion contribution.

The one-electron bond is very weak unless the two nuclei are identical. If they are not identical, either ψ_A or ψ_B will give a lower E for the system. Let us suppose that ψ_A gives the lower energy. In semiclassical terms, we say that the electron spends most of the time in the region of nucleus A, the resonance motion of the electron between the two nuclei is less frequent, and the bond is weak. In accurate quantum terms, we again describe this situation by a distortion of a combination of the wave functions ψ_A and ψ_B. Since ψ_A gives the lower energy, this combination is not $\psi_A + \psi_B$ but $a\psi_A + b\psi_B$, where $a \neq b$. Making $a \gg b$ "takes advantage" of the lower energy that can be obtained if the electron is in the ψ_A state and produces a lower energy for the system than would be obtained if $a = b$. Distortion of this wave function produces very little lowering of the energy, since the region between the two nuclei is not much more "attractive" than other regions at the same distance from nucleus A. Therefore very little distortion occurs, and the one-electron bond is weak if one nucleus has an appreciably stronger attraction than the other.

7-4 THE HYDROGEN MOLECULE AND COVALENT BINDING

The hydrogen molecule problem is very similar to the H_2^+ problem, but the hydrogen molecule has two electrons. The H_2 problem can be solved by means of approximations. The wave function illustrated in Figs. 7-6 and 7-7 is a good starting point for an approximate solution of this problem. We can put one electron with spin $+\frac{1}{2}$ into this quantum state and one with spin $-\frac{1}{2}$ without violating the Exclusion Principle. The contribution of each electron to the decrease in potential energy is a curve like the dashed line of Fig. 7-8, and therefore this effect produces a strong binding. But there is an additional repulsion energy, namely, the electrostatic repulsion between the electrons. Because of this energy (which was, of course, not included in the H_2^+ problem) the wave function for H_2^+ is not correct for H_2. The actual wave function keeps the two electrons well separated, on the average. For this reason the $|\psi|^2$ contours are somewhat fatter near $x = 0$ than those plotted in Fig. 7-7b. This electrostatic repulsion energy is not so large as the attractive energy resulting from putting two electrons (on the average) into the low-potential-energy region. Thus binding occurs, and the *nature* of this binding in H_2 is just the same as in H_2^+. The internuclear separation R_0 and dissociation energy $|E_d|$ can be calculated to as high a precision as desired. The theoretical values agree satisfactorily with the values 0.74 Å and 4.48 e.V. from molecular-spectra experiments.

This type of binding is called *covalent* or *electron-pair* binding. It is responsible for the binding in all organic and many inorganic molecules. Even an ionic molecule like MgO has some binding of this type superimposed on the ionic binding.

One feature of covalent binding is that the bond is strongest when exactly two electrons participate. We have already seen that the hydrogen bond ($E_d = -4.48$ e.V.) with two electrons is stronger than the hydrogen-ion bond ($E_d = -2.65$ e.V.) with only one electron. If three electrons participate the bond becomes much weaker, because only two electrons can be put into the wave function of lowest energy. Therefore HeH does not exist as a molecule, for example. The Exclusion Principle compels us to put the third electron into another quantum state, which has a much higher energy. The repulsion arising from the increased energy of the system from this cause dominates over the attraction described above, and there is no net binding energy. For some combinations of atoms the three-electron bond is stable, but it is always weaker than the electron-pair bond.

The binding between dissimilar atoms by the electron-pair bond is as strong as that between similar atoms. This is not true for the one-electron bond, as was noted in connection with H_2^+. When only one electron is present between dissimilar atoms it spends most of its time in whichever atom gives it the lowest energy level. It therefore is only rarely in the low-potential-energy region between the atoms. When two electrons are present, this behavior cannot occur, since if both electrons were on the same atom their mutual repulsion would raise the energy considerably. (An exception to this statement occurs when one atom has an electron affinity, but this is the ionic binding already discussed in Sec. 7-2.)

Another way of explaining the strength of the electron-pair bond between dissimilar atoms is to note that the resonance in the electron-pair bond does *not* require identical energy levels in the two atoms being joined. The resonance is between the state with electron 1 near nucleus

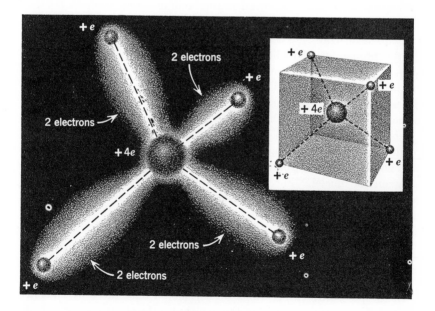

Fig. 7-9. The geometry of the methane molecule. The hydrogen nuclei are at the corners of a regular tetrahedron, or at 4 of the 8 corners of a cube. The electrons are concentrated along the carbon-hydrogen lines.

A, electron 2 near nucleus B and the state with electron 1 near nucleus B, electron 2 near nucleus A. The energies of these two states are identical regardless of the energy levels of A and B.

Another feature of covalent binding is that the electron-charge density around each nucleus is not at all spherically symmetrical. An example of this fact is illustrated in Fig. 7-9 for the methane molecule, CH_4. Carbon has 4 *valence* (outer-shell) electrons, and each hydrogen atom has 1. The 8 electrons are concentrated chiefly along the lines joining the protons to the center of the C atom: the electron density along and near these lines is very similar to that plotted in Fig. 7-7. (Of course Fig. 7-9 should not be viewed as a literal picture of the electron density, since $|\psi|^2$ is actually zero only at a few special planes in this figure.) The *tetrahedral* arrangement of the hydrogen nuclei shown in Fig. 7-9 minimizes the mutual repulsion of electrons. If the 4 hydrogen nuclei were in one plane, for example, the electron-electron repulsion would be larger than in this tetrahedral arrangement.

In the hydrogen molecule the nuclei do not go closer together because of their electrostatic repulsion. This is a different situation from the closed-shell, Exclusion-Principle repulsion described in the discussion of the ionic bond. Hydrogen is a rather special case, since the only electrons present are those involved in the binding. The covalent bond in the Na_2 molecule is more typical. Here there are two electrons outside the two neon cores. These electrons constitute the bond (there is a greater probability of finding them in the region between the two cores than far away from this region). The resistance to interpenetration of the neon cores (Exclusion Principle) is the chief reason the sodium atoms do not go closer together. The Exclusion-Principle repulsion is the principal repulsive effect in all molecules except H_2.

This section and Sec. 7-3 may be summarized as follows: Covalent binding occurs if the ground state of the molecule has a wave function that concentrates electronic charge in the region of low potential energy between the nuclei. For a one-electron bond, this concentration by resonance occurs only if the two atoms are identical. For a two-electron bond, this concentration occurs whether or not the two atoms are identical.

7-5   ELECTRON ENERGY LEVELS IN MOLECULES

In the preceding two sections we considered only the lowest energy level of the system composed of the nuclei plus the electrons. We therefore considered only the lowest energy level of the electrons involved in the bonds. (All the inner-shell electrons were in their lowest energy

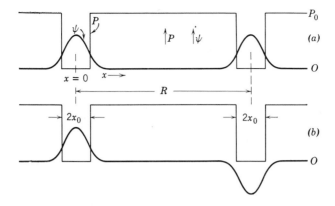

Fig. 7-10. Wave functions for two square wells superimposed on $P(x)$ plots. When R is large, the two types of ψ give nearly the same E.

levels before the atoms were brought together and remained in these levels, which are not appreciably affected by the association of two atoms.) To see how the higher levels are affected by the binding together of atoms into molecules it will be worth while first to consider a square-well problem. Study of this problem will show that an energy level of an atom is split into a pair of levels when this atom is bound to another in a molecule. In addition, the concept of resonance is illustrated by this model.

Two identical, one-dimensional square-well potentials are illustrated in Fig. 7-10. The wave function ψ for the lowest quantum state is superimposed on the same plot; this is the solution of the Schrödinger equation for one electron moving in the potential energy $P(x)$ that is plotted. In Fig. 7-10 the wells are separated enough so that ψ becomes practically zero between them. The solution in this case is just like the ψ of Fig. 5-5b in each well, except that normalization reduces the magnitude $|\psi|^2$ by a factor of 2; ψ is plotted in Fig. 7-10a.

Another solution of the Schrödinger equation, plotted in Fig. 7-10b, is possible for this problem. The only difference between this solution and the ψ of Fig. 7-10a is the different sign of ψ in one of the wells. This change in sign means that the wave function Ψ (including the time dependence) in one well is 180° out of phase with respect to the Ψ in the other well. There is practically no difference in the energy E for the two solutions (a) and (b), which is apparent from the fact that the shapes of $\psi(x)$ are the same, and therefore the average kinetic energy ($\sim |d\psi/dx|^2$) and average potential energy are the same for the solutions. It is also apparent from the fact that either wave function (a) or (b) gives almost

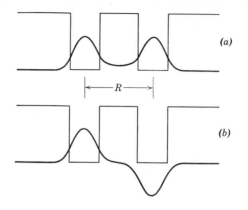

Fig. 7-11. Wave functions for two square wells with a smaller separation R.

exactly the energy for a single square well. The wave function (a) is said to be *symmetric,* and the wave function (b) is said to be *antisymmetric.*

In Fig. 7-11 the wells have been brought closer together, and now a significant change in shape is developing. This change becomes even more obvious in Fig. 7-12. It should be apparent that the ψ of (a) gives the lower E, since the average P is about the same as in (b) but the average K is lower because the average $|d\psi/dx|^2$ is smaller. The limiting case of small separation of wells is illustrated in Fig. 7-13. Here the wells just touch, and the barrier between them has zero width. We

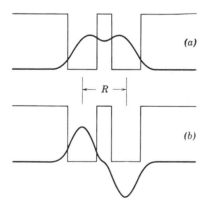

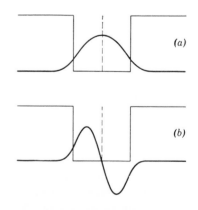

Fig. 7-12. Wave functions for the case where R is so small that the barrier separating the wells has almost disappeared.

Fig. 7-13. Wave functions for the limiting case where the barrier has just disappeared. R equals the width of one of the original wells.

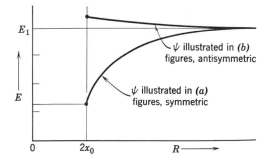

Fig. 7-14. Energy as a function of R for symmetric and antisymmetric states of two square wells.

recognize ψ of (a) as the $n = 1$ wave function for a well of width $4x_0$. Since from eq. 5-26 $E \sim n^2/(\text{width})^2$, this E is only $\sim \frac{1}{4}$ of the E for Fig. 7-10. The ψ of Fig. 7-13b is also recognizable as the $n = 2$ wave function for a well of width $4x_0$. The E associated with this ψ is about the same as the E for the ψ of Fig. 7-10, since n and the width have both increased by a factor of 2 (it would be exactly the same if the square well had infinite sides).

In Fig. 7-14 the energies of the two kinds of wave functions are plotted as functions of the distance R between the centers of the wells. They both start from E_1 at $R = \infty$. E_1 is the energy of an electron in the $n = 1$ state of a square well with finite sides. It is approximately $h^2/32mx_0^2$ by eq. 5-26, which applies as an approximation here but applies accurately only to the square well with infinite sides. The limiting values when R has decreased to a value ($R = 2x_0$) such that the barrier disappears were estimated as about E_1 and $E_1/4$ in the preceding paragraph. The important features of Fig. 7-14 are that for any R *the level E_1 for a single well is split into two levels* and that *the splitting becomes greater as the separation of the wells decreases.*

We should note parenthetically that the square-well wave functions of Figs. 7-11a and 7-12a also illustrate the concept of resonance. The electron described by one of these wave functions is moving back and forth between the two wells, and there is a relatively large $|\psi|^2$ between the wells. Thus more charge is concentrated in the region between the wells than at the same distance from an isolated well, and this is the phenomenon called resonance in Sec. 7-3. The resonance concentration of charge does not lower the potential energy of the square-well system (unlike the H_2^+ ion), since the region between the wells is a region of *large P*, but it does lower the kinetic energy. The juxtaposition of the two wells gives a larger region for the electron's motion, and therefore

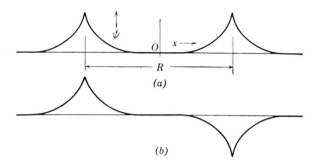

Fig. 7-15. Symmetric and antisymmetric ψ's for the $H_2{}^+$ ion with a large internuclear separation.

the kinetic energy decreases for the same reason it decreases in a square well when the sides are spread apart.

Next we consider the energy levels of the $H_2{}^+$ ion, for which the $P(x)$ curve was plotted in Fig. 7-5. The ψ's plotted in Fig. 7-15 are the ψ's for the hydrogen atom, and they are nearly correct for the two-nucleus problem when R is very large. As the two protons are brought closer together, the ψ's of Fig. 7-16 result. This figure is drawn for the observed value $R = 1.06$ Å for $H_2{}^+$; curve (a) of Fig. 7-16 is the same ψ that was plotted in Fig. 7-6a. It is useful to plot also the extreme case $R = 0$ as an aid in sketching the energies as a function of R; this is illustrated in Fig. 7-17. Since $R = 0$, the two protons coincide to give

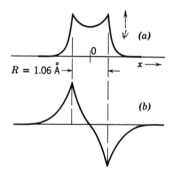

Fig. 7-16. Symmetric and antisymmetric ψ's for the $H_2{}^+$ ion for $R = 1.06$ Å, the observed internuclear separation.

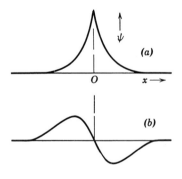

Fig. 7-17. Symmetric and antisymmetric ψ's for the limiting case $R = 0$. The ψ's are the $1s$ and $2p$ ψ's of the He^+ ion.

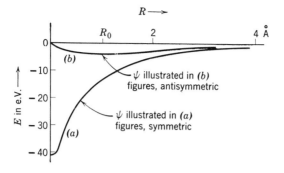

Fig. 7-18. Energy as a function of R for the symmetric and antisymmetric states of the H_2^+ ion. The zero of E is defined by setting $E = 0$ for an H^+ ion and an H atom (in its ground state) at infinite separation ($R = \infty$).

essentially the problem of the helium ion He^+ ($Z = 2$). The (b) wave function has developed continuously into the $2p$ wave function of He^+. This is the lowest-energy wave function that has the characteristic $+$ and $-$ feature of the antisymmetric (b) curves of Figs. 7-15 and 7-16. (See Fig. 6-3a.) Its energy happens to be the same as that of the hydrogen atom ground state (see eq. 6-12, with $n = 2$ and $Z = 2$). The symmetric wave function (a) is the $1s$ wave function of He^+, which gives an energy a factor of 4 lower than the ground state of hydrogen (see again eq. 6-12, but with $n = 1$ and $Z = 2$).

In Fig. 7-18 the energies of the two types of quantum states are plotted as functions of R. The splitting of one level into two is again apparent.

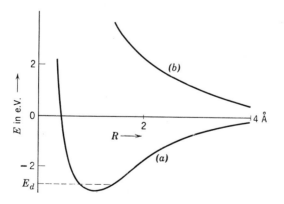

Fig. 7-19. Energy as a function of R for the symmetric and antisymmetric states of the H_2^+ ion, including the nuclear-repulsion energy. E_d is slightly above the minimum in the curve because of the zero-point energy to be discussed in Sec. 7-6.

Curve (*a*) appeared before in Fig. 7-8. As in Fig. 7-8, we let $E = 0$ represent the energy of an H atom and an H^+ ion at $R = \infty$. We repeat the sum of curve (*a*) and the nuclear-repulsion energy in curve (*a*) of Fig. 7-19. The similar sum of curve (*b*) and the nuclear repulsion is also plotted. The (*b*) curve has no minimum, and therefore a molecule with the electron in this state would be unstable.

The energy levels in the hydrogen molecule are similar to those in the H_2^+ ion. The 1*s* state of the hydrogen atom is split into two states, and the energy difference between these states increases as the internuclear separation decreases. The total energy E of the molecule consists of the energies of the electrons and the (repulsion) energy of the two protons. Curves of E as a function of R are given in Fig. 7-20. The lower curve is for electrons distributed in space as explained in Sec. 7-4 in connection with the binding energy of the H_2 molecule. The $|\psi|^2$ for this state has large values in the region between the protons and is roughly the same as that illustrated in Fig. 7-7 for the H_2^+ ion; the ψ for this state is similar to the ψ of Fig. 7-16*a*. The upper curve in Fig. 7-20 is for electrons distributed so that ψ and $|\psi|^2$ are zero midway between the two protons; ψ for this state is quite similar to Fig. 7-16*b*. If the electrons are in this state, a stable molecule is not formed since there is no minimum in the upper curve of Fig. 7-20. In this state there is no concentration of charge in the low-potential-energy region and no binding.

Before the two atoms composing the H_2 molecule were brought together there was a total of four 1*s* states, all with practically the same energy. There were only two electrons, but four could have been accommodated. After the atoms are brought together these 1*s* states are

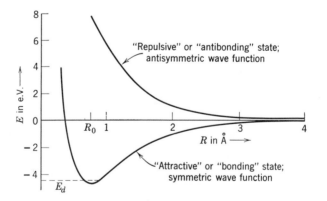

Fig. 7-20. The total energy of the H_2 molecule as a function of the internuclear separation R.

greatly modified and are split into two types of states, symmetric and antisymmetric. Two electrons can be accommodated in the lower state, one with plus spin and one with minus spin. Any additional electrons would have to go into the upper state. We shall see an important consequence of this feature of covalent binding in the properties of silicon and germanium crystals, which will be studied in Chapter 11.

The preceding discussion of the splitting of atomic energy levels when atoms are bound into molecules was developed for the H_2 molecule, but this same effect occurs for *all* molecules and for the same reasons. Splitting occurs for the inner-shell electrons as well as for the outer, but the magnitude of the splitting is much less than for the outer (valence) electrons. The relatively small splitting occurs because the ψ functions for the inner electrons do not extend very far from the nuclei. At internuclear separations R such that the valence electron ψ's look like Fig. 7-16, the inner shell ψ's are still well separated, as in Fig. 7-15.

The splitting of energy levels in molecules is quite analogous to the splitting of the resonant frequencies of two resonant circuits when they are coupled together. The analogy is apparent if we compare (a) the Schrödinger equation in one dimension for $P = 0$ (eq. 5-22) with (b) the differential equation for the current I in an L-C circuit without resistance, written in terms of the natural frequency of oscillation $\nu_0 = 1/(2\pi\sqrt{LC})$:

$$(a) \quad \frac{d^2\psi}{dx^2} + \frac{8\pi^2 mE}{h^2}\psi = 0; \qquad (b) \quad \frac{d^2 I}{dt^2} + 4\pi^2\nu_0^2 I = 0 \qquad (7\text{-}6)$$

The solutions $\psi(x)$ must therefore be analogous to the solutions $I(t)$, and the analogue of the frequency is the square root of the energy.

For two identical resonant circuits well separated (Fig. 7-21a), equations like eq. 7-6b apply to each, and the resonant frequencies ν_0 can be computed. As the circuits become coupled, through a mutual inductance M between the two inductors, for example, eq. 7-6b becomes more complicated and the resonant frequency ν_0 is split into two frequencies. If the circuit is excited at one of these frequencies, the currents in the two parts are in phase; if it is excited at the other, they are out of phase. This result is analogous to the signs of ψ in the two wells of Fig. 7-11. If the electron is in the state with the lower energy E (Fig. 7-11a), the ψ's are in phase in the two wells; if the electron is in the other state, they are out of phase.

7-6 MOLECULAR SPECTRA AND DISSOCIATION

All the binding-energy curves of Figs. 7-2, 7-8, 7-19, and 7-20 have the characteristic shape of the solid line in Fig. 7-22. Stable equilibrium

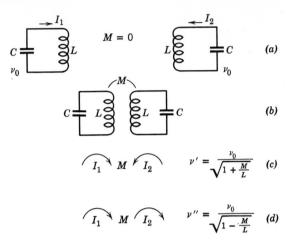

Fig. 7-21. Coupled-circuit analogy to energy-level splitting in molecules. The coupled circuits have two resonant frequencies. The splitting is greater for tighter coupling (larger M/L).

occurs with the nuclei separated by a distance R_0. If the nuclei are separated by a slightly greater or smaller distance R, the energy E is raised, and there is effectively a restoring force. The electrons follow

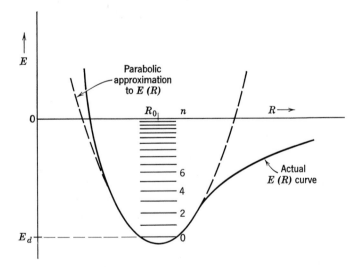

Fig. 7-22. Vibrational energy levels of a diatomic molecule. The depth of the well is $|E_d| + h\nu_0/2$. The number of levels is typical for molecules containing hydrogen; many more levels occur in heavier molecules.

the nuclei as R varies, since the electron velocities are much higher than the nuclear velocities.

The dashed curve in Fig. 7-22 is a parabola fitted to the $E(R)$ curve. It is a good approximation for small vibrations about the equilibrium position R_0. As explained in Sec. 5-5, small-amplitude vibrations about a position of stable equilibrium always lead to a force like eq. 5-37. In terms of the parameters of Fig. 7-22, the force is

$$F = -\left(\frac{d^2E}{dR^2}\right)_{R=R_0} (R - R_0) \tag{7-7}$$

The classical frequency of vibration ν_0 is

$$\nu_0 = \frac{1}{2\pi\mu^{1/2}} \left[\left(\frac{d^2E}{dR^2}\right)_{R=R_0}\right]^{1/2} \tag{7-8}$$

Here μ is the *reduced mass* of the vibrating system

$$\mu = \frac{M_1 M_2}{M_1 + M_2} \tag{7-9}$$

The use of the reduced mass reduces the problem of two particles M_1 and M_2 vibrating about their common center of mass to the problem of a single mass μ vibrating about a fixed point. (See prob. 7-22; the reduced-mass concept is applicable to both classical and quantum mechanics.)

The energy levels for a problem with a force like eq. 7-7 were given in Sec. 5-5:

$$E_n = (n + \tfrac{1}{2})h\nu_0 \tag{5-43}$$

This energy E_n is measured relative to $E = 0$ for the system at rest at the equilibrium position $R = R_0$, and ν_0 is the classical oscillator frequency, as computed in eq. 7-8. These levels are sketched on Fig. 7-22. For *large*-amplitude vibrations, the parabola does not fit the actual energy curve and the force from eq. 7-7 is not correct. Equation 5-43 no longer holds, and the upper energy levels therefore are not equally spaced. The energy well is wider than the parabola for the higher E_n's, and therefore these energy levels are closer together than the lower E_n's.

The ground state of the vibrating molecule is the state $n = 0$.* We learned in Chapter 5 that *all* systems have kinetic energy in the ground state, and of course eq. 5-43 predicts such a result for $n = 0$. Thus even at the absolute zero of temperature, the atoms are vibrating with kinetic energy $\tfrac{1}{2}h\nu_0$, which is therefore called the *zero-point energy*. (It should be noted that if there were no zero-point energy the internuclear separation R would be fixed, and the Indeterminacy Principle would be vio-

---

* The symbol n is used for the vibrational quantum number in this section; *electronic* quantum states will be referred to without specifying quantum numbers.

lated.) In the discussion of dissociation energies in Secs. 7-2, 7-3, and 7-4 we really should have included this energy, which reduces slightly the dissociation energy of the molecule (see prob. 7-25).

The vibrational energy levels can be experimentally determined for many molecules from absorption spectra. Most of the experiments involve transitions between electronic, as well as vibrational, energy levels, but there is a simpler situation that illustrates the way observations of spectra provide information about molecular binding, namely, the absorption spectra of *polar* molecules. Photons *cannot* be absorbed by transitions from one vibrational level to another of the same electronic state in a perfectly symmetrical molecule like H_2, as explained for symmetrical electronic distributions in Sec. 5-8 (there is no oscillation of charge, no coupling to the electromagnetic field). But in a polar molecule like HCl, light of the proper frequency *can* produce the transition from $n = 0$ to $n = 1$; most of the molecules are in the state $n = 0$ at room temperature, and the harmonic oscillator selection rule $\Delta n = \pm 1$ applies. The wavelength for this absorption in HCl, for example, is about 33,500 Å, which is in the infrared. Since the force law of eq. 7-7 is not exactly correct for any except infinitesimal-amplitude oscillations, the wave functions for the vibrations of the molecule are not exactly the simple-harmonic oscillator wave functions. They therefore do not have the symmetry necessary to make the transitions $n = 0$ to $n = 2$, $n = 0$ to $n = 3$, etc., *exactly* impossible. Absorption caused by these forbidden transitions can be measured even though they are much weaker than the allowed transition $n = 0$ to $n = 1$. All the energy levels, and thus the depths and shapes of binding energy curves, of such molecules can thereby be determined experimentally.

Dissociation of the molecule into its constituent atoms occurs whenever the vibrational energy becomes larger than $|E_d|$. Experimental determination of the energy levels such as those of Fig. 7-22 determines also the dissociation energy $|E_d|$. The maximum vibrational energy is $|E_d|$, and this maximum can be readily inferred from a series of levels that terminates at a particular quantum number ($n = 15$ in the example of Fig. 7-22).

Another form of energy of the molecule is the kinetic energy of *rotation*

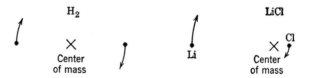

Fig. 7-23. Rotation of two typical diatomic molecules about their centers of mass.

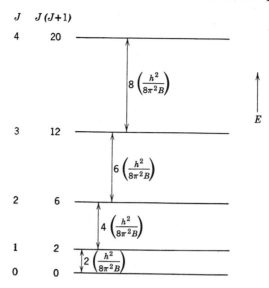

Fig. 7-24. Rotational energy levels of a diatomic molecule with moment of inertia B.

of the nuclei about their common center of mass, as sketched in Fig. 7-23. The Schrödinger equation can be solved for the rotation of a diatomic molecule and gives the following energy levels:

$$E_J = \frac{h^2}{8\pi^2 B} J(J+1) \qquad (7\text{-}10)$$

Here J is the rotational quantum number and takes on the values $0, 1, 2, \cdots$; the factor $J(J+1)$ originates from the angular dependence of the wave function in the same way that the factor $l(l+1)$ originated in the hydrogen atom. The constant B is the moment of inertia of the two nuclei about their common center of mass. In terms of the masses M_1 and M_2 and their internuclear separation R_0:

$$B = \frac{M_1 M_2}{M_1 + M_2} R_0{}^2 \qquad (7\text{-}11)$$

(Compare eqs. 7-11 and 7-9.)

The energy levels of a rotating molecule are plotted in Fig. 7-24. The selection rules for transitions are $\Delta J = \pm 1$. Thus in absorption or emission a set of lines with $h\nu$ values equal to $2(h^2/8\pi^2 B)$, $4(h^2/8\pi^2 B)$, etc., occur. The measurement of the frequencies of such lines permits the measurement of B and hence R_0 (see prob. 7-36). The rotational energies are very small, and therefore the lines of the *pure rotation spec-*

trum are in the far infrared or microwave regions of the electromagnetic-wave spectrum.

We have discussed the vibrational and rotational energies of molecules separately, but a molecule can have both forms of energy at the same time. Furthermore, the electrons can be in excited states. The total energy of a molecule is, therefore,

$$E = E_{\text{electronic}} + E_{\text{vibration}} + E_{\text{rotation}} \qquad (7\text{-}12)$$

The separation between the lowest state and the first excited state is about 2 to 10 e.V. for electronic energies, about 0.2 to 2 e.V. for vibration energies, and about 10^{-5} to 10^{-3} e.V. for rotation energies. These energies are illustrated schematically in Fig. 7-25.

Each transition involving a change in electron energy produces a whole series of emission or absorption lines, since many combinations of changes in vibration and rotation energy are possible. If such a system of lines is observed under low resolution conditions, it appears to be a *band* with practically a continuous distribution of frequencies. With higher resolution, the individual lines can be resolved and the energy differences

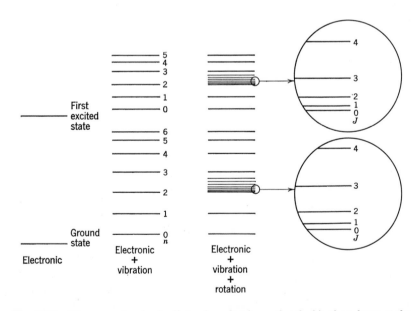

Fig. 7-25. The energy levels of a diatomic molecule are sketched in the column at the right. Every level of electronic+vibration energy has the rotation fine structure, but this structure is illustrated only for the $n = 2$ states.

measured.   In this way the energies of the vibrational states (eq. 5-43) and of the rotational states (eq. 7-10) can be measured.

It is apparent in Fig. 7-22 that in the higher vibration states the average internuclear distance R is larger because of the asymmetry of the binding energy curve.   This fact produces a higher moment of inertia B when the vibrational quantum number is large and a correspondingly smaller separation in the rotational energy levels.   Thus precision measurement of the rotational levels as a function of vibrational quantum number permits the study of the asymmetry of the binding energy curve; it is this asymmetry that is responsible for the thermal expansion of molecules.

The applications of molecular spectroscopy to the determination of R_0 and the binding energy curves for diatomic molecules have been described above.   The geometrical arrangement of atoms in polyatomic molecules can also be learned from such measurements, since the observed moments of inertia can be compared with theoretical moments based on trial structures such as linear or tetrahedral arrangements of atoms.   Molecular spectroscopy is also useful in analyzing mixtures of complex molecules, especially organic molecules, since the complete infrared absorption spectrum of a particular complex molecule is unlike that of any other molecule.   The force constants of the various bonds give rise to characteristic absorption frequencies, and the measurement of these frequencies identifies the molecule.   Although no two different molecules have the same absorption spectrum, there are important similarities.   For example, an OH group in any molecule gives rise to approximately the same absorption frequency.   The occurrence of an absorption peak at this frequency therefore reveals the presence of an OH group in a complex molecule.

REFERENCES

J. C. Slater, *Quantum Theory of Matter*, McGraw-Hill, New York, 1951, Chapters 8 and 9.

L. Pauling and E. B. Wilson, *Introduction to Quantum Mechanics*, McGraw-Hill, New York, 1935, Chapters 12 and 13.

F. O. Rice and E. Teller, *The Structure of Matter*, Science Editions, New York, 1961, Chapter 7.

L. Pauling, *The Nature of the Chemical Bond*, Cornell University Press, Ithaca, 2nd Ed., 1940.

G. Herzberg, *Molecular Spectra and Molecular Structure*, Van Nostrand, New York, 2nd Ed., 1950, Chapters 2, 3, and 4.

R. B. Leighton, *Principles of Modern Physics*, McGraw-Hill, New York, 1959, Chapter 9.

PROBLEMS

7-1. In all but the last section of this chapter we have assumed that the nuclei were fixed at a separation R. Show that the electron motion is so fast compared to that of the nuclei that such an assumption is justified; to do this, calculate the ratio of the velocity of an electron to that of a potassium ion with the same kinetic energy.

7-2. Prove the assertion in the third paragraph of Sec. 7-2 that the force between spherically symmetrical charge distributions whose centers are a distance R apart and that do not overlap is the same as the force between point charges (each equal to the total charge in one of the spheres) a distance R apart. *Hint:* Use of Gauss' law is much simpler than direct integration.

7-3.* Calculate the repulsive energy E_r at $R = R_0$ for NaCl by using the fact that $R_0 = 2.51$ Å and $E_d = -4.24$ e.V. Neglect zero-point and van der Waals energies.

7-4.* Assume a repulsive energy for the KCl molecule of the form $E_r = CR^{-n}$, where C and n are constants. Use the results of eqs. 7-2 and 7-3 and the value (0.24 e.V.) of E_r at $R = R_0$ to find n. Make an enlarged, labeled plot of the region of the binding energy curve near its minimum. [There is little significance to the precise value of n, since it is sensitive to the small energy terms (zero-point and van der Waals) that have been neglected. The significant fact is that $n \gg 1$, as explained below eq. 7-3.]

7-5. Find the exponent n of the repulsive energy of NaCl (as in prob. 7-4; use the result of prob. 7-3).

7-6.* Predict the binding energies of the two possible structures for the calcium oxide molecule, Ca^+O^- and $Ca^{++}O^{--}$, using the observed value $R_0 = 1.822$ Å. The electron affinity of the first electron added to the neutral oxygen atom is 2.2 volts, and of the second is about -9.0 volts. The first and second ionization potentials of calcium are 6.11 and 11.87 volts. Which structure do you predict will occur? (The observed dissociation energy is 5.9 e.V.; neglect the repulsive energy.)

7-7.* Show that penetration of the energy hump (Fig. 7-4) by tunneling is very unlikely because of the large mass of the *atom* involved. Do this by replacing the hump by a square barrier 1 e.V. high (i.e., $P - E = 1$ e.V.) and 2 Å wide, by using the mass of the potassium ion as the mass of the particle attempting to penetrate the barrier, and by calculating the tunneling probability as in prob. 5-28.

7-8. Draw a sketch like Fig. 7-6 for the wave function for a system consisting of a proton, a helium nucleus about 1 Å from the proton, and an electron. Do you expect the $(HeH)^{++}$ molecular ion to be a stable system?

7-9. Why is the intercept $R = 0$ of the dashed curve in Fig. 7-8 equal to -40.8 e.V.? *Hint:* By the definition of $E = 0$, this corresponds to a total energy of the electron moving in the field of two protons of $-13.6 - 40.8 = -54.4$ e.V.

7-10.* Calculate the potential energy (in electron volts) of the repulsion of two protons separated by $R = 1.06$ Å. Let $P = 0$ at $R = \infty$.

7-11.* Calculate the potential energy of an electron assumed fixed at the point $x = 0$ in Fig. 7-5. Let $R = 1.06$ Å.

7-12. Predict the geometrical arrangement of the ethane molecule, C_2H_6, by studying Fig. 7-9 and the associated discussion in the text. Draw a sketch of the three-dimensional structure of ethane.

7-13. The CO_2 molecule has no dipole moment. What must be the arrangement of the atoms? Discuss the type or types of binding.

7-14. Predict the geometrical arrangement of the carbon tetrachloride molecule (CCl_4). Draw a sketch of the three-dimensional structure. Are the carbon and chlorine atoms electrostatically neutral? Should this molecule have a dipole moment?

7-15. Sketch the electron distribution in the Na_2 molecule. Include all the electrons but pay particular attention to the electrons involved in the binding.

7-16. No electron spin resonance (Sec. 6-5) will be observed if all the electron spins in an atom or molecule are *paired*, i.e., if for each electron with $m_s = +\frac{1}{2}$ there is another with $m_s = -\frac{1}{2}$. Explain why unpaired spins are rare in molecules. (Since almost all atoms are commonly found in molecular form at ordinary temperatures, electron spin resonance is generally confined to systems like free radicals.)

7-17. In the square-well problem illustrated in Fig. 7-11, assume that at time $t = 0$ the electron was inserted into one of the wells (i.e., probability that it is in well at the left is 1, probability that it is in well at the right is 0). Sketch $\psi(x)$ at $t = 0$. Describe the subsequent behavior.

7-18. Draw sketches like Figs. 7-11, 7-12, and 7-13 but for the $n = 2$ wave functions of an electron in two identical square wells.

7-19. Draw sketches like Figs. 7-15, 7-16, and 7-17, but start with $2s$ wave functions on each proton. Show that the splitting of energy levels becomes appreciable at larger values of R than with the $1s$ wave functions plotted in those figures.

7-20. Sketch the lowest energy ψ for an electron moving in the electrostatic field of three protons lying on a line with nearest neighbor separation equal to 1 Å. After considering the proton-proton repulsion (which makes the difference between the energies plotted in Figs. 7-18 and 7-19), comment on the likelihood of a stable molecular ion H_3^{++}.

7-21. Let each of the circuits of Fig. 7-21b contain the same small series resistance (i.e., high but not infinite Q) and assume that a source of power at a variable frequency ν is very weakly coupled to them. Sketch $I(\nu)$ for several choices of M/L. (This is only a qualitative problem.)

7-22. In classical mechanics the problem of the interaction of two particles, masses M_1 and M_2, is solved by introducing the reduced mass μ defined in eq. 7-9. Let the two masses be moving along the OX-axis, let $x = 0$ be the position of the center of mass, let x_1 be the position of M_1 and x_2 of M_2, let $u = x_1 - x_2 = |x_1| + |x_2|$ be their separation, and let $F(u)$ be the force between them. Show that the equation of motion for the separation u is $\mu(d^2u/dt^2) = F(u)$. Thus the two-body problem is reduced to a one-coordinate problem through the use of the reduced mass μ.

7-23. Once the differential equation in the preceding problem has been solved, $u(t)$ is known. How are $x_1(t)$ and $x_2(t)$ then obtained? If the atom A, mass M_A, in the diatomic molecule AB executes simple harmonic motion of amplitude a and frequency ν about the center of mass, describe the motion of atom B, mass M_B.

7-24.* In the problem of the hydrogen *atom*, how much different are the reduced mass μ and the electron mass? Does this difference increase or decrease the 13.60-e.V. binding energy, and by how many electron volts? (Equation 6-2 with μ in place of m is the precisely correct form and agrees with experiment.)

7-25. The ultraviolet emission spectrum of the H_2 molecule consists of several thousand lines. Nevertheless, pairs of lines can be identified in which each line of the pair arises from the same electronic transition, the same rotational transition, the same vibrational level in the initial states, but from vibrational levels differing by unity in the final state (in the present problem, the ground electronic state). Several observed pairs, with the vibrational quantum numbers in the final state, are: $n = 0$, 9,405,300 m.$^{-1}$, and $n = 1$, 8,989,600 m.$^{-1}$; $n = 1$, 8,864,400 m.$^{-1}$, and $n = 2$, 8,472,100 m.$^{-1}$; $n = 2$, 8,344,000 m.$^{-1}$, and $n = 3$, 7,974,400 m.$^{-1}$. (The units are wavelengths per meter, from which the energies or energy differences can be easily calculated.) Why are the energy differences between neighboring vibrational states not exactly constant? Show that the zero-point energy is about 0.265 e.V.

7-26.* Compare quantitatively the vibrational energy levels of H_2, HD, and D_2 (D is deuterium, the heavy isotope $_1H^2$ of hydrogen). What is the difference (in electron volts) between the dissociation energy of H_2 and of D_2? (Use the zero-point energy result from prob. 7-25.)

7-27. Why is the average R for H_2 slightly different from that for D_2?

7-28.* Use the result of prob. 7-25 to determine the ratio of the number of hydrogen molecules in the state $n = 1$ to the number in $n = 0$ in thermal equilibrium at 0°C. (The statistical weights of the two states are equal.)

7-29. From the result of prob. 7-25, calculate the shape of the binding energy curve near its minimum and compare with Fig. 7-20.

7-30. Electron diffraction experiments show that the internuclear separation R of diatomic molecules increases as the temperature increases. Why does R increase? What can be learned about the binding energy curves from a knowledge of $R(T)$?

7-31. What rôle, if any, do excited electronic states play in the thermal expansion of diatomic molecules?

7-32.* Suppose that an energetic electron strikes an H_2 molecule in its ground state. It excites the H_2 molecule to the first excited electronic state. Assume that Fig. 7-20 is drawn to scale, and calculate the minimum electron energy for this excitation process. *Hint:* Show first that the internuclear separation R cannot change much from R_0 during the short time of the electron collision.

7-33. The excited state formed as in prob. 7-32 will not decay to the ground state in a time shorter than about 10^{-8} sec. Show that in less time than this the molecule dissociates.

7-34. The minimum electron energy for the dissociation process of probs. 7-32 and 7-33 is greater than $|E_d|$. What happens to the extra energy?

7-35. Show that the sum of the moments of inertia of two particles of mass M_1 and M_2 rotating about their common center of mass as a fixed point is eq. 7-11.

7-36.* Calculate the internuclear distance R_0 for the HCl molecule from the fact that some of the lines of its pure rotation spectrum occur at wavelengths $\lambda = 120.3\ \mu$, $\lambda = 96.0\ \mu$, $\lambda = 80.4\ \mu$, $\lambda = 68.9\ \mu$, $\lambda = 60.4\ \mu$. (1 $\mu = 1$ micron $= 10^{-6}$ m.) *Hint:* Calculate the energy differences between the $h\nu$ values for these lines, and compare with Fig. 7-24. Then use eq. 7-11. Assume that all the molecules are $_1\text{H}^1\ _{17}\text{Cl}^{35}$.

7-37.* Calculate B and the first four rotational energy levels for the hydrogen molecule (in its lowest electronic and vibrational state).

7-38. Calculate B and the first four rotational levels for the HD and D_2 molecules. ($D = _1\text{H}^2$.)

7-39.* Calculate from the Boltzmann factor and from the data of prob. 7-36 the relative numbers of HCl molecules in the $J = 0$ and $J = 1$ rotational states at 17°C; the statistical weight of the state J is $2J + 1$. (This result is quite different from the results for the comparison of populations of electronic and vibrational levels.)

7-40. In Fig. 7-25 the separation between the first and second vibrational levels is smaller for the excited electronic state than the similar separation for the ground electronic state. Why?

7-41. In Fig. 7-25 the separation between the first and second rotational levels is smaller for the excited electronic state than the similar separation for the ground electronic state. Why?

7-42.* Show that the Correspondence Principle holds for the emission or absorption of radiation as a diatomic molecule with constant $R = R_0$ changes from one rotational state to another. Calculate the classical frequency of rotatation ν_c in terms of B and the energy E (this is the classical radiation frequency). Calculate the wave mechanical frequency $\nu_w = \Delta E/h$ for $\Delta J = 1$ for large J values in terms of B and E, and compare ν_w with ν_c.

7-43. Discuss by referring to Fig. 7-25 the specific heat of a diatomic gas as a function of temperature. Use the results of probs. 7-25 and 7-37 to estimate the temperatures at which interesting changes occur in the specific heat of hydrogen.

8

Binding and Energy Bands in Solids

8-1 INTRODUCTION

There are two immediate problems when the aggregation of atoms to form solids is considered. The first problem may be presented as a question: What is the origin of the forces holding the atoms together? This problem is analogous to the similar problem in molecular binding, and the answer is very nearly the same. The attractive forces are the electrostatic forces between charged particles that are distributed in space in the manner prescribed by quantum mechanics; a stable array of atoms is formed because these forces are opposed by the electron-electron repulsive forces described by the Exclusion Principle. As in molecular binding, we speak of several different kinds of binding, depending on how the electrons involved in the binding are distributed. Binding in solids will be discussed in Secs. 8-2 and 8-3.

The second problem may also be presented as a question: When atoms are brought together into a solid, what are the allowed energy levels of electrons? This problem is analogous to the similar problem in molecules, and the answer is an extension of the answer for molecules. In the hydrogen molecule the energy levels of the individual atoms split into pairs when the atoms come together. A similar process takes place in the solid and splits each level into a large number of closely spaced levels. The *energy bands* arising in this way are of the greatest importance in

254

understanding the properties of solids and will be considered in Secs. 8-4 and 8-5.

All the solids considered in this chapter and the following chapter are *crystals*. The atoms in a crystal are arranged in a perfectly symmetrical, periodic array. This symmetry permits many calculations of properties and phenomena that otherwise would become entangled with impossibly large numbers of atoms in large varieties of relative positions. The understanding of single crystals attained through such calculations can then be applied to most of the properties of common polycrystalline aggregates, such as metals and alloys. Crystal boundaries play a minor role in most of these properties but a major role in some (notably in mechanical properties); boundaries will be considered briefly in Chapter 10.

No materials other than crystals will be discussed in detail in this book, but there are many important non-crystalline or partially crystalline materials, such as concrete, glass, and plastics. These solids lack a perfectly repeated order over long distances, but the neighborhood of any atom exhibits considerable local order. Quantum physics has been less valuable in the understanding of these materials than in the understanding of crystals, but much of the physics of Chapters 8 to 10 is applicable to non-crystalline solids.

8-2 IONIC AND COVALENT CRYSTALS

The physics of *ionic* binding is almost identical with that of the binding in ionic molecules (Sec. 7-2). The KCl molecule will be used as an example in the following discussion, but the principles apply to other alkali halides and also to other ionic crystals. The same combinations of atoms that produce ionic molecules produce ionic crystals.

The crystal structure of KCl has already been described; it is the *NaCl structure* illustrated in Fig. 4-20. The potassium atoms are arranged at the corners and the centers of the faces of an array of cubes. Such an array is called a *face-centered cubic* array. The chlorines are also arranged in a face-centered cubic array. Hence the *NaCl structure* can be described as two interleaved arrays, one of potassium atoms, the other of chlorine atoms, each in a face-centered cubic arrangement.

X-ray diffraction experiments (Sec. 4-4) have established the fact that NaCl and most other alkali halide crystals have this structure. These experiments have also determined the closest spacing between the centers of ions of unlike sign, which is 2.81 Å in NaCl and 3.14 Å in KCl

It is instructive to estimate the *binding energy* or *cohesive energy* of solid KCl. Large cohesive energy means that a large energy must be supplied in order to break up the solid into its constituent atoms. Thus large cohesive energy (e.g., diamond or tungsten) implies high melting temperatures and low vaporization rates, since the atoms will remain in the solid until the thermally produced energy fluctuations are large enough to overcome the binding forces. The cohesive energy can, in principle, be determined from the observed latent heat of sublimation of a solid, but actual determinations are usually made by combining data for the latent heat of fusion, the latent heat of vaporization from the liquid, and the heat capacity. We shall define the cohesive energy as the energy given off when the crystal is formed from the neutral atoms in the form of monatomic gases.* This energy is commonly expressed in electron volts per molecule (or in kilocalories per gram molecular weight).

The first step in estimating the cohesive energy is to calculate the energy required to produce a K^+ and a Cl^- ion from the neutral atoms. This step requires a net expenditure of 0.52 e.V. per pair, which is the ionization potential of K (4.34 e.V.) minus the electron affinity of Cl (3.82 e.V.).

The next step is to compute the energy $-E_L$ per ion pair that is given off when these ions are brought together to form the crystal lattice. This energy is composed principally of the same two terms as in the KCl molecule, namely, an electrostatic attraction and a repulsion originating from the difficulty of interpenetration of ion cores. But there is a difference in the electrostatic term because of the interaction of a given ion with *all* the other ions of the crystal. A positive ion is surrounded closely by six negative ions, and the attraction is strong to these nearest neighbors. It is also surrounded by twelve next nearest neighbors of positive sign (which give a repulsion), and so on. Therefore the effects of all the other ions on the one in question must be summed to learn how tightly it is held in the crystal.

This summation depends only on the geometrical arrangement of ions in the crystal and on the charge of the ions, not upon other properties of the ions. The electrostatic energy per molecule for diatomic structures with ionic charge q is

$$E_L = - \frac{\alpha q^2}{4\pi\epsilon_0 R} \tag{8-1}$$

---

* The *lattice energy* is more frequently calculated for ionic crystals; it is the energy given off when the crystal is formed from the separated *ions*. We could also start with separated KCl molecules or K_2 and Cl_2 molecules in our example. Of course translation from one definition to another is possible because the atomic ionization energies and molecular binding energies are known. We prefer neutral atoms to ions as the reference condition because the "cost" of forming the ions must be considered when deciding, for example, whether a particular solid will be ionic or covalent.

where α is the *Madelung constant*. Multiplication by the Madelung constant has modified the expression for the interaction of two ions so that it is correct for the interaction of *all* the ions in the lattice. The value of α for the KCl structure is 1.748; for other simple structures α lies between 1.6 and 1.8. Since q equals e in KCl (singly charged ions), the electrostatic energy per ion pair is

$$E_L = -\frac{1.748e^2}{4\pi\epsilon_0 R}\text{ joules} = -\frac{1.748 \times 14.40}{R\text{ (\AA)}}\text{ e.V.}$$

This energy is plotted as the dashed line in Fig. 8-1.

The repulsive energy (dotted line in Fig. 8-1) has precisely the same cause (the Exclusion Principle) as in the molecule. As in the molecule, the repulsive energy rises so steeply as R decreases that its contribution to E is relatively small at the minimum of $E(R)$. Also as in the molecule, there are minor contributions to E from van der Waals attraction

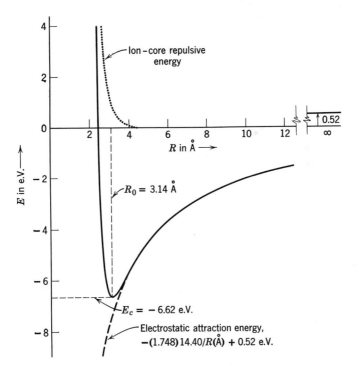

Fig. 8-1. Energy of a lattice of K^+ and Cl^- ions as a function of the nearest-neighbor spacing R.

(binding, negative) and from zero-point vibration energy (repulsive, positive).

The cohesive energy $|E_c|$ can be estimated by neglecting the minor terms and by using the experimental value of the nearest-neighbor distance ($R_0 = 3.14$ Å):

$$E_c = -\frac{1.748 \times 14.40}{3.14 \text{ Å}} + 4.34 - 3.82$$

$$= -7.50 \text{ e.V. per ion pair} \tag{8-2}$$

The experimental value of $|E_c|$ is 6.62 e.V. The difference 0.88 e.V. is caused by the neglect of the minor contributions to the energy, principally the Exclusion Principle repulsion (probs. 8-4 and 8-5).

The chemical compounds that form ionic crystals are the same as those that form ionic molecules. As in molecules, a small ionization potential for the electropositive atom and a large electron affinity for the electronegative atom favor ionic binding. Compounds of elements in groups I and II with elements in groups VI and VII of the periodic system produce solids in which the binding is almost completely ionic. Other compounds have partially ionic, partially covalent binding. Of course monatomic solids cannot possess ionic binding.

Experimental proof that KCl, for example, is ionic is available in several forms. Precision comparison of the intensities of the X-ray reflections from various atomic planes permits the measurement of the electron density in the crystal as a function of position. Such measurements on compounds of light elements (e.g., LiF) have shown that electron transfer from the positive to the negative ions has occurred and therefore that the crystals are ionic. The existence of electrical conductivity by positive and negative ions in a crystal is also evidence for ionic binding (see Sec. 10-3).

Another way of determining whether a crystal is ionic is by comparing the *low frequency* (d-c through microwave frequencies) dielectric constant κ_e (relative electric permittivity) with the *optical* dielectric constant κ_{eo}, which is the square root of the optical index of refraction in the visible region of the spectrum. If a solid is not ionic, κ_e equals κ_{eo}. If it is ionic, κ_e is greater than κ_{eo}. At frequencies of visible light only the *electron* distribution can be distorted by an electric field, since the motion of the heavy ions is too sluggish to follow the rapidly oscillating electric field. At lower frequencies (less than about 10^{13} cycles/sec) the *ions* in an ionic crystal can follow the oscillations of the field, a larger polarization occurs because of the oppositely directed displacements of the positive and the negative ions, and this polarization makes κ_e greater than κ_{eo}. Of course no such displacement occurs in a covalent crystal since

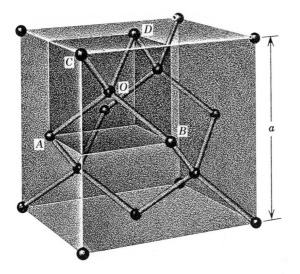

Fig. 8-2. The diamond structure. The atom O is in the center of the small cube and is closely bound to the four nearest neighbors A, B, C, and D at alternate corners of that cube. Every atom has four nearest neighbors.

the atoms in such a crystal are not charged, and κ_e equals κ_{eo}.* Two examples of ionic crystals are: KCl, $\kappa_e = 4.7$, $\kappa_{eo} = 2.1$; BaO, $\kappa_e = 34$, $\kappa_{eo} = 4$. An example of a non-ionic crystal is: Ge, $\kappa_e = 16$, $\kappa_{eo} = 16$.

The physics of *covalent* binding in solids is the same as in molecules. The attractive force originates from the concentration of electronic charge along the lines joining adjacent nuclei. The repulsive force originates from the Exclusion Principle interaction, as in all solids. Typical examples of crystals with nearly pure covalent binding are diamond, silicon, germanium, and silicon carbide. Examples of crystals with part covalent, part ionic binding are quartz (silicon dioxide), tungsten carbide, and aluminum antimonide. Direct experimental proof of covalent binding is not generally available. We rely on theory and on the elimination of other forms of binding by the use of experimental information to show that a particular crystal is covalent.

Carbon (diamond), silicon, germanium, and gray tin are an especially interesting group of covalent crystals. All have the *diamond structure* shown in Fig. 8-2, which should be compared with the structure of the

* Molecular crystals composed of molecules with dipole moments (like HCl) also exhibit a difference between κ_e and κ_{eo}. These crystals can be distinguished from ionic crystals by the fact that the transition from κ_e to κ_{eo} occurs in the far infrared or radio regions of the spectrum for molecular dipolar crystals.

methane molecule illustrated in Fig. 7-9.   Each of these elements has four valence electrons.   The diamond structure is characterized by the fact that in it each atom has *four* nearest neighbors.   The covalent binding consists of electron pair bonds between a given atom and each of its nearest neighbors.   The atom in question contributes one of its valence electrons to each bond, and this exhausts its supply of valence electrons. Each of its neighbors contributes one electron to each bond.   Thus all valence electrons are involved in the binding, and each is held rather tightly between a pair of atom cores.

In addition to ionic, covalent, and metallic crystals, there is a fourth class called *molecular* crystals.   In these crystals the elementary unit of which the solid is built is the molecule (rather than the atom).   For example, hydrogen atoms are very strongly bound into the hydrogen molecule, but hydrogen molecules are bound together only by the weak van der Waals force.   Solid hydrogen is therefore a weakly joined crystal of molecules, and its melting point is accordingly very low (14°K).

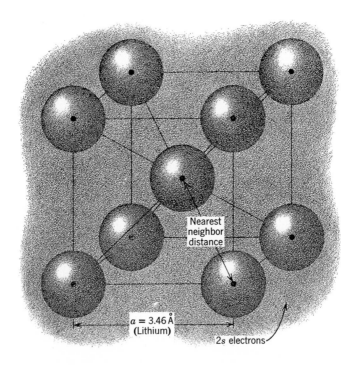

Fig. 8-3. The *body-centered cubic* structure, a common metallic structure.   The valence electrons (2s electrons in the example of lithium) are *not* localized on individual atoms.

Other typical molecular crystals are oxygen, carbon dioxide, and most organic crystals.

8-3 METALLIC CRYSTALS

Metallic binding has no counterpart in diatomic molecules.* Binding occurs in metals because when the atoms come together in a metal it is possible for the outer, *valence* electrons to be near positive ions at all times (hence low potential energy) and at the same time to have "spread-out" (not sharply peaked) wave functions (hence low kinetic energy). This type of binding will be studied by discussing the example of lithium. The structure of lithium metal is shown in Fig. 8-3.

A rough sketch of the wave functions of the lithium *atom* is presented in Fig. 8-4. The $1s$ electrons are close to the nucleus, but the presence of the $2s$ electron increases the size of the atom by about a factor of 5 over the Li^+ ion. The energy of the system could be lowered if the $2s$ electron could be brought closer to the nucleus without increasing its kinetic energy, and this is what happens when lithium atoms come together to form a solid. Suppose that lithium atoms are brought together so closely that their cores (the $1s$ electrons) almost overlap. The valence electrons ($2s$) can no longer be considered as "belonging" to particular cores, since the cores are closer together than the size of the $2s$ atomic wave functions. The valence electrons form a "gas" and belong to the solid as a whole. No valence electron ever gets so far from a lithium nucleus as its average distance from the nucleus in the lithium atom. The potential energy of the valence electrons is therefore much reduced. They are making better use of the positively charged cores by staying closer to them.

The valence electrons thus have lower potential energies than in lithium atoms, but what has happened to their kinetic energies? To a first, crude approximation, the array of lithium ion cores can be considered to be a single square well, a region of low potential energy surrounded by barriers (the edges of the crystal). The kinetic energy of the lowest state is very low indeed, since in eq. 5-26 the width $2x_0$ of the "box" is very large compared to atomic dimensions. But all the valence electrons cannot have this minimum kinetic energy without violating the Exclusion Principle, since there are only two quantum states in the whole crystal with this energy. In order to dispose of all the valence electrons and yet have at most one electron per quantum state, states

---

* If the reader is familiar with the benzene molecule and similar organic molecules he will note a considerable resemblance between the nature of the binding in the benzene ring and in metals.

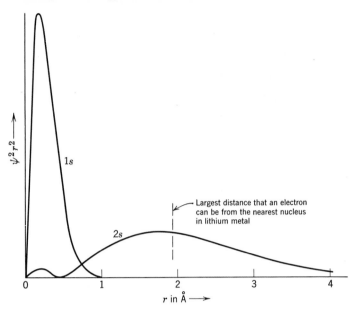

Fig. 8-4. Comparison of the 1s and 2s probability densities of the lithium atom. The ordinates are proportional to the probability that an electron in the 1s or 2s state will be at a radius between r and $r + dr$.

with a large range of momenta and therefore of kinetic energies must be occupied. The range of kinetic energies required in order that there should be at most one electron per quantum state will be calculated in Sec. 9-4. The required range is from zero to 4.7 e.V. for lithium, and the average electron kinetic energy is 2.8 e.V. per atom when the internuclear separation R has its equilibrium value R_0; this energy increases as R decreases, since then more electrons per unit volume must be accommodated without violating the Exclusion Principle. The average kinetic energy is of the order of 2 to 10 e.V. per atom for other metals and alloys.

Another way of estimating the kinetic energy difference between separated lithium atoms and atoms bound into a solid is to consider the relative magnitudes of $|d\psi/dx|^2$ in the two situations. Throughout most of the crystal, ψ varies relatively slowly from point to point. Near the ion cores (Fig. 8-3), ψ exhibits the "wiggles" of the 2s wave function shown in Fig. 8-4, but elsewhere it behaves like the sinusoidal ψ's of a square well, with wavelengths for the ψ's of the various electrons all considerably longer than the internuclear spacing. The behavior near the ion cores is nearly the same in separated atoms and in the solid and thus neither enhances nor suppresses binding. But the $|d\psi/dx|^2$ away from the cores is less in the solid, since ψ need not fall to zero a few angstroms from each nucleus (as it does in the atom, Fig. 8-4). Thus the kinetic energy is reduced from its atomic value by bringing the atoms together.

Permitting the valence electrons to be bound to an extended array of ion cores, rather than to individual separated atoms, thus permits lowering both the potential and kinetic energies. The potential energy becomes more negative as the ion cores are brought closer together, but the kinetic energy (though still less than in the isolated atoms) becomes larger. The actual internuclear separation R_0 is thus a compromise between these two effects, as shown schematically in Fig. 8-5.

The relative importance of potential and kinetic energy in metallic binding varies from metal to metal; the potential energy effect is larger in lithium, but the kinetic energy effect is larger in most other metals. In addition, ion core repulsion becomes significant in some metals, and covalent binding involving inner shell electrons becomes significant in metals like the transition metals with incomplete d-shells. Furthermore,

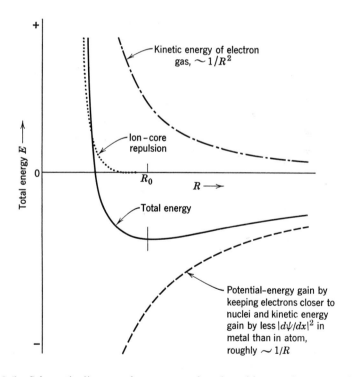

Fig. 8-5. Schematic diagram of energy as a function of internuclear separation in lithium metal.

the details of the wave functions are such that electrons on the average avoid one another to minimize the electron-electron repulsive energy (*correlation* and *exchange* effects). Thus calculation of the cohesive energy of a metal is quite complicated, but the basic binding process in all metals is the process described above.

Metallic binding is favored over other forms if the number of valence electrons per atom is small, since only the lowest kinetic energy states need be occupied if there are relatively few valence electrons per unit volume. The elements at the left side of the periodic table therefore form metals. It is difficult to predict whether a particular atom near the middle of the periodic table will form a metal or a covalent crystal. Sometimes the competition between these structures is very close. For example, tin exists in two forms: metallic (white) tin is stable above 13°C; covalent (gray) tin is stable below this temperature (the latent heat of transformation from one structure to the other is much less than 1 e.V., and therefore the two structures have almost exactly the same binding energy).

Several properties of metals that depend upon the fact that the valence electrons are like particles in a gas will be discussed in later chapters. Meanwhile it should be remarked that the above interpretation of metallic binding suggests that atomic mixtures of metals ought to behave much like the metals themselves. Atomically dispersed solid solutions of metals are in fact formed by the elements at the left side of the periodic table. An alloy of, say, copper and nickel has a cohesive energy and other properties much like those of copper or nickel. Furthermore, these properties vary continuously as the percentage of nickel is changed from 0 to 100. Such behavior does not occur with solids in which the binding is ionic or covalent, since specific arrangements of electrons are required in both these types of binding ($Na_{20\%}Cl_{80\%}$ could scarcely exist as an ionic solid, for example, since four electrons would be required from each Na atom and the ionization potentials for all but one of these are very high). The metallic type of binding, on the other hand, predicts the smooth dependence of properties on composition of alloys. In metallic binding the details of the atomic wave functions are relatively unimportant, and what matters most is how closely the cores can be packed together and what the average electron kinetic energy is.*

---

* Many alloys show discontinuities in their properties as a function of composition. Much of this behavior is caused by differences in crystal structure. If the ion cores are much different in sizes or not spherical, different space arrangements are preferred at different concentration ratios of the constituents. Additional quantum effects also occur, and quantum theory has been successful in explaining the changes in structure and properties of many alloys as functions of composition (*Hume-Rothery rules*).

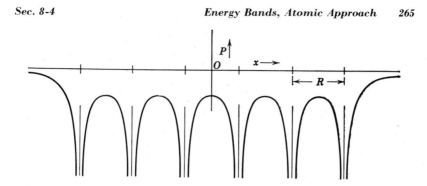

Fig. 8-6. Potential energy as a function of distance along a line of six evenly spaced protons.

8-4 ENERGY BANDS, ATOMIC ENERGY LEVEL APPROACH

In this section the possible energy levels of electrons in solids are discussed by the same approach used in Sec. 7-5 for molecules. We could begin, as in Sec. 7-5, with a square-well example, but we shall leave that for a problem (prob. 8-28). We shall consider first a linear array of six evenly spaced hydrogen atoms. This forms a transition example between diatomic molecules and solids.

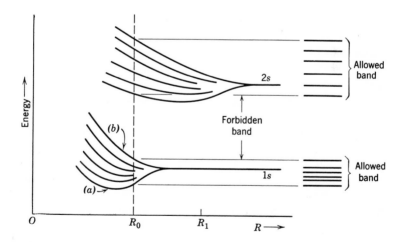

Fig. 8-7. Energy levels as functions of internuclear separation for a row of six hydrogen atoms. The levels are shown at the right for the separation R_0. (Schematic, from W. Shockley, *Electrons and Holes in Semiconductors*, Van Nostrand, New York, 1950.)

The potential energy of an electron as a function of the distance x along the line of nuclei is illustrated in Fig. 8-6. The energies for all the possible 1s and 2s wave functions are presented in Fig. 8-7 as functions of the internuclear spacing R. Six energy levels are present where only one 1s level occurred in an individual atom. The number six is not surprising in view of the production of two levels in the two-atom molecule. There are actually two quantum states ($m_s = +\frac{1}{2}$ or $m_s = -\frac{1}{2}$) in the original 1s level of hydrogen, and there are twelve quantum states in the 1s group for the array of six atoms. Pairs of these wave functions differ only in spin, however, and this makes such a small difference in energy that the twelve levels appear to be only six.

This is an example of the general principle (which we shall not attempt to prove here) that *bringing atoms together leaves the total number of quantum states with a given quantum number unchanged.* For example, there are eight possible $n = 2$ states in an atom. If N atoms are brought together, there are *exactly* 8N states, even though the energies of the states may be altered considerably. As another example, since there are two electronic wave functions with $n = 2$, $l = 0$ (2s states) in the atom, a solid of N atoms has 2N of the 2s states. This principle will be of great importance when we consider in Sec. 9-2 the difference between conductors and non-conductors of electricity.*

As the number of atoms in this one-dimensional model of a solid is increased, the additional levels appear in the regions labeled *allowed bands* in Fig. 8-7. This fact can be verified by first considering the wave functions for the (a) and (b) states for the six-atom problem. These wave functions are plotted schematically in Fig. 8-8. They are the obvious extensions of wave functions like those of Fig. 7-16 for a two-atom problem. The (a) function never changes sign whereas the (b) function changes sign between each pair of atoms. These represent the extreme cases of low energy (a) and high energy (b). The other four states change sign 1, 2, 3, and 4 times. As the number of atoms in the array increases, the (a) and (b) type wave functions do not change their energies appreciably. Practically the same kinetic and potential energies occur as in the array of six, since the shapes of the wave functions are very nearly the same. As the number of atoms increases, the new energy states appear as additional fine structure between the two extremes.

By the time a milligram of matter is assembled there are $\sim10^{19}$

---

* This principle must be expressed in a somewhat different form if the quantum states involved *overlap* in energy; the concept of overlap will be explained in conjunction with Fig. 8-10. If, for example, the 3s and 3p states of sodium overlap, we can no longer distinguish 3s states from 3p states but must count only the *total* number of 3s plus 3p states (which equals 8N).

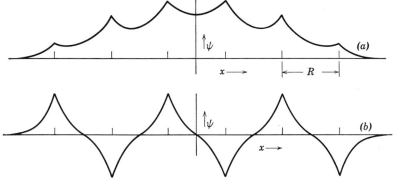

Fig. 8-8. Schematic diagram of the real part of the wave functions at the bottom (*a*) and at the top (*b*) of the 1*s* band for the array of six hydrogen atoms.

energy levels, and the spacings between them are so small that no experiment can provide any meaning or significance to the spacing between levels, which is of the order of 10^{-19} e.V. We thus speak of this group of levels as an *allowed band*, and treat it as if a *continuous* distribution of energies were allowed for electrons within such a band. It is worth repeating that the width of an allowed band does not grow as the number of atoms in the aggregation is increased. Therefore the *forbidden band* of Fig. 8-7 remains an energy region without any electronic energy states, regardless of the number of atoms in the solid.

The one-dimensional theory should, of course, be replaced by a three-dimensional one. This replacement does not alter the basic conclusion that allowed and forbidden bands exist. The mathematical description becomes complicated since the energy associated with an electron wave function is a function of the direction of motion of the electron. The bands must be described in terms of the *vector* momentum. This *Brillouin zone* treatment is necessary to make quantitative theoretical predictions of many of the properties of solids but is not necessary to understand the *nature* of energy bands and the properties they affect.

A useful way of illustrating the bands in a solid is to combine potential energy curves like Fig. 8-6 with the information about the bands at the observed lattice spacing from curves like those in Fig. 8-7. Such an illustration is shown in Fig. 8-9 for lithium. The curve is the potential energy of an electron along a line of atom centers and was obtained by adding together curves like Fig. 6-7*a* for each lithium atom.

Since the ordinate in Fig. 8-9 is energy, we can also plot on this same graph the total energy of possible electron states. The 1*s* levels are shown

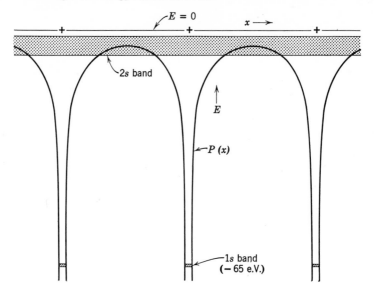

Fig. 8-9. Energy bands for lithium superimposed on a plot of the potential energy as a function of the distance x along a line of nuclei.

as a narrow band.   This band is narrow because the atomic $1s$ wave function is concentrated near the nucleus, and the value of $|\psi|^2$ of such a function is very small at a distance from the nucleus equal to half the separation between nuclei.   Thus, when a ψ of one atom is joined to that of the next, it makes very little difference in energy whether the ψ's have the same or opposite signs, since both ψ's are very nearly zero at the point of joining. This is the same situation revealed in Fig. 8-7 at an R like R_1, which is a value large enough compared to the $1s$ state of hydrogen to make the interaction of wave functions of electrons on adjacent atoms slight (see also Fig. 7-15).

The $1s$ band in Fig. 8-9 is drawn as if it did not exist outside the classical turning points of the $1s$ electrons.   It is drawn so to emphasize that the electrons in this band are on the average very close to the nuclei. Thus the plot gives a rough idea of the spatial extension of wave functions as well as the allowed energies.   Of course the $1s$ wave functions *do not* fall abruptly to zero at the point where the kinetic energy is zero (and at larger R becomes negative), but they begin to decrease sharply here, and the probability of finding an electron much farther away is very small.

The $2s$ electron states are much more strongly affected by the interaction of neighboring atomic states than the 1s states are, since the radial extent of the 2s wave functions is much larger (see Fig. 8-4). The 2s band is therefore much wider. We could draw it as if it stopped at the classical turning points, but the wave functions spread out so much farther beyond these points that these points have lost any real significance. (Compare the large penetration into the negative kinetic energy region shown in Fig. 5-5c.) The electrons in this band are essentially shared by all the atoms in the crystal. An electron that was initially near a certain atom core is just as likely to be near any of a large number of other atoms after a short time. (The "short time" is required only because of the finite velocity of the electron, and a time of the order of 10^{-15} sec is ample.)

Of course, there are also the 2p, 3s, 3p, 3d, etc., states of the lithium atoms (all ordinarily empty). In the solid all these states are broadened into wide bands, since they have wave functions extending at least as far from the nucleus as the 2s wave functions. These bands overlap, since they are each several volts wide and there are only a few volts difference between the 2s band and the continuum (the energy $E = 0$ line). Above this energy, electron states are no longer bound states, and electrons in these states are not confined to the solid crystal.

We have used lithium as an illustration, but the same general behavior occurs in all solids. Low-lying levels in atoms become narrow bands in the solid, since they are relatively little affected by joining the atoms together into the solid. High-lying levels are broadened into wide bands. Electrons occupying states in these upper bands move relatively freely throughout the crystal; each electron is bound to the solid, but not to any particular atom of the solid. If there is more than one kind of atom in the solid (e.g., KCl), bands originate from the energy levels of both of the atomic constituents, but otherwise the analysis is the same as above.

It should now be apparent why X-ray emission lines can be observed using solids as the targets in X-ray tubes, although optical spectra can be obtained only with gaseous sources. Consider sodium, for example. The energy bands for sodium have been calculated and are illustrated in Fig. 8-10. If a high-energy electron strikes a sodium crystal it can give enough energy to a 1s electron to remove it from the crystal or to excite it to a vacant energy band in which it moves away to another region of the crystal. In either case an empty state in the 1s shell is produced, and a 2p electron can make a transition to this state with the emission of an X-ray photon. Such photons have energies of 1010 e.V. with only a tiny spread in energies. The X-ray line is sharp (small spread in photon energy) because both the 2p band and the 1s band have

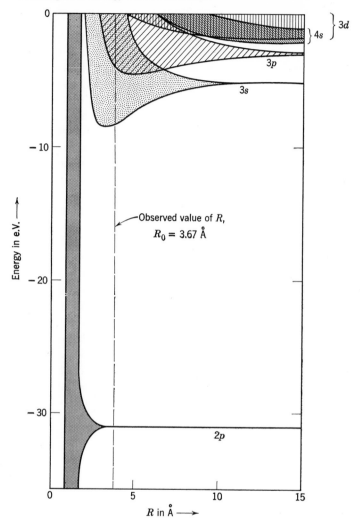

Fig. 8-10. Energy bands in sodium as functions of the internuclear spacing R. The 2s band is at -63.4 e.V., and the 1s band is at -1041 e.V. Both of these bands are even narrower than the 2p band. [From J. C. Slater, *Phys. Rev.*, **45**, 794 (1934).]

very small widths compared to 1010 e.V. In elements with larger atomic number (like tungsten) even the $n = 3$ and $n = 4$ levels are very narrow in the solid, since even these levels are inner levels; therefore sharp X-ray lines can be observed even from $n = 3$ and $n = 4$ to $n = 1$ transitions in heavy elements.

The levels involved in the optical spectra of atoms, on the other hand, are so broadened in the solid that line spectra are quite impossible. Exciting the electrons in a solid by heating it to incandescence gives rise only to the continuous, black-body spectrum. All solids show essentially the same spectrum with only minor modifications in intensity as a function of wavelength, which modifications depend on the energy band structure of the particular solid. The spectral emissivity ϵ_λ introduced in Sec. 4-8 is thus somewhat different for different solids, but the main features of light emission are the same for all solids. Electron bombardment gives rise to luminescence in some solids, and in a few solids the luminescence appears in relatively narrow wavelength regions. But the electron transitions that produce this light usually involve imperfections in the solid and are *not* transitions from one energy band to another. This type of light emission will be discussed in Sec. 10-6.

Some experimental proof of the existence and relative widths of energy bands in solids thus comes from experiments on the X-ray and optical properties of solids. Much additional, but sometimes indirect, proof comes from experimental information described in the next three chapters.

8-5 ENERGY BANDS, MATHEMATICAL MODEL APPROACH

The existence of allowed and forbidden energy bands for electrons in solids is of such importance that it is worth demonstrating their existence by an additional method. In this section we shall describe a one-dimensional mathematical model of the electrons in the higher-lying energy bands in a solid called the *Kronig-Penney model*. This model uses a regular array of the square-well potentials of Sec. 5-4. It thus is the same type of crude approximation to a solid that the model of Sec. 5-4 was to an atom. As in that problem, we seek an understanding of the *nature* of the allowed energy levels rather than a quantitative calculation of their

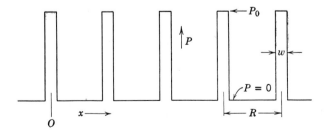

Fig. 8-11. Square well model of a one-dimensional solid.

position. It will be recalled that the model of Sec. 5-4 showed that a bound electron could have only one of a discrete set of energies. Similarly the model of the present section will show that the energy states for electrons in a structure like a crystal (where the same potential vs. distance curve is repeated over and over) are confined to *bands*.

In Fig. 8-11 a set of square-well potential functions is repeated in a one-dimensional array. This plot is similar to Fig. 8-9 in that regions of low and high potential energy alternate with perfect regularity, and it is this property of the solid that produces the allowed and forbidden bands. Except for this point of similarity, the $P(x)$ curve of Fig. 8-11 was chosen solely to make the problem solvable in terms of elementary mathematics. The algebra is still rather involved, even with this simplified model, and an additional approximation makes solution somewhat easier. The approximation consists of letting $w \rightarrow 0$ and $P_0 \rightarrow \infty$ in such a way that wP_0 remains constant. The classically forbidden regions of x-values shown in Fig. 8-11 are thereby replaced with very narrow but very high regions. This approximation does not change the result appreciably but makes the work less tedious.

The solutions of the Schrödinger equation for this problem are of two types, and each type of solution is valid over a particular range of values of the electron energy E. One type of solution is

$$\Psi = u_\lambda(x)e^{2\pi i\left(\frac{x}{\lambda} - \frac{Et}{h}\right)} \tag{8-3}$$

The exponential part of this Ψ is a plane wave with wavelength λ traveling toward $+x$ if $\lambda > 0$ or toward $-x$ if $\lambda < 0$. The factor $u_\lambda(x)$ is a periodic function with the same period R as the lattice. In any one well, $u_\lambda(x)$ has approximately the sinusoidal form of eq. 5-24 (it has precisely this form only for $\lambda = \infty$). Ψ functions like eq. 8-3 are called *Bloch functions;* with the appropriate $u_\lambda(x)$ they are the solution of *any* periodic potential problem.

The continuity conditions on Ψ and $\partial\Psi/\partial x$ are just as important in this problem as they were in the square-well problems of Sec. 5-4. These conditions restrict the validity of eq. 8-3 to values of E that satisfy the following equation:

$$\left(\frac{4\pi^2 mRP_0w}{h^2}\right)\frac{\sin \beta R}{\beta R} + \cos \beta R = \cos \frac{2\pi R}{\lambda} \tag{8-4}$$

where $\beta = (2\pi/h)\sqrt{2mE}$. If the strength P_0w of the barrier between wells is very small, there is a solution $\beta(\lambda)$, and hence $E(\lambda)$, for every E (see prob. 8-30). But if P_0w is not small, eq. 8-4 has solutions only for certain ranges of values of E, the allowed bands. The right side of eq.

8-4 has a maximum value of 1 and a minimum of -1. The left side can oscillate over a much larger range as E, and hence β, varies. Whenever the left side is >1 or <-1, no solution of eq. 8-4 is possible and no traveling waves of the type described by eq. 8-3 exist. Such values of E constitute the forbidden bands.

The other type of solution of the Schrödinger equation, which is valid for values of E in the forbidden band, is similar to eq. 8-3 in other respects but has a *real* constant multiplying x in the exponent:

$$\Psi = u_\lambda(x)e^{2\pi\left(\frac{x}{\lambda} - \frac{iEt}{h}\right)} \tag{8-5}$$

where λ can be either >0 or <0. This is a rapidly damped function (the rapidly growing alternative can never fit the boundary conditions) that can have appreciable values only near a source of electrons, such as at the ends of the array (surfaces of the crystal). If an electron is injected into the solid with an energy in a forbidden band, it will be reflected; its Ψ near the surface will be an exponentially decreasing function of distance into the crystal. Although in the next three chapters we shall be dealing primarily with electrons described by wave functions

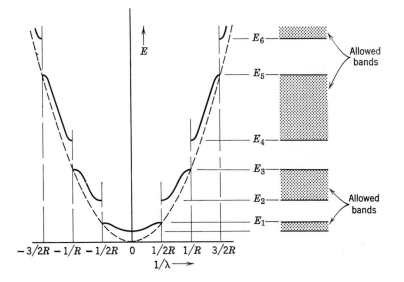

Fig. 8-12. Energy as a function of the reciprocal of the wavelength for the problem of Fig. 8-11 with $P_0 w = 3h^2/8\pi mR$. The *allowed bands* are energy regions in which Ψ has the form of a traveling wave (eq. 8-3). The dashed line is the relation between E and $1/\lambda$ for free electrons. [From A. Sommerfeld and H. A. Bethe, *Handbuch der Physik*, Vol. 24 (2nd part), J. Springer, Berlin, 1933.]

like eq. 8-3, there are important applications of eq. 8-5, notably to the tunnel diodes to be discussed in Chapter 11.

Figure 8-12 shows the relation between E and $1/\lambda$ for the traveling wave solutions (eq. 8-3) in the allowed bands and shows the allowed and forbidden bands alternating as E increases. This figure is drawn for a particular strength $P_0 w$ of the barriers. If this strength were much larger, the allowed bands would be very narrow and the forbidden bands very broad. In the limit $P_0 w \rightarrow \infty$, the segments of curves reduce to horizontal lines, the energy levels of the square well of Sec. 5-4 (see prob. 8-29). In the limit $P_0 w \rightarrow 0$, the forbidden bands disappear, and $E(\lambda)$ approaches the $E(\lambda)$ characteristic of free electrons, namely,

$$E = \frac{p^2}{2m} = \frac{h^2}{2m} \left(\frac{1}{\lambda}\right)^2 \tag{8-6}$$

where p is the momentum. This $E(\lambda)$ is the dashed curve in Fig. 8-12.

We can now verify the analysis of band widths that was given in Sec. 8-4. In order to approximate the behavior of electrons in low-lying (e.g., 1s) bands by our model, we should have to use a large P_0 and a large w, since these electrons are strongly bound to nuclei and the region of x where the classical kinetic energy is negative is very large. Hence we expect narrow allowed bands separated by wide forbidden bands. The reverse is true for high-lying levels. These conclusions agree with the analysis in Sec. 8-4.

The physics associated with the striking behavior whenever $1/\lambda$ equals an integer times $1/2R$ is worth studying. This condition can be written

$$n\lambda = 2R \tag{8-7}$$

where n is an integer, and this equation is precisely the condition for Bragg reflection. An electron wave is partially reflected at each barrier. If these walls are separated by $\lambda/2$, or $2\lambda/2$, or $\cdots$, the wave reflected at $x = R$ returns to $x = 0$ in phase with the wave reflected at $x = 2R$, $x = 3R$, $\cdots$. Even if the individual reflections are weak, the total effect is 100% reflection if there are enough walls (i.e., long enough array of atoms). Hence there are no traveling waves when λ satisfies the condition for Bragg reflection.

The Ψ functions are quite different for $1/\lambda$ a little smaller than $n/2R$ than they are for $1/\lambda$ a little larger than $n/2R$. This difference produces the discontinuity in E as a function of $1/\lambda$ and produces the gap between allowed bands. For example, the Ψ for $1/\lambda$ a little smaller than $1/2R$ is a series of $n = 1$ wave functions (Fig. 5-5b and Fig. 7-12b) that change phase from each well to the next. The Ψ for $1/\lambda$ a little larger than $1/2R$

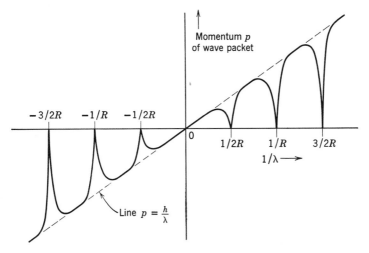

Fig. 8-13. Momentum of electron waves in the problem of Fig. 8-11 as a function of reciprocal wavelength.

is a series of $n = 2$ wave functions (Fig. 5-5c) with the same phase in each well.

Figure 8-13 illustrates another feature of the solution of this problem. The average momentum of an electron wave packet is plotted as a function of $1/\lambda$. The momentum was computed from the wave functions by the method of Appendix F. It is the product of the electron mass and the group velocity of a wave packet. If the electron were completely free, its momentum would be exactly h/λ, the de Broglie relation. The actual curve approaches this behavior except near the values of $1/\lambda$ satisfying eq. 8-7. At those points the momentum drops to zero, since for those values of $1/\lambda$ the total reflection of the traveling waves produces a set of standing waves, carrying no momentum. The decrease of the momentum to zero at the bottom and top of an allowed band will be encountered again in the explanation of the properties of semiconductors (Sec. 11-2).

The problem examined in this section is mathematically identical with the problem of the transmission of an electromagnetic wave through a wave filter or through a transmission line with periodically repeated susceptances: Figure 8-14 shows a general wave filter or transmission line; the only restriction imposed is that there be no resistive terms, and hence pure reactances are shown instead of impedances. This structure has the same periodic repetition that is the principal feature of the solid.

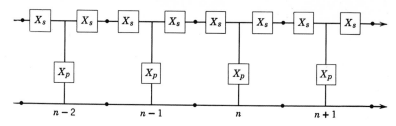

Fig. 8-14. Wave filter. The pass bands of frequency are the analogue of the allowed bands of energy in the problem of Fig. 8-11.

Furthermore, if we substitute the voltage V in the filter for Ψ, the differential equation for V has the same form as the Schrödinger equation, as was noted at the end of Sec. 7-5. It differs only in that, instead of potential energy and atomic constants, the electric-circuit equation has the magnitudes of the reactances. Also, the requirements on continuity of Ψ and $\partial\Psi/\partial x$ at the boundary between adjacent atoms have a parallel in the requirements that the voltage and current at the output of the section n must be the same as at the input of the section $(n + 1)$. Where the total energy E appears in the Schrödinger equation, we find the square of the frequency ν in the filter equation.

Since the differential equations and the boundary conditions are the same, the solution of the problem of energy levels in solids must be the same as the solution for the allowed frequencies in wave filters. It will be recalled * that in general a filter has one or more *pass bands*, which are regions of frequency ν in which the voltage V has the form

$$V = V_0 e^{-in\gamma - 2\pi i\nu t} \tag{8-8}$$

Here γ is the propagation constant and n is the number of the section at which V is measured, and hence n measures distance through the array in units of one filter section. V is measured at the same point (e.g., the middle) of each section. If V were measured at different points within each section, an additional factor would multiply eq. 8-8. This factor is obtained from the solution of the circuit equation within a given section and is precisely analogous to the factor $u_\lambda(x)$ in eq. 8-3. The solution eq. 8-8 is thus the exact analogue of eq. 8-3. The pass bands of frequencies that give traveling-wave solutions are the analogues of the allowed energy bands.

* See T. E. Shea, *Transmission Networks and Wave Filters*, Van Nostrand, New York, 1929; an introduction to wave filters is provided by L. C. Jackson, *Wave Filters*, Wiley, New York, 1944.

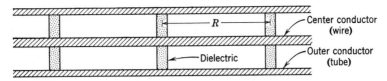

Fig. 8-15. Coaxial transmission line with regularly spaced dielectric supports.

It will also be recalled that a filter has one or more *stop bands*, which are regions of frequency in which V has the form

$$V = V_0 e^{-\alpha n - 2\pi i \nu t} \tag{8-9}$$

This is the same as the behavior of Ψ in a forbidden energy band. Another point of similarity is that the group velocity of propagation of a pulse (or packet of waves) through a filter decreases to zero at the edges of a pass band, which is analogous to the behavior of the electron momentum exhibited in Fig. 8-13.

An interesting special case of a wave filter is a coaxial line with a periodic arrangement of dielectric supports for the center conductor (Fig. 8-15). This problem can be handled by the general theory, but the important features of the result are almost obvious. A wave coming from the left is partially reflected at the first dielectric support, which acts like a shunt capacitance across the line. If the support is thin and of not too high a dielectric constant, the reflection is weak. Reflections also occur at all the other supports. Whenever $n\lambda = 2R$ these reflected waves add in phase and behave like one large reflection, no traveling wave to the right occurs, and this situation corresponds to a stop frequency and complete reflection of the incident wave. If the individual reflections were stronger, a stop *band* of frequencies would be produced.[*] Of course this condition is the same as eq. 8-7; constructive interference of small reflections to make total reflection is what is called Bragg reflection in the solid.

We have presented the solution of an artificial, one-dimensional model of a solid. This bears the same relation to the problem of interest (the actual solid) as the square-well problems of Sec. 5-4 bear to the hydrogen atom. In order to determine the energy bands in the actual solid we should have to use Bloch functions like eq. 8-3 but with the vector **r** re-

---

[*] It probably does not have to be noted that in practice such a coaxial line would not be a good broad-band transmission line. Either R is made very much less than any λ for which the line might be used or else the spacing between supports is varied enough to destroy the periodicity of the reflections.

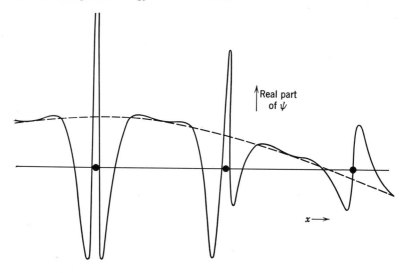

Fig. 8-16. Real part of ψ plotted as a function of distance along a line of atom centers in sodium (three centers are shown). [From J. C. Slater, *Phys. Rev.*, **45**, 794 (1934).]

placing x and with a *wave vector* **k** replacing the scalar $1/\lambda$. Furthermore, the functions $u_\lambda(x)$ would be replaced by the actual solutions for the atoms composing the solid. It would take us far afield even to present the complete solutions for actual solids.*

It is instructive, however, to plot a particular example of a wave function for sodium metal. The curve in Fig. 8-16 is one of the wave functions of the $n = 3$ band of sodium plotted as a function of distance x along a line of centers of sodium atoms. Only the real part of the amplitude ψ has been plotted:

$$\psi = u_\lambda(x) \cos (2\pi x/\lambda) \tag{8-10}$$

whereas the actual wave function has the form

$$\Psi = u_\lambda(\mathbf{r})e^{2\pi i\left(\mathbf{k} \cdot \mathbf{r} - \frac{Et}{h}\right)} \tag{8-11}$$

The ψ plotted oscillates near each atom core just like the $3s$ wave function of sodium, but is modulated by the cosine factor (dashed line in

---

* See F. Seitz, *Modern Theory of Solids*, McGraw-Hill, New York, 1940, Chapter 8; C. Kittel, *Introduction to Solid State Physics*, Wiley, New York, 2nd Ed., 1956, Chapter 11; J. C. Slater, *Quantum Theory of Matter*, McGraw-Hill, New York, 1951, pp. 275–286.

Fig. 8-16), which depends on λ. If the electrons were free (no atoms present), the $u_\lambda(x)$ term would be absent. Thus the Bloch function combines the properties of free electrons with the properties of electrons bound to the atoms composing the solid.

REFERENCES

C. Kittel, *Introduction to Solid State Physics*, Wiley, New York, 2nd Ed., 1956, Chapters 3 and 11.

W. Shockley, *Electrons and Holes in Semiconductors*, Van Nostrand, New York, 1950, Chapter 5.

F. Seitz, *Modern Theory of Solids*, McGraw-Hill, New York, 1940, Chapters 1, 2, and 8.

J. C. Slater, *Quantum Theory of Matter*, McGraw-Hill, New York, 1951, Chapters 9 and 10.

W. Hume-Rothery and G. V. Raynor, *The Structure of Metals and Alloys*, Institute of Metals, London, 1954.

F. O. Rice and E. Teller, *Structure of Matter*, Science Editions, New York, 1961, Chapter 8.

L. V. Azároff, *Introduction to Solids*, McGraw-Hill, New York, 1960.

A. J. Dekker, *Solid State Physics*, Prentice-Hall, Englewood Cliffs, N.J., 1957.

PROBLEMS

8-1.* In chemistry it is customary to express the latent heats of vaporization and fusion in kilocalories per *gram* molecular weight. One kilocalorie (1000 calories) equals 4180 joules. How many kilocalories per gram-molecular weight equal 1 e.V. per molecule?

8-2.* Calculate the cohesive energy of NaCl for which $R_0 = 2.81$ Å. Neglect the Exclusion Principle repulsion, zero-point energy, and van der Waals energy.

8-3. The crystal structure of BaO is the NaCl structure, and the cohesive energy is 10.4 e.V. per molecule. Predict the cohesive energies per molecule of the hypothetical crystals Ba^+O^- and $Ba^{++}O^{--}$. The observed nearest-neighbor internuclear distance is $R_0 = 2.76$ Å, the first and second ionization potentials of Ba are 5.19 and 9.96 e.V., and the electron affinities of the first and second electrons added to the neutral oxygen atom are 2.2 and about -9.0 volts. Which structure do you predict will occur?

8-4.* Assume that the Exclusion Principle repulsive energy in KCl is of the form CR^{-n}, where C and n are constants. The experimental value of the cohesive energy is 6.62 e.V. per molecule. Assume that the difference between this value and the value of eq. 8-2 is caused entirely by the neglect in eq. 8-2 of the Exclusion Principle repulsion. Compute n.

8-5. The experimental value of the cohesive energy for NaCl is 6.61 e.V. (which happens to be approximately the same as for KCl). Compute n as in prob. 8-4.

8-6. The internuclear distances R_0 of nearest neighbors in a group of alkali halides with the NaCl structure are: LiF, 2.01 Å; LiCl, 2.57 Å; LiBr, 2.75 Å; NaF, 2.31 Å; NaCl, 2.81 Å; NaBr, 2.98 Å; KF, 2.67 Å; KCl, 3.14 Å; KBr, 3.29 Å. Investigate with these data the hypothesis that the repulsive energy changes so rapidly as R changes that a fairly definite *ionic radius* can be deduced for each ion. Make a table of ionic radii based on the assumption of a radius of 1.33 Å for F^-.

8-7. How many *third-nearest neighbors* are there in the NaCl structure? Evaluate the first three terms in the infinite series for the Madelung constant α for this structure (nearest, second-nearest, and third-nearest neighbors). (Because the convergence of this direct summing approach is so slow, α is actually evaluated in practice by more elegant, indirect methods.)

8-8. A *lattice* may be defined as the set of points (*lattice points*) in space given by the vectors $\mathbf{R} = m_1\mathbf{a}_1 + m_2\mathbf{a}_2 + m_3\mathbf{a}_3$, where the m's are allowed to take on all $\pm$ integer values or zero and where the $\mathbf{a}$'s are vectors giving translations along arbitrary, non-coplanar coordinate axes. A crystal may be defined as a repetition in three dimensions of a fixed group of atoms. Thus a crystal may be regarded as a fixed group of atoms placed in the same arrangement about each of the lattice points given by $\mathbf{R}$. It follows that in any given crystal structure, the immediate (and extended) surroundings of each lattice point are identical. Show that in the diamond structure (Fig. 8-2) the points at the corners and face-centers of the large cube (e.g., points like C and D) may be taken as lattice points, but that points such as O do not have the same surroundings as C and D and hence are *not* lattice points. (It is only in the face-centered and body-centered cubic structures that atoms occur at and only at lattice points.)

8-9. Sketch the NaCl structure in three dimensions. Show that the volume per molecule, Ω, is $2R_0^3$. To do this, choose any cubical volume V. Count as belonging entirely to V any ion wholly inside V. Count as belonging $\frac{1}{2}$ to V any ion located on a face, as $\frac{1}{4}$ to V any on an edge, and as $\frac{1}{8}$ to V any on a corner of the cube.

8-10. As in prob. 8-9, sketch the diamond structure and show that the volume per atom is $\Omega = a^3/8$, where a is the length of the side of the large cube in Fig. 8-2. Show that the nearest-neighbor distance is $R = a\sqrt{3}/4$.

8-11.* In Sec. 8-2 it was noted that ions can follow an electric field at frequencies as high as about 10^{13} cycles. Make a crude verification of this statement as follows: Assume that a potassium ion in a KCl crystal has a natural frequency of oscillation about its equilibrium position equal to 10^{13} cycles. Calculate the potential energy as a function of the distance $R - R_0$ from the equilibrium position (compare eq. 7-8), and compare with Fig. 8-1.

8-12. The crystal structure of solid GaAs is the *zincblende* structure. Each Ga is surrounded by four nearest neighbors, all As atoms. Each As is surrounded by four nearest neighbors, all Ga atoms. Except for the distinction between the two kinds of atoms, the structure is the same as the diamond structure illustrated in Fig. 8-2. Discuss the type or types of binding in GaAs. (The melting point of germanium is 958°C and of GaAs is 1240°C.)

8-13. A new crystal form of boron nitride (BN) has been made artificially through use of high-temperature, high-pressure techniques similar to those used in making diamonds. The structure is the zincblende structure (prob. 8-12). Discuss the binding type or types and whether the binding energy should be less than or greater than that of diamond.

8-14. Make models of the following crystal structures from corks or gumdrops and toothpicks: (*a*) NaCl; (*b*) face-centered cubic (like NaCl with all the Cl's removed); (*c*) body-centered cubic (Fig. 8-3); diamond; zincblende (prob. 8-12).

8-15. Hooke's law for volume changes states that compressional stress equals a constant times volume strain. Show how this law follows from E vs. R curves like Fig. 8-1 or 8-5. *Hint:* Use analysis similar to the first few paragraphs in Sec. 5-5.

8-16.* The compressibility β of KCl is 4.8×10^{-11} m.^{2}/Newton. Use the method of the preceding problem to determine and plot to scale the shape of $E(R)$ near the minimum in Fig. 8-1 (i.e., let $|R - R_0| \ll R_0$). Compute the frequency of vibration of a K^+ ion in this force field. Calculate the wavelength of electromagnetic waves of this frequency. In what region of the spectrum are such waves? [β is the ratio of volume strain (dV/V) to compressional stress (pressure, Newtons/m.^{2} or joules/m.^{3}) and is the reciprocal of the bulk modulus. Since the pressure $p = -dE/dV$ and $\beta = -(1/V)(dV/dp)$, we can find $d^2E/dV^2 = 1/V\beta$ and from this compute d^2E/dR^2.]

8-17. Pressures up to $\sim 10^9$ Newtons/m.^{2} are commonly employed (and up to $\sim 10^{10}$ Newtons/m.^{2} occasionally employed) in order to decrease R from its equilibrium value R_0 and thereby to investigate $E(R)$ curves for solids. Use the magnitude of compressibility given in prob. 8-16 to discuss the extent to which the $E(R)$ curve in Fig. 8-1 can be explored by such techniques. To what extent can $E(R)$ for $R > R_0$ be determined?

8-18.* Titanium carbide, TiC, is a promising high-temperature material with the NaCl structure and nearest-neighbor distance 2.15 Å. What is the carbon-carbon distance? Compare with the C—C distance in diamond (1.54 Å) and in graphite (1.42 Å). Do you expect the C—C bond to contribute appreciably to the binding energy of TiC?

8-19. Is it possible to separate metallic from non-metallic solids by a single, jogged line on the periodic system chart (Appendix B)? If this is possible, indicate where such a line crosses each of the six complete rows of the chart. (Use tabulated electrical resistivity values as the principal indication of metallic behavior.)

8-20. Most metals have *close-packed* structures (like face-centered cubic, body-centered cubic, or hexagonal close-packed). Show from the nature of metallic binding that arranging the atoms so that there are as many as possible per unit volume is likely to produce the lowest energy for the system.

8-21. Gold, silver, copper, and many other metals have a *face-centered cubic* structure. Sketch this structure and compare with the arrangement of the potassium ions in Fig. 4-20.

8-22.* Let a be the edge length of the basic cube in either the body-centered cubic (Fig. 8-3) or face-centered cubic (prob. 8-14 or 8-21) structures. What is the nearest-neighbor internuclear separation in each structure in terms of a?

8-23.* What is the location of the point in the body-centered cubic structure that is farthest from the nearest nucleus? What is the maximum distance from any nucleus that an electron can ever be in lithium metal? *Hint:* The point in question is on a cube face; compare your result with Fig. 8-4.

8-24. The compressibilities of the alkali metals are much larger (a factor of 10 or more) than those of other metals or ionic crystals. Discuss the implication of this fact for the relative importance of the ion core repulsion in alkali metals and in other solids.

8-25. Sketch the (a) and (b) type wave functions as in Fig. 8-8 but for a linear array of *ten* hydrogen atoms. Determine by an analysis of average energies $\bar{P}$ and $\bar{K}$ (as in Sec. 5-7) the answers to the following questions: Is the energy of an electron in an (a) type state in the array of ten much different from its energy in the same type state in the array of six? Is there much difference in the energies of the (b) states for the array of ten and of six?

8-26. Sketch the real part of a $2s$ wave function for the array of six hydrogen atoms, like Fig. 8-8a. Study Fig. 8-16 and eq. 8-3 before attempting this.

8-27. Sketch the real parts of the other four wave functions, with energies lying between the extreme cases of Fig. 8-8a and 8-8b, for the array of six hydrogen atoms. Use wave functions like eq. 8-3.

8-28. Consider a linear array of four equally spaced square wells, spaced about as in Fig. 7-12. Sketch the four wave functions with lowest energies, all derived from the $n = 1$ state of a single well (as was done in Fig. 8-8 for two ψ's for the problem of six H atoms). Plot the energies E for these four as functions of $1/\lambda$ (as in Fig. 8-12) for $0 < 1/\lambda < 1/2R$; you will have to use an arbitrary, qualitative E scale, but the location of points on your $1/\lambda$ scale can be approximately quantitative. Compare the shapes of your ψ's with eqs. 8-3 or 8-10, namely, $\psi = u_\lambda(x) \sin (2\pi x/\lambda)$, where $u_\lambda(x)$ is the same as ψ for $n = 1$ of the square well. [The x in $u_\lambda(x)$ need not be measured from the same position $x = 0$ as the x in $\sin (2\pi x/\lambda)$; also, $\cos (2\pi x/\lambda)$ can be substituted for $\sin (2\pi x/\lambda)$.]

8-29. In Fig. 8-12 the energies at the tops of the allowed bands are marked E_1, E_3, $\cdots$. Show that these values of E are the set of allowed energies that was obtained in the square-well problem of Sec. 5-4. *Hint:* Consider eq. 8-4 for the case $\beta R = n\pi$, where n is an integer.

8-30. Show that, if the strength $P_0 w$ of the barrier between wells is very small, eq. 8-4 reduces to the relation between wavelength and energy that is appropriate for a free electron in a region of zero potential energy.

8-31.* Find the energy E_2 at the bottom of the second allowed band for the particular strength $P_0 w$ of the barrier for which Fig. 8-12 is drawn. *Hint:* The value of βR lies between π and 2π.

8-32. Devise a *low-pass* wave filter composed of resistanceless elements (i.e., all frequencies up to a cut-off frequency should be passed and higher frequencies rejected). Present your filter in the form of a diagram like Fig. 8-14 but with L's and C's replacing X's.

8-33. Devise a *band-pass* filter, as in prob. 8-32 (i.e., only frequencies within a certain band are passed).

8-34. Draw a sketch like Fig. 8-16 but for the 3s wave function with the lowest energy (bottom of the $n = 3$ band).

8-35. Draw a sketch like Fig. 8-16 but for the $n = 1$ wave functions at the top and at the bottom of the $n = 1$ band.

8-36. For a *two*-dimensional lattice, the interference condition eq. 8-7, and hence the gaps in the allowed E values in Fig. 8-12, will appear at different $1/\lambda$ values for different directions of motion of the electron wave. Show that this fact leads to the possibility of *overlapping* of allowed bands, as in Fig. 8-10. (The three-dimensional, Brillouin-zone treatment provides the same possibility.)

9

Electrical, Thermal, and Magnetic Properties of Solids

9-1 INTRODUCTION

The basic theory of binding and energy bands is applied
in this chapter to some of the properties of solids. The
band theory will be used in Sec. 9-2 to show why the electrical conductivity of some solids is very much greater than
that of others. The physical distinction between conductors and non-conductors is readily accomplished with the
theory of the preceding chapter and impossible to accomplish without that theory. The distribution of electron
energies in the partially filled band of a metal will be developed in Secs. 9-3 and 9-4. The study of these energies
is necessary to complete the theory of metallic binding
(Sec. 8-3), but the principal reason for this study is that
the results are required for the discussion of the properties
of metals in the present chapter. The thermal properties
of solids will be studied in Sec. 9-5 because they are of interest in themselves and because the thermal vibrations
of the atoms of a solid profoundly affect the electrical conductivity. Electrical conduction in metals will be discussed in Secs. 9-6 and 9-7. The discussion of the electrical
conduction in non-metals is delayed until Chapters 10
and 11, since such conduction is very sensitive to the
presence of the imperfections studied in Chapter 10. The
physics of the magnetic properties of materials, especially
ferromagnetic materials, will be presented in Sec. 9-8.

284

9-2 CONDUCTORS AND NON-CONDUCTORS OF ELECTRICITY

The best conductors, metals, have an electrical conductivity greater than that of the best insulators by more than a factor of 10^{24}. No other property of solids has such an enormous range of values; indeed, it is difficult to conceive of such an enormous range of values for *any* property, whether of solids, atoms, nuclei, or macroscopic objects. Clearly there must be some interesting physics behind the electrical conductivity of solids. In addition to the magnitude of the conductivity, there is another distinction between good and poor conductors, namely, that the conductivity of a pure crystal of a metal or alloy is greater than that of an impure crystal. The poor conductors (*semiconductors* and *insulators*) have larger conductivities when impure than when pure.

Our task in this section is to distinguish conductors from non-conductors on the basis of the band theory of electron energy states. We have seen in Secs. 8-4 and 8-5 that the energies of electrons in solids can have only certain values, namely, those values lying within the allowed energy bands. For energies within such a band, electrons have wave functions like eq. 8-3 and are traveling waves that move through the crystal. In all solids the low-lying energy bands are completely *filled;* that is, there is one electron occupying each allowed energy state. It was noted in Sec. 8-4 that bringing atoms together into a solid does not change the total number of quantum states of any one kind (e.g., $1s$ or $3p$) provided that the bands do not touch one another. Thus if N lithium atoms are brought together, each with two $1s$ states, there are $2N$ of these states in the solid. There were $2N$ electrons occupying these states in the individual atoms, and these are just sufficient to fill all the $1s$ states in the resulting solid. Electrons in these states are going in all directions and with all values of energy that lie within the band.

No electric current can be carried by electrons in a filled band. We can show this by first considering the solid without an applied electric field. There is, of course, no electric current, no net momentum of electrons, in any one direction. Although electrons are moving in all directions, the vector sum of all the momenta is zero. For each quantum state representing, for example, motion of an electron in the $+x$-direction with a certain speed, there is a similar state giving motion in the $-x$-direction with the same speed. Furthermore, both states are occupied by electrons. This is an obvious result—we could not expect a current without an electric field to urge the electrons predominantly in one direction.

Now suppose that an electric field is applied to the solid. If the band remains filled, there are still just as many electrons traveling just as fast in one direction as in the opposite direction. The field does not change the quantum states (see prob. 9-2) or the fraction of quantum states filled, since all are filled. The action of an applied field is illustrated in Fig. 9-1, which is the lowest band part of Fig. 8-12 (the same result would be obtained with any other band). It should be recalled from Sec. 8-5 that a positive λ means an electron wave traveling toward $+x$. If ε is directed so as to accelerate electrons toward $+x$, it can increase the energy and momentum of an electron like A or B. But some of the electrons (like C) will be accelerated sufficiently to reach a critical value of λ, to suffer Bragg reflection, and to appear at negative λ's (like C'). Thus the distribution of electron momenta is just the same as before the application of a field, and the field is unable to produce a current.

At first sight it might appear that the electric field could give an electron enough energy to excite it to another band. If so, the argument of the preceding paragraph would not hold, since we should no longer have a filled band. But the electric field cannot do this because the spacing between bands is too large (usually several volts). The electric field accelerates an electron, but the electron ultimately suffers a collision. When this occurs, the effect of the field on the electron's velocity is destroyed, and the acceleration process must start all over again. The *mean free path* between collisions is only of the order of 10^{-8} m., and the electron can gain only a tiny energy from even a very strong field. Even if the field were as great as 10^4 volts/m., an energy of only 10^{-4} volt would be attained, and this is far too small to permit excitation to the next higher energy band.* [There is one interesting but rare exception to the conclusion of this paragraph. It is possible to produce a very high concentration of electric field inside a semiconductor crystal, and this high field produces excitation to a higher band. This is the Zener effect (also called tunneling), which will be discussed in Sec. 11-4.]

Of course it might be possible to have *thermal excitation* of electrons from a filled band to an empty band above it. This process does produce appreciable conduction in solids that have a narrow forbidden band separating a filled and an empty allowed band. Such solids—*intrinsic*

---

* The description of the collision process is necessarily vague at this point; it will be discussed in more detail in Sec. 9-6. It is worth noting that there is also an experimental way of demonstrating that the field does not excite an electron from one band to the next. Ohm's law holds with as small voltages as can be measured; the conductivity of a specimen with only microvolts of applied voltage is the same as that of the same specimen with several volts applied. Certainly in the former case no electron could receive an energy from the field sufficient to excite it to another band. Since the conductivity is the same at higher voltages, this process must not be occurring at any voltage for which Ohm's law holds.

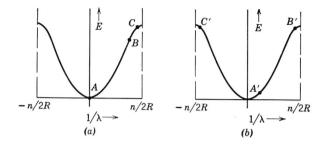

Fig. 9-1. Energy as a function of reciprocal wavelength for a filled band. (a) No electric field. (b) Electric field accelerating electrons toward $+x$, but still no current.

semiconductors—will be discussed in Sec. 11-2. The fraction of electrons so excited is very small, however, even when the forbidden band is only a volt or two in width. According to eq. 6-10, this fraction is determined by the Boltzmann factor $e^{-(E_2-E_1)/kT}$. Since kT equals $\frac{1}{40}$ e.V. at room temperature, the Boltzmann factor for an $E_2 - E_1$ of the order of 1 e.V. is very small. Thus the number of thermally excited electrons is only a tiny fraction of the total number of electrons. In good insulators the band gap is many volts, and what little conduction there is comes from other processes (see Chapter 10).

If a solid has a *partially filled band*, it is a good conductor of electricity. The effect of the electric field in this case is illustrated in Fig. 9-2. The field accelerates electrons toward the right, increasing $1/\lambda$ and the mo-

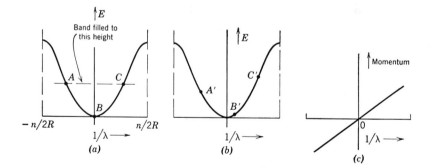

Fig. 9-2. Energy and momentum as functions of reciprocal wavelength for a partially filled band. (a) No electric field. (b) Electric field producing a net current of negative charge toward $+x$. (c) Momentum vs. $1/\lambda$, plotted here to recall that if more electrons have positive λ's than negative there is a net current of electrons toward $+x$.

mentum of those with $\lambda > 0$ and decreasing $|1/\lambda|$ and the magnitude of the momentum of those with $\lambda < 0$. A net total momentum is thus attained, and a current flows. Such a partly filled band is called the *conduction band*. Of course there is at most one such band in a particular solid.

The problem of distinguishing conductors from non-conductors has now reduced to the question: "Does the solid have a partially filled band?" If it does not, it cannot be a good conductor because there are no electrons (or very few) that can give a net current when an electric field is applied. If it does, it has an abundance of electrons that can participate in conduction and is a good conductor.

We now consider some examples of particular solids, beginning with sodium. The band structure is shown in Fig. 8-10. Since all the $n = 1$ and $n = 2$ states are filled in sodium atoms, they will also be filled in the solid. N electrons, which are the $3s$ electrons of the N atoms, remain. There are $2N$ states in the $3s$ band, and therefore this band is only partially filled. In addition, there are the $6N$ states in the $3p$ band that overlaps the $3s$ band. Thus the $n = 3$ band is only partially filled, and sodium is a good conductor of electricity.

Another interesting example is KCl. We saw in Sec. 8-2 that the KCl crystal is composed of K^+ and Cl^- ions. The band structure (Fig. 9-3)

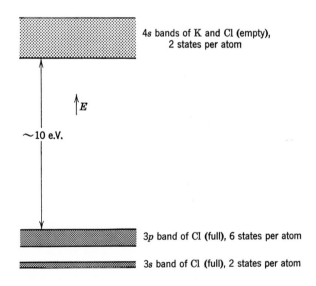

4s bands of K and Cl (empty),
2 states per atom

E

$\sim$ 10 e.V.

3p band of Cl (full), 6 states per atom

3s band of Cl (full), 2 states per atom

Fig. 9-3. Schematic diagram of KCl energy bands. The $3p$ and $3s$ bands of K are considerably below those of Cl.

is hence the energy bands produced by the broadening of the levels of these two ions, both of which have complete rare-gas shells. The $3d$ and $4s$ states lie very much higher. There are just enough electrons to fill the $3s$ and $3p$ bands. Therefore there are no partially filled bands, and KCl is an insulator.

A generalization from the last two paragraphs is justified: All the alkalies form metallic solids; all ionic crystals with rare-gas ions form insulators.

Beryllium is another interesting example. Both the $2s$ states in the atom are filled and all the $2p$ are empty; therefore we might expect beryllium to be an insulator. The $2p$ band overlaps the $2s$, however, and the combined band has four times as many states as electrons and hence satisfies our criterion for a metal. In general, atoms with 1, 2, or 3 valence electrons outside a rare-gas shell produce metallic solids.

Considerable caution must be exercised in trying to decide whether other solids will be metals or non-metals. For example, consider hydrogen. We might expect hydrogen to form a metal, and it might if it were not for the strength of the hydrogen *molecule* bond. As pointed out in Sec. 8-2, the binding energy of this molecule is large enough to keep the hydrogen atoms bound in pairs, even in the solid, and the hydrogen molecules are weakly bound together. The energy bands are therefore very nearly just the energy levels of the hydrogen molecule. In discussing this molecule we have noted that the $1s$ levels split into two groups, with a considerable energy difference between them (Fig. 7-20). The two electrons fill this lower pair of states in the molecule. Hence, in the solid, the lower band is filled, the upper band is empty, and solid hydrogen is an insulator.

A similar effect occurs in other solids with covalent binding (such as diamond, silicon, germanium, and gray tin). All these have a band structure like that shown in Fig. 9-4. Unlike solid hydrogen, these solids are not molecular solids, and the binding is strong. Detailed calculations have shown that, with this crystal structure and with covalent binding, the bands *cross* (rather than overlap as they do with metallic binding), leaving an energy gap E_g. There are four states per atom in the lower branch and four states per atom in the upper branch.

This splitting of the band into halves is characteristic of covalent binding and is related to the splitting of the levels in the hydrogen molecule. The lower branch corresponds to wave functions (like Fig. 7-16a) that concentrate electronic charge along the lines joining the atoms; these functions have the four-pronged symmetry of the charge distribution shown in Fig. 7-9. The upper branch corresponds to wave functions (like Fig. 7-16b) that have a very small average charge along these lines,

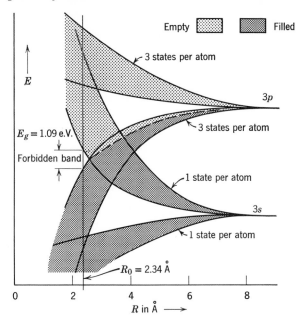

Fig. 9-4. Schematic diagram of the energy bands of silicon as a function of nearest-neighbor distance R. At the observed value R_0 of R the bands have crossed and leave an *energy gap* E_g of 1.09 e.V. The bands for germanium are similar, but $R_0 = 2.43$ Å and $E_g = 0.72$ e.V.

which are regions of low potential energy. The wave functions of both these branches at $R = R_0$ are distinctly different in symmetry from the $3p$ and $3s$ wave functions for large R. The two types of wave functions differ appreciably in energy, as explained in connection with Fig. 7-20, and therefore an energy gap occurs. At the observed lattice constant, this gap is 7 e.V. for diamond, 1.09 e.V. for silicon, 0.72 e.V. for germanium, and about 0.2 e.V. for gray tin. Since there are four valence electrons per atom, the lower band is filled and the upper is empty at low temperatures. For all these except diamond, the energy gap is small enough for an appreciable number of electrons to be thermally excited to the upper band at room temperature. These materials, semiconductors, will be described further in Chapter 11.

Experimental evidence that the analysis in this section is correct will be presented at various places in later sections, since the details of the electron energy distribution in solids must be described before most of these experiments can be made intelligible.

9-3 FERMI DISTRIBUTION OF ELECTRON ENERGIES

We must now digress to study the distribution of energies of electrons in solids. The theory developed will be general, but it will be immediately applied only to electrons in metals. In later sections it will be applied to semiconductors and to problems in physical electronics.

The relative numbers N_1 and N_2 of atoms in the energy states E_1 and E_2 of an assembly of hydrogen atoms was stated in Sec. 6-2 to be

$$\frac{N_2}{N_1} = \frac{w_2}{w_1} \, e^{-(E_2 - E_1)/kT} \tag{6-10}$$

The number of different wave functions in the atom with energy E_1 was called w_1, and similarly for w_2. There was no necessity in that problem to consider the Exclusion Principle, since there was only one electron in each atom. Equation 6-10 can also be applied to the higher excited states of any atoms or to the quantum states near the *top* of a conduction band in a solid, since those states are almost always empty. Suppose that for some E the fraction of conduction band states that are occupied is 10^{-6}. It would then be extremely unlikely for *two* electrons to be in the same state, and so the predictions of eq. 6-10 are not in appreciable disagreement with the Exclusion Principle when applied to states that are almost always empty.

The situation is quite different for the lower states of many-electron atoms or for conduction band electrons in metals, since the Exclusion Principle must be considered in these problems. (In Sec. 6-4 the Exclusion Principle was included from the beginning.) The classical Maxwell distribution of energies of free particles predicts an average kinetic energy of $\frac{3}{2}kT$. If all the conduction band electrons in a metal were crowded into the energy states lying within $\sim \frac{3}{2}kT$ (~ 0.04 e.V. at room temperature) of the bottom of the band, there would be more than 1000 electrons per quantum state. Therefore this distribution is quite wrong for electrons in a metal.

We shall present the correct distribution of electron energies in the present section; it is called the *Fermi distribution*. We shall not derive * it, but only show that it agrees with the Exclusion Principle and with eq. 6-10 in the region of high energies, where eq. 6-10 should apply.

The Fermi distribution is

$$N(E) \, dE = S(E) \, dE \, [f(E)] = S(E) \, dE \left[\frac{1}{e^{(E - E_0)/kT} + 1} \right] \tag{9-1}$$

* A brief derivation can be found in C. Kittel, *Introduction to Solid State Physics*, Wiley, New York, 2nd Ed., 1956, pp. 251–255.

$N(E)\ dE$ is the number of *electrons* per unit volume with total energy between E and $E + dE$. $S(E)\ dE$ is the number of *quantum states* per unit volume with total energy between E and $E + dE$; it is different for different solids and can be computed from the energy band calculations described in Secs. 8-4, 8-5, and 9-4. Of course $S(E)$ equals zero for values of E lying in the forbidden bands. $f(E)$ is called the *Fermi factor;* it is the probability that a quantum state with energy E is occupied. It is a universal function applicable to all solids and to many other systems (e.g., heavy nuclei and very high-pressure gases, such as are encountered in shock waves). E_0 is a parameter with the dimensions of energy that is called the *Fermi level*, to be explained below.

Equation 9-1 satisfies the Exclusion Principle. This is apparent since $f(E)$ is never greater than unity, because the first term in the denominator is never negative. Thus $N(E)$ cannot be larger than $S(E)$, and the Fermi distribution does not put more than one electron into a single quantum state.

Equation 9-1 also agrees with eq. 6-10 in the region of high energies $(E \gg E_0)$ where eq. 6-10 should apply. If E is greater than E_0 by more than a few kT, the first term in the denominator of eq. 9-1 is very large. The second term $(+1)$ can therefore be neglected, and

$$f(E) \cong e^{-(E-E_0)/kT} \tag{9-2}$$

In order to compare eq. 9-1 with eq. 6-10 it is necessary to compute the ratio of the number of electrons per unit energy interval for one value E_2 to the number for another value E_1. This ratio can be computed by using eq. 9-2:

$$\frac{N(E_2)}{N(E_1)} = \frac{S(E_2)e^{-(E_2-E_0)/kT}}{S(E_1)e^{-(E_1-E_0)/kT}} = \frac{S(E_2)}{S(E_1)} e^{-(E_2-E_1)/kT} \tag{9-3}$$

Since $S(E_2)/S(E_1)$ is the ratio of the numbers of quantum states at the

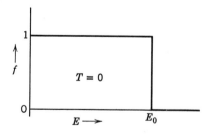

Fig. 9-5. Fermi factor f as a function of energy at absolute zero. All states with $E < E_0$ are filled, and all others are empty.

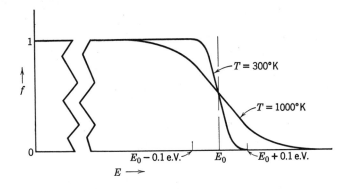

Fig. 9-6. Fermi factor f as a function of energy at two temperatures. Note that the variation of f occurs within a few kT of the Fermi level E_0. The zero of E may be a few volts or many volts below E_0, depending upon the arbitrary choice of $E = 0$.

two energy values, this ratio is the same as w_2/w_1, and eq. 9-3 agrees with eq. 6-10.

It is instructive to examine the Fermi factor

$$f(E) = \frac{1}{e^{(E-E_0)/kT} + 1} \qquad (9\text{-}4)$$

in the limit $T \rightarrow 0$. If $E < E_0$, the first term in the denominator approaches $e^{-\infty} = 0$. Therefore $f(E) = 1$ for $E < E_0$. If $E > E_0$, this term approaches $e^{+\infty} = \infty$. Therefore $f(E) = 0$ for $E > E_0$. The factor $f(E)$ is plotted in Fig. 9-5 for $T = 0°\text{K}$. Every quantum state with $E < E_0$ is filled, and every other quantum state is empty. (This is the occupation probability that was tacitly assumed in Sec. 6-4.)

The Fermi function is plotted in Fig. 9-6 for two values of T other than zero. In this plot the zero of E may be many electron volts to the left of the region plotted.* At any temperature the function $f(E)$ goes from nearly 1 to nearly 0 when E changes from E_0 minus a few kT to E_0 plus a few kT. Since kT at any temperature below the melting points of solids is of the order of 0.1 e.V. or less, we cannot plot $f(E)$ on a plot covering several electron volts without losing the details of the variation in $f(E)$ near $E = E_0$. (Such a plot would look almost like Fig. 9-5 at *any* ordinary temperature.) Further interesting properties of $f(E)$ are explored in the problems.

The significance of the Fermi level E_0 may already be apparent. At

---

* Of course the zero of E is arbitrary since the zero of P is arbitrary. In Sec. 9-4 it will be convenient to define $E = 0$ at the bottom of the conduction band, and E_0 will be several electron volts.

$T = 0$, the lowest quantum states are filled with electrons until there are no more electrons. The value of E that is reached at this point is E_0, and thus E_0 is frequently called the *Fermi brim*. At any temperature, E_0 can be determined by equating the total number of electrons N to the total number of occupied quantum states $\int_0^\infty S(E) f(E) \, dE$, but this equation is not easy to use when $T \neq 0$. In the next section E_0 will be calculated for a particularly important problem, and additional examples of the calculation of E_0 will be given in Chapter 11.

One of the reasons for the importance of the Fermi level is its significance in the problem of two solids in contact. If two solids are in contact and the system is in thermal equilibrium, E_0 in one solid must be at exactly the same energy as in the other solid. To verify this statement we first suppose that it were not true, and that upon placing the solids in contact, E_0 in solid A were lower than in B. Then some electrons from B would spill into A, charging A negatively and B positively, since the energy of the system would be lowered by the transfer of these electrons to lower energy states. This charging process in turn would change the electrostatic potential of B relative to A, which raises the energies of all the bands and the energy E_0 in A and lowers them in B. This process stops only when E_0 in A is aligned with E_0 in B. Thus the Fermi level, rather than the electrostatic potential, is constant across a boundary between dissimilar solids in equilibrium. Since the position of the Fermi level governs the supply of electrons at any value of E, it is not surprising that E_0 has this property. It should also be noted that in thermodynamic language E_0 is the electrochemical potential for electrons, and a rigorous thermodynamic argument can be applied to demonstrate the alignment of Fermi levels at a contact.

At the contact between dissimilar solids, there is always a double layer of positive and negative charges and a step in the electrostatic potential (an important example of this will be studied in Sec. 11-4). At the external surfaces of two dissimilar solids in contact, there is always a difference in electrostatic potential. This is the *contact difference in potential*, which will be explained in Sec. 12-3. If the two solids are not in thermal equilibrium but are connected to a voltage source, the Fermi levels no longer coincide. The difference in Fermi levels is the product of the electronic charge and the voltage of the battery (or other source) to which the solids are connected.

9-4 CONDUCTION BAND ELECTRONS IN METALS AND ALLOYS

To apply the Fermi distribution to electrons in the partially filled band (conduction band) of a metal we must determine the density of

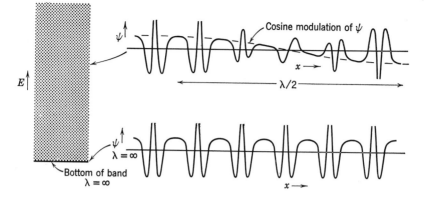

Fig. 9-7. The real part of ψ plotted as a function of the distance along a line of atom centers for a wave function somewhat above the bottom of the band and for a wave function at the bottom of a band ($\lambda = \infty$). Note that the rapid oscillations near the atom cores contribute approximately the same energy in each case. The higher energy of the upper ψ comes from the kinetic energy of the cosine modulation, which is the kinetic energy of translation that a free electron would have.

states $S(E)$ in the conduction band. We shall first show that $S(E)$ for these electrons is approximately the same as for completely free electrons if E is not too close to the top of the band. Then we shall compute $S(E)$ and $N(E)$ and compare the computed values with the results of X-ray experiments.

The energy E of an electron can be separated into two terms to a good approximation: (1) The kinetic energy of translation of the electron through the crystal; this is the energy derived from the

$$e^{2\pi i\left(\frac{x}{\lambda} - \frac{E t}{h}\right)} \tag{9-5}$$

term of eq. 8-3 (or the similar term in eq. 8-11). (2) The kinetic and potential energy of the electron's interaction with each ion core; this is the energy derived from the $u_\lambda(x)$ term of eq. 8-3 (or the similar term in eq. 8-11). The rapid oscillation of $u_\lambda(x)$ near each atom core produces a large $|d\psi/dx|^2$ and hence a large kinetic energy. There is also a large negative potential energy from this term since it makes $|\psi|^2$ large in the region of large negative P. These two energies derived from $u_\lambda(x)$ are almost the same for all values of λ, as illustrated in Fig. 9-7. We can thus set $E = 0$ at the bottom of the conduction band, equivalent to assuming that the $u_\lambda(x)$ energies are constant, as we investigate the distribution of the translational kinetic energies in the conduction band. With this definition of $E = 0$, E equals the translational kinetic energy,

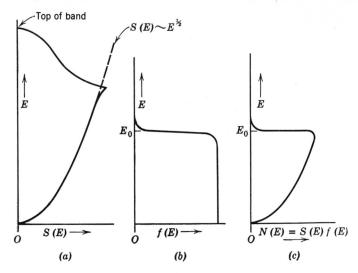

Fig. 9-8. Quantities involved in the distribution of electrons in the conduction band of a metal. (*a*) Number of quantum states per unit energy range (density of states). (*b*) Probability that a state is occupied. (*c*) Number of electrons per unit volume per unit energy range.

and the relations among the wavelength, momentum, and energy of conduction band electrons are approximately the same as for free electrons moving in a region of constant potential, that is, $E = \frac{1}{2}mv^2$ and $\lambda = h/mv$.*

$S(E)$ can be calculated in this approximation by determining the number of quantum states per unit energy interval for nearly free electrons, electrons bound to a region of space (much larger than atomic sizes) in which the electrostatic potential is constant. This calculation must be performed in three dimensions to be relevant to real solids; it is carried out in Appendix G. The result (eq. G-9) is

$$S(E) = \left(\frac{2^{1/2}m^{3/2}\pi}{h^3}\right)E^{1/2} \tag{9-6}$$

* The free electron approximation outlined here is not the best approximation to the actual behavior of conduction band electrons. A better approximation can be obtained by replacing the electron mass by an *effective mass* m^* which is different for each solid and for each energy in the band. In the lower-energy part of bands of simple metals, m^* equals approximately the free electron mass. The effective mass approximation is quantitatively better but predicts no results that are qualitatively different from the free electron approximation. We shall use the free electron approximation.

This function is plotted in the lower part of Fig. 9-8a, with $S(E)$ as abscissa and E as ordinate in order to compare this figure more readily with the energy band diagrams of past and future sections. Figure 9-8a also shows schematically the actual $S(E)$ curve in the upper part of the conduction band, where it departs from eq. 9-6.

$S(E)$ approaches zero at the bottom of the band because there are only a few combinations of quantum numbers that give nearly zero energy; a geometrical representation of this effect is given in Fig. G-1, and the effect is similar to that explored in Fig. 2-4. Another way of seeing that $S(E) \to 0$ as $E \to 0$ is by studying the eq. 6-11 statement of the Exclusion Principle; there are only a few combinations of the momentum components p_x, p_y, and p_z that will permit $E = p^2/2m$ to be nearly zero.

$S(E)$ also approaches zero at the *top* of a band. This property, distinctly different from the properties of free electrons, is of less interest for metals but becomes of central importance in the study of semiconductors. As E increases, electrons moving in a particular direction eventually have a small enough $1/\lambda$ to suffer Bragg reflection, and their momenta decrease to zero as shown in Fig. 8-13. As E approaches the top of the band, there are fewer and fewer possible directions of motion (directions of the **k** vector of eq. 8-11) that do *not* involve Bragg reflection, the magnitude of the momentum p decreases toward zero, and $S(E)$ decreases accordingly. In fact, $S(E)$ near the top of the band is given simply * by eq. 9-6 with $(E_{\text{top}} - E)$ replacing E.

To return to metals, the number of electrons $N(E)\,dE$ per unit volume with energies between E and $E + dE$ can now be calculated by inserting eq. 9-6 into eq. 9-1. The Fermi factor $f(E)$ is plotted in Fig. 9-8b, and the product $S(E)f(E) = N(E)$ is plotted in Fig. 9-8c. The example plotted is typical of monovalent and divalent metals. The Fermi level E_0 is *not* close to the top of the band, and therefore eq. 9-6 applies to the whole occupied part of the band. For example, if Fig. 9-8 is drawn for an s band (2 states per atom) of a monovalent metal (1 electron per atom), the area under the curve of Fig. 9-8c is one-half the area under the curve of Fig. 9-8a, and E_0 is near the middle of the allowed band.

The vertical position of the $f(E)$ plot relative to the $S(E)$ plot in Fig. 9-8 was determined by calculating E_0. To learn how this is done we shall consider the example of the conduction band of a monovalent metal. Since E_0 is very insensitive to temperature, we shall assume for convenience that T equals zero. E_0 is determined by setting the integral of $S(E)\,dE$ from 0 to E_0 equal to the total number N of electrons per unit volume in the band. (This operation amounts to moving Fig. 9-8b

---

* Again, the effective mass m^* should replace the free-electron mass m, and the m^* at the top of the band is different from that at the bottom.

up or down until the area under the curve of Fig. 9-8c just equals the number of electrons per unit volume.)

$$N = \int_0^{E_0} S(E) \, dE = \frac{2^{7/2} m^{3/2} \pi}{h^3} \int_0^{E_0} E^{1/2} \, dE = \frac{2^{7/2} m^{3/2} \pi}{h^3} \left(\tfrac{2}{3} E_0^{3/2}\right)$$

$$E_0 = \frac{h^2}{8m} \left(\frac{3N}{\pi}\right)^{2/3} \tag{9-7}$$

If E_0 is expressed in electron volts,

$$E_0 = 3.65 \times 10^{-19} N^{2/3} \text{ e.V.} \tag{9-8}$$

E_0 depends, therefore, on the number of conduction electrons per unit volume (it also depends on the effective mass if the more accurate approximation is used). Some values of E_0 are given in Table 9-1. It should be noted that in metals E_0 is always much greater than kT, which means that the Maxwell distribution of velocities would predict far more than one electron per quantum state.

The experimental values of E_0 in Table 9-1 were obtained from X-ray emission spectra. After an electron has been ejected from the $1s$ or $2s$ or $2p$ bands by electron bombardment in an X-ray tube, an electron from a higher band can make a transition to the vacant state. If conduction band electrons make this transition, the resulting X-ray photons will have a distribution of energies over a width E_0. Hence the $N(E)$ distribution can be measured by studying the widths and shapes of the K- or L-emission lines. Such investigations * have led to the results shown in Fig. 9-9 and in the right-hand column of Table 9-1. In the example of the $3s \rightarrow 2p$ transition plotted in Fig. 9-9, the transition

TABLE 9-1

Fermi Level E_0 of Metals

Metal	Calculated E_0	Experimental E_0
Ag	5.5 e.V.	
Au	5.5	
Cu	7.1	
K	2.1	
Li	4.7	3.2 e.V.
Na	3.2	3.2

* D. H. Tomboulian in *Encyclopedia of Physics*, Vol. 30, edited by S. Flügge, Springer, Berlin, 1957, pp. 246–304; F. Seitz, *Modern Theory of Solids*, McGraw-Hill, New York, 1940, pp. 436–441.

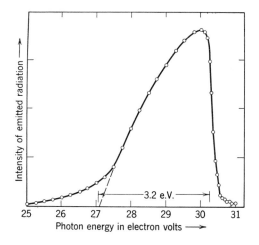

Fig. 9-9. *L*-emission band of sodium. From D. H. Tomboulian. [See also W. M. Cady and D. H. Tomboulian, *Phys. Rev.*, **59**, 381 (1941).]

probability is nearly independent of E, and the width of the $2p$ band is negligible compared to E_0. Therefore the X-ray intensity as a function of energy gives the number of conduction band electrons per unit energy, and this is found to be in very good agreement with the theory given in eqs. 9-1, 9-6, and 9-8. (The rounding at the left end of the experimental curve appears to be caused by more complicated transitions similar to the Auger transitions briefly described in Sec. 6-6.)

The example illustrated is for X-ray transitions of the type $s{\rightarrow}p$, which are allowed in the individual atoms and allowed in the solid. In addition to such transitions, others that would be forbidden in the atoms in a gas are allowed in the solid. For example, the $3s{\rightarrow}2s$ and $3s{\rightarrow}1s$ are permitted for all electrons except those at the very bottom of the $3s$ band (i.e., electrons with $\lambda = \infty$). The reason for this can be seen by consulting Fig. 8-16, where ψ for the $3s$ band of sodium is plotted (see also Fig. 9-7). Note that ψ is an s function near the left-hand atom core, since it is symmetric about this atom center. But at other positions ψ does not have this symmetry, and at the right-hand atom it resembles a p function. There is therefore a possibility of radiation, since the condition of eq. 5-54 is not satisfied unless the symmetry of the wave function for the initial state is identical with that for the final state.

The calculation of the distribution of energies of conduction band electrons permits the completion of the theory of metallic binding (Sec. 8-3). The average kinetic energy $\bar{E}$ of the conduction band electrons was needed for that theory, but the calculation of $\bar{E}$ had to be delayed until the present section. $\bar{E}$ can be shown to equal $\frac{3}{5}E_0$ (see prob. 9-13). As the internuclear spacing R varies, the number N of conduction band electrons per unit volume varies as $1/R^3$. Equation 9-7 therefore predicts that E_0 and $\bar{E}$ will be proportional to $1/R^2$. The repulsion energy curve labeled "kinetic energy of electron gas" in Fig. 8-5 is thus proportional to $1/R^2$. This variation with R is more rapid than that of the attraction term, which varies roughly as $1/R$. Therefore a minimum occurs in the total energy, and a stable crystal is formed. It should be noted that the repulsion energy from the kinetic energy of conduction band electrons, like the ion-core repulsion, is ultimately derived from the Exclusion Principle.

The theory of this section applies to many alloys as well as to metals. The value of E_0 (and hence the kinetic energy of the electron gas) depends on the number of valence electrons per unit volume. As the composition of a homogeneous alloy with a given crystal structure is changed, E_0 varies continuously with composition. If E_0 becomes very large, another crystal structure may become more favorable in that a lower energy for the crystal would be obtained, or a non-metallic form of binding may become more favorable. For example, the transition from α brass (face-centered cubic) to β brass (body-centered cubic) can be understood in terms of the difference between the $S(E)$ curves for the two structures and the fact that the addition of zinc (two valence electrons) to copper (one valence electron) increases the average number of valence electrons per atom.*

9-5 THERMAL PROPERTIES OF SOLIDS

The heat capacity, thermal conductivity, and thermal expansion of solids are considered in this section. The study of thermal properties precedes the study of electrical conductivity of metals since the thermal vibrations of the atoms are the principal cause of electrical resistance in metals at ordinary temperatures. Another reason for studying thermal properties at this point is that additional experimental evidence for the validity of the Fermi distribution is provided by studies of the heat capacity of solids.

(a) Heat capacity

In an ideal monatomic gas, each atom has three degrees of freedom, and on the average each has a kinetic energy equal to $\frac{3}{2}kT$. A kilogram

---

* See the books by W. Hume-Rothery listed at the end of this chaper and of Chapter 8 for many applications of the electron theory of solids to the properties of metals and alloys.

atomic weight of such a gas has an energy equal to $\frac{3}{2}N_0 kT = \frac{3}{2}RT$, where R is the gas constant and equals 8314 joules per kilogram atomic weight per degree centigrade (or 2 calories per mole per degree centigrade). The heat capacity per kilogram atomic weight at constant volume C_v is defined as

$$C_v = dE/dT \tag{9-9}$$

that is, as the rate of change with temperature of the total energy E per kilogram atomic weight. It is understood that the volume is kept constant as T changes. (Frequently we wish the *specific heat*, or heat capacity per kilogram; it can be obtained from C_v by dividing by the atomic weight.) For the ideal gas, eq. 9-9 becomes

$$C_v = \frac{d}{dT}\left(\tfrac{3}{2}RT\right) = \tfrac{3}{2}R \tag{9-10}$$

If there were any contribution to the energy other than the kinetic energy of translation, C_v would be increased; such contributions occur in gases of *molecules*, in which there is energy of rotation and vibration. The energies of electrons in atoms or molecules are not involved in C_v at ordinary temperatures, since the energies of electron transitions are very much greater than kT, and hence the electronic energy does not change with T at ordinary temperatures.

Before the development of the quantum theory, it was thought that a somewhat similar simple theory should apply to solids. Experiments showed that at sufficiently high temperatures the heat capacity of any solid is

$$C_v = 3R \tag{9-11}$$

This relation, known as the *law of Dulong and Petit*, is a good approximation for most materials even at room temperature. It is just what was expected on the basis of classical theory. Each atom in the solid has $\frac{3}{2}kT$ kinetic energy. It is not, however, a free particle but is bound into a position of stable equilibrium and executes simple harmonic oscillations about the equilibrium position. It is possible to show that the average potential energy of such an oscillator equals its average kinetic energy (see prob. 9-24). The total energy of each atom is hence $3kT$ and of a kilogram atomic weight is $3RT$. Equation 9-11 follows by differentiation.

The exact classical theory is actually somewhat more complicated than the theory explained in the preceding paragraph, but the result is the same. We really cannot accurately look upon the individual atoms in the lattice as *independent* simple harmonic oscillators. If a particular atom is displaced toward $+x$ at a particular instant, the equilibrium

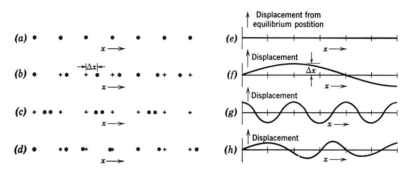

Fig. 9-10. Displacements of atom cores (nuclei plus inner electrons) in the thermal vibrations of a crystal lattice. (*a*), (*b*), (*c*), and (*d*) show actual positions of atoms (•) and equilibrium positions (+). (*e*), (*f*), (*g*), and (*h*) are corresponding plots of displacements from equilibrium positions as functions of *x*.

positions of its neighbors are affected. The motions of the atoms are hence correlated and should not be treated as if they were independent. This correlation can be taken into account by expressing the vibrations of the atoms by the superposition of vibrational waves. In Fig. 9-10 a few waves of various wavelengths are sketched, with amplitudes greatly exaggerated. Any positions of the atoms of the lattice can be described by the superposition of lattice waves, as in Fig. 9-10*h*. (This is just the same process that is used to represent an arbitrary current as a function of time in an electric circuit by the superposition of sinusoidal current-time functions, as is done in Fourier synthesis.) These waves are simply sound waves, but of very high frequencies (see prob. 9-25).

The shortest-wavelength wave is illustrated in Fig. 9-10*c* and 9-10*g*, in which the oscillations of neighboring atoms are 180° out of phase. Waves with wavelengths shorter than this are not different from longer-wavelength waves already counted. For example, a wave with half the wavelength of Fig. 9-10*g* would give every atom identical displacements, which is indistinguishable from the situation of Fig. 9-10*e*. Counting the number of independent waves is just like the counting of wave functions in Sec. 9-4, and the result is that the total number of independent waves is three times the number of atoms. The energy of the system can be expressed as the superposition of the energy of all these acoustic waves. The classical theory assumes that each wave has $\frac{1}{2}kT$ kinetic energy and $\frac{1}{2}kT$ potential energy. Equation 9-11 is thus obtained.

Although this simple theory works well at high temperatures, it fails badly at low temperatures (i.e., much below room temperature for common substances). The reason for its failure is essentially a quantum effect. The energy of the acoustic waves should be quantized, like the energy of electrons or electromagnetic waves. The energy of a wave can change only in units of $h\nu$, where ν is its frequency. Because of the similarity of this situation to the photon description, the quantum $h\nu$ of acoustic energy is called a *phonon*. At very low temperatures kT is so small compared to the $h\nu$ of most of the waves that there are very few phonons (very few waves have one quantum of energy). At somewhat higher temperatures, there are many low-frequency phonons but few high-frequency ones. Finally, at high temperatures kT becomes greater than $h\nu$ for even the highest-frequency waves (Fig. 9-10c or 9-10g). At such a temperature the discreteness of wave energy becomes of little importance and eq. 9-11 becomes valid. (Compare this with the discussion of black-body radiation in Sec. 4-8.)

The mathematical calculation, *Debye theory*, of the total phonon energy as a function of temperature produces the following result for the heat capacity of any solid:

$$C_v = 9\mathsf{R}\left(\frac{T}{\Theta}\right)^3 \int_0^{\Theta/T} \frac{e^x x^4 \, dx}{(e^x - 1)^2} \tag{9-12}$$

The parameter Θ is called the *Debye temperature* and is different for different solids. It is defined in terms of the frequency ν_m of the shortest-wavelength wave (Figs. 9-10c or 9-10g; see also prob. 9-31):

$$\Theta = h\nu_m/k \tag{9-13}$$

Hence Θ is a function of the velocity of sound in the solid and of the interatomic spacing. Table 9-2 gives values of Θ for several solids, which were determined by comparing observed values of C_v with eq. 9-12.

TABLE 9-2

Debye Temperature Θ of Solids

Substance	Θ	Substance	Θ
Ag	215°K	Ge	290°K
Al	390	K	100
Au	170	Na	150
Be	1100	Ni	370
C (diamond)	1860	Pb	88
Cu	315	Si	620
Fe	420	W	310

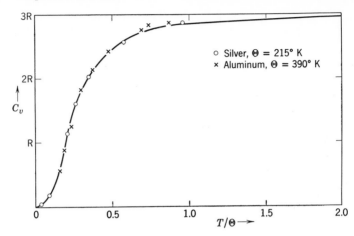

Fig. 9-11. Heat capacity at constant volume as a function of temperature. The solid curve is the Debye function (eq. 9-12). The curve was fitted to the data points for each metal in order to determine the Debye temperature Θ for the metal, and then the data were replotted as a function of T/Θ. (From F. Seitz, *Modern Theory of Solids*, McGraw-Hill, New York, 1940.)

C_v is plotted as a function of T/Θ in Fig. 9-11. Experimental values for silver and aluminum are presented for comparison.* The theory gives good agreement with experiment for all solids. Although eq. 9-11 is nearly correct for $T > \Theta$, it is quite wrong for $T < \Theta$. Thus $C_v(T)$ measurements on solids show that quantum theory applies to acoustic (as well as electromagnetic) energy, and such measurements might well have been discussed in Chapter 4 as one of the experiments demonstrating the necessity of quantum physics.

It is striking that the Debye theory agrees well with experiment even though no contribution of the *electrons* to the heat capacity was considered. There are many times as many electrons as nuclei present, but the inner-shell electrons should not contribute to the heat capacity because their excitation energies are so high (the same argument as that used for gases near the beginning of this section). But the conduction electrons in a metal certainly would contribute if they behaved like a gas with a Maxwellian distribution of velocities. Each electron would have $\frac{3}{2}kT$ energy, and the electronic contribution to the heat capacity

---

* Ordinarily the heat capacity at constant pressure, C_p, is the quantity measured. C_p is about 3 to 5% larger than C_v for common solids at ordinary temperatures. C_v can be computed from C_p, the coefficient of linear expansion, and the compressibility.

of a metal would be $\frac{3}{2}R$ for monovalent metals and proportionally larger for polyvalent metals. Yet the agreement of experiments with the Debye theory for ordinary temperatures shows that the electronic contribution to the heat capacity must be negligible except at very high and very low temperatures.

One of the greatest accomplishments of the quantum theory of metals (Secs. 9-3 and 9-4) is the demonstration that the electronic heat capacity, C_{ve}, of a metal should be negligibly small at ordinary temperatures. This demonstration is easily carried out by studying Fig. 9-6 again. Note that very few electrons change their energy as the temperature changes. Only those electrons with energies within about kT of the Fermi level E_0 are affected by changes in T. It follows from the definition of heat capacity in eq. 9-9 that most of the electrons make no contribution to C_{ve} since their energy is constant. Therefore C_{ve} is very much less than $\frac{3}{2}R$ per kilogram atomic weight.

The amount of the electronic heat capacity can easily be estimated by considering Fig. 9-12, in which $N(E)$ is plotted as a function of E near the Fermi brim. The shaded area is a rough estimate of the number of electrons that can change energy as the temperature changes. This estimate is $2kTS(E_0)$, which is a fraction $2ktS(E_0)/N$ of all the valence

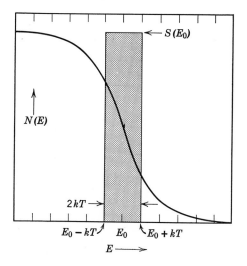

Fig. 9-12. $N(E)$ as a function of E near the Fermi level E_0. The shaded area is the approximate number of electrons that can change their energies as T changes.

electrons.   If all N participated, we should have an electronic heat capacity $C_{ve} = \frac{3}{2}R$, as in eq. 9-10; hence

$$C_{ve} \cong \frac{2kTS(E_0)}{N} \left(\tfrac{3}{2}R\right) \tag{9-14}$$

Although the numerical factor is only approximate, this expression is valid as an approximation whether or not the Fermi level E_0 lies low enough in the conduction band for $S(E)$ to have the form of eq. 9-6 at $E = E_0$.

The magnitude of C_{ve} can be visualized if we now assume that eq. 9-6 gives $S(E)$ correctly for $E \leqq E_0$.   Then by comparing eqs. 9-6 and 9-7 we see that $N = \frac{2}{3}E_0 S(E_0)$.   If this N is substituted into eq. 9-14,

$$C_{ve} \cong 3 \left(\frac{kT}{E_0}\right) \left(\tfrac{3}{2}R\right)$$

The result of the actual calculation using the Fermi distribution is

$$C_{ve} = \frac{\pi^2}{3} \left(\frac{kT}{E_0}\right) \left(\tfrac{3}{2}R\right) \tag{9-15}$$

where kT and E_0 are in the same units.   Since $kT \ll E_0$, this heat capacity is only a small addition to the heat capacity of the lattice. It can be observed near $0°K$ (where the lattice term is very small, being proportional to T^3 near absolute zero) and at very high temperatures (where the lattice term is constant).   Good agreement with experiment is found for the simpler metals.

(b) Thermal conductivity

The thermal conductivity σ_t of a solid is defined in a manner similar to the electrical conductivity: σ_t equals the heat conducted across unit area per second per unit temperature gradient.   The units of σ_t are watts meter^{-1} (degree centigrade)$^{-1}$.   The lattice vibrations considered in the discussion of heat capacity conduct heat in all solids.   In metals, electrons in the conduction band also conduct heat.   The electronic heat conduction process is roughly one hundred times as effective as the lattice vibration process, and therefore pure metals are much better heat conductors than non-metals.   But the ratio of σ_t for a metal to σ_t for a non-metal is by no means as great as the ratio of electrical conductivities for the two kinds of solids.

The conduction of thermal energy by lattice vibrations occurs in a rather obvious way.   If an atom is vibrating about its equilibrium position with an amplitude characteristic of temperature T, it exerts periodic forces on its neighbors and will increase their amplitudes of vibration if they were initially vibrating with smaller amplitudes, characteristic of a lower temperature. If the ends of a solid specimen are maintained at different temperatures, there will be a flow of heat by this process. Each atom will be vibrating a little less energetically than its neighbor toward the hot end, and will on the average receive energy of vibration from it.

The quantitative theory of heat conduction by these lattice vibrations is very difficult. As in the heat-capacity analysis, it is necessary to treat the process in terms of phonons (quanta of acoustic wave energy). These phonons transfer heat just as visible-light photons transfer heat from hot bodies to cold. The phonons do not travel very far before they are scattered—the longer this distance, the higher the conductivity. The calculation of the *mean free path* of phonons (which is of the order of 10 to 100 Å at room temperature) is very involved and will not be presented here.

In metals, the electrons in the partially filled band are the principal carriers of heat. The elementary theory of this process is very similar to the theory of the conduction of electricity, which will be given in Sec. 9-6.

Some experimental values of σ_t at room temperature are given in Table 9-3. It should be noted that alloys with a disordered array of different kinds of atoms have a lower conductivity than well-annealed crystals of pure metals, because the electron mean free path is shorter when the electrons are moving through a disordered solid. The relatively small ratio of the thermal conductivity of the best solid conductors to the thermal conductivity of the best solid insulators means that

TABLE 9-3

Thermal Conductivity σ_t of Solids

Substance	σ_t	Substance	σ_t
Aluminum	210	Nichrome	14
Copper	390	Brass	100
Gold	290	Steel	50
Iron	63	NaCl	7.0
Nickel	70	KCl	7.0
Silver	420	AgCl	1.1
Germanium	59	Window glass	0.8

σ_t is expressed in watts/(meter °C). In order to obtain σ_t in calories per (cm sec °C), divide the numbers given by 418. (Room temperature.)

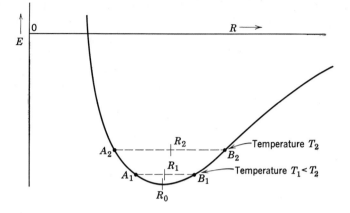

Fig. 9-13. Energy as a function of internuclear spacing. At the higher T, the vibrational energy is larger and the average R is larger because of the asymmetry of the curve. (The amplitudes of vibration are exaggerated.)

practically useful insulators cannot be real solids at all; good insulators are highly porous substances with voids constituting a large fraction of their volumes.

(c) Thermal expansion

All solids increase their dimensions with a rise in temperature; the fractional increase in dimensions per degree centigrade temperature rise is of the order of 1 part in 10^5 for most solids.*

In Fig. 9-13 we have drawn an energy vs. internuclear spacing curve similar to Figs. 8-1 and 8-5. At some low temperature the atoms in the solid vibrate so that the interatomic spacing varies from A_1 to B_1, with an average value R_1. At a higher temperature there is a higher energy of vibration E_2 and a larger amplitude. The interatomic spacing varies from A_2 to B_2 with a mean value R_2. Since $R_2 > R_1$, the solid has expanded with rising temperature. This result occurred because of the asymmetry of the energy vs. internuclear spacing curve: If the curve were symmetrical, R_2 would be directly above R_1. It was explained in Chapter 8 that the repulsive energy varies rapidly with R for all solids.

* These statements do not hold for a few non-crystalline solids or for some crystals near temperatures at which phase transformations occur, in which changes of the geometrical arrangement of atoms are more important than internuclear spacing in determining the volume per atom. They do not hold quantitatively for some ferromagnetic metals and alloys, where magnetic effects combine with thermal expansion. For example, Invar, an iron (64%) nickel (36%) alloy, exhibits virtually no thermal expansion over the temperature range 0°C–200°C.

TABLE 9-4

Coefficient α of Linear Expansion of Solids

Substance	α	Substance	α
Aluminum	25 ($\times 10^{-6}$)	Invar (0°–150°C)	0.6 ($\times 10^{-6}$)
Copper	17	Invar (300°–400°C)	15
Gold	14	Germanium	5.5
Iron	12	Silicon	2.3
Nickel	13	Window glass	10
Potassium	83	Concrete	12
Sodium	62	Al_2O_3	9

α is per degree centigrade, and all entries in this table should be multiplied by 10^{-6}, as indicated in the first row. (Room temperature unless noted.)

Since this energy is responsible for the rise at the left side of Fig. 9-13, and since the right side is a slowly varying function (like $-1/R$), it is apparent that the asymmetry of these curves is a universal property of solids. Thermal expansion is therefore a universal property of solids.

The coefficient of linear expansion is defined by $\alpha = (1/L)(dL/dT)$, where L is the length of a specimen of the solid. A quantitative theory shows that the coefficient of thermal expansion α is proportional to the heat capacity C_v. This result is to be expected, since the heat capacity measures the rate of change with T of the vibrational energy. In turn, the increase of vibrational energy increases R as shown in Fig. 9-13. The constant of proportionality between C_v and α depends on the shape of the binding energy curve, Fig. 9-13. Thus we expect and find that α is nearly constant for $T > \Theta$ but falls rapidly toward zero at lower temperatures. Some representative values of α for temperatures near room temperature are tabulated in Table 9-4.

9-6 ELECTRICAL CONDUCTIVITY OF METALS AND ALLOYS

The electrical conductivity σ is the current density (amperes per square meter) per unit electric field (volts per meter), and its units are mhos per meter. It is the reciprocal of the resistivity ρ (ohm-meters). Suppose that there were n_c electrons per cubic meter that could participate in conduction, each carrying a charge of magnitude e. If each were given the same velocity $\mathbf{v}$ by the electric field $\mathcal{E}$, the current density (charge crossing unit area per second) would be $n_c ev$, where v (the speed) is the magnitude of $\mathbf{v}$. This description is applicable to electrons in a

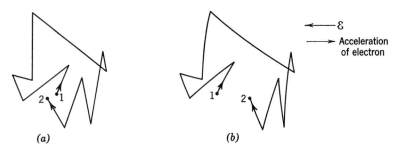

Fig. 9-14. Electron free paths, (*a*) without an electric field, and (*b*) with an electric field. The field produces a small *drift component* of velocity toward the right. The drift is greatly exaggerated if the field and mean free path have values typical of metals.

vacuum tube, but not to electrons in gas discharges or solids. If a conductivity $\sigma = n_c ev/\mathcal{E}$ were to be defined, it would be a function of position and electric field since $v/\mathcal{E}$ would be a function of these. Ohm's law would not be valid.

In a gas discharge or a solid the velocities of the electrons are in all directions. The electrons suffer collisions, and after each collision the new velocity is nearly independent of what the velocity was before the collision. The application of an electric field alters these velocities only slightly, but in a systematic way. If the field is $\mathcal{E}$, each electron is given an acceleration $e\mathcal{E}/m$ in the $-\mathcal{E}$ direction. The *change* in velocity produced by this acceleration is never a substantial fraction of the velocity itself because the electron soon suffers a collision after which its velocity is again random. The electric field has to "start all over again" to change the velocity of the electron. This situation is illustrated schematically in Fig. 9-14. The electric field produces a systematic component of velocity, the *drift velocity*. We shall learn later (from the values of the mobility presented in Table 9-6) that this drift velocity is very small compared to the speeds of the electrons; the drift has been greatly exaggerated in Fig. 9-14.

The drift current per square meter is called j, the current density. It is the product of the number of electrons participating per cubic meter, the magnitude of the charge on each, and the drift velocity:

$$j = n_c e \, \Delta v \tag{9-16}$$

which follows from consideration of the flow of electrons across a plane normal to the direction of the electric field (similar to the argument associated with Fig. 2-1). The conductivity σ is defined as $j/\mathcal{E}$, and the *mobility* μ as $\Delta v/\mathcal{E}$, the drift velocity per unit field (meter^2/volt second). Therefore

$$\sigma = n_c e \mu \qquad (9\text{-}17)$$

which holds for *any* conduction process if the correct n_c and μ are inserted.

Before the theory of the mobility in metals is presented it will be useful to consider a theory of mobility that is applicable to semiconductors, in which the electron velocities have a Maxwellian distribution. It is not applicable to metals because it ignores the Exclusion Principle and the Fermi distribution. This theory is presented because it will be useful later in the discussion of semiconductors and because it introduces the concept of an external electric field producing slight, systematic changes in the velocities of electrons and thereby producing an electric current.

First we estimate the magnitude of the drift velocity by a crude theory in which we assume that all the electrons have the same speed v, the same free time between collisions t_c, and therefore the same free path length l between collisions. The electric field gives each electron an acceleration $e\mathcal{E}/m$, and therefore the increment in the electron's velocity at the end of a time t_c is $e\mathcal{E}t_c/m$. The average increment in velocity Δv is $e\mathcal{E}t_c/2m$, since the velocity increment increases linearly from 0 to $e\mathcal{E}t_c/m$ during the free time t_c. Our crude estimate of the mobility is therefore $\mu \cong et_c/2m = el/2mv$.

This theory can be refined by averaging over the distribution of free path lengths, and the mobility can be expressed in terms of the *mean free path* $\bar{l}$ and the average value of $1/v$:

$$\mu = \frac{2e\bar{l}}{3m}\overline{\left(\frac{1}{v}\right)} \qquad (9\text{-}18)$$

By combining eqs. 9-17 and 9-18 with the expression $(2m/\pi kT)^{1/2}$ for the average value of $1/v$ from the Maxwellian distribution (prob. 9-40), we find

$$\sigma = \frac{4n_c e^2 \bar{l}}{3(2\pi mkT)^{1/2}} \qquad (9\text{-}19)$$

Since the drift velocity Δv is very small compared to v, the electric field makes only a small change in the velocity distribution that existed before the field was applied. Therefore the mean value of the free time t_c and $\bar{l}$, which depend on the distribution of velocities, are not appreciably changed by the field. The mobility and conductivity are thus independent of $\mathcal{E}$, and Ohm's law holds.

The treatment so far is applicable to situations such as semiconductors in which the Exclusion Principle is not important, that is, in situations

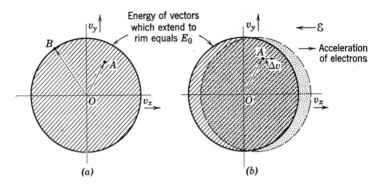

Fig. 9-15. Velocity distribution of electrons in the conduction band of a metal: (*a*) without electric field; (*b*) with electric field.

where the probability that a conduction band state is filled by an electron is very small. In the foregoing theory we have tacitly assumed that an electron's velocity and energy could be changed by the electric field and by collisions without regard for the availability of empty quantum states. In a metal this is clearly wrong, since almost all the conduction electrons have energies many kT below the Fermi brim E_0, and all quantum states at such energies are filled.

The situation in a metal is illustrated in Figs. 9-2 and 9-15. Figure 9-15*a* is a plot of the x and y components of the velocities of electrons. The vector terminating at A represents a typical electron velocity vector. The vector terminating at B represents an electron moving in a different direction with an energy equal to E_0, the Fermi brim energy. The circular area thus contains all the electron velocities (except for those few electrons with $E \gg E_0$). All possible quantum states for energies less than E_0 are filled (as in Fig. 9-8).

In Fig. 9-15*b* an electric field has been applied and is accelerating electrons toward $+x$. Let the solid circle represent the velocity distribution of a large number of electrons just after each has suffered a collision. Before the next collision, the electric field skews the whole distribution to the right, and the new distribution of velocities is given by the dashed line. Each electron has been given a tiny velocity increment Δv (grossly exaggerated in size in Fig. 9-15). The new distribution gives a current of electrons toward $+x$. Of course collisions will soon occur, and, if all the electrons suffered collisions at the same time, the symmetrical distribution (solid line in Fig. 9-15*b*) would be restored at that time. We speak of the skewed distribution *relaxing* to the equilibrium distribution with a characteristic *relaxation* time. Of course the collisions do not occur at the same time, and the skewed distribution will continue with an average drift velocity Δv. The averaging over the velocity distribution is now quite different from the averaging over the Maxwellian distribution. It is apparent from Fig. 9-15b that the $\bar{l}$ and v to use in eq. 9-18 should now be the $\bar{l}$ and v appropriate to electrons with energy E_0, since the amount of the displacement of the distribution is proportional to the changes of velocities of electrons near the Fermi brim. It is thus not surprising that the correct expression for σ for a metal is

$$\sigma = \frac{n_c e^2 \bar{l}}{mv} \tag{9-20}$$

where $\bar{l}$ and v are evaluated for electrons with energy E_0.*

The collisions mentioned in the foregoing analysis must now be explained. An electron wave in an allowed energy band would travel through a *perfect* crystal without scattering, reflection, or attenuation. The traveling-wave functions of Sec. 8-5 represent electrons moving without change in momentum through the crystal lattice. This was found in Sec. 8-5 to be analogous to an electromagnetic wave traveling through a perfect (lossless) filter. In the absence of thermal vibrations and crystal imperfections, the mean free path $\bar{l}$ of the electron wave in the crystal would be ∞. Suppose, however, that in the wave filter the inductances and capacitances varied somewhat from section to section, or that occasionally a quite different L or C was included by mistake. In the first case a large number of small reflections of the waves would occur. In the second, a small number of almost complete reflections would occur. Similar reflections (or, in three dimensions, scattering) of the electron waves in a crystal occur if because of thermal motion the lattice ions are not quite in their correct positions or if an impurity or other concentrated imperfection is present.

Let us first consider the thermal vibration of atoms in an otherwise perfect crystal. The scattering of electron waves by lattice ions vibrating about their perfect lattice positions (or, in other words, the collisions of electrons with phonons) increases with increasing T. In the temperature region $T > \Theta$ the heat capacity is constant and the lattice vibrational energy is proportional to T. The square of the amplitude A of the vibrations of the atoms about their equilibrium positions is proportional to this vibrational energy, since these are small-amplitude vibrations for which the harmonic oscillator serves as a good approximate model (see prob. 9-47). It is true, though not obvious, that the scatter-

---

* W. Shockley, *Electrons and Holes in Semiconductors*, Van Nostrand, New York, 1950, pp. 187–211, 250–282.

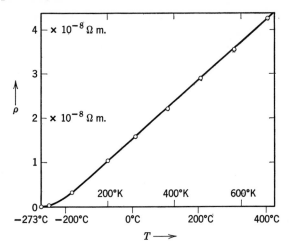

Fig. 9-16. Resistivity of copper as a function of temperature.

ing cross section for electrons is proportional to A^2 and hence to T; thus since $1/\bar{l}$ is proportional to this cross section (see Sec. 2-5), we find that $\rho \sim 1/\bar{l} \sim T$. For $T < \Theta$, the lattice vibrational energy can be computed from the Debye function (eq. 9-12), which is the derivative with respect to T of this energy. This vibrational energy hence increases rapidly as T increases from zero, and ρ also increases rapidly, as illustrated in Fig. 9-16. The similarity of $d\rho/dT$ to the Debye curve (Fig. 9-11) should be noted; these two functions are not identical, however, since some wavelengths of lattice vibrations scatter electrons more effectively than others.

Let us next consider the effect of impurities or other crystal imperfections. (Imperfections in solids will be studied systematically in Chapter 10.) All imperfections increase the electrical resistivity of metals. An atom displaced from the position it would occupy in a perfect lattice or an impurity atom constitutes an incorrect value of L or C in the wave filter analogy noted above or, more directly, a square well deviating in depth or width from the other wells in the model of Fig. 8-11. Thus strong scattering of electrons occurs and adds a contribution to ρ that is almost completely independent of temperature. (See Fig. 9-18a, in which the upper curve is for a less pure specimen than the lower.)

An extreme example of imperfections occurs in a disordered alloy in which the various atomic constituents are distributed almost at random. Electrons move in a region of fluctuating (instead of regularly repeating) potential energy, and the mean free path is therefore very short. Since the reduction of $\bar{l}$ by this effect is much greater than by the lattice vibra-

TABLE 9-5

Electrical Conductivity σ of Metals and Alloys

Substance	Measured σ	Calculated σ	Substance	Measured σ
Aluminum	35 ($\times$ 10^6)		Mercury (liquid)	1.0 ($\times$ 10^6)
Copper	59	161 ($\times$ 10)^6	Nichrome	0.9
Iron	10		347 Stainless Steel	1.4
Sodium	22	22	Constantan	2.3

σ is given in mhos per meter at 20°C, and all entries in the table should be multiplied by 10^6 as indicated in the first row. The calculated values are from J. Bardeen, *J. Applied Physics,* **11,** 88 (1940).

tions, σ for such an alloy does not vary rapidly with T. For example, Nichrome is a disordered alloy of nickel, iron, and chromium; its σ is much less than the σ of a pure metal and changes by only about 9% from room temperature to 1000°C. A less extreme example of imperfections is a pure metal in the form of a hard-drawn wire; its resistivity is higher than that of the same wire after annealing with consequent healing of many of the defects.

The quantitative calculation of σ for metals is very complicated because of the difficulty of calculating $\bar{l}$. Some experimental values of σ for metals and alloys are given in Table 9-5, together with theoretical calculations.

Observation of the *Hall effect* permits experimental determination of μ and $\bar{l}$. In this experiment a rectangular block of a solid is subjected to electric and magnetic fields at right angles to each other and to the faces of the block, as illustrated in Fig. 9-17. The applied field is in the

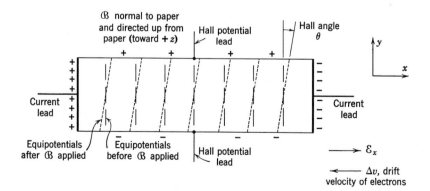

Fig. 9-17. The Hall effect experiment on a rectangular block of a solid.

$+x$-direction, and the current flow of electrons is in the $-x$-direction. When the magnetic field is turned on there is an additional force

$$\mathbf{F} = -e\mathbf{v} \times \boldsymbol{\mathcal{B}} \qquad (1\text{-}6)$$

on each electron. This force is in a different direction for each electron, since each vector velocity is different, and the net deflection of electrons by it would be zero if it were not for the small drift component of velocity $\Delta\mathbf{v}$, which is in the same direction $(-x)$ for each electron and has the magnitude $\mu\mathcal{E}_x$. The force $\mathbf{F}$ is in the $-y$-direction, and

$$F_y = -e|\boldsymbol{\mathcal{B}}|\,|\Delta\mathbf{v}| = -e\mathcal{B}\mathcal{E}_x\mu \qquad (9\text{-}21)$$

This force deflects electrons toward the bottom of the block and quickly produces a net $(+)$ charge at the top and a $(-)$ charge at the bottom. The charge builds up until the electrostatic force

$$-e\mathcal{E}_y \qquad (9\text{-}22)$$

is equal and opposite to F_y from eq. 9-21. The net force in the y-direction is again zero, but a measurable difference in potential now exists between the top and bottom surfaces. This potential difference is called the *Hall effect* and is measured by instruments connected to the Hall potential leads.

When the transverse field has been established the sum of eq. 9-21 and eq. 9-22 must be zero:

$$\mathcal{E}_y = -\mathcal{B}\mathcal{E}_x\mu \qquad (9\text{-}23)$$

The *Hall constant R* is defined by

$$R = \mathcal{E}_y/\mathcal{B}j_x$$

where $j_x = \sigma\mathcal{E}_x$ is the current density (amperes/meter^2). Using eq. 9-16, which applies to all materials, and eq. 9-23, we obtain:

$$R = -\frac{\mathcal{E}_x\mu}{\sigma\mathcal{E}_x} = -\frac{1}{n_c e} \qquad (9\text{-}24)$$

Thus the Hall-effect experiment permits the measurement of the number n_c of conduction electrons per unit volume.

Another way of describing the results of the Hall-effect experiment is in terms of the *Hall angle θ*, which is the angle through which the equipotential surfaces are rotated when $\mathcal{B}$ is applied (Fig. 9-17). This angle can be calculated from eq. 9-23

$$\tan \theta = \mathcal{E}_y/\mathcal{E}_x = -\mathcal{B}\mu$$

TABLE 9-6

Hall Constant R and Mobility μ of Metals

Metal	Observed R	Calculated R	Observed μ
Al	-0.30×10^{-10}	-0.35×10^{-10}	0.0012
Cu	-0.55	-0.74	0.0032
Li	-1.70	-1.35	0.0018
Na	-2.50	-2.46	0.0053

R is in volt meter³/ampere weber. μ is in meter²/volt second and is for 20°C.

Since θ is almost always very much less than 1 radian, $\tan \theta \cong \theta$ and

$$\theta = -\mathscr{B}\mu \tag{9-25}$$

It should be apparent from the theory that R and θ would be positive if the sign of the charge carriers were positive.

Some observed values of R and μ are tabulated in Table 9-6. The calculated values of R were obtained by assuming one conduction band electron per atom in copper, lithium, and sodium, and three per atom in aluminum. Mean free times and mean free paths can be calculated from these values. Further application of Hall-effect experiments will be found in the discussion of semiconductors in Chapter 11.

In concluding this section it is wise to point out again that *any* metal (or alloy) is a much better conductor than a non-metal. The differences in conductivity among metals are primarily caused by differences in the mean free paths of conduction electrons, but *any* metal has a large number (one or more per atom) of conduction electrons. Non-metals in the pure state at low temperatures have only a very small number of electrons that can participate in conduction. Hence we find an enormous difference in conductivity between the best metals and all pure non-metals at low temperatures (and good insulators at ordinary temperatures). The differences in conductivity among non-metals are primarily caused by differences in the number of conduction electrons.

9-7 SUPERCONDUCTIVITY

One of the most striking phenomena in physics is the sudden vanishing of the electrical resistance of many solids as the temperature is lowered through a critical temperature T_c. This phenomenon of superconductivity, although discovered in 1911 by Kamerlingh Onnes, is still actively studied, and new understanding and applications are developing rapidly.

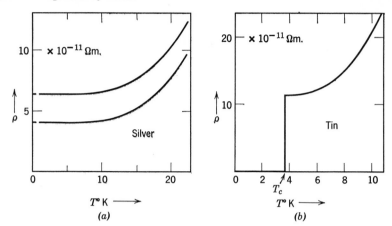

Fig. 9-18. Resistivities of metals as functions of temperature. (a) Silver, a normal metal; curves are for two samples with different impurity contents. (b) Tin, a superconductor; the resistivity drops exactly to zero as T is decreased through $T_c = 3.72°$K. [Data from W. J. de Haas, G. J. van den Berg, and J. de Boer, *Physica*, **2**, 453 (1935) and **3**, 440 (1936).]

Because of this rapid development and because of the complexity apparently inherent in the phenomenon, we shall be able to give here only a brief and qualitative introduction.

The difference between superconductivity and the reduction toward zero of the ordinary temperature-dependent resistivity as $T{\to}0$ should be made clear at the outset. Figure 9-18a shows the resistance $\rho(T)$ of silver at very low temperatures. The temperature-dependent part of the resistance continuously approaches zero as $T{\to}0$ but is finite at any finite temperature; furthermore, the temperature-independent part (impurities and structural imperfections, different from sample to sample) remains as $T{\to}0$. In contrast, the resistance of tin, a superconductor, drops to zero almost abruptly (within a fraction of a degree) at $T_c = 3.72°$K, as shown in Fig. 9-18b. Although a part per million of impurity substantially affects ρ of a normal metal (Fig. 9-18a), ρ remains *exactly* zero for $T < T_c$ in a superconductor even when heavily "doped" with impurities. A *persistent current* can in fact be established in a superconductor: A current once established continues as long as T is maintained less than T_c. The continued existence of such a current can be verified by measuring the magnetic field it produces, and such currents have been observed to persist for years.

Superconductors are by no means rare. More than twenty elements, all metals, are superconducting, and so are innumerable alloys and intermetallic compounds. Curiously enough, the best conductors at ordinary temperatures (silver, copper, and gold) are *not* superconductors; strong electron-phonon interaction (which makes normal metals *poorer* conductors) seems to be essential for superconduction.

Perhaps equally as striking as the vanishing of electrical resistance is the *Meissner effect*, the complete exclusion of magnetic induction from the interior of a superconductor (in the language of Sec. 9-8 we should say that a superconductor exhibits perfect diamagnetism). It might be thought that this effect is simply the result of a superconducting current on the surface of the specimen, but arguments from Maxwell's equations show that this is not true; the Meissner effect and vanishing resistance appear to be different aspects of superconductivity.

All other properties that depend on the motion of conduction-band electrons also change as a metal becomes superconducting. Such properties as the electronic heat capacity, the thermal conductivity (which is *smaller* for most superconductors below T_c than above it!), and thermoelectric and magnetic resonance effects are different in the superconduct-

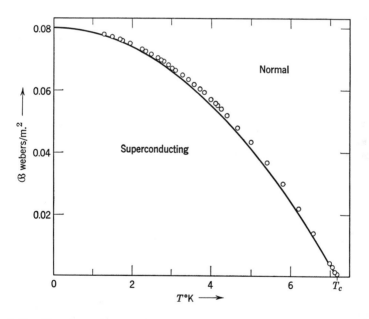

Fig. 9-19. The effect of a magnetic field $\mathfrak{3C}$ on the temperature at which bulk lead becomes superconducting. The magnetic induction $\mathfrak{B} = \mu_0\mathfrak{3C}$ of the externally applied field is plotted as a function of temperature. The data are from D. L. Decker, D. E. Mapother, and R. W. Shaw, *Phys. Rev.*, **112**, 1888 (1958). The curve is a fit of the expression $\mathfrak{3C} = \mathfrak{3C}_0(1 - T^2/T_c^2)$ to the data.

TABLE 9-7

**Transition Temperatures and Critical Fields for
Some Superconductors**

Substance	T_c,°K	$\mathfrak{B}_0$		Substance	T_c,°K	$\mathfrak{B}_0$
Al	1.175	0.0106		Ti	0.39	0.010
Sn	3.72	0.031		V	5.13	0.134
Nb	8.7	0.20		$Ti_{0.42}V_{0.5\acute{5}}$	7.49	
Pb	7.18	0.0802		Nb_3Sn	17.9	*

The magnetic induction (in webers/m.²) in vacuum $\mathfrak{B}_0 = \mu_0 \mathfrak{K}_0$ equivalent to the critical field $\mathfrak{K}_0$ is tabulated in order to make the magnitudes easier to visualize.

* Depends on microstructure; can be approximately 10 webers/m.²

ing state. Evidently normal and superconducting states correspond to a qualitative difference in the electronic structure, even though the fact that the phase transition between the two structures occurs at such a low temperature indicates that there cannot be substantial energy or entropy differences between the two phases.

An external magnetic field affects the temperature of the transition between these two phases, as illustrated in Fig. 9-19, which gives the conduction-electron phase diagram for bulk lead. The fact that a magnetic field influences the temperature at which superconductivity occurs is not surprising in view of the Meissner effect, which is evidence of the important role magnetism plays in superconductivity. Other metals follow diagrams similar in shape but with different values of T_c and $\mathfrak{K}_0$ (see Table 9-7); the shape of such curves is very nearly given by $\mathfrak{K} = \mathfrak{K}_0 (1 - T^2/T_c^2)$.

There are many additional items of experimental information that give guidance in the creation of a theoretical understanding of super-conductivity (such as high-frequency and size effects), but we shall mention only two. The first of these is the *isotope effect*, the dependence of T_c of a pure metal on the isotopic mass M of the metal nuclei. It is observed that T_c is proportional to $M^{-\frac{1}{2}}$. Since phonon processes (velocity of sound, Debye temperature) depend on $M^{-\frac{1}{2}}$, the isotope effect suggests that an electron-phonon interaction (just what *increases* the resistance in a normal metal!) is vital to superconductivity. The second is the *superconductor energy gap*, a thin region near E_0 on a plot of the density of states $S(E)$ as a function of E where $S(E)$ equals zero. The existence of this gap, which is about $3.5kT_c$ in width as $T \rightarrow 0$ and approaches zero as $T \rightarrow T_c$, has been established by the abrupt change in the absorption of far infrared or microwave radiation when $h\nu$ equals the gap energy. Many other properties of superconductors, such as the electronic heat capacity, can be interpreted in terms of this same energy gap. A direct demonstration (and incidentally a possible application) of the energy gap comes from *tunneling* experiments, in which the current through a thin insulating film separating two superconductors is studied. These experiments will not be described here but are identical in principle with the tunnel diode experiments to be described in Sec. 11-4.

We can provide here very little theoretical understanding of the complex phenomena of superconductivity, but we shall sketch an outline of the current theory. Clearly superconductivity is a process in which many electrons cooperate, unlike the processes we have hitherto considered in which electrons were weakly coupled to one another and could individually lose momentum by collisions. The superconducting state is a quantum state on a large scale in which the motions of all the electrons in the conduction band are locked together. This state can have zero momentum or a net momentum, depending on whether a current has been established in it by an external source. If it has a net momentum, the current persists indefinitely in the superconducting state. A finite amount of energy, corresponding to the gap energy, must be supplied to excite the electron distribution as a whole to the next higher (non-superconducting) state. Thus the current cannot be attenuated by a large number of infinitesimal interactions, as in a normal metal.

The interaction that locks the electrons together appears to be an electron-electron attraction with a phonon serving as a "mutual friend"; that is, an electron momentum state is coupled to a phonon, which in turn is coupled to the momentum state of another electron. If this attraction exceeds the normal electron-electron repulsion (Sec. 8-3), a superconducting state is possible. Hence the criteria for superconductivity include a large electron-phonon interaction, and hence T_c depends on the isotopic mass. The ground state is made up of coupled pairs of electrons of opposite spin and oppositely directed momenta; the momenta can be equal in magnitude (no current) or slightly different (a net momentum, a superconducting current).

This theory, which has been created in part by many physicists but is ultimately the work of Bardeen, Cooper, and Schrieffer, has been rather successful in explaining the wealth of experimental facts about superconductivity. Much remains to be clarified and extended (and the whole theoretical structure may be replaced by a more general one), but good agreement is found between theory and a large number of experiments.

There are many possible and intriguing applications of superconductivity. At first sight, the data of Fig. 9-19 and Table 9-7 seem to exclude the application of superconductivity to the production of strong magnetic fields. But thin films, filamentary structures (inhomogeneous alloy wires containing tiny filaments of a superconducting phase), and highly strained metals and alloys exhibit much larger values of $\mathfrak{IC}_0$ than do pure metals. Magnetic inductions of the order of 10 webers/m.^{2}, far above the largest value obtainable with iron-core electromagnets, have been obtained in superconducting solenoids. The Meissner effect can be exploited to support a superconductor in free space by repulsion from a permanent magnet. Other applications of superconductivity are based on the effect of an applied magnetic field on the transition between normal and superconducting states. For example, at a constant temperature below T_c, changes back and forth from normal to superconducting behavior can be effected by varying the external magnetic field, which thereby can control a current in a circuit connected to the superconductor. Thus amplifiers, oscillators, control systems, and (especially) the logic and information storage functions of a large-scale computer can be provided by the control a magnetic field exercises on superconductivity.

9-8 MAGNETIC PROPERTIES OF SOLIDS

There are three principal types of magnetic behavior of solids. Of these, ferromagnetism is the most complicated but the most important for engineering applications. The other two are diamagnetism and paramagnetism. We shall discuss these two briefly and then concentrate on ferromagnetism and other forms of magnetism closely related to it.

(a) Diamagnetism and paramagnetism

First we review the definitions of some magnetic parameters. $\mathfrak{IC}$ is the magnetic field intensity (ampere-turn/meter); $\mathfrak{B}$ is the magnetic induction (webers/meter^2). The relation between $\mathfrak{B}$ and $\mathfrak{IC}$ is expressed by $\mathfrak{B} = \mu\mathfrak{IC} = \mu_0(1 + \chi)\mathfrak{IC}$. μ is the magnetic permittivity; χ is the magnetic susceptibility, the magnetization per unit $\mathfrak{IC}$. The magnetization $\chi\mathfrak{IC}$ is zero if there is no material in the field, and hence in a vacuum $\chi = 0$. Thus $\mu = \mu_0$ in a vacuum, and μ_0 is therefore called the magnetic permittivity of a vacuum. Our problem in studying the magnetic properties of solids is to learn what χ is for different solids and how it varies with $\mathfrak{IC}$ and T. When we consider ferromagnetism it is more convenient to use the relative magnetic permittivity κ_m as a measure of the magnetization. κ_m is defined as $\kappa_m = \mu/\mu_0$, and since $\mu = \mu_0(1 + \chi)$ we can calculate χ from κ_m, or κ_m from χ. κ_m is also (and usually) called the permeability.

Most solids have a very small magnetization when placed in a magnetic field. There are two contributions to χ. The first is a *diamagnetic*, negative contribution that is independent of $\mathcal{3C}$ and T and originates from the motion of the electrons in the solid. When a magnetic field is applied, the quantum states of all the electrons in a solid are modified slightly by the magnetic force on the moving charges (eq. 1-6). The modified motion of each electron in turn produces a local magnetic moment that opposes the applied field. The magnetization is therefore negative, and we call this effect *diamagnetism*. The negative sign of this contribution to χ can be shown to be in agreement with Lenz's law, which is that the magnetic flux produced by the current induced in a circuit by a changing magnetic induction is in a direction to decrease that induction.

The second contribution to χ is a *paramagnetic*, positive contribution. It is a consequence of the fact that many atoms have permanent magnetic moments that partially line up with the applied magnetic field and thereby produce magnetization; the magnetization is the magnetic moment per unit volume. These moments come from the orbital motion of the electrons and from the magnetic moment intrinsic to each electron and associated with its spin (the two types of moments are of the same order of magnitude). The spin magnetic moment was expressed in Sec. 6-5 as $\pm\mathfrak{M}_B$, where $\mathfrak{M}_B$ is the Bohr magneton *

$$\mathfrak{M}_B = \frac{eh}{4\pi m} = 0.927 \times 10^{-23} \text{ joule m.}^2/\text{weber} \qquad (6\text{-}14)$$

Both the diamagnetic and paramagnetic effects produce a magnetization that is very small compared to $\mathcal{B}$ (except in superconductors, in which $\chi = -1$ and $\mu = 0$). That is, μ very nearly equals μ_0 for diamagnetic or paramagnetic solids. We shall demonstrate this fact for paramagnetism since the demonstration prepares the way for the study of ferromagnetism. In any atom the inner, closed shells have a net magnetic moment of zero, since all quantum states are filled and there are just as many electron spin moments and orbital moments in one direction as in the opposite direction. Electrons in the outermost shell may or may not produce a moment. For example, s states have no angular

---

* The magnetic moment has been defined here by writing the energy of a moment $\mathfrak{M}$ in a magnetic induction $\mathcal{B}$ as $\mathfrak{M}\cdot\mathcal{B}$. The moment of a current I flowing in a closed plane circuit of area A is IA. This definition is not unique; $\mathfrak{M}$ is frequently defined by writing the interaction energy as $\mathfrak{M}\cdot\mathcal{3C}$ (instead of $\mathfrak{M}\cdot\mathcal{B}$). The moment of a current is then $\mu_0 IA$, and $\mathfrak{M}_B = \mu_0 eh/4\pi m$.

momentum and therefore no orbital magnetic moment. If there is a single s electron in a shell, the total magnetic moment of the atom is $\mathfrak{M}_B$, the moment associated with the spin of one electron. If there are two s electrons, the total moment is zero since the two electron spins $(+\frac{1}{2}$ and $-\frac{1}{2})$, and therefore their magnetic moments, are oppositely directed. The angular momentum of a p or d electron is not zero since the quantum number l is not zero, and such electrons can produce an orbital magnetic moment. This moment is of the same order of magnitude as $\mathfrak{M}_B$.

In a magnetic induction $\mathfrak{B}$, a dipole of magnetic moment $\mathfrak{M}_B$ has a potential energy that depends on its orientation with respect to the magnetic induction. The difference between minimum potential energy (dipole aligned with the induction) and maximum potential energy (dipole and induction oppositely directed) is $2\mathfrak{M}_B\mathfrak{B}$. In a magnetic induction of 1 weber/m.^{2}, this energy difference is

$$2 \times 0.927 \times 10^{-23} \times 1 \cong 1.8 \times 10^{-23} \text{ joule} \cong 10^{-4} \text{ e.V.}$$

which is very much less than kT at room temperature ($\frac{1}{40}$ e.V.). Therefore the thermal energy is so large that there are nearly as many moments opposite to the field as parallel to it. Furthermore, the field taken for this example is nearly as large as can be obtained in practice. By using the Boltzmann factor to calculate the relative numbers of moments in the low-energy and high-energy orientations, we could find that in this example the average moment was $\frac{1}{250}\mathfrak{M}_B$ per atom. If there are about 2.5×10^{28} atoms/m.^{3} in a solid, then the magnetization (magnetic moment per unit volume) is

$$\chi\mathfrak{K} \cong 2.5 \times 10^{28} \times \tfrac{1}{250} \times 0.927 \times 10^{-23} \cong 10^3 \text{ joule/weber m.}$$

and

$$\mu_0\chi\mathfrak{K} \cong (4\pi \times 10^{-7} \text{ weber/m. joule})$$

$$\times (10^3 \text{ joule/weber m.}) \cong 10^{-3} \text{ weber/m.}^2$$

Since for this example we have taken $\mathfrak{B} = 1$ weber/m.^{2}, the second term on the right in $\mathfrak{B} = \mu_0\mathfrak{K} + \mu_0\chi\mathfrak{K}$ is only 1/1000 of the first, $\chi \cong 10^{-3}$,[*] and $\kappa_m \cong 1.001$.

Thus paramagnetic materials have a permeability κ_m not much different from unity at ordinary temperatures. We state without proof that the diamagnetic susceptibility is of the same order of magnitude. In some materials diamagnetism dominates; in others, paramagnetism is the stronger effect. (The simple, monovalent metals can be shown

---

[*] χ was estimated here for a particular field strength, but χ is independent of the field. The average magnetic moment per atom ($\frac{1}{250}\mathfrak{M}_B$ in this example) is in general proportional to the field, and therefore the magnetization per unit field is a constant.

to be paramagnetic by the application of the energy band theory and the Fermi statistics.)

(b) *Atomic theory of ferromagnetism*

Ferromagnetic solids are quite different from the solids considered above in two respects: (1) The permeability attains very much higher values, as high as 10^5 in some solids. (2) A magnetization remains after the field is removed, and the familiar hysteresis loop occurs in the relation between $\mathfrak{B}$ and $\mathfrak{3C}$. These are really different aspects of the same fundamental phenomenon, namely, the spontaneous magnetization of small regions called *domains* of the ferromagnetic solid. We shall first consider a single domain and show how a magnetization can occur even if no external field is applied. Then we shall show how the $\mathfrak{B}$ vs. $\mathfrak{3C}$ curves characteristic of ferromagnetic solids can be explained in terms of these domains.

The commonest ferromagnetic elements are iron, cobalt, and nickel. There are two other, relatively rare, ferromagnetic elements and many ferromagnetic alloys. Our first problem is to understand why iron, cobalt, and nickel exhibit spontaneous magnetization. Atoms of all these elements have partially filled $3d$ shells, and each atom has two $4s$ electrons. Iron, for example, has six $3d$ electrons (a full d shell contains ten electrons). Line-spectra experiments show that these six electrons are divided as follows: five with one spin, one with opposite spin. This makes the spin magnetic moment of the atom $4\mathfrak{M}_B$.

Why should the electron spins be divided this way rather than, say, 3-and-3? The answer lies in the *exchange energy*, which is a change in the electrostatic energy caused by the Exclusion Principle. Two $3d$ electrons will stay farther apart, on the average, if they have the same spin than if they have opposite spins. This behavior occurs because they have the same momentum, and only one electron with each spin is permitted by the Exclusion Principle in the volume $(h/2)^3$ of momentum and coordinate space (eq. 6-11). Since they stay farther apart, their electrostatic repulsion energy is smaller. Thus it happens that in iron the lowest energy state consists of electrons in all five of the states with one spin, the sixth electron appearing with the opposite spin. (The exchange energy here constitutes an attraction, but keeping the electrons farther apart may *increase* the energy of a state, since the electrons may be farther from the nucleus as a result. Thus the exchange energy can be a repulsion energy.)

When the iron atoms, each with magnetic moment $4\mathfrak{M}_B$, are brought together to form a small section of a solid crystal, all the magnetic moments are in the same direction. We might at first think that the alignment was produced by the magnetic interaction of the "atomic magnets."

The magnetic interaction energy is, however, far too small to produce alignment. We can infer this indirectly from the calculation above of the paramagnetic χ. In that calculation a $\mathfrak{B}$ was used that is of the same order of magnitude as that produced by alignment of all the atomic moments, namely, $\mathfrak{B} = 1$ weber/m.² This is about as large a $\mathfrak{B}$ as can be obtained by complete alignment of the atomic moments in a ferromagnetic solid. Yet this $\mathfrak{B}$ was far too small to produce alignment. Hence the magnetization that would be produced by alignment is far too small to maintain alignment.

The interaction that actually aligns the magnetic moments is the exchange interaction. The $3d$ electrons in one atom move slightly away from the $3d$ electrons in an adjacent atom if these electrons have the same spin. Thus the electrostatic repulsion energy is smaller for the magnetized (spins and magnetic moments aligned) than for the unmagnetized state. The distortion of the $3d$ wave functions by the exchange interaction therefore makes the energy of the system lower for the magnetized than for the unmagnetized state.

We can also describe this effect in terms of the energy bands. The $3d$ band is a narrow band compared to the $4s$ band, since it is a band formed from wave functions that do not extend so far out from the nuclei as the $4s$ wave functions. This $3d$ band would be filled if there were ten electrons per atom in it, five with one spin and five with the other. These two types of states are pictured separately in Fig. 9-20a. In the unmagnetized state of iron (for example), there would be three electrons per atom in each of the two partial bands. Iron is never found in this form, however, because the configuration shown in Fig. 9-20b has a lower

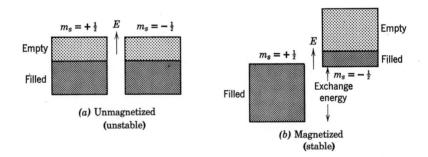

Fig. 9-20. The $3d$ energy band of iron separated into the partial bands with $+$ and with $-$ spin (and therefore $+$ and $-$ magnetic moment). The magnetized state is the stable state because the exchange interaction lowers the total energy more than the Fermi energy ($\frac{3}{5}E_0$) raises it when 5 of the 6 electrons are put into one of the partial bands.

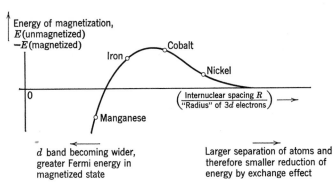

Fig. 9-21. Criterion for ferromagnetism. The difference in energy between the magnetized and unmagnetized states is plotted as a function of the ratio of the internuclear separation R to the "radius" of the $3d$ shell. Only iron, cobalt, nickel, and alloys with abscissa values similar to those of these elements exhibit ferromagnetism. [See W. Shockley, *Bell Sys. Tech. J.*, **18**, 645 (1939).]

energy. Putting as many electrons as possible (five per atom) into the band with one spin has lowered the energy of that band by the exchange energy. Of course the Fermi energy is now larger, since the states now occupied at the top of the $m_s = +\frac{1}{2}$ band have larger kinetic energies than the states made vacant (near the middle of the $m_s = -\frac{1}{2}$ band). In the example illustrated (iron), the effect of the exchange energy is greater than the increase in Fermi energy and the stable, observed condition is the magnetized state.*

We are now in a position to learn why some elements are ferromagnetic and others are not. If in a particular solid the lowering of the energy by the exchange interaction is greater than the increase in Fermi energy, one partial band will be filled and there will be a large magnetization. This situation occurs for only a few elements and alloys. Even in these solids, the energy difference favoring spontaneous magnetization is only a few tenths of an electron volt per atom. The criterion for spontaneous magnetization (and hence for ferromagnetism) is illustrated in Fig. 9-21. The difference in energy between the magnetized and unmagnetized states is plotted as a function of the inter-

---

* In order to concentrate on the principal processes we have ignored the two $4s$ electrons and the $4s$ band, which partially overlaps the $3d$ band. It is much wider than the $3d$ band and is full when only two electrons per atom occupy it, and therefore the density of states $S(E)$ in the $4s$ band is very much smaller than in the $3d$ band. There are eight electrons per atom of iron to be accommodated in these two bands, and about 7.4 of these appear in the $3d$ band and 0.6 in the $4s$ band (in the above we assumed six in the $3d$ band and ignored the $4s$ band).

atomic separation divided by the "radius" (as defined in Sec. 6-2) of the $3d$ wave functions. The magnetized state has the lower energy for iron, cobalt, and nickel. The unmagnetized state has the lower energy for the other elements of this group. At a large separation of atoms, the exchange interaction is too weak to produce alignment. At a small separation of atoms the widening of the $3d$ band produces too great an increase in the Fermi energy for the magnetized state.

We should expect on the basis of this criterion for ferromagnetism that alloys like copper-manganese alloys would be ferromagnetic if the atomic separation was in the iron-cobalt-nickel range, and this prediction is found to be correct. It is not necessary that the elements composing the alloy be ferromagnetic.

Since the energy difference between the magnetized and unmagnetized states is only a few tenths of a volt, we should expect that increasing the temperature would ultimately destroy the spontaneous magnetization. At a temperature called the *Curie temperature* the spontaneous magnetization disappears. This temperature is 1043°K for iron, 1400°K for cobalt, and 631°K for nickel. (The other two known ferromagnetic elements, gadolinium and dysprosium, have Curie temperatures of 289°K and 105°K, respectively, and therefore are of less practical interest. In these elements the magnetism arises from an incomplete $4f$ shell, rather than an incomplete $3d$ shell.)

(c) Ferromagnetic domains

The region of a ferromagnetic solid in which the spins are aligned is called a ferromagnetic domain. The magnetization within a single domain is large, corresponding to a magnetic induction of the order of 1 weber/m.^{2} (see prob. 9-61). If a solid were a single domain, it would have a large magnetic moment even in the absence of an applied magnetic field. There would be a large external magnetic field and therefore a large energy of this field. A lower energy for the system can be obtained if it is divided into four domains, as shown in Fig. 9-22a, since there is now a much smaller external magnetic field and therefore less energy in that field. However, it takes some energy to create a wall between domains. Furthermore, there is a difference in energy between a crystal magnetized in one direction relative to the crystal axes and in another direction; this energy difference is called the *anisotropy energy*. The actual configuration of domains in a particular solid is determined by the minimization of the magnetic field energy, domain wall energy, and anisotropy energy.

The existence of domains can be experimentally verified by depositing

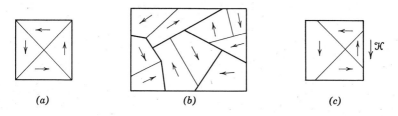

Fig. 9-22. Ferromagnetic domains.   (*a*) Single crystal, macroscopically unmagnetized but each domain magnetized.   (*b*) Polycrystalline solid, otherwise the same as (*a*); the boundaries between crystals are the heavy lines, which also bound domains (if the crystals are very small, the domains can be larger than the individual crystals).   (*c*) Single crystal in an external magnetic field, partially macroscopically magnetized.   (The arrows indicate the directions of magnetization.)

ferromagnetic particles of very small (colloidal) size onto an etched surface of the ferromagnetic solid.   The ferromagnetic particles concentrate at the places where the domain walls intersect the surface, since there is a strong local magnetic field in these places.   The size and shape of domains can therefore be studied by microscopic examination of the surface.   This technique also permits study of the motion of domain walls during the application of an external magnetic field.   In this way the statements of the following paragraphs have been verified experimentally.

A polycrystalline specimen is illustrated in Fig. 9-22*b*.   The domain pattern is, of course, much more complicated than the single crystal of Fig. 9-22*a*, since each little crystal has crystal axes oriented differently and the preferred (low-energy) orientation of the domain magnetization is therefore also oriented differently in each little crystal.   Neither of the two specimens of Fig. 9-22*a* and Fig. 9-22*b* has a net magnetic moment, and we would call both specimens "unmagnetized."   It should be emphasized, however, that in each domain of each specimen the magnetization is the large and constant value produced by the spin alignment. Dimensions of domains vary widely, but a typical linear dimension is 10^{-5} m.

If the sample of Fig. 9-22*a* is placed in an external magnetic field the domain walls move and a net macroscopic magnetization is produced (Fig. 9-22*c*).   A similar effect occurs in the polycrystalline sample, the domains favorably oriented relative to $\mathcal{H}$ growing at the expense of those unfavorably or neutrally oriented.   At still larger values of $\mathcal{H}$, the direction of magnetization of the remaining domains rotates into alignment with $\mathcal{H}$.   Thus, at very high $\mathcal{H}$ values, all the magnetic moments of the

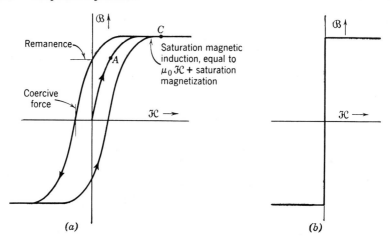

Fig. 9-23. ℬ-ℋ curves. (*a*) A typical curve for a polycrystalline specimen. (*b*) Curve for a rectangular ring of an ideal single crystal, with ℋ along the preferred (low-energy) directions in the crystal.

individual domains are aligned and the magnetization of the specimen is the same as the magnetization of a single domain.

The familiar ℬ vs. ℋ curve of a ferromagnetic solid is illustrated in Fig. 9-23*a*. As ℋ increases from zero, the domain walls move, produce a net magnetization, and increase ℬ. This part of the process stops at about the point *A*, and the further increase of ℬ is caused by rotation of the magnetic moments of the remaining domains. At *C* all the domains are in the direction of ℋ, and ℬ has reached its saturation value. The return curve as ℋ is decreased does not retrace the initial curve because some of the domain wall motion is irreversible. This would not be true in a perfect single crystal, which under ideal conditions exhibits a ℬ-ℋ.curve like Fig. 9-23*b*. In the actual polycrystalline specimen, the motion of domain walls is impeded by strains, impurities, and other crystal imperfections. A large imperfection (like a precipitated colloidal particle) creates spike-shaped domains around it which are very effective in retarding the motion of a domain wall.

For transformers and rotating electric machinery we seek as small as possible an area of the hysteresis loop of Fig. 9-23*a*. Relatively pure metals, without precipitated impurities, free from strain, and composed of large crystals, are therefore used. For permanent magnets, a large remanence and a large coercive force are required. Permanent-magnet materials are therefore alloys with precipitated phases and small crystals (see Fig. 12-31).

(d) Antiferromagnetism and ferrimagnetism

In a few elements and in many compounds (NiO, FeF_2, and MnS are examples) the exchange interaction has a sign opposite to its sign in ferromagnetic solids. Magnetic moments of neighboring magnetic atoms in these materials are aligned *anti*parallel, a phenomenon called *anti*ferromagnetism. This antiparallel alignment commonly occurs through the intervention of a negative ion, commonly oxygen, between the two positive ions possessing magnetic moments. The way the oxygen atom serves to create the alignment is sketched in Fig. 9-24. The strength and even the sign of this interaction, frequently called *superexchange*, depend on whether the positive ions are diametrically opposed or arranged in some other fashion relative to the oxygen. The susceptibility of antiferromagnetic solids is positive, as in paramagnetism, but the susceptibility *decreases* as the temperature is lowered below the Curie point, the point at which alignment of moments of equal magnitude occurs.

Another class of solids, more interesting from the point of view of applications, is the class of *ferrites* and related magnetic oxides, and the type of magnetism exhibited by these compounds is called *ferri*magnetism. Neighboring atomic moments in these materials are aligned antiparallel by the antiferromagnetic exchange interaction, but since these crystals contain ions with at least two different magnitudes of moments, the net magnetization is in general *not* zero. These compounds therefore exhibit spontaneous magnetization, like ferromagnetic materials.

We shall first consider the example of magnetite, which has the formula Fe_3O_4 and the spinel crystal structure, a complicated structure with

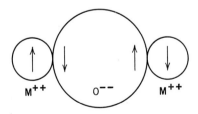

Fig. 9-24. Antiferromagnetic exchange interaction. The ground state of the oxygen ion when surrounded by two ions with magnetic moments is lowest if its electron spins are, on the average, paired on each side with those of the positive ions. Thus the two positive ions are aligned antiparallel. (See J. Smit and H. P. J. Wijn, *Ferrites*, Wiley, New York, 1959, pp. 30 ff.)

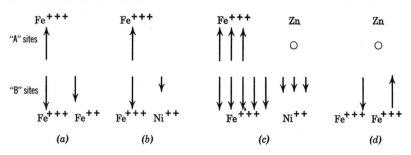

Fig. 9-25. Magnetic moment arrangements in magnetite and some ferrites. The upper row gives the moments of ions in A sites and the lower in B sites; each arrow represents an ion, and the length of the arrow is proportional to the magnetic moment of the ion. (a) Magnetite. (b) Nickel ferrite. (c) Divalent zinc (zero moment) substituted for one-fourth of the nickel to give $Ni_{0.75}Zn_{0.25}Fe_2O_4$ (four times as many ions are shown as in the other parts of this figure). (d) Zinc ferrite; the B-B interaction is so weak that the ordering shown disappears above a Curie temperature of $\sim 4°K$.

eight molecules of Fe_3O_4 per cubic unit cell. Measurements of saturation magnetization help to verify the assertions about the arrangements of magnetic moments made below. The most powerful tool for verifying these assertions and learning about ferrites generally is, however, neutron diffraction; the magnetic moment of the neutron interacts strongly with the moments of atoms, and the Bragg reflection of neutrons from a regular array of magnetic moments reveals the structure of this array.

One-third of the iron ions in Fe_3O_4 are doubly charged (six $3d$ electrons remaining per atom) and two-thirds are triply charged (five $3d$ electrons), and the chemical formula is therefore frequently written $Fe^{++}Fe_2^{+++}O_4^{--}$. There are two types of positions for the metal ions: They can be located so that their nearest neighbors are four oxygen ions ("A" or "tetrahedral" sites) or six oxygen ions ("B" or "octahedral" sites). All the Fe^{++} ions are in B sites, and half the Fe^{+++} ions are in A and half in B sites. All of the iron ions interact (through the oxygens) by the antiferromagnetic exchange interaction, but the A-B interaction is much stronger than the A-A or B-B interaction. Thus the net magnetic moment of the two sets of Fe^{+++} ions is zero, and the magnetization is simply that of the Fe^{++} ions, each with a net magnetic moment of $4\mathfrak{M}_B$ as noted in (b) above. This magnetic structure is illustrated schematically in Fig. 9-25a.

Many chemical modifications of this Fe_3O_4 structure can be created, and desired magnetic and other properties can be varied over wide ranges. Figure 9-25b illustrates the magnetic moment arrangement in nickel ferrite, $NiFe_2O_4$, which is very similar to magnetite. As zinc is substituted for some of the nickel in nickel ferrite, the magnetization at first increases, even though the zinc ions have no moments. This is because the zinc substitutes for Fe^{+++} in A sites and thus prevents the perfect cancellation of the Fe^{+++} moments. Figure 9-25c shows the structure of $Ni_{0.75}Zn_{0.25}Fe_2O_4$ in which four of these basic units are shown in order to have integral numbers of Ni and Zn ions (but we continue to refer to the "molecule" as a single MFe_2O_4 unit, where "M" is a divalent ion). Further increase of the Zn content until Ni and Zn are in about equal concentration increases the magnetic moment per molecule, but eventually the number of Fe^{+++} ions in A sites becomes too small to align all the Fe^{+++} ions in B sites; in the limit as all the Ni ions are replaced, the structure is that shown in Fig. 9-25d and is paramagnetic except at very low temperatures. There are many additional ferrite structures, most of them more complicated than these examples.

Most of the applications of ferrites depend on the fact that, unlike ferromagnetic materials, they are ionic crystals with very small electrical conductivity.* Thus eddy current losses are negligible, and ferrites can be used to quite high radio or even microwave frequencies. Thousands of tiny ferrite rings are used as information storage elements in a typical electronic digital computer; each ring is magnetized to saturation in either the clockwise or counterclockwise sense, corresponding to the binary digits 0 or 1. The absence of appreciable eddy currents permits rapid (<1 μsec) storing and reading-out of the information stored by current-carrying wires threaded through the rings. Other ferrites can be designed with large values of the coercive force, for permanent magnet applications.

REFERENCES

General

C. Kittel, *Introduction to Solid State Physics*, Wiley, New York, 2nd Ed., 1956.

F. Seitz, *Modern Theory of Solids*, McGraw-Hill, New York, 1940, Chapters 3, 4, 15, 16.

W. Shockley, *Electrons and Holes in Semiconductors*, Van Nostrand, New York, 1950, Chapters 5–8.

---

* Magnetite has a large conductivity because of the transfer of electrons from Fe^{++} to Fe^{+++}. Each Fe^{++} contains one more electron than Fe^{+++}. These electrons can move from one iron ion to the next in the narrow $3d$ energy band, which is only partially filled. This process does not occur in the ferrites, such as nickel ferrite, since the Ni^{++} energy levels (though broadened somewhat to make an energy band in the solid) make a *filled* band and do not overlap the Fe^{+++} levels.

J. C. Slater, *Quantum Theory of Matter*, McGraw-Hill, New York, 1951, Chapters 10, 11, 12, 14.

W. Hume-Rothery, *Atomic Theory for Students of Metallurgy*, The Institute of Metals, London, 2nd Ed., 1952, Parts IV, V, and VI.

W. Hume-Rothery, *Electrons, Atoms, Metals and Alloys*, Philosophical Library, New York, 1955, Parts II and III.

The Science of Engineering Materials, edited by J. E. Goldman, Wiley, New York, 1957.

Superconductivity

J. Bardeen and J. R. Schrieffer in *Progress in Low Temperature Physics*, Vol. 3, edited by C. J. Gorter, North Holland, Amsterdam, 1961.

B. Serin in *Encyclopedia of Physics*, Vol. 15, edited by S. Flügge, Springer, Berlin, 1956.

Ferromagnetism

R. M. Bozorth, *Ferromagnetism*, Van Nostrand, New York, 1951.

E. C. Stoner, *Magnetism*, Methuen, London, 5th Ed., 1954.

C. Kittel in *Modern Physics for the Engineer*, edited by L. N. Ridenour, McGraw-Hill, New York, 1954, Chapter 4.

W. Shockley, *Bell System Technical Journal*, **18**, 645–723 (1939).

"Action Pictures of Ferromagnetic Domains," 16-mm silent motion picture loaned by Publications Department, Bell Telephone Laboratories, 463 West St., New York 14, N. Y.

Ferrites

J. Smit and H. P. J. Wijn, *Ferrites*, Wiley, New York, 1959.

C. L. Hogan, *Scientific American*, **202**, No. 6, 92–104 (June, 1960).

PROBLEMS

9-1. Find from tables the electrical conductivities of four common metals and four solid insulators. Are the ratios of conductivities of one metal to another greater or less than the ratios among the insulators?

9-2.* In Sec. 9-2 it was stated that the application of an electric field to a solid does not appreciably alter the quantum states. Consider a field in the $+x$-direction of 100 volts/m. applied to a lithium crystal. Sketch the modification to the potential energy curves of Fig. 8-9 produced by this field, and compute the difference in height (in electron volts) of successive maxima of the $P(x)$ curve. The nearest-neighbor distance in lithium is 3.0 Å. What order of magnitude of field would be required to alter the wave functions appreciably?

9-3. Are the solid noble gases conductors or non-conductors of electricity? Why?

9-4. Use analysis like that of Sec. 9-2 to comment on the likelihood that the following solids are conductors or non-conductors: NaCl, BeO, Mg, and Sc.

9-5. Do you expect compounds such as GaAs and InSb to be conductors or non-conductors? Why? (See prob. 8-12 for the crystal structure of these compounds.)

9-6. Devise, if you can, a function $f(E)$ other than eq. 9-4 that has the properties (a) always ≤ 1, (b) like Fig. 9-5 at $T = 0$, and (c) like eq. 9-2 in the region $E \gg E_0$.

9-7.* In the Fermi factor, write $E = E_0 + \delta$. Compute δ for $f = \frac{1}{4}$ and for $f = \frac{3}{4}$, and express δ in units of kT. (This problem shows that values of f differing appreciably from 1 or 0 occur only for energies within a few kT of E_0.)

9-8. Again write $E = E_0 + \delta$. Show that, for any δ, $f(\delta) = 1 - f(-\delta)$, where "$f(\delta)$" means f evaluated at $E = E_0 + \delta$ and similarly for $f(-\delta)$. (This property of f will be useful in discussing semiconductors.)

9-9.* Again write $E = E_0 + \delta$. Evaluate f for $\delta = 0.1, 0.2$, and 1.0 e.V. for room temperature ($kT = 0.025$ e.V.).

9-10.* How accurate is eq. 9-2 as an approximation to $f(E)$ when $E - E_0 = 4kT$? when $E - E_0 = 10kT$?

9-11. Figure 9-7 is drawn for an s-band; what is the quantum number n of this band? Sketch $\psi(x)$ for a single atom with this n and $l = 0$. Sketch $\psi(x)$ as in the upper part of Fig. 9-7, but with λ equal to twice the distance between atom centers.

9-12. Maxwell-Boltzmann statistics are generally used to describe the properties of an electron gas just outside of and in equilibrium with a hot thermionic emitter. Show that this approximation is appropriate if (as is almost always true) the work function $e\phi$ (Fig. 4-1) is several electron volts. (For definiteness, assume $\phi = 1.5$ volts and $T = 1000°K$.)

9-13. Show that the average kinetic energy of conduction band electrons in a metal is $\frac{3}{5}E_0$. (See Sec. 2-3 for the method of computing an average when a distribution function is given; let $T = 0°K$.)

9-14. In prob. 9-13 it was convenient to set $T = 0°K$. Estimate the percentage difference between $\frac{3}{5}E_0$ and the actual $\bar{E}$ at $T = 300°K$ by plotting $S(E)$ and $f(E)$ near $E = E_0$ and graphically determining the difference in $\bar{E}$ that is made by replacing the $f(E)$ of Fig. 9-5 with that of Fig. 9-6, keeping the total number of electrons constant. Let $E_0 = 10$ e.V.

9-15.* Calculate the Fermi energy E_0 in electron volts for sodium and copper. The density of sodium at room temperature is 970 kg/m.[3], and the density of copper is 8920 kg/m.[3] In each metal there is one conduction electron per atom.

9-16.* Calculate E_0 for an alloy of 10% zinc in copper. Assume the same structure and interatomic spacing as copper.

9-17.* In order to obtain a crude estimate of the temperature at which Fermi statistics are no longer necessary, one frequently calculates a T from $E_0 = kT$. Calculate this T for copper. What do you conclude from this calculation and the order of magnitude of the E_0's in Table 9-1?

9-18.* What effective mass m^* would have to be assumed in eq. 9-7 in order to produce agreement between the theoretical and experimental values of E_0 for lithium?

9-19. A recent theory by Callaway of electron wave functions in lithium concludes that $E_0 = 3.39$ e.V. What is the ratio of the effective mass m^* to the free electron mass m according to this theory?

9-20.* Assume that the only cause for a change in E_0 with temperature is the thermal expansion of the lattice. Calculate the difference (in electron volts) between E_0 for copper at $100°C$ and at $0°C$.

9-21. In connection with Fig. 4-9 we spoke of "an electron with maximum initial kinetic energy"; we see now that strictly speaking there is no such electron. Describe as quantitatively as possible the effect that replacing the electron energy distribution assumed in Fig. 4-9 (the same as Fig. 9-5) by that of Fig. 9-8c has on the $I(V)$ curve of Fig. 4-3 near $V = -V_0$.

9-22. The $K\beta$-line of X-rays from sodium arises from the transition $3s{\to}1s$. In what way is the shape of this line different from the sodium L-line illustrated in Fig. 9-9? *Hint:* The intensity of photons emitted with any energy E is equal to the product of the number of electrons that can make the transition giving this $h\nu$ and the probability of a transition actually occurring; the dependence on E of the transition probability was discussed qualitatively in Sec. 9-4.

9-23. In a heavy metal like tungsten, even the $5p$ band is filled, and therefore the X-ray transitions $2p{\to}1s$, $3p{\to}1s$, $4p{\to}1s$, and $5p{\to}1s$ are all possible and produce the $K\alpha$, $K\beta$, $K\gamma$, and $K\delta$ lines, respectively. Compare the intensities of these four lines. *Hint:* The transition probability is proportional to the integral on the right in eq. 5-53, which in turn depends on the "overlap" in space of the wave functions for the initial and final states.

9-24. Show that the average kinetic energy (averaged over one cycle) of a classical simple harmonic oscillator (eq. 5-37) equals the average potential energy. (This result is used in Sec. 9-5.)

9-25.* Calculate the frequency of an acoustic wave in copper with a wavelength equal to twice the nearest-neighbor distance R_0. In copper $R_0 = 2.55$ Å, and the sound velocity is 3600 m./sec. Calculate also the frequency of a wave with λ equal to twice the specimen dimension (let this dimension be 1 mm). Estimate the temperatures at which these two waves will become thermally excited to an appreciable extent $(kT = h\nu)$ and hence contribute to the lattice heat capacity.

9-26. Show that C_v from eq. 9-12 becomes $3R$ in the limit $(\Theta/T){\to}0$. *Hint:* Expand the integrand in powers of x, and, since x will have only very small values in the interval of integration, make the appropriate approximations.

9-27.* At room temperature, the specific heat at constant pressure of silver is 234 joules/kg and of iron is 450 joules/kg. Should the law of Dulong and Petit hold in both cases? What are the heat capacities at constant pressure per *atom* (expressed in terms of k) of silver and of iron?

9-28.* The specific heat of lead at constant volume is 123 joules/kg°C and of beryllium is 1980 joules/kg°C at 20°C. Calculate the heat capacity per kilomole (C_v) for each metal, and compare with the law of Dulong and Petit. Consider the Debye temperature of each substance; should the law hold for each metal?

9-29.* Devise an approximate expression $C_v{'}$ for C_v that is valid at very low temperatures by letting the upper limit of the integral in eq. 9-12 approach ∞. The value of the integral is then $4\pi^4/15$.

9-30. To learn the accuracy of our approximation in the preceding problem, estimate the approximate value of C_v'/C_v for $T = \Theta/20$. (*Hint:* The error in our approximation of the integral is the integral from 20 to ∞ of the integrand in eq. 9-12, and in this range of x, e^x is always $\gg 1$.)

9-31. The frequency ν_m of the shortest wavelength wave in the Debye theory of heat capacity, like the vibration frequency of a diatomic molecule (eq. 7-8), is larger the smaller the mass and the sharper the binding energy curve near its minimum. Discuss likely values of Θ for the metal Cs and for the ionic crystal BeO. Why are the Debye temperatures of the noble metals in the order $\Theta_{Au} < \Theta_{Ag} < \Theta_{Cu}$? Why do the alkali metals have smaller Θ's than other metals of comparable atomic mass?

9-32.* Calculate the electronic heat capacity C_{ve} for copper at 1000°K, and compare with the lattice heat capacity C_v.

9-33.* Calculate the electronic heat capacity C_{ve} for copper at 1.6°K, and compare with the lattice heat capacity, using C_v' from prob. 9-29 for the latter.

9-34. Why is the electronic heat capacity of non-metals equal to practically zero?

9-35. Why do the electrons in non-conductors of electricity not conduct heat?

9-36. What is the form of the observed heat capacity C_{obs} as a function of T at very low temperatures? Devise a method of plotting some function of $C_{obs}(T)$ vs. some other function of T in order to get a straight line and to separate the lattice and electronic contributions to the total heat capacity. *Hint:* Use the form of eq. 9-15 and the result of prob. 9-29.

9-37. Why are the coefficients of linear expansion α of germanium, silicon, and diamond near room temperature ordered as follows: $\alpha_C < \alpha_{Si} < \alpha_{Ge}$? ($\alpha$ for diamond is 1.0×10^{-6} per degree C.)

9-38. It was pointed out at the beginning of Sec. 9-2 that values of the electrical conductivity of solids varied over a tremendous range. Why is the range of thermal conductivity values much smaller?

9-39. Draw a diagram like Fig. 2-1 but for the flow of electrons across a plane perpendicular to the direction of the electric field. Show that eq. 9-16 follows from consideration of this diagram.

9-40. The distribution function $\phi(v)$ of speeds v in a Maxwellian distribution of N particles is

$$\phi(v) = N \left(\frac{2m^3}{\pi k^3 T^3} \right)^{1/2} v^2 e^{-mv^2/2kT}$$

where $\phi(v)\, dv = dN$ is the number of particles of mass m with speeds between v and $v + dv$ (see prob. 2-9). Show that the average value of $1/v$ is $(2m/\pi kT)^{1/2}$.

9-41. Estimate the order of magnitude of the relaxation time in metals (discussed in connection with Fig. 9-15). Comment on the possibility of a direct observation of this time.

9-42.* Calculate the speed of an electron with kinetic energy E_0, in terms of E_0 expressed in electron volts. Calculate the speed of an electron with kinetic energy E_0 in copper.

9-43.* Evaluate l for conduction electrons for sodium and copper from eq. 9-20 and Table 9-5.

9-44.* Calculate the drift velocity Δv for an electron with energy E_0 in copper when an electric field of 100 volts/m. is applied (use the observed mobility). Calculate the ratio of Δv to the speed computed in prob. 9-42.

9-45. The speeds calculated in probs. 9-42 and 9-44 are much less than the speed of light, but, if a current pulse flows into one end of an isolated copper wire of length L, it is observed to flow out of the other end at a time L/c later. Why does the pulse move with the speed c of light rather than with the speed of the conduction electrons?

9-46. Use Table 9-5 to compare the electrical conductivity of liquid mercury with other pure metals and with disordered alloys. Why does its conductivity resemble disordered alloys rather than pure metals?

9-47. Show that the total energy of a classical harmonic oscillator (eq. 5-37) is proportional to the square of the amplitude of oscillation. What is the comparable statement for this oscillator treated by quantum mechanics? Verify the latter statement qualitatively by studying Figs. 5-12 and 5-13, eq. 5-43, and the definition of a.

9-48. Look up in tables the value of ρ and $d\rho/dT$ for copper and two other metals near room temperature. (See also Fig. 9-16.) Calculate $(1/\rho)(d\rho/dT)$ for each. Why do you expect this function to be practically the same for almost all metals at room temperature?

9-49. The resistivity of a metal can be written as the sum of the contributions from lattice scattering and from impurities. *Matthiessen's rule* states that the lattice scattering is approximately independent of purity and the impurity scattering is approximately independent of temperature. Investigate the validity of Matthiessen's rule for the samples of Fig. 9-18a. Use this rule to estimate the ratio of the resistivities of these two samples at room temperature.

9-50. Why should a certain small percentage of zinc in copper increase the resistivity more than the same percentage of silver?

9-51. It is impossible to locate the Hall potential leads (Fig. 9-17) exactly symmetrically, and the Hall potential is usually a tiny fraction of the potential difference applied at the current leads. What difficulty do these facts cause? How would you suggest that the experiment be performed in order to eliminate the effects of lack of perfect placement of the Hall leads?

9-52.* A Hall-effect experiment is being performed on a rectangular block of copper 0.1 m. long (in the direction of $\mathbf{j}$), 0.001 m. thick (in the direction of $\mathbf{\mathcal{B}}$), and 0.01 m. wide. Hall potential leads are connected to the narrow sides, and the potential difference V_1 between these is measured when $\mathcal{B}$ is 1.5 webers/m.^{2} and when a total current of 40 amp is flowing. $\mathcal{B}$ is then reversed in sign (but with the same magnitude and the same current), and the new potential difference is V_2. Compute $V_2 - V_1$ in volts.

9-53.* Calculate the Hall constant R for copper.

9-54.* How would the *normal* resistivity of $Ti_{0.42}V_{0.58}$ be expected to compare with that of pure titanium or of pure vanadium? Compare with Table 9-7, which gives the behavior of the *superconducting* alloy.

9-55. Ginsberg and Tinkham have measured the transmission of far infrared radiation by thin films of superconductors at 1.5°K and compared the transmission t_s when in the superconducting state with the transmission t_n when in the normal state (the normal state was obtained by applying a sufficiently strong $\circledB$). t_s/t_n as a function of λ shows a peak at $\lambda = 0.454$ mm for lead, at $\lambda = 0.745$ mm for mercury ($T_c = 3.79°$K), and at $\lambda = 0.825$ mm for indium ($T_c = 3.88°$K). Calculate $h\nu$ at the peak in units of kT_c for each metal. (Current theory indicates that this quantity should be the same for all superconductors.)

9-56. Biondi, Garfunkel, and McCoubrey have measured the attenuation of microwaves in aluminum wave guides at low temperatures. As the temperature is raised at a fixed wavelength λ, an additional absorption of the microwaves begins when the superconducting energy gap ΔE, a function of T, equals the $h\nu$ of the microwaves. The threshold temperatures measured were: 0.646°K at $\lambda = 4.03$ mm, 0.934°K at $\lambda = 5.17$ mm, 1.069°K at $\lambda = 7.91$ mm, and 1.105°K at $\lambda = 9.56$ mm. Plot ΔE as a function of T. Are these data consistent with the theoretical expectations that $\Delta E = 0$ at $T = T_c$ and $\Delta E = 3.5kT_c$ at $T = 0$?

9-57.* Calculate from the Boltzmann factor the total magnetic moment of an assembly of N moments, each of magnitude $\mathfrak{M}_B$, at room temperature in a magnetic induction of 1 weber/m.^{2}

9-58.* It was asserted in Sec. 9-8 that the orbital motion of an electron produced a magnetic moment of the same order of magnitude as $\mathfrak{M}_B$. Demonstrate this fact by a classical calculation of the magnetic moment produced by the motion of an electron in a circular orbit. Express this moment in terms of the angular momentum, and obtain the order of magnitude of the orbital angular momentum from Sec. 6-2.

9-59. Why should the alkali halides be diamagnetic rather than paramagnetic?

9-60. Gadolinium and dysprosium are ferromagnetic at low temperatures. Why do those rare earths that are ferromagnetic appear near the *middle* of the group, rather than at either end (near lanthanum or lutetium)? What are the electronic structures of the gadolinium and dysprosium atoms?

9-61.* Estimate the magnetic induction (μ_0 multiplied by the magnetic moment per unit volume) in a single domain of iron, of cobalt, and of nickel after consulting Table 6-1 for the electronic structures of these elements. Neglect any effect of the 4s bands and the 4s electrons. Compare your results with handbook data for the maximum magnetic inductions in these three metals. (The difference between experiment and the simple theory is produced by the canceling effect of the 4s electrons, which reduces the net magnetizations by about the same amount in each of these three elements.)

9-62. How should we arrange four rectangular bar magnets, each of length L and cross-sectional dimensions $L/2 \times L/2$, in such a way as to minimize the energy in the external magnetic field? (Magnetic domains arranged like this are observed.)

9-63. The largest saturation magnetization of any ferrite is only about one-third that of pure iron. Why?

9-64.* The magnetic moment alignment in $CoFe_2O_4$ is the same as in nickel ferrite. What is the magnetic moment per molecule in $CoFe_2O_4$ (in units of $\mathfrak{M}_B$)? Multiply by the number of molecules per unit volume and by μ_0 to obtain the magnetic induction. (The density is 5290 kg/m.^{3})

9-65.* What would be the magnetic moment per molecule in magnetite (in units of $\mathfrak{M}_B$) if all the moments were aligned in the same direction? (The observed value is 4.1 Bohr magnetons per molecule.)

10

Imperfections in Solids

10-1 INTRODUCTION

Some properties of solids are practically the same for perfect crystals as for crystals with small concentrations of chemical impurities or with other deviations from perfection. Examples of such properties are density, heat capacity, and thermal expansion. On the other hand, minute deviations from crystal perfection control many of the properties of solids, such as the strength of metals and the electrical conductivity of semiconductors, and cause interesting phenomena, such as the light emission by insulators. In this chapter we shall study a few examples of the ways in which imperfections determine the behavior of solids. The analysis in Chapters 8 and 9 provided the basic understanding required to investigate these and many additional imperfection phenomena.

Section 10-2 will survey the types of imperfections present in solids. Diffusion will be described briefly in Sec. 10-3. The absorption of light in insulating crystals will be considered in Sec. 10-4, and the change in electrical conductivity produced by this absorption of light will be studied in Sec. 10-5. The emission of light by solids will be considered in Sec. 10-6. The effect of dislocations on the mechanical properties of metals and alloys will be treated in Sec. 10-7. The electrical conductivity of semiconductors, another important application of imperfection theory, will be considered in detail in Chapter 11.

341

10-2 TYPES OF IMPERFECTIONS

In a perfect crystal every atom is in precisely the correct place in the crystal lattice, there are no atoms missing from sites in the lattice, and there are no foreign atoms. Furthermore, if the perfect solid is not a metal, each of the energy bands is either completely *empty* (no electrons) or completely *full* (every possible quantum state occupied by an electron). Such a perfect crystal does not exist. Any real crystal always contains chemical and physical imperfections. In the following paragraphs the various types of deviations from perfection are listed. Undoubtedly every crystal has all these imperfections. It frequently happens, however, that a single type of imperfection is responsible for a particular physical property of the real crystal. This may be because one type is present in a much greater concentration than other types, or it may be because the property of interest is not appreciably altered by imperfections of other types.

(a) Phonons

The vibrations of the atoms in a solid about their equilibrium positions have already been described in terms of phonons in Sec. 9-5 and were illustrated in Fig. 9-10. The energy of these vibrations increases rapidly with temperature, and they strongly scatter electrons moving through the crystal. The electron transfers energy and momentum to the vibrational waves in the lattice; in the quantum language, the electron collides with a phonon, a quantum of energy of vibration of the atoms in the crystal. As was explained in Sec. 9-6, phonons are responsible for the electrical resistance of pure metals at ordinary temperatures. They are also important imperfections in the phenomena of semiconduction and luminescence, and in many other phenomena. Unlike the other imperfections we shall study, the number or properties of phonons cannot be changed appreciably by changes in the preparation of the solid. Only by changing the temperature can the concentration and distribution of wavelengths of phonons be changed appreciably.

(b) Chemical impurities

No crystal is really chemically pure. A crystal with a concentration of 0.1% of an impurity is considered to be quite pure, but no point in such a crystal is more than a few lattice constants (d_0) from an impurity atom (each cube with an edge length of 10 lattice constants would contain about 1000 atoms and hence on the average would contain 1 impurity atom). Impurity atoms can be incorporated in the *host* crystal

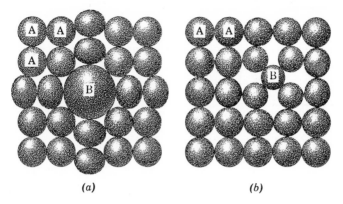

(a)   (b)

Fig. 10-1. Impurity atoms (B) in a host crystal of atoms (A). (a) Substitutional impurity. (b) Interstitial impurity.

in either of two ways: (1) *Substitutional* impurities; the impurity atom takes the place of a host crystal atom at a regular lattice site, as shown in Fig. 10-1a. (2) *Interstitial* impurities; the impurity is squeezed into one of the interstices in the crystal, without replacing a host crystal atom, as illustrated schematically in Fig. 10-1b.

Some distortion of the lattice is produced by either type of impurity. Substitutional atoms may be either larger or smaller than the host atoms they replace. An interstitial atom generally causes greater distortion, and therefore large atoms are rarely interstitial impurities. In fact, only a few very small atoms (such as hydrogen, lithium, nitrogen, and

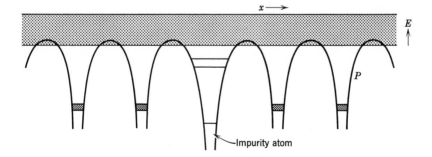

Fig. 10-2. Localized energy levels introduced by an impurity. The energy levels of the impurity atom are altered in position when it is dissolved in the solid but are not broadened into bands.

carbon) can enter metals (close-packed structures) interstitially. It should be noted that the two-dimensional plot of Fig. 10-1b exaggerates the distortion around an interstitial impurity. In open crystal structures, like the diamond structure, there is enough space in interstices for even relatively large atoms like copper to be found interstitially, but substitutional impurities are also common in such structures.

An impurity atom introduces electron energy levels that are not present in a perfect crystal. This fact is illustrated schematically in Fig. 10-2, which is a plot of the potential energy P of an electron as a function of distance along a line of atoms. The energy levels of the impurity atom are *not* broadened into bands because the impurity atoms are at least several lattice spacings apart. The interaction between impurity atoms is therefore weak, and sharp energy levels are preserved (compare Fig. 8-7 or Fig. 8-10 for large values of the internuclear spacing R). The electrons of the impurity atoms move in a region of potential energy that is quite different from the potential energy when the atom is isolated (such as in a gas, rather than immersed in the host crystal). Therefore the energy levels of the impurity atom are not the same when it is dissolved in the solid as they are for such atoms in a gas.

The wave filter analogy used in Sec. 8-5 provides another way of showing that sharp, atomic-like energy levels are introduced by an impurity atom. If the repeated array of identical inductances and capacitances in a wave filter is interrupted by a different L or C, a *local resonance* occurs at that point in the filter. That is, a sharp resonant frequency occurs that can be observed only by exciting the filter near that imperfection.

The energy levels of an impurity are usually broadened by the thermal vibrations (phonons) of the lattice. Since the change in an energy level depends sensitively upon the distances from the impurity atom to its neighbors, a level is different for different instantaneous positions of the neighbors. At high temperatures the vibrations of the atoms in the lattice therefore broaden the energy levels of the outer electrons of the impurity. Note that this broadening is different in origin from the broadening of energy levels into bands in a perfect crystal.

In addition to providing localized energy levels, an impurity atom scatters the traveling waves of electrons in the energy bands. Here, too, the action is similar to the partial reflection produced by an L or C different from the usual L and C in a wave filter. The mean free path of electrons in a crystal is thus decreased as the impurity content is increased. We have already noted (Sec. 9-6) that scattering by impurities limits the conductivity of metals at low temperatures. We shall see in Sec. 11-3 that it is also important in semiconductors.

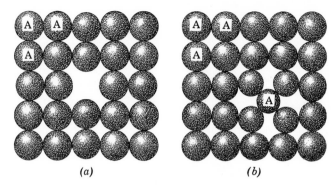

(a)   *(b)*

Fig. 10-3. (*a*) Vacancy. (*b*) Interstitial.

(c) *Vacancies and interstitials*

The imperfection produced by removing an atom from a lattice site is called a *vacancy;* it is illustrated in Fig. 10-3*a*. The imperfection produced by inserting an atom of one of the chemical elements composing the crystal into one of the interstices between ordinary lattice sites is called an *interstitial;* it is illustrated in Fig. 10-3*b*. Vacancies and interstitials behave like impurities in that they have localized energy levels and scatter electron waves.

These imperfections can be produced in several ways. One way is by the displacement of atoms from their normal positions upon bombardment by energetic neutrons or heavy ions in a nuclear reactor. Another way is by quenching a crystal from a high temperature. In thermal equilibrium at a high temperature there are appreciable disorder and appreciable concentrations of these crystal defects. (In thermodynamic terms, the state of the crystal is determined by minimizing the *free energy*, which equals $E - TS$, where S is the entropy. S is greater for larger concentrations of defects. At high temperatures, $E - TS$ is smaller when there is a large concentration of defects even though the energy E would be smaller if every atom were in its proper place.) In most crystals there are many more vacancies than interstitials. In a crystal with two kinds of atoms (e.g., KCl), there are two kinds of vacancies and interstitials (e.g., a K^+ or Cl^- vacancy, and a K^+ or Cl^- interstitial).

(d) *Electrons, holes, and excitons in non-metals*

In a perfect non-metal there are no partially filled energy bands. A free electron in a normally empty band is therefore a type of imperfec-

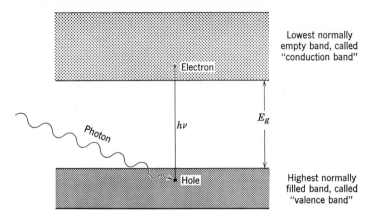

Fig. 10-4. Schematic diagram of the production of a conduction electron and a hole by a photon.

tion. Another type is a quantum state in a normally filled band that is not occupied by an electron. Such a quantum state is called a *hole*. The concept of a hole will be explained further in Sec. 11-2, but meanwhile we should note that a hole behaves like a positive charge if an electric field is applied to the crystal. The electrons in the normally filled band are urged in one direction by the field, which means that the one quantum state not occupied by an electron (i.e., the hole) is urged in the opposite direction. The hole (a vacant electron state) should not be confused with the vacancy (a missing *atom or ion* from a lattice site).

Electrons and holes can be produced by *thermal excitation* and are responsible for the electrical conductivity of non-metals. At any temperature above 0°K there are some electrons in the lowest normally empty band (*conduction band*) and some holes in the highest normally filled band (*valence band*). The number of electrons and holes is small if the *energy gap* E_g (the difference in energy between the top of the valence band and the bottom of the conduction band) is more than a few tenths of an electron volt, as explained in Sec. 9-2 and calculated in Sec. 11-2.

Another way of producing free electrons and holes is by the absorption of light, as illustrated schematically in Fig. 10-4. An incident photon with an energy $h\nu$ greater than E_g, the energy gap, can excite an electron from the valence band to the conduction band. This process is just the same in principle as the ionization of an atom by a photon with $h\nu$ greater than the ionization energy. Both the electron and the hole produced by this process are free to migrate through the crystal. Immediately after such an absorption process, the vacant state in the valence band is generally not at the top of the band, nor does the free electron appear at the bottom of the conduction band (except when the incident photon has barely the threshold energy $h\nu = E_g$). The electron quickly loses its kinetic energy of motion in the conduction band by collisions with phonons until its kinetic energy is about $\frac{3}{2}kT$. Similarly, the hole rises to the top of the valence band, like a bubble in a tube carrying water. What really happens to the hole is, of course, that an electron with energy nearer the top of the valence band makes a transition to lower energy, thereby losing energy to the lattice vibrations, filling the hole, and creating a hole nearer the top of the valence band.

Photons with $h\nu < E_g$ can excite an electron without giving it enough energy to leave the hole. This process is the same in principle as the excitation of an atom by a photon with $h\nu$ equal to the energy difference between the ground state and one of the excited states. The electron and hole produced in the solid are bound together. This electron-hole pair is called an *exciton*. It cannot conduct electricity, since its total charge is zero. It can nevertheless transfer energy from one point to another, since both electrons and holes are mobile. The exciton is therefore quite different from an excited state of an impurity atom in a crystal, which is fixed in position. The minimum photon energy required to produce an exciton is usually between $\frac{3}{4}$ and $\frac{9}{10}$ of E_g, the

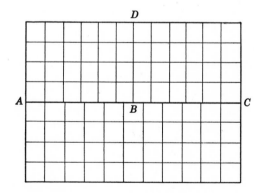

Fig. 10-5. An edge dislocation in an otherwise perfect crystal. The part of the crystal above the line ABC has one more plane of atoms than the part below ABC. The upper part is under compression, and the lower part is under tension. The line normal to the paper at B is the *dislocation line*, and the symbol $\perp$ placed at B is a shorthand way of indicating the dislocation illustrated in full here. (Each block represents an atom.)

energy required to produce free electrons and holes. Additional properties of excitons will be considered in Secs. 10-4 and 10-5.

(e) Dislocations

A dislocation is a region of a crystal in which the atoms are not arrayed in the perfect crystal lattice. The simplest type of dislocation, an *edge dislocation*, is illustrated in Fig. 10-5. An incomplete atomic *plane* is normal to the paper at *BD*. The distortion of the crystal is greatest near the edge *B* of this plane, which is called the *dislocation line* and is normal to the plane of the figure. A two-dimensional bubble model of a dislocation is illustrated in Fig. 10-6. The individual bubbles (representing atoms) are compressed on one side and dilated on the other side of the dislocation line. The distortion of the individual bubbles is so slight that it is difficult to determine the position of the dislocation line unless we look at the page at a glancing angle.

Dislocations are produced during the solidification of the original crystal from the melt. They can also be produced by plastic deformation of the cold crystal. Dislocations are most significant in determining the strength of ductile metals (Sec. 10-7) but are involved in many other processes.

Dislocations can be observed by any of several techniques. Disloca-

Fig. 10-6. Bubble model of a dislocation. [From W. L. Bragg and J. F. Nye, *Proc. Roy. Soc.*, **A190,** 474 (1947).]

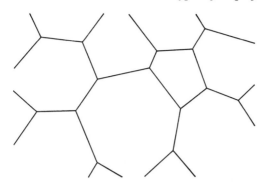

Fig. 10-7. Grain boundaries in a polycrystalline metal. Each area is a single crystal with an orientation different from its neighbors.

tions in specimens thin enough (a few hundred angstroms) for appreciable transmission of 100-kilovolt electrons can be studied by electron microscopy (see Fig. 12-32). The region of dilation along a dislocation line is an especially favorable place for the precipitation of impurities. Thus dislocations in transparent crystals can be seen in the optical microscope if they are first decorated by precipitating metallic impurities along the dislocation lines (e.g., silver decoration of alkali halides). The intersections of dislocation lines with the surface of a crystal can be revealed by the etch-pit technique, to be discussed in the last paragraph of this section (see also the Frontispiece).

(f) Grain boundaries

Up to this point we have been considering a single crystal, with the same orientation at all points. Most materials are not single crystals but aggregates of little single crystals called *grains*. Each grain is a crystal oriented in a different direction. The surface separating one grain from another is called a *grain boundary*. A schematic diagram of grain boundaries is presented in Fig. 10-7. The freezing of any melt produces polycrystalline material, with many grain boundaries, unless great pains are taken to have the freezing proceed slowly from a single point. Grain boundaries are of most importance in the plastic properties of metals, but they must also be considered in many other areas of solid-state physics.

A grain boundary is not usually listed with the primary imperfections discussed above for two reasons: (1) Unlike the others, this imperfection can be avoided; large single crystals can be grown without grain boundaries. (2) A grain boundary can be considered as merely a great con-

centration of dislocation lines, as illustrated in Fig. 10-8. It is nevertheless fitting to include the grain boundary in this list of types of imperfection because it is such a common imperfection.

The connection between grain boundaries and dislocations has been
experimentally verified for very low-angle grain boundaries separating
crystals of nearly the same orientation. Examination of Fig. 10-8
shows that, for a small angle between orientations, the dislocations are
separated by many lattice constants. Since dislocations are regions of
less than perfect order, if a crystal is etched by an acid the regions near
dislocations are etched more rapidly than the perfectly ordered regions.
The crystal therefore shows *etch pits* at the places where dislocation lines
intersect the surface. A photograph of the surface of a germanium
slab containing a low-angle grain boundary is shown in Fig. 10-9. The
regular spacing of etch pits is just what the dislocation theory predicts.
Furthermore, from the spacing of the pits and the size of the blocks of
Fig. 10-8, which is 3.98 Å for the germanium crystal orientation of

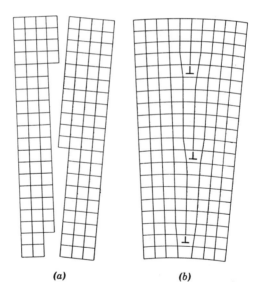

(a)   *(b)*

Fig. 10-8. Dislocation interpretation of a simple grain boundary. (*a*) Two grains
with a common crystal axis perpendicular to the paper. (*b*) The two grains joined
together to form a bicrystal. The plane boundary between the grains is normal to
the paper. Note that the spacing between the edge dislocations (⊥) can be determined from the angle θ between the grains and the size of the blocks (each block
represents an atom). (From W. T. Read, Jr., *Dislocations in Crystals*, McGraw-Hill,
New York, 1953.)

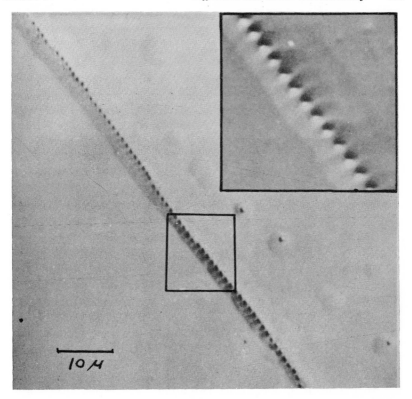

Fig. 10-9. Etch pits along a low-angle grain boundary in germanium. The intersection of an edge dislocation with the surface is a region of disorder that is preferentially attacked by the acid etchant. The length of the scale at the lower left is 10^{-5} m. since $1\mu = 10^{-6}$ m. [From Vogel, Pfann, Corey, and Thomas, *Phys. Rev.*, **90**, 489 (1953).]

Fig. 10-9, we can calculate the angle between the two crystals. This calculation agrees well with the angle determined by X-ray investigation of the relative orientation of the two crystal grains.

10-3 DIFFUSION AND IONIC CONDUCTIVITY

Diffusion is the transport of matter, especially atoms, ions, or electrons, under the driving force of a concentration gradient. If there is a higher concentration of a particular chemical impurity at one side of a crystal than at another, and if these impurities are at all mobile, there will be a net flow of impurity atoms or ions in the direction to equalize the con-

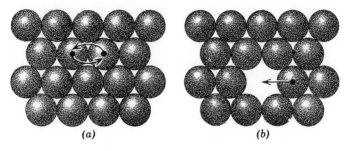

(a)   (b)

Fig. 10-10. Diffusion mechanisms. The direct interchange mechanism (a) is unlikely because of the great distortion of the lattice required. The vacancy mechanism (b) requires much less distortion and therefore has a smaller activation energy.

centration. *Self-diffusion*, the diffusion of atoms or ions of the host
(pure) crystal, is also an important process and can equalize the concentrations of vacancies and interstitials throughout a solid. The most
powerful way of studying all these diffusion processes is by depositing a
layer of radioactive atoms (such as Fe^{59} on an iron crystal, in order to
study self-diffusion) and then following the motion of these atoms by
examining the radioactivity of slices of the solid after a known time at a
known temperature.

Diffusion in solids necessarily involves vacancies or interstitials.
It is very difficult to interchange two atoms of a perfect lattice, as illustrated in Fig. 10-10a, because a great distortion of the lattice is required in order to squeeze them past each other. On the other hand, if a
vacancy is present, a lattice atom can jump into the vacancy with relatively little distortion, as shown in Fig. 10-10b. This jump moves the
vacancy to the right. Interstitials and interstitial impurities can also
diffuse through the lattice.

The flow of atoms J (particles/m.^{2} sec) is proportional to the gradient
of the particle density, dN/dx, and is of course in the direction in which
N decreases. Thus a diffusion constant D can be defined by

$$J = -D(dN/dx) \tag{10-1}$$

In the vacancy diffusion process illustrated in Fig. 10-10b, any of the
atoms adjacent to the vacancy can jump into the vacancy, thereby moving the vacancy to the position where that atom had been. Each atom
is vibrating about its equilibrium position with a frequency ν_0 of the
order of 10^{14} sec^{-1}, and thus the neighbors of a vacancy attempt to fill
it about 10^{14} times each second. But there is a potential energy barrier
ΔE arising from the difficulty of making room for the jump between

adjacent atoms; only if an atom happens to have enough energy to surmount that barrier is an attempt successful. The probability of having the energy ΔE is proportional to the Boltzmann factor $e^{-\Delta E/kT}$, and therefore D is also proportional to this factor. The activation energy for common vacancy or interstitial diffusion processes is of the order of magnitude 1 e.V. (In addition to the temperature dependence of D occurring in this way, the concentrations of vacancies and interstitials may also be functions of temperature, which produces an even more rapid exponential dependence of D upon T.)

A process closely related to diffusion is ionic conductivity, but (unlike diffusion) it occurs only in ionic crystals and in some other solids containing ionic impurities. An ionic crystal contains both positive-ion and negative-ion vacancies, but usually one is so much more mobile than the other that it alone need be considered. If there is no concentration gradient of vacancies, there will be no diffusion. But if an electric field is applied, the positive-ion vacancies are more likely to move toward

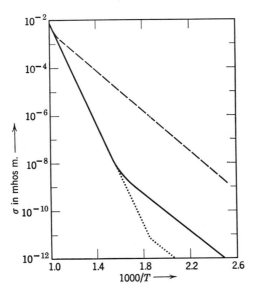

Fig. 10-11. The logarithm of the ionic conductivity plotted as a function of reciprocal absolute temperature for three crystals of KCl. In the steep-slope region at the left of each curve, the concentration of K^+ vacancies is changing with temperature and is the same for all samples. In the region to the right of each curve, the K^+ concentration is fixed and equal to the concentration of divalent impurities; this slope gives the activation energy ΔE for motion of K^+ ions. The dotted curve is for the purest crystal that has been grown. (Data from Anthony N. Taylor.)

the positive side of the crystal than toward the negative, because a vacancy is more likely to be filled by a positive ion jumping into it from that side of the vacancy which is at a higher electrostatic potential. Such an ion is aided in the jump process by the electric field. There is therefore a net current of positive ions toward the negative side of the crystal; similarly, the negative-ion vacancies will move toward the negative side and the negative ions toward the positive side of the crystal. This process depends on the same activation energy ΔE and the same attempt frequency as does diffusion, and hence the ionic mobility (drift velocity per unit electric field) is proportional to the diffusion constant (see probs. 10-19 and 10-20 and Fig. 10-11).

Diffusion is by no means confined to solids but is a common process in liquids and gases. In gases, the mean free path plays the role of the jump distance introduced above. For the diffusion of electrons and holes in semiconductors, the mean free path introduced in the discussion of electrical conductivity enters in the same way. We shall find in Sec. 11-4 that diffusion is a most important process in semiconductor devices.

10-4 OPTICAL ABSORPTION

The mathematical description of the absorption of light by a solid is very similar to the description of the attenuation of a molecular beam, which was described in Sec. 2-5 (see also eq. 6-16). Photons are removed from the incident beam by collisions that produce excited states or free electrons and holes. If L_0 is the original light intensity of a beam and L is the intensity after traversing a thickness x of the solid, then

$$L = L_0 e^{-Ax} \tag{10-2}$$

where A is the *optical absorption constant*. This equation can be derived by an argument similar to that leading to eq. 2-19.

The absorption of light by a pure insulator was described briefly in conjunction with Fig. 10-4 and in the discussion of excitons. No absorption can occur ($A = 0$) for photon energies below the energy required to make the first excitation, which results in an exciton with the electron and hole tightly bound together. As the $h\nu$ of the photons increases, there may be several higher-energy excitation processes (producing less tightly bound excitons), and finally the photon energy will be large enough to produce a free electron and a hole. The experimentally observed absorption constant of BaO is plotted as a function of photon energy in Fig. 10-12. Note the exciton absorption bands, which are absorption lines broadened primarily by the imperfect environment

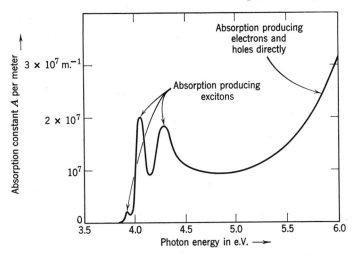

Fig. 10-12. Optical absorption constant of BaO as a function of photon energy at 80°K. Note the large magnitude of A. [From R. J. Zollweg, *Phys. Rev.*, **97**, 288 (1955).]

of the exciton because of the vibration of the surrounding atoms (in other words, because of phonons). The absorption beginning at $\sim$5 e.V. produces free electrons and holes. For pure BaO, the absorption constant equals zero throughout the visible part of the spectrum.

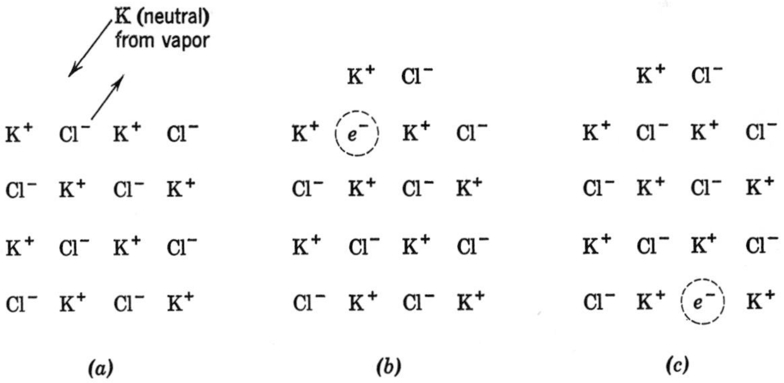

Fig. 10-13. Formation of an F-center in KCl by treating a KCl crystal in potassium vapor. In (a) an atom from the vapor combines with a Cl$^-$ ion to form another pair of ions in the crystal and an F-center. In (c) the F-center has diffused into the crystal.

If the solid has chemical impurities or other imperfections with localized energy levels, absorption of much lower-energy photons is possible. We shall consider the example of chlorine-ion vacancies in KCl, which can be produced by heating a crystal in potassium vapor. Chlorine-ion vacancies are produced at the surface and diffuse into the crystal. For each K atom added to the crystal, a K^+ ion, a Cl^- vacancy, and an electron are produced. The electron is *trapped* at the Cl^- vacancy, since this is a region in the crystal that has a net $+$ charge and therefore binds an electron. The electron bound to a negative-ion vacancy is called an *F-center*. Figure 10-13 illustrates the process of formation. *F*-centers can also be formed in ionic crystals by bombardment with energetic charged particles, X-rays, or ultraviolet light, but the processes of formation by such bombardment are not yet well understood.

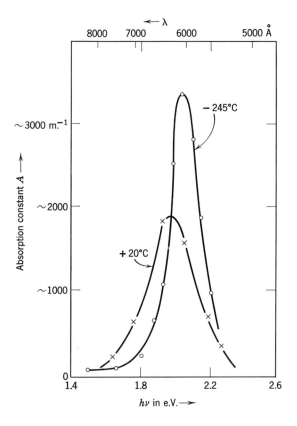

Fig. 10-14. The *F*-absorption band of KBr at two different temperatures. [From R. W. Pohl, *Physik. Z.*, **39,** 36 (1938).]

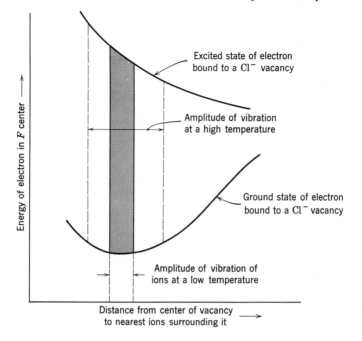

Fig. 10-15. Electron energy levels of an *F*-center as a function of lattice expansion in its neighborhood.

The electron of the *F*-center is in a discrete energy state (like the state of a chemical impurity described in Sec. 10-2*b*), and an incident photon can excite this electron to a higher-energy state. The optical absorption arising in this way is much weaker than the fundamental absorption of Fig. 10-12, since there are many fewer electrons in *F*-centers than electrons in the valence band. The absorption constant for *F*-center absorption is plotted in Fig. 10-14.

The absorption band is broader at higher temperatures because of the dependence of the photon energy required for absorption on the precise positions of neighboring atoms, as illustrated schematically in Fig. 10-15. The energies of the ground state and the excited state are plotted as a function of the distance from the site of the vacancy to any one of the surrounding six K^+ ions. During the very short time required for excitation of an electron by a photon, the heavy atom cores do not move appreciably. Therefore a transition is represented by a vertical line on this figure. At a low temperature, the interatomic distance is nearly constant, and only the transitions within the shaded region shown in Fig.

10-15 can occur. At a higher temperature, with a larger amplitude of the variation of the interatomic distance, a wider range of photon energies can be absorbed.

We have discussed a particular example (electron bound to a Cl^- vacancy in KCl) of a *color center* in a solid. Absorption bands like those of Fig. 10-14 give color to solids, since more light is removed from an incident white-light beam at some wavelengths than at others. Other color centers are produced by electrons bound to aggregates of $+$ and $-$ ion vacancies and by holes bound to $+$ ion vacancies. Chemical impurities are the commonest source of localized electron energy levels in solids. The colors of almost all non-metals are caused by photon absorption by localized energy levels, produced either by chemical impurities or by vacancies. In a few solids the absorption to produce excitons occurs at the blue edge of the visible region of the spectrum (and hence colors the crystal orange-red). The exciton absorption usually occurs, however, only in the ultraviolet, and the crystal without imperfections is transparent.

Before leaving the subject of optical absorption, we should draw attention to the meaning of transparency of a solid. All solids have large concentrations of electrons; at first sight we might predict that the electric field of a light beam would always accelerate some of these, and thus the beam would have some of its energy absorbed. Yet some solids, such as optical glass with $A < 0.01$ m.$^{-1}$, are (within the limits of measurement) completely transparent, and all pure non-metals have wavelength regions of great transparency. This transparency is, of course, a quantum effect: If the $h\nu$ of incident photons is insufficient to excite electrons from a filled to an empty band, exactly *no* absorption can occur.

10-5 PHOTOCONDUCTIVITY

Photoconductivity is the increase in the electrical conductivity of a non-metal that occurs if photons excite electrons to the conduction band or create holes in the valence band. The simplest way in which this can occur is if the photon energy is large enough to produce the free charge *carriers* (electrons and holes) directly (for example, if $h\nu > \sim 5.5$ e.V. in Fig. 10-12). But there are other ways: An exciton (produced by a lower-energy photon) can transfer its excitation to an electron at an F-center or at a chemical impurity, giving this electron enough energy to excite it to the conduction band. Also, a photon can excite an electron from a color center or chemical impurity to the conduction band.

A simple example of the photoconductive process is illustrated in Fig.

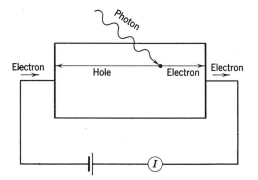

Fig. 10-16. Photoconductivity experiment. The situation illustrated is the simple but rare case in which the electric field is strong enough that it can cause both carriers to move to the electrodes without recombination or permanent trapping.

10-16. An electron and hole are created by the absorption of each photon, and one electronic charge flows through the external circuit for each photon absorbed. Such a simple process can occur only with nearly perfect crystals, with large electric fields, and with low light intensities. Some of the complicating effects usually present are space charge in the crystal, localized fields at the electrodes, and carrier *trapping*.

The trapping process for electrons is illustrated schematically in Fig. 10-17a. The trap is a localized empty quantum state a few hundredths or tenths of an electron volt below the conduction band. It can be introduced by a chemical impurity, by an interstitial, or by a negative-ion vacancy (which becomes an *F*-center after trapping an

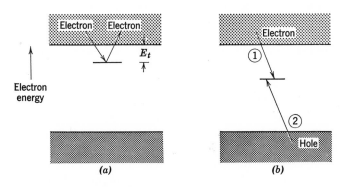

Fig. 10-17. (*a*) Trapping of an electron and subsequent release. (*b*) Recombination of an electron and a hole at a recombination center.

electron). The electron can lose a small amount of energy by exciting lattice vibrations. If it loses energy while in the neighborhood of the trap it will remain in the trap until it again acquires the energy necessary to permit escape. A trapped electron can be thermally released into the conduction band and can then continue its progress through the crystal. If the *depth* E_t of the trap (energy difference between the energy level of the trap and the bottom of the conduction band) is $\gg kT$, the electron will spend a long time in the trap before release.

Some impurity atoms produce localized electron and hole states that permit another process to take place. These atoms are called *recombination centers*, and the process is *recombination*. Such centers provide deep traps for electrons. The atom with a trapped electron has an attraction for a hole, which can then be trapped (process 2 in Fig. 10-17*b*). The *trapping of a hole* means, of course, that an electron from the impurity has made a transition into the vacant energy state in the valence band; it has "filled up the hole." This sequence of electron and hole trapping has removed the two free carriers (electron and hole) and *reset* the center so that the whole cycle can recur at the same center. The energy and momentum brought to the trap by the carriers are readily transferred to the lattice, since the impurity atom interacts strongly with the lattice (if the atom becomes larger because of the addition of an electron, for example, it pushes aside its neighbors, transferring energy and momentum to them). Not all impurities are effective as recombination centers. The transition elements iron, cobalt, and nickel appear to be particularly effective in many crystals.

Recombination at an impurity is much more likely than direct recombination between free electrons and holes, because momentum must be conserved in the process, and this is rarely possible for free carriers. The photon emitted in recombination has very little momentum, and the free carriers are coupled too weakly to the lattice to transfer momentum to it during the short time they are close to one another. Therefore direct recombination is possible only if an electron and a hole are moving in very nearly the same direction with very nearly the same magnitude of momentum, and this condition is rarely met.

Experimental data from a photoconductivity experiment are plotted in Fig. 10-18. A crystal of KCl containing *F*-centers is illuminated with light of constant intensity, and a constant electric field is applied. The photoconduction currents are measured for different temperatures of the crystal. At very low temperatures (below 140°K), a free electron is *not* produced for each photon absorbed; an electron in an *F*-center is raised to an excited state by the incident photon, but some thermal energy is necessary to raise the electron from this excited state to the conduction

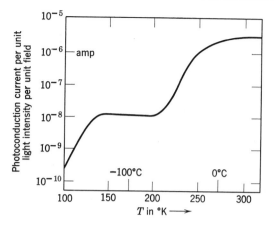

Fig. 10-18. Photoconductivity of KCl as a function of temperature. [From Rögener, *Gött. Nacht.*, **3**, 219 (1941).]

band. At intermediate temperatures ($140° < T < 200°$K), all excited electrons can be thermally excited into the conduction band and the photoconduction current is independent of T. Each electron moves until trapped. The most effective traps are unexcited F-centers, and a larger concentration of F-centers causes a proportional decrease in photoconduction current per absorbed photon in this temperature range. At higher temperatures ($T > 200°$K), the trapped electrons are thermally released and the photoconduction current is larger. In this temperature range the current does not fall to zero as soon as the light is turned off, since the thermal release of electrons from traps continues for some time. The *time constant* of the release is determined by the Boltzmann factor $e^{-E_t/kT}$ and therefore becomes shorter at higher temperatures. The experimental data presented in Fig. 10-18 thus confirm the importance of trapping in photoconductors and provide information on the excited state of the F-center.

The photoconductive processes described up to this point have yielded at most the flow of a single electronic charge in the external circuit for each photon absorbed, but processes yielding much larger currents are possible. One such process is the following: A hole and an electron are produced by the incident photon. The electron quickly moves through the crystal and into the external circuit. The hole becomes trapped, it remains in the crystal, and it creates a positive space charge. An electron flows into the crystal from the negative electrode to neutralize this space charge and flows through the crystal. The electron flow con-

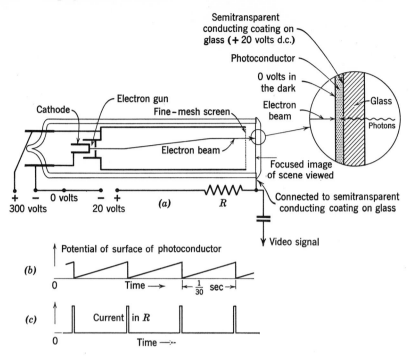

Fig. 10-19. "Vidicon" television pick-up tube using photoconduction. The tube and connections are shown in (*a*), but the magnetic focusing and deflection coils are not shown. The potential as a function of time at the surface of an illuminated spot of the photoconductor is plotted in (*b*). The signal transmitted to the video amplifier is shown in (*c*), where it is assumed that only a single spot of the photoconductor is illuminated.

tinues as long as the hole remains trapped; the process stops when an electron recombines with the hole in the crystal or at an electrode. The current gain of this photoconductive process (compared to the simpler one described above) equals the ratio of the lifetime of a trapped hole to the transit time of an electron.

One interesting application of photoconductivity is in a television camera tube called the *vidicon* (Fig. 10-19). The electron beam is made to scan across the thin photoconductor target by magnetic deflection coils. It adds electrons to any spot of the target that is at a potential greater than zero (the beam electrons cannot reduce the potential below zero because they would be turned back before reaching the target if the target potential were less than zero). Hence in the dark there is a potential difference of 20 volts across the target. If a spot on the target is illuminated, a photoconduction current flows, and the potential of the beam side of the target rises. The next time the scanning beam strikes this spot, electrons from the beam return the potential of the spot to zero. The charge deposited by the beam is proportional to the integrated photoconduction current during the time between scans and hence to the light intensity. Because of the large capacitance between the beam side of the photoconductor and the conducting layer on the other side, a charge flows in R just equal to the charge deposited by the beam. Therefore a signal is transmitted to the amplifier. This amplified signal controls the intensity of the electron beam in a picture tube (kinescope); this beam is deflected in synchronism with the beam in the camera tube. Thus a bright spot appears on the screen of the picture tube at the same relative position as the bright spot on the camera tube, and the intensity of the spot on the picture tube is proportional to the intensity of the spot on the camera tube.

Another interesting application of photoconductivity is the photographic process. A photographic film is composed of tiny crystals (*grains*) of silver bromide suspended in gelatin. Visible light incident on an AgBr crystal excites electrons from the valence band to the conduction band, and these electrons wander until trapped. The trap acquires a negative charge when it acquires an electron. AgBr contains appreciable numbers of interstitial silver ions, and furthermore these interstitials have a low value of the activation energy for jumping from one site to the next. AgBr is thus an ionic conductor of electricity. A positive ion is attracted to the negatively charged trap and bound there. The trap is thereby reset and can trap another electron, which attracts and neutralizes another Ag^+ ion. Thus, for each photon absorbed, a silver atom is added to the trapping region. This aggregation of silver forms the *latent image* (see Fig. 10-20).

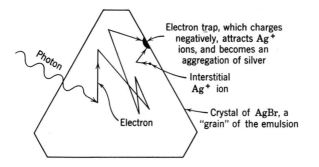

Fig. 10-20. Schematic diagram of the formation of a photographic latent image.

An AgBr crystal that has a silver particle with more than about four silver atoms in it can be converted entirely to silver by means of chemical reducing agents called *developers*. The silver particle forms a nucleus from which the conversion of the whole grain proceeds. If the tiny AgBr crystal is not exposed to light, no nucleus of silver exists; the developer is not sufficiently strong to create a nucleus and hence cannot convert the AgBr crystal to Ag. The film is *fixed* after development by dissolving away the remaining AgBr. Thus a negative image of the original illumination is formed. The exposed regions are opaque because of the light absorption by the silver crystals, and the unexposed regions are transparent.

It should be noted how the photographic process depends on the interaction of three imperfections: (1) Free *electrons* are produced by photons. (2) These electrons are concentrated at a *trap*. (3) *Interstitial ions* provide the mechanism for storing the ordinarily transient photoconductive effects.

Another important application of photoconductivity is in printing and copying, where photoconduction is used in a variety of ways. One way is the *xerographic* process of copying, in which an image of the document to be copied is focused on a thin photoconducting plate (frequently vitreous selenium) across which a potential difference has previously been established. Photoconduction at illuminated regions discharges the surface potential at these regions. Fine black powder particles are then attracted to the areas that have retained their charge, and this powder pattern is next transferred by electrostatic attraction to a sheet of paper placed close to the plate. The powder image on the paper is made permanent by heating or spraying. The photoconducting plate can of course be re-charged and used again and again.

10-6 LUMINESCENCE

The emission of light by a crystal is called luminescence. The two commonest ways of exciting luminescence in a solid are by high-energy electron bombardment and by irradiation with ultraviolet light; but high-energy nuclear particles can be used, and an intense alternating electric field also produces some light (*electroluminescence*). For both practical and physical reasons, two kinds of luminescence are distinguished: (1) Fluorescence, which is light emitted practically simultaneously with the introduction of the excitation energy and ceases as soon as the exciting radiation ceases. (2) Phosphorescence, which is light that persists some time after excitation is removed. The division between these two classes is usually made by stating that, if most of the light is

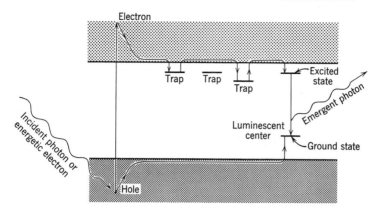

Fig. 10-21. Schematic diagram of the process of phosphorescence.

emitted within $\sim 10^{-8}$ sec after the excitation ceases, the solid is fluorescent; otherwise it is phosphorescent. The light from various phosphorescent materials (*phosphors*) persists for times from $\sim 10^{-7}$ sec ("short persistence phosphors") to minutes or even hours ("long persistence phosphors"). We shall discuss such phosphors in some detail and then briefly consider fluorescent solids.

There are three aspects of the process of light emission by a phosphor: (1) the absorption of the energy of the primary bombarding electron or photon; (2) the transfer and storage of this energy; (3) the conversion of the energy into light. These processes are illustrated schematically in Fig. 10-21 and will be discussed in the order stated.

An energetic primary electron loses its energy chiefly by the excitation of electrons from filled bands to the conduction band. The bombarding electron interacts with the electrons in the solid by the electrostatic force, and an electron with an energy of a few thousand electron volts can give an energy E_g or greater to many such electrons in succession. Electrons and holes can also be produced by a sufficiently energetic incident photon, as explained in Secs. 10-2, 10-4, and 10-5.

The electrons and holes wander * through the crystal until they encounter traps or *luminescent centers*. (In Fig. 10-21 and in the discussion below we assume that only the electrons are trapped, but if the holes are

---

* Free electrons and holes are not involved in luminescence in some phosphors, and no photoconductivity is observed in these. The electron traps are shallow wells adjacent to the luminescent center that are introduced by the same chemical impurity that creates the center. Absorption, storage, and conversion to light of the primary energy all take place within a few lattice constants of the luminsecent center.

also trapped the analysis can be extended to include this process.)   An electron in a trap is immobilized until by chance it acquires enough energy from the lattice vibrations to be again excited to the conduction band.   The rate of release is $se^{-E_t/kT}$ per second.   Here E_t is the depth of the trap and s is a factor related to the classical frequency of vibration of an electron in the trap potential well; s is about 10^8 sec^{-1}.   The electron can be considered as making s "tries" per second to surmount the energy barrier E_t.   The probability of success on each try is given by the Boltzmann factor $e^{-E_t/kT}$.   Electrons can thus be trapped for long or short times, depending on E_t (which varies with the type of impurity producing the traps), on T, and on the relative number of traps and luminescent centers.   Traps thus store the energy of the incident particles or photons and are responsible for the persistence of light emitted by phosphors.

An electron eventually is trapped in an excited state of a luminescent center or *activator*, which is an imperfection intentionally added to the pure crystal.   Activators are usually chemical impurities but may be interstitials or vacancies.   For example, a common phosphor is zinc sulfide to which a tiny amount (of the order of 1 part per million) of copper is added to produce luminescent centers.   Another phosphor is zinc oxide activated with zinc; a small excess of zinc is present as interstitial ions.

The process that occurs at a luminescent center is the recombination process described in Sec. 10-5.   (When we discussed recombination we were not concerned with the changes occurring in the trap after an electron was trapped.)   In some traps an electron is caught in an excited state of the trap and later makes a transition to the ground state.   This transition can involve the emission of a photon (*radiative* recombination, luminescence), or it can produce only phonons (*non*-radiative recombination).   At the present state of knowledge of solid-state physics we cannot predict which chemical impurities will give luminescence with a useful efficiency in a particular host crystal and which will merely dissipate the electron-hole energy into the heat associated with lattice vibrations.   The latter process is much more common, and therefore phosphors must be highly purified to reduce the concentration of such impurities.

Energy diagrams of a luminescent center and of a non-radiative recombination center are presented in Fig. 10-22.   The electron from the conduction band is trapped in the excited state.   The impurity atom is vibrating about its equilibrium position (the minimum in the curve of total energy).   The excited state of a luminescent center can also be produced by the transfer of energy from an exciton.

The electron in the luminescent center (Fig. 10-22a) can make a transi-

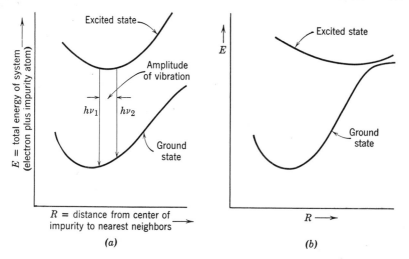

Fig. 10-22. Distinction between a luminescent center (*a*) and a non-radiative recombination center (*b*). Either type of impurity can trap an electron, and the resulting system has the $E(R)$ plotted in the upper curves. Luminescence is more likely in (*a*) because the non-radiative transition to the ground state would require the simultaneous emission of many phonons, which is as improbable as the simultaneous collision of many phonons. Recombination without radiation is more likely in (*b*), since only a few phonons need be emitted.

tion to the ground state by emitting a photon. The energy of the photon may be any value between $h\nu_2$ and $h\nu_1$ and depends on the phase of the vibration of the impurity atom at which the photon emission takes place. Thus a broad *band* of light is emitted. The impurity atom is in a high vibrational state after the electron transition, and in a higher state for the transition $h\nu_2$ than for the transition $h\nu_1$. Thus the energy that does not appear in $h\nu$ appears as lattice vibrations (i.e., is dissipated into heat).

The non-radiative center is illustrated in Fig. 10-22*b*. The transition to the ground state is made by the collision with phonons or perhaps by the production of a very long-wavelength infrared photon. In either event no visible light is produced. This is the common type of impurity and must be avoided to produce high-efficiency phosphors.

The *color* of the light emitted by the luminescent center depends, of course, on the range of emitted $h\nu$ values. This range in turn depends on the details of the two curves in Fig. 10-22*a* and is therefore different for different activators. The emission spectra of three phosphors are

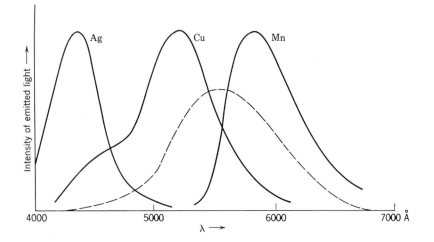

Fig. 10-23. Luminescent emission intensity as a function of the wavelength of emitted light. The three solid curves are for hexagonal ZnS phosphors with the three different activators labeled on the curves (the peak heights were arbitrarily set equal). (From H. W. Leverenz, *Introduction to Luminescence of Solids*, Wiley, New York, 1950.) The dashed curve is the relative sensitivity of the light-adapted human eye. The curve for the dark-adapted eye is almost identical in shape but is displaced ~500 Å toward the left (toward the violet).

presented in Fig. 10-23. The host crystal for each spectrum is ZnS, but three different activators have been added and produce quite different colors. More than one kind of activator can be added to a host crystal in order to give a broader spectrum or to approximate a white color. Also, mixtures of the host crystals can be used, since the same chemical impurity has different energy levels when immersed in different host crystals. Both approaches are common in the application of phosphors in "fluorescent" lamps (Fig. 10-25), where a common luminescent coating is a mixture of calcium chlorophosphate and fluorophosphate activated with antimony and manganese.

In *fluorescent* materials the excited state of the luminescent center is produced either directly by the incident photon, by receiving the energy of an exciton that was produced by the incident photon, or by electrons and holes. No trapping with attendant delay of emission occurs in such materials. Many organic solids such as stilbene and anthracene fluoresce without addition of activators, and many organic liquids fluoresce. In these compounds the luminescent centers are the individual molecules of the pure materials. Since these materials are invariably molecular crystals, the molecules are loosely bound together and interact weakly.

Fluorescence can occur within a particular molecule if an excited state of the molecule has a minimum at a different position of the ions from the minimum of the ground state, as illustrated in Fig. 10-24.

It should be noted that in all luminescent processes the emitted photons have a different frequency (almost always a smaller frequency) than the frequency of photons that excite luminescence. Thus the crystal is transparent to its emission spectrum, a necessary condition if light is to be emitted from the interior of the crystal.

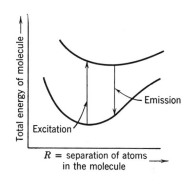

Fig. 10-24. Fluorescence in a molecule. The minimum in the excited state is at a larger interatomic spacing than the minimum in the ground state.

The principal practical applications of the luminescence of solids are in television picture tubes and in "fluorescent" lamps, but there are other applications (an application to the detection of nuclear particles will be explained in Sec. 14-6). In a television picture tube (kinescope), an electron beam excites the phosphor coated on the inside of the end of the tube. The intensity of the beam is varied as it is deflected from one point of the screen to another, and the light output varies accordingly. A medium-persistence phosphor is required to reduce flicker and yet to permit portrayal of rapid motion in the picture. The use of a phosphor in a "fluorescent" lamp is illustrated in Fig. 10-25. The low-pressure discharge in a mixture of argon gas and mercury vapor produces very little visible light yet produces $\lambda = 2536$ Å ultraviolet light with high efficiency. The phosphor converts the ultraviolet to visible light. The production of visible light by this two-stage process is much more efficient than by an incandescent lamp.

A different kind of emission of electromagnetic waves from solids is becoming increasingly important (atoms and molecules also emit in this way, but we shall not explore that aspect here). This is *stimulated emission*, to be distinguished from the *spontaneous emission* described above. Spontaneous emission, as the name suggests, does not require any external influence after an upper energy level of an impurity atom has been populated. For stimulated emission, a strong electromagnetic wave with frequency such that $h\nu$ precisely equals the energy difference between two states, the upper occupied and the lower vacant, is incident upon an atom or ion; the prompt emission of a photon is induced, and the outgo-

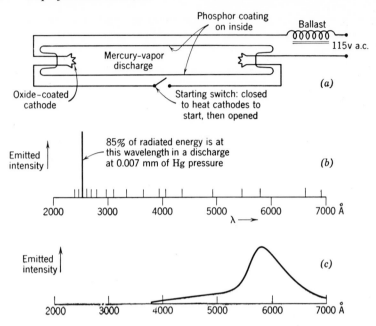

Fig. 10-25. "Fluorescent" lamp. The mercury discharge produces largely ultraviolet radiation, as shown in (*b*). The phosphor converts this to visible light with the emission spectrum shown in (*c*). The particular phosphor has a "warm white" color and is a mixture of calcium halophosphates. (The halogens are F and Cl and are present in a concentration ratio of 7 to 2 in this particular phosphor; the activators are Sb and Mn.)

ing wave so generated is in phase with (and thus strengthens) the stimulating wave. We could verify these statements by extending the arguments preceding eq. 5-53: The electric field of the incoming wave *perturbs* the atomic system and enhances the strength of the oscillating dipole and establishes its phase. (Stimulated, rather than spontaneous, emission is the true converse of absorption.) Since the stimulating wave must have precisely the frequency of the transition to be induced, either the energy levels involved must be quite sharp or at least the *difference* in energies between levels must be sharply defined. If broad levels (Figs. 10-22 and 10-23) were used, most of the stimulating wave would be wasted since it would be at the correct frequency for only a small fraction of the atoms or ions. We shall sketch two of a large and rapidly growing number of applications of stimulated emission.

One useful set of sharp levels is the set of levels into which a single level of an ion with a magnetic moment is split when the ion is in a magnetic field (Sec. 6-5). The separation $h\nu$ of these in a typical magnetic induction (0.1 to 1 weber/m.^{2}) gives a frequency ν in the microwave region of the spectrum. The energy difference between two such spin states is precisely defined and practically unaffected by the other ions in the solid and their thermal motion. For example, in a synthetic ruby crystal, a Cr^{+++} impurity ion in the Al_2O_3 host crystal has a magnetic moment, and its energy levels are split into several components in a magnetic field; since the Al^{+++} and O^{--} ions have no moments and since the Cr^{+++} ions are far apart (a typical concentration is a few hundredths of 1%), there is no significant broadening of the levels by the interaction of the Cr^{+++} ion with its environment.

A device that exploits stimulated emission and these magnetic energy levels is the *maser* (*m*icrowave *a*mplification by *s*timulated *e*mission of *r*adiation). In Fig. 10-26a are illustrated the energy levels of a three-level maser, which could, for example, be constructed of the synthetic ruby described above. The population of level E_3 is greatly increased over the thermal-equilibrium population by absorption of energy at a microwave frequency $\nu_{13} = (1/h)(E_3 - E_1)$ supplied by a local oscillator (*pump*). The population in level E_3 can thus be made greater than in level E_2, instead of substantially less, as in thermal equilibrium. The signal to be amplified, at frequency $\nu_{23} = (1/h)(E_3 - E_2)$, induces transitions from level E_3 to level E_2, and such transitions add photons in phase with this signal, thereby amplifying it. This amplifier can be operated at very low temperatures and contributes an extremely small amount of noise (Sec. 4-8) to the signal. It is hence useful for amplifying

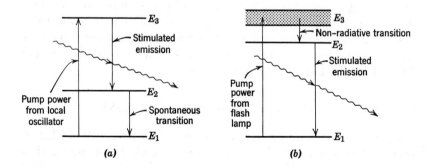

Fig. 10-26. Examples of stimulated emission of radiation. (*a*) Three-level maser. (*b*) Three-level laser; the non-radiative transition disperses some of the energy of state E_3 into lattice vibrations and leaves the impurity ion in the state E_2. The state E_2 can also be populated in some solids (e.g., GaAs) by injection of carriers from the *p-n* junctions to be discussed in Sec. 11-4.

the weak signals from satellites and space vehicles. By feeding back some of the output, this amplifier can be converted to an oscillator with remarkable frequency stability.

The energy levels of some impurity ions are properly spaced and sufficiently sharp to make possible a device similar to the maser but operating at the frequencies of visible light (it is sometimes called a *laser*, the "*l*" originating from *light*). Cr^{+++} in Al_2O_3 (synthetic ruby) is again a good example. The energy differences between pairs of states of the Cr^{+++} ion, whereas not so sharply defined as the differences produced in a single level by a magnetic field, are precisely enough defined that a typical emission line $\lambda = 6943$ Å has a width at half-maximum of 4 Å. These energy differences arise from transitions within the $3d$ shell of the small, isolated Cr^{+++} ion; the long distance to the next such ion and the tight binding of the $3d$ electrons into the Cr^{+++} core protect these energy levels from broadening.

Figure 10-26*b* illustrates the levels of a ruby laser that emits light in short pulses. Intense bursts of light from a gas-discharge flash lamp "pump" chromium ions into the level E_3; the absorption band for this process is a broad band in the green part of the spectrum. After a non-radiative transition, the Cr^{+++} ion is left in the excited state E_2 which is 1.7450 ± 0.0005 e.V. above the ground state. The cylindrical crystal, ~0.02 m. long, has plane-parallel, silvered ends. Light emitted in the transition $E_2 \rightarrow E_1$ is reflected back and forth, and a standing wave is produced with wavelength such that $n\lambda/2$ is equal to the length of the crystal, where n is an integer (see Figs. 5-8 and 5-10). Other ions are stimulated to emit in the same phase, the same direction, and at the same λ (i.e., the same value of n). Any other radiation (for example, radiation not precisely perpendicular to the faces of the crystal) does not build up a strong standing wave and hence does not induce further transitions. Thus a light pulse that is extremely homogeneous in frequency and in direction is generated and can be extracted through a small hole or partially silvered region at one end of the crystal.

10-7 SLIP AND STRENGTH OF METALS AND ALLOYS

A perfect crystal should have a strength much greater than that observed for any real crystal. It should be possible to produce a shear strain of the order of $10°$ in a perfect crystal without permanent deformation when the shear stress is removed. No real crystal of appreciable size has such great elasticity. Some crystal imperfection must cause the disagreement between perfect-crystal theory and real-crystal experiments. The imperfection responsible is the dislocation.

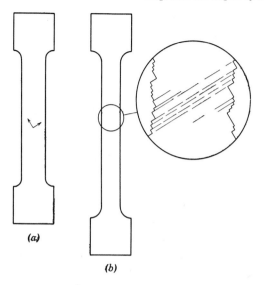

Fig. 10-27. Tensile test experiment on a single crystal of a ductile metal. The original crystal is shown in (*a*) with the arrows indicating the principal lattice directions. (*b*) is the same crystal after pulling. Slip has occurred on a set of parallel planes.

We shall first briefly explain the statements about the behavior expected for the perfect crystal. Then we shall show how the presence of dislocations modifies this behavior in the correct way to give agreement with experiments. Finally, we shall note briefly the origin of dislocations and some other effects of dislocations on the mechanical properties of metals, both single cyrstals and the common polycrystalline metals of engineering practice.

The tension test of a single-crystal specimen is illustrated in Fig. 10-27. For small elongations of the specimen this test measures Young's modulus of the crystal. As the tension force is increased, eventually plastic deformation of the crystal occurs. A brittle crystal (like quartz, Al_2O_3, and most metals at sufficiently low temperatures) breaks in two along a plane of atoms, the *cleavage plane*. Metals of most engineering interest are ductile rather than brittle, and ductile crystals exhibit *slip* as shown in Fig. 10-27*b*. The tension force has a component in the *slip planes* illustrated that acts to produce a shearing of one part of the specimen past the other part in a particular direction in these planes. The stress on the tensile specimen can be resolved into a *resolved shear stress*, which is the component of stress in the direction of slip. The *critical* *shear stress* for a particular slip direction is the resolved shear stress that is just sufficient to initiate slip. The critical shear stress is a property of the crystal and of the slip direction.

The critical shear stress of a *perfect* crystal can be roughly estimated by studying Fig. 10-28. The two rows of atoms pictured are on opposite sides of the slip plane. If a small shearing stress is applied, an elastic shear strain is produced, and the upper plane is moved slightly to the right relative to the lower plane. The binding forces between atoms oppose this motion, since it results in a larger separation between atomic centers than the equilibrium separation (Sec. 8-3 and Fig. 8-5). This opposing force increases with strain for small strains, but it must decrease for large strains since the shear strain shown in Fig. 10-28b must have a zero opposing force. The maximum opposing force occurs for $\theta \cong 10° \cong \frac{1}{6}$ radian, as illustrated in Fig. 10-28c. If a larger stress is applied than the stress necessary to produce this strain, the planes slip past each other. In other words, the critical shear stress of a perfect crystal should be the stress that produces an elastic shear strain of $\sim\frac{1}{6}$ radian.

The experimental critical shear stress of real crystals is smaller than this prediction by a factor of as much as 30,000. Real crystals contain dislocations, and a crystal with dislocations can slip with a much smaller

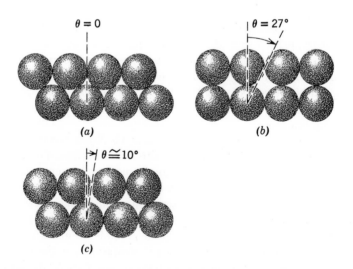

(a)

(b)

(c)

Fig. 10-28. Shear of a *perfect* crystal. (a) Equilibrium positions of two adjacent planes of atoms. (b) A shear strain so large that the shear stress has reduced to zero. (c) A shear strain for the maximum shear stress. A *real* crystal begins to slip (to deform plastically) long before an elastic strain as large as 10° is attained.

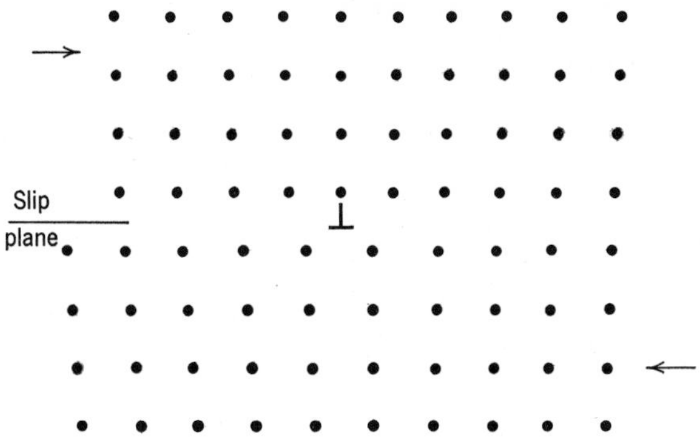

Fig. 10-29. Region of a crystal in which a slip of one atomic plane spacing has occurred for a small distance from the edge of the crystal (at the left). The edge dislocation ⊥ is perpendicular to the plane of the paper.

resolved shear stress than is necessary in an ideal crystal. This results from two features of a dislocation: (1) It is easily moved with only moderate stresses. (2) Its motion produces slip.

The motion of an edge dislocation is illustrated in Fig. 10-29. The symbol ⊥ is used to identify the position of the dislocation. The atoms above and to the left of the dislocation have slipped one lattice spacing to the right. If the dislocation moves one lattice constant farther to the right, the atoms near it move only slight distances. Furthermore, for each interatomic force resisting the motion there is a force favoring it because of the symmetrical displacements from their equilibrium positions of the atoms on either side of the dislocation. To a first approximation these forces cancel, and there is no net force resisting the motion of a dislocation. Although second-order effects do give a resisting force, this force is very much smaller than the force required to cause slip in a perfect crystal.

The deformation of the crystal produced by the motion of a dislocation is shown in Fig. 10-30. The motion of the dislocation through the crystal produces a slip of one lattice spacing along the slip plane. Thus the motion of this imperfection, unlike the other imperfections considered in Sec. 10-2, produces a change in the exterior shape of a crystal. The deformation in Fig. 10-30 is produced by a large number of atomic displacements *in sequence,* quite different from slip in a perfect crystal (Fig. 10-28), in which all the displacements must be produced *simultaneously.* The critical shear stress is much less for the dislocation process because of this fact and because of the cancellation effect described in the preceding paragraph.

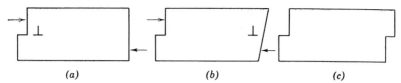

Fig. 10-30. The shear stress produces slip by the motion of a dislocation along the slip plane. (*a*) The dislocation is beginning to cross the crystal. (*b*) The dislocation has almost completely traversed the slip plane. (*c*) The passage of the dislocation has produced slip over the whole plane. (The amount of slip has been greatly exaggerated, and many dislocations must traverse the slip plane in order to produce measurable slip.)

Dislocations are inevitably present in a real crystal. Even a very small crystal may be able to grow from the melt only if it contains a dislocation. Crystals of ordinary size contain many small regions that do not fit perfectly together because of strains or impurities present during the solidification of the crystal from the melt, even though the long range order of atoms is characteristic of a single crystal. The regions of misfit necessarily contain dislocations, as explained in Sec. 10-2 for grain boundaries (a more extreme misfit, requiring a larger density of dislocations). Additional dislocations are produced during slip, by a multiplication process called the *Frank-Read source*. This process cannot be described in terms of the two-dimensional picture of dislocations we have presented. One type of Frank-Read source consists of a dislocation line bent in a right angle (instead of straight like the edge dislocation we have discussed). When the area of the crystal containing this source is stressed it generates dislocations. The passage of each dislocation through the crystal produces a slip of one lattice constant. The multiplication of dislocations during slip permits a macroscopic (many lattice constants) slip along a slip plane.

As slip proceeds, the dislocations produced do not continue to move so readily as they did at first. Relatively little is known about the way dislocations become "tangled up," but it is thought that they do and that their interference increases the stress that must be applied to continue slip after some slip has occurred. Thus the stress in a tension test continues to rise after slip has begun. The crystal is said to be *work hardened*, or hardened by *cold working*.

The resistance to slip, and therefore the strength, of a metal can be increased by impeding the motion of dislocations. There are two common ways of doing this, both of which are used in all engineering materials. The first of them is the use of polycrystalline metals rather than single crystals. We have described single crystals in the foregoing discussion because the fundamental processes are clearer in a single crystal, but a structural engineer would not use single crystals, because of their high cost and their weakness. The small crystals (grains) in a rolled, extruded, or forged metal exert constraints on each other. Slip in one grain cannot occur without deforming the adjacent grains. Since these do not generally have slip planes in the same direction, a polycrystalline specimen cannot be deformed so readily as a single crystal.

The second way of impeding the motion of dislocations is by alloying. Consider, for example, the addition of copper to aluminum, which produces an alloy known as Duralumin, a typical *precipitation hardening* alloy. At high temperatures a few per cent of copper is in solid solution in the aluminum; the copper is not a separate phase but is atomically dispersed. As the temperature is lowered, the solubility of copper in aluminum decreases, and copper precipitates in tiny $CuAl_2$ plates. A dislocation encountering such a plate is *pinned* and cannot continue to move until a much greater stress is applied. Thus slip requires much greater stress than in pure aluminum, and Duralumin is a stronger material than aluminum. The growth of the precipitate requires some time. If an alloy of this type is quenched from a high temperature it is quite soft and weak. It becomes hard and strong upon aging at room temperature, which permits the precipitated particles to grow by diffusion of copper. These alloys are therefore also called *age hardening* alloys.

Very tiny metal crystals ("whiskers") have been grown that are so small and so nearly perfect that they contain no dislocations that can facilitate slip. These crystals are observed to have the strength characteristic of a perfect crystal. Such observations help to confirm the explanation of slip in terms of dislocations.

We have pointed out the success of the dislocation theory in explaining the fact that real materials are not so strong as a perfect crystal should be. Dislocations are called upon to explain many other phenomena and processes in physical metallurgy, such as *creep* and work hardening. In only a few problems, such as low-angle grain boundaries and crystal growth, can the theory be called really successful at the present stage of our understanding. Much work remains to be done, particularly in the quantitative aspects of dislocations.

REFERENCES

General

C. Kittel, *Introduction to Solid State Physics*, Wiley, New York, 2nd Ed., 1956. Chapters 17, 18, and 19.

N. F. Mott and R. W. Gurney, *Electronic Processes in Ionic Crystals*, Clarendon Press, Oxford, 2nd Ed., 1948.

W. Shockley, Editor, *Imperfections in Nearly Perfect Crystals*, Wiley, New York, 1952.

The Science of Engineering Materials, edited by J. E. Goldman, Wiley, New York, 1957.

H. G. van Bueren, *Imperfections in Crystals*, North Holland, Amsterdam, 1960.

Optical Absorption

F. Seitz, "Color Centers in Alkali Halide Crystals," *Revs. Modern Phys.*, **18**, 384–408 (1946); **26**, 7–94 (1954).

Photoconductivity

R. H. Bube, *Photoconductivity of Solids*, Wiley, New York, 1960.

T. H. James and G. C. Higgins, *Fundamentals of Photographic Theory*, Wiley, New York, 1948.

Luminescence

H. W. Leverenz, *An Introduction to Luminescence of Solids*, Wiley, New York, 1950.

G. F. J. Garlick, *Luminescent Materials*, Clarendon Press, Oxford, 1949.

Dislocations

W. T. Read, Jr., *Dislocations in Crystals*, McGraw-Hill, New York, 1953.

A. H. Cottrell, *Dislocations and Plastic Flow in Crystals*, Clarendon Press, Oxford, 1953.

Dislocations and Mechanical Properties of Crystals, edited by J. C. Fisher, W. G. Johnston, R. Thomson, and T. Vreeland, Jr., Wiley, New York, 1957.

W. L. Bragg and J. F. Nye, "Bubble Model of a Metal," Cinegraph 2015, distributed by Kodak, Ltd., Kingsway, London, W. C. 2, England (16-mm. silent motion picture).

PROBLEMS

10-1.* A cubic crystal contains 1 part per million of a chemical impurity. Consider any point in the crystal. Estimate the distance in lattice constants, on the average, from this point to the nearest impurity.

10-2. Boron, aluminum, gallium, indium, or thallium can be incorporated into a germanium host crystal as *substitutional* impurities. Which of these would you expect to expand the lattice in the neighborhood of the impurity atom, and which would you expect to contract the lattice?

10-3. Suppose that the atoms of the preceding problem formed *interstitial* impurities. Would they expand or contract the lattice? How would you use precision lattice constant determinations (by means of X-rays) to learn whether these impurities formed substitutional or interstitial imperfections?

10-4. Where would you expect an interstitial impurity to be located in the diamond structure (Fig. 8-2)? (Specify the position relative to the edges of the small cube in that figure.)

10-5. How can we use a Hall effect experiment to tell whether the electrical conduction in a particular non-metal is by electrons or by holes?

10-6.* A crystal of germanium has 10^{22} free electrons per m.^{3} and very many fewer holes. The electron mobility μ is 0.39 m.^{2}/volt sec. What is the conductivity σ?

10-7. Compare the minimum photon energy to produce an exciton (expressed as a fraction of the energy E_g to produce free electrons and holes) with the minimum photon energy to produce an excited hydrogen atom (expressed as a fraction of the ionization energy).

10-8. What is the position of the intersection of the dislocation line with the plane of the paper in Fig. 10-6? (Measure the x and y coordinates in inches using the lower left-hand corner of the figure as the origin.) If a vacancy is to be formed anywhere inside the region of this figure, what is the most likely position for this vacancy? (Give coordinates as before.)

10-9. Refer to Fig. 10-5 and consider a geometrical figure made up of the following vectors connected head-to-tail: n interatomic distances to the right, m such distances up, n to the left, and m down (n and m are integers, and the distances are not constant in magnitude or direction in a distorted region such as this). Show that such a figure is closed unless it encloses the dislocation line, in which case the vector required to close it (the *Burgers vector*) is perpendicular to the edge dislocation line.

10-10. A *screw dislocation* is defined by a Burgers vector (see preceding problem) *parallel* to the dislocation line. Sketch in three dimensions a layer (not a plane!) of atoms perpendicular to the dislocation line; the origin of the name "screw" should be apparent from your sketch.

10-11.* Sketch a boundary between two crystal grains of the same simple cubic material with the angle between the cubic planes of one grain and those of the other equaling 2.9°. What is the spacing (in lattice constants d_0) between edge dislocations along the grain boundary?

10-12.* Measure the spacing of etch pits in Fig. 10-9. Calculate the angle between the two crystals by dislocation theory, using $d = 3.98$ Å for the size of the blocks of Fig. 10-8. (The value measured by X-ray crystallography is 65.0 ± 2.5 seconds of arc.)

10-13. Explain why the disordered region at an edge dislocation should be etched more rapidly by an acid than a perfect region of the same crystal.

10-14. Grain boundaries have been called the "garbage cans of solids" because impurities frequently segregate at grain boundaries. Explain why they are likely to do this.

10-15. The diffusion process in a low-pressure gas is similar to diffusion in solids in that eq. 10-1 holds with D a constant. But in gases D depends on the mean free path of the gas atoms (instead of on the jump distance), and gaseous diffusion is not a thermally activated process. For diffusion in gases: (a) Tell how D depends on T at constant gas density. (b) Tell how D depends on pressure p at constant T. (c) Show how measurements of D can reveal the sizes of atoms or molecules (Sec. 2-5).

10-16.* Show that tunneling is *not* a likely diffusion process in solids by computing the tunneling probability (see prob. 5-28) for a copper atom through a square potential barrier 1 e.V. high and 2 Å thick.

10-17.* If the attempt frequency is 10^{14} sec^{-1} and the activation energy is 0.75 e.V. for an interstitial diffusion process at $T = 290°K$, what is the average jump time τ? Why is not an interstitial displaced a distance $a\tau$ per second on the average, where a is the jump distance?

10-18. Let a be the distance that an interstitial jumps in diffusion, ν_0 be the attempt frequency, and ΔE the activation energy for a jump. Show that the diffusion constant is then $D = a^2\nu_0 e^{-\Delta E/kT}$. *Hint:* Let the number per unit area of interstitials in one atomic plane normal to the concentration gradient dN/dx be N, and let the number per unit area in a plane a distance a away be $N + a(dN/dx)$.

10-19.* Draw a sketch of the energy as a function of distance for an interstitial positive ion (charge q) moving from one stable position in the lattice to the next (a distance a away). Label the sketch with the activation energy ΔE for a jump. Draw a similar sketch but with an applied electric field $\mathcal{E}$. By what amount is ΔE for a jump less in the direction of the field than in the opposite direction? Calculate the jump probability in each direction, the net jump probability in the direction of the field, and the ionic mobility $\mu = $ (average velocity)/$\mathcal{E}$. *Hint:* Since the change produced by $\mathcal{E}$ in ΔE is so small compared to ΔE, the exponential terms can be approximated after factoring out $e^{-\Delta E/kT}$.

10-20. Show from the results of probs. 10-18 and 10-19 that $D/\mu = kT/q$, the *Einstein relation*. Show that this is a *general* result for all diffusion processes by considering the steady state of a density $N(x)$ of charged particles (each with charge q) in an electric field $\mathcal{E}_x$. There are just as many particles moving toward $+x$ as toward $-x$, and therefore the flow (eq. 10-1) of particles by diffusion must just balance the flow by the electric field. From the Boltzmann statistics, $N = $ (constant) $e^{-qx\mathcal{E}_x/kT}$.

10-21.* What does the observation that a particular crystal is transparent and colorless (in the visible region of the spectrum, of course) demonstrate about the energy gap E_g of this crystal?

10-22.* Would you expect the optical absorption constant of a *metal* to have the order of magnitude of A in Fig. 10-12 or of A in Fig. 10-14? Why? About how thick a coating of a metal on a transparent glass would be required to transmit only 50% of the incident light?

10-23. Estimate the concentration of F-centers in the crystal on which the data of Fig. 10-14 were obtained by assuming that the contribution to A of each atomic absorption act is the same when the atom is an impurity as when it is a host atom (a crude but useful approximation).

10-24. Glass that is carefully manufactured for use in optical instruments has an A in the visible part of the spectrum of less than 0.1 m.$^{-1}$. Use the method of the preceding problem to estimate for this glass the maximum concentration of impurity atoms with energy levels that permit visible light to be absorbed.

10-25. Why will an appreciable width of the *F*-absorption band remain even at $T = 0°K$?

10-26.* What is the color of a crystal of KBr, 0.0005 m. thick, with *F*-centers in the same concentration as that in the crystal of Fig. 10-14, when viewed by transmitted light at room temperature? How would its color change if it were observed at 28°K? (See the dashed curve in Fig. 10-23.)

10-27. Suppose that a Cl^- vacancy is created inside a KCl crystal and that no electron has yet been trapped at this position. The neighboring K^+ and Cl^- ions move to new equilibrium positions. Are these closer to or farther from the Cl^- vacancy than their original positions (before the Cl^- ion was removed)?

10-28.* If 10^{-11} joule of light energy of wavelength 2000 Å is absorbed by a small crystal of BaO, how many electrons and holes are produced? If the electric field across the crystal is large enough to draw all free carriers to the electrodes, how many coulombs flow in the external circuit? (Study Fig. 10-16 to avoid making an error of a factor of 2.)

10-29.* If the light energy of the preceding problem is provided in a short pulse of 1-μ sec duration, how large a voltage pulse can be transferred to a pulse amplifier? Assume that the capacitance between the electrodes is 30 $\mu\mu$f. Sketch the coupling of the crystal electrodes to the amplifier, and determine the resistances and capacitances in order to preserve the pulse shape.

10-30. The most effective carrier traps have a depth E_t of the order of kT. Show that both shallower traps ($E_t \ll kT$) and deeper traps ($E_t \gg kT$) are less effective. *Hint for the deeper traps:* If an electron is to be trapped, it must lose energy by colliding with phonons while in the field of a trap. Consider the phonon energies and the probability of a large number of nearly simultaneous collisions.

10-31. Describe in words and in a diagram the phenomena of trapping and thermal release of holes, but do not use the concept of a hole. Describe the process entirely in terms of valence band electrons.

10-32. An electron and a hole arrive at a recombination center. How is their energy (E_g plus a small kinetic energy) converted into heat? What happens to their momentum?

10-33.* Consider the direct radiative recombination at $T = 290°K$ of a free hole and a free electron, each with a kinetic energy kT, in a solid with a band gap $E_g = 2$ e.V. What is the maximum value of the vector sum of their momenta? How much momentum does the radiated photon have? Compare the answers to these two questions and comment on the probability of direct recombination (without the agency of a recombination center).

10-34. Photoconductors with high sensitivity are desired for detectors of infrared radiation. If such detectors utilize the high-current-gain photoconductive process, why should high sensitivity be accompanied by sluggish response (inability to distinguish two pulses of radiation close together in time)? What should the relation be between sensitivity and resolving power (in time)?

10-35. The *speed* of a photographic emulsion is a measure of the fraction of the tiny AgBr crystals (grains) rendered developable by a given exposure. In order to increase the speed it is desirable to have the electron traps on the surface of the grain. Why?

10-36. Coarse grain (large crystal size) photographic emulsions can be made with greater sensitivity (speed) than fine grain emulsions. Why? [Speed increases with increasing grain size up to a size of about 1 μ; in grains larger than this not all of the electrons can reach the surface before being trapped.]

10-37. The *reciprocity law* for photographic emulsions, valid for moderate light intensities, states that the optical density of the developed image is a function only of the *product* of intensity and time and not an explicit function of either variable by itself. Show from the theory of latent image formation why this law fails at too high intensities. *Hint:* The diffusion of Ag$^+$ ions is a rather slow process; the law fails whenever electrons are released faster than about 10^4 per sec within a single grain.

10-38. Show that the particular xerographic process described in the text produces a *positive* copy (unlike the photographic process, which produces a negative).

10-39.* For a particular phosphor, one-half of the light is emitted in $\frac{1}{30}$ sec (at $T = 290°$K) after excitation. The rate of release of electrons from traps is $Nse^{-E_t/kT}$, where N is the number trapped and s for this phosphor is 10^8 per sec. Compute the trap depth E_t in electron volts.

10-40. How does the decay time ($\frac{1}{30}$ sec at room temperature) of the phosphor of prob. 10-39 depend on temperature?

10-41. A particular phosphor has two different chemical impurities, each with the same concentration, inserted into it to produce traps. One set of traps has the depth of the traps in prob. 10-39 and one has twice this depth. Sketch the logarithm of the light output at room temperature as a function of time after excitation by a short pulse of high energy electrons.

10-42. The phosphor of prob. 10-41 is excited at a very low temperature ($-190°$C) and warmed in the dark at a constant rate of $10°$ per sec to $400°$C. Sketch qualitatively the light output as a function of time; assume that the expression for the release rate given in prob. 10-39 applies to each set of traps.

10-43.* What are the colors of the three phosphors of Fig. 10-23?

10-44. List all the physical and chemical requirements that a phosphor must satisfy if it is to be practically useful in "fluorescent" lamps (Fig. 10-25).

10-45. How can a solid-state maser be *tuned* (i.e., its signal frequency be changed)?

10-46. Why should the signal and pump frequencies of a maser be different? *Hint: Two*-level masers have been constructed, but they are suitable only for intermittent operation.

10-47. In a *four*-level maser, the transition $E_3 \rightarrow E_2$ is the stimulated emission. Show that this maser requires less pumping power than the three-level maser. *Hint:* Compare the extents to which the ground state E_1 must be emptied in the two arrangements.

10-48. * The length l of a ruby laser is 3 cm, the emitted wavelength λ is about 6943 Å, and the index of refraction of Al_2O_3 is 1.76. The standing wave builds up only if the emitted wave is so monochromatic that $n\lambda/2 = l$ with the same value of the integer n for all photons. The higher the reflectivity of the ends, the more precisely this equation must be obeyed. What is a possible value of n? Calculate an upper limit to the wavelength spread $\Delta\lambda$ by assuming that sufficient reflections interfere with each other to require that $n = n_0 \pm 0.01$, where n_0 is an integer. What is $\Delta\lambda/\lambda$? (This is an upper limit since the phase agreement between stimulating and emitted radiation produces *coherent* radiation, like the emission from a vacuum-tube oscillator circuit. For long pulses or continuous waves, $\Delta\lambda$ approaches 0 as the pulse length approaches ∞, as explained in connection with Figs. 4-41 and 4-42.)

10-49. Find in tables for each of two metals or alloys: (*a*) A measure of the shear modulus (shear stress divided by shear strain); although not equal to this, Young's modulus is of the same order of magnitude and is more frequently tabulated. (*b*) A measure of the shear strength; "yield strength at 0.2% offset" or "tensile strength" are adequate approximations for the purpose of this problem. Compare the values in (*a*) with those in (*b*) for each material, and discuss this comparison with reference to the text associated with Fig. 10-28.

10-50. Make a tracing of the central region (containing about 40 atoms) of Fig. 10-29, using dots for the positions of the atoms, as in that figure. On this tracing draw crosses for the new positions of the atoms after the dislocation has moved one lattice spacing to the right.

10-51. Why are Duralumin rivets that have been quenched from a high temperature refrigerated if they must be stored for some time before use?

10-52. During annealing under no external stress after dislocations have been introduced into a crystal by deformation, edge dislocations move perpendicularly to the slip plane by the arrival of vacancies at the dislocation line. Sketch this process of *dislocation climb* on the atomic scale of sizes, and explain how the rate of this process should depend on the annealing temperature.

11

Semiconductors

11-1 INTRODUCTION

The study of semiconductors makes use of the concepts of energy bands, Fermi statistics, mobility of current carriers, holes, and energy levels of impurity atoms, which were introduced in Chapters 8, 9, and 10. Thus the study of semiconductors provides an opportunity to apply and to test much of the theory of solids that was developed in previous chapters.

The study of semiconductors is also important because of the extensive applications of semiconductor devices such as rectifiers and transistors to physical instrumentation and to engineering.

The behavior of intrinsic semiconductors will be explained in Sec. 11-2. These are crystals with no important imperfections other than phonons, electrons, and holes. Semiconductors with intentionally added impurities will be discussed in Sec. 11-3. The *p-n* junction, an internal boundary between regions of a semiconductor with different impurities, will be analyzed in Sec. 11-4; it is the basic element of diode rectifiers and transistors. Transistors will be discussed in Sec. 11-5. Thermoelectric effects in semiconductors and their application to thermoelectric cooling and power conversion will be described in Sec. 11-6.

It is interesting to note that the junction transistor was a product of the ingenious application of solid-state physics to semiconductors. It was predicted by theory and later demonstrated experimentally. It should also be noted that the contributions of the chemist and metallurgist have been indispensable to the development of this and other semiconductor devices.

The physics developed in this chapter is applicable to any semiconductor, but we shall frequently use germanium and silicon as examples. There are many important semiconductors, most of which are semiconducting compounds rather than single elements. Research in the physics, chemistry, and metallurgy of new materials and engineering ingenuity in developing new devices are rapidly expanding the field of semiconductor applications.

This chapter differs in purpose from the others in this book. It presents the physics of one class of solids, semiconductors, in considerable detail. The presentation of this small part of modern physics thus covers the full range from fundamental theory (Chapter 5) to devices and applications (the present chapter). The subject of semiconductors thereby provides an illustration of the power and importance of quantum mechanics in modern science and technology. We could carry out a similar demonstration in many other areas of modern physics, but one such chapter should suffice as an illustration of the connection between quantum physics and technology.

11-2 INTRINSIC SEMICONDUCTORS

We shall now investigate the electrical conductivity produced in a chemically pure semiconductor by thermally excited charge carriers. This is called *intrinsic* conduction since it is a property of the pure crystal. We shall use silicon and germanium as examples to illustrate the typical magnitudes of physical properties such as the conductivity, but the theory applies to any semiconductor.

Both silicon and germanium are covalent crystals with the diamond structure (Fig. 8-?). Because of the tight covalent binding, the energy band structure illustrated in Fig. 9-4 occurs. The four valence electrons from each atom are most probably near the lines joining a germanium atom to its nearest neighbors. A schematic representation in two dimensions of the joining of atoms is presented in Fig. 11-1, but the actual arrangement of atoms is, of course, the three-dimensional arrangement of Fig. 8-2. The minimum energy required to remove an electron from one of these bonds and to permit it to move throughout the crystal is, by definition, the *energy gap* E_g between the top of the *valence* (filled) band and the bottom of the *conduction* (empty) band. E_g is 1.09 e.V. for silicon and 0.72 e.V. for germanium at room temperature. These are

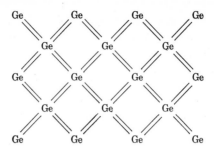

Fig. 11-1. Schematic diagram of a germanium crystal. Each pair of lines represents a two-electron bond.

smaller energy gaps than are typical for non-metals and are small enough to permit appreciable conduction by thermally excited carriers at room temperature.

Our first task is to find the Fermi level E_0. Figure 11-2a presents the $S(E)$ functions near the edges of the valence and conduction bands of germanium. The two $S(E)$ functions have the same shape * because the electron momentum decreases to zero at the top of a band in the same way that it rises from zero at the bottom of a band (see Fig. 8-13). The density of states $S(E)$ is proportional to the momentum, as indicated by eq. 9-6. We set E equal to zero at the top of the valence band.

The Fermi function $f(E)$ is plotted in Fig. 11-2b with the same vertical scale as in Fig. 11-2a. As in Fig. 9-8, drawn for a metal, the purpose of this plot is to locate E_0 by an argument about $N(E)$. The situation is now different from the metal, however, and we cannot use an argument for the special case $T = 0$ in order to find E_0. At $T = 0$ the valence band is full ($f = 1$) and the conduction band is empty ($f = 0$). There are no energy levels in the forbidden energy band, and hence $N(E)$ is zero there regardless of the value of $f(E)$. Thus from an argument for $T = 0$, all we can learn is that E_0 lies somewhere in the energy gap.

At any temperature above zero it is possible to determine E_0. The number of electrons in the conduction band must be the same as the number of holes (empty quantum states) in the valence band, since one empty quantum state in the valence band is produced for each electron that is thermally excited to the conduction band. The concentration of electrons in the conduction band, $N(E) = S(E)f(E)$, and the concen-

---

* This statement is not precisely true because the effective mass m^* of an electron is not exactly the same in the two regions, nor is m^* the same for electrons moving in different directions relative to the crystal axes. We shall, however, set m^* equal to m in this chapter. This approximation introduces an error of less than 0.03 e.V. in the calculation of E_0 in germanium.

tration of holes in the valence band, $N'(E) = S(E)\,[1 - f(E)]$, are plotted in Fig. 11-2c. (The horizontal scale has been greatly expanded from Fig. 11-2a.) The areas under these two curves are the total electron concentration and the total hole concentration, and these two areas must therefore be equal. Because the two $S(E)$ curves have the same shape, this requirement means that

$$f(E_g) = 1 - f(0) \tag{11-1}$$

This equation has the solution $E_0 = E_g/2$, which can be inferred from prob. 9-8 or can be demonstrated by inserting into eq. 11-1 the explicit expressions for f:

$$\frac{1}{e^{(E_g - E_0)/kT} + 1} = 1 - \frac{1}{e^{-E_0/kT} + 1} = \frac{e^{-E_0/kT}}{e^{-E_0/kT} + 1}$$

$$e^{-E_0/kT} + 1 = e^{-E_0/kT}\{e^{(E_g - E_0)/kT} + 1\}$$

$$1 = e^{(E_g - 2E_0)/kT}$$

$$E_0 = E_g/2 \tag{11-2}$$

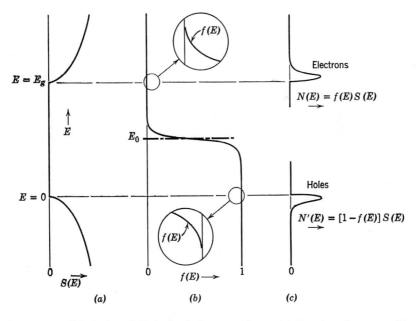

Fig. 11-2. Calculation of E_0 in intrinsic germanium. (a) Density of states. (b) Fermi function; the circles show $f(E)$ on a greatly enlarged scale. (c) Densities of holes and electrons as functions of E, with the same E scale as in (a) and (b); the horizontal scale of (c) is greatly enlarged compared to (a).

In an intrinsic semiconductor the Fermi level is therefore halfway from the top of the valence band to the bottom of the conduction band. The density (number per unit volume) of electrons (or holes) is therefore proportional to $e^{-E_g/2kT}$.

The dependence of the densities of holes and electrons upon T can also be obtained by a simple argument in terms of the rates of generation and of recombination of free electrons and holes. Let N_n be the density of electrons in the conduction band and N_p the density of holes in the valence band. We assume that both these densities are small compared to the densities of quantum states in the bands. The number of electrons per unit volume per second that are excited to the conduction band is proportional to the product of the density of electrons in the valence band and the Boltzmann factor, $e^{-E_g/kT}$. The number of electrons per unit volume per second that recombine with holes is proportional to the product of the density N_n of electrons in the conduction band and the density N_p of holes in the valence band. These *generation* and *recombination* rates must be equal when the crystal is in equilibrium. Since one hole is produced for each electron in the conduction band, N_n equals N_p, and

$$N_n N_p = N_n{}^2 = \text{(constant)} \times e^{-E_g/kT}$$

Therefore N_n is proportional to $e^{-E_g/2kT}$, which is the same result cited in the preceding paragraph.

The generation-recombination argument is less abstract but is also incomplete: In order to complete the argument we should have to show that the same result was obtained regardless of the process of recombination. In the foregoing argument it was tacitly assumed that direct recombination occurred between free electrons and holes. The recombination process in all real crystals actually occurs through the agency of recombination centers, as explained in Sec. 10-5. We shall not complete the present argument, since the proof of eq. 11-2 was general and did not depend on the nature of the recombination process.

It is convenient to define an *effective density of states in the conduction band*, which we shall call N_c. This quantity is defined by writing

$$N_c f(E_g) = \int_{E_g}^{\infty} N(E)\, dE = \int_{E_g}^{\infty} S(E) f(E)\, dE$$

In other words, the total number of electrons per unit volume in the conduction band is written as the product of N_c and the Fermi factor evaluated at the bottom of the band. Evaluation of N_c is facilitated by using the approximate $f(E)$ from eq. 9-2,* which is valid here since $(E_g - E_0) \gg kT$:

$$N_c e^{-(E_g-E_0)/kT} = \int_{E_g}^{\infty} \left(\frac{2^{7/2}m^{3/2}\pi}{h^3}\right)(E - E_g)^{1/2}e^{-(E-E_0)/kT} \, dE$$

N_c can be found by dividing both sides by the exponential term on the left (which does not contain the variable E) and by making the substitution $u = (E - E_g)/kT$:

$$N_c = \frac{2^{7/2}m^{3/2}\pi}{h^3}\int_{E_g}^{\infty}(E - E_g)^{1/2}e^{-(E-E_g)/kT} \, dE$$

$$= \frac{2^{7/2}\pi}{h^3}(mkT)^{3/2}\int_{0}^{\infty}u^{1/2}e^{-u} \, du = \frac{2^{7/2}\pi}{h^3}(mkT)^{3/2}\left(\frac{\pi^{1/2}}{2}\right)$$

$$= 2\left(\frac{2\pi mkT}{h^2}\right)^{3/2} = 4.83 \times 10^{21}T^{3/2} \tag{11-3}$$

The total number of electrons per cubic meter in the conduction band is the product of eq. 11-3 and $f(E_g)$:

$$N_n = 4.83 \times 10^{21}T^{3/2}f(E_g) \tag{11-4}$$

The total number of holes per cubic meter in the valence band is the product of eq. 11-3 and $[1 - f(0)]$; this statement can be proved by a development precisely equivalent to that of the preceding paragraph. The densities of free carriers can therefore be conveniently calculated. The introduction of the concept of N_c has permitted the foregoing integration to be made only once instead of each time a different semiconductor or temperature is encountered.

A schematic illustration of the population of the allowed energy bands in an intrinsic semiconductor is given in Fig. 11-3. The Fermi factor is plotted as a function of the reciprocal of the wavelength of an electron (compare Figs. 8-12 and 8-13). The Fermi factor equals unity for all the lower bands, and therefore in Fig. 11-3b only the region of the $1/\lambda$ scale near the top of the valence band and the bottom of the conduction band is plotted. Figure 11-3c differs only in that T is large enough for some electrons to be excited to the conduction band. The effect on $f(E)$ is greatly exaggerated in the figure; at any ordinary temperature the difference between Figs. 11-3b and 11-3c would not be apparent in such plots.

---

* $S(E)$ has been inserted from eq. 9-6; since $E = E_g$ at the bottom of the band, instead of $E = 0$ as in eq. 9-6, the term $(E - E_g)^{1/2}$ appears instead of $E^{1/2}$.

The effect of an electric field in the $-x$-direction is illustrated in Fig. 11-4a. The electrons receive positive momentum increments, and so there are now more electrons moving toward $+x$ in each band. Figure 11-4b is the same as Fig. 11-4a but is interpreted in terms of holes in the valence band. What is actually happening in the valence band is that there is a net current of electrons to $+x$, but it is much more convenient and fruitful to view this as a net current of holes to $-x$. Both the electrons in the conduction band and the holes in the valence band carry current, and both currents move positive charge toward $-x$.

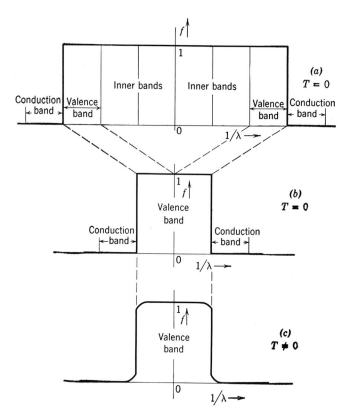

Fig. 11-3. Fermi factor plotted as a function of $1/\lambda$. (*a*) At $T = 0$, all the inner bands are full and the conduction band is empty. (*b*) Same as (*a*) except that the inner bands have been removed from the center of the plot since they play no part in conduction. (*c*) Same as (*b*) except drawn for an appreciable T; the deviations from $f = 1$ or $f = 0$ which are drawn are much larger than those appropriate to room temperature.

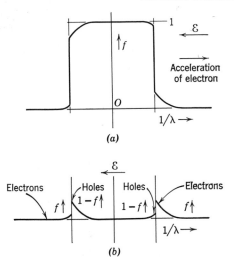

Fig. 11-4. (*a*) Plot like Fig. 11-3*c* except that an electric field is applied. (*b*) Conditions identical with (*a*), but interpreted in terms of holes; there are more holes moving to the left than to the right.

The total conductivity can therefore be expressed as

$$\sigma = N_n e\mu_n + N_p e\mu_p \tag{11-5}$$

where N_n and N_p are the number of negative (electrons) and positive (holes) carriers per cubic meter, and μ_n and μ_p are their respective mobilities. Equation 11-5 is the generalization of eq. 9-17 to the case of simultaneous conductivity by electrons and holes. The conductivity of an intrinsic semiconductor can be expressed by combining eqs. 9-2, 11-2, 11-4, and 11-5:

$$N_n = N_p = 4.83 \times 10^{21} T^{3/2} e^{-E_g/2kT}$$

$$\sigma = 4.83 \times 10^{21} T^{3/2} e(\mu_n + \mu_p) e^{-E_g/2kT} \tag{11-6}$$

Both the $T^{3/2}$ term and the mobilities * vary much more slowly with T than the exponential factor, and hence the logarithm of the conductivity is very nearly a linear function of $1/T$. The energy gap E_g can therefore be measured approximately from the slope of a log σ vs. $1/T$ plot.

---

* At low temperatures the mobilities are limited by scattering of electrons by impurities. At high temperatures the mobilities are limited by lattice scattering. See Sec. 9-6, Fig. 11-7, and prob. 11-4.

Intrinsic semiconductors find applications chiefly in the measurement of temperature and in the compensation of electronic circuits for changes in temperature; both applications exploit the rapid dependence of σ on T.

11-3   *n*- AND *p*-TYPE SEMICONDUCTORS

The most interesting semiconductors are those whose electrical properties have been changed by intentionally incorporating chemical additives into the crystals. Such impurity atoms are called *donors* if they introduce occupied energy levels from which electrons can easily be excited to the conduction band. They are called *acceptors* if they introduce vacant energy levels to which electrons can easily be excited from the valence band, thereby producing holes. The typical donors in germanium and silicon are elements from group V of the periodic table: P, As, and Sb. The typical acceptors are from group III: B, Al, Ga, and In. These additives have been shown by X-ray and other experiments to be substitutional impurities (Sec. 10-2*b*) in germanium and silicon. We shall first show why these impurities act as donors or acceptors and then investigate the properties of germanium with donors (*n-type*) and with acceptors (*p-type*).

A schematic diagram of an arsenic atom inserted substitutionally into the germanium lattice is given in Fig. 11-5. The arsenic atom has five valence electrons, and only four of them are used in the covalent bonds.

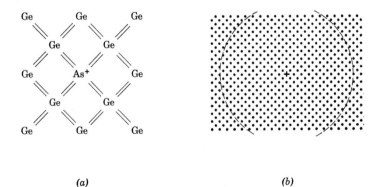

(a)   (b)

Fig. 11-5. An arsenic atom substituted for a germanium atom in a germanium crystal. (*a*) Only four of the five arsenic valence electrons are localized in the bonds. (*b*) The extra electron is usually many lattice constants from the arsenic ion (+); the electron is inside the circle only one-fourth of the time, even at $T = 0$ (the circle has been drawn on the assumption that the effective mass of an electron in germanium is $\frac{1}{5}m$).

The region of the crystal near the arsenic atom thus has a net positive charge $+e$, and there is one extra electron bound to this region. At a sufficiently high temperature (20°K is high enough for germanium) this electron may be thermally released into the conduction band and may move throughout the crystal, which is a process like the ionization of an atom in a gas.

The binding energy of the extra electron to the positive charge can be estimated by a simple theory. We assume tentatively that the average distance of the electron from the arsenic ion is at least several lattice spacings. If this is true, then to a good approximation the germanium crystal can be replaced by a continuous medium of the same dielectric constant as the crystal, namely 16. The calculation of the binding energy of the extra electron is now like the calculation of the binding energy of an electron in the ground state of hydrogen, except that κ_e equals 16 instead of unity. The potential energy is

$$-\frac{e^2}{4\pi\kappa_e\epsilon_0 r} \quad \text{instead of} \quad -\frac{e^2}{4\pi\epsilon_0 r}$$

The energy levels are then

$$-\frac{e^4 m}{n^2 h^2 8\kappa_e^2 \epsilon_0^2} \quad \text{instead of} \quad -\frac{e^4 m}{n^2 h^2 8\epsilon_0^2}$$

The latter expression (eq. 6-2) gave a binding energy of 13.6 e.V. (E for $n = 1$ equals $- 13.6$ e.V.). Hence the binding energy of an electron to a donor in germanium is estimated to be $13.6 \div (16)^2 = 0.05$ e.V. This is the energy difference $E_g - E_d$, where E_d is the energy of the donor levels.

Our tentative assumption that the average distance of the electron from the arsenic ion was large can now be verified. The average distance is about $\frac{3}{2}\rho$, according to Fig. 6-2b, and ρ for our problem is 16 times ρ for hydrogen, which was 0.53 Å (see eq. 6-5, with $\kappa\epsilon_0$ in place of ϵ_0). Thus the replacement of the periodic array of atoms by a continuum was a good approximation.

If the theory is refined by replacing m by the effective mass of an electron in germanium, the binding energy predicted is about 0.01 e.V. The experimental values are 0.0127, 0.0120, and 0.0097 e.V. for arsenic, phosphorus, and antimony, respectively. Both the quantitative agreement and the fact that these three donors give almost identical binding energies testify to the validity of the simple theory. We shall set $E_d = E_g - 0.01$ e.V. in our calculations below.

A precisely similar argument could be carried through for a substitutional impurity with a valence of 3. Such an acceptor impurity has one too few electrons to complete the covalent bonds. The vacant state, or hole, is not adjacent to the acceptor, for the same reason the extra electron was not close to the donor. The problem of calculating the binding energy of the hole (charge $+e$) to the acceptor ($-e$) is in fact the same as the problem of calculating the binding energy of the extra electron to a donor. Experimental values of this energy are 0.0104, 0.0102, 0.0108, and 0.0112 e.V. for boron, aluminum, gallium, and indium, respectively, in germanium. Accordingly, we shall set $E_a = 0.01$ e.V. in problems involving acceptors. (A free hole with zero kinetic energy is at the top of the valence band, where $E = 0$, and hence E_a is simply the binding energy of a hole to an acceptor.)

The method of calculating the position of the Fermi level E_0 for germanium containing donors or acceptors will be illustrated by calculating E_0 at room temperature for a crystal with a particular concentration $(5 \times 10^{22}$ m.$^{-3})$ of donor atoms. The process, illustrated in Fig. 11-6, is very similar to the analysis in Sec. 11-2. In the present example

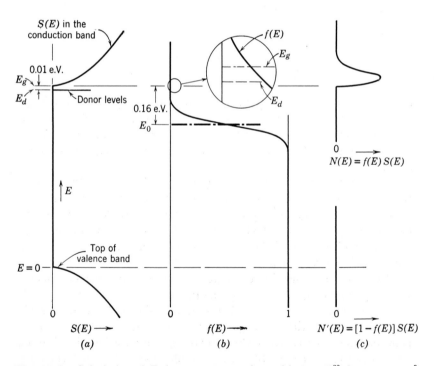

Fig. 11-6. Calculation of E_0 in n-type germanium with 5×10^{22} donors per m.^{3} The circle in (b) is magnified to show the relative magnitudes of $f(E_g)$ and $f(E_d)$. There are practically no holes in the valence band and practically no electrons bound to donors. See Fig. 11-2 for the meaning of (a), (b), and (c).

there is an *additional* source of electrons at the donor levels, and the supply of electrons from the valence band is negligible (this statement will be verified after the calculation is completed). From eq. 11-4 the density of electrons in the conduction band is

$$4.83 \times 10^{21} T^{3/2} e^{-(E_g - E_0)/kT}$$

since eq. 9-2 is a good approximation to $f(E)$ here. From the definition of the Fermi factor $f(E)$ the number of vacant (ionized) donor states is

$$N_d[1 - f(E_d)] = N_d[1 - e^{-(E_g - 0.01 - E_0)/kT}]$$

where N_d is the number of donor atoms per m.^{3}, and $E_d = E_g - 0.01$ e.V. is the energy of the donor levels. The density of free electrons must equal the density of ionized donors. If we equate these two densities and substitute $T = 300°$ and $N_d = 5 \times 10^{22}$, we obtain

$$2.51 \times 10^{25} e^{-(E_g - E_0)/kT} = 5 \times 10^{22}[1 - e^{-(E_g - 0.01 - E_0)/kT}]$$

Since the coefficient on the left is much larger than the coefficient on the right, the exponential on the left must be very much less than unity. The exponential on the right is larger than the exponential on the left by a factor of only $e^{0.01 \times 40} = e^{0.4} = 1.5$, and hence it too must be very much less than unity. Neglect of the exponential on the right with respect to unity is therefore permissible and gives an easily solvable equation:

$$e^{-(E_g - E_0)/kT} = 2 \times 10^{-3}$$

At $T = 300°$K,

$$E_g - E_0 = 0.16 \text{ e.V.}$$

which is the position of E_0 sketched on Fig. 11-6.

The density of holes in the filled band can now be calculated for $T = 300°$K; it is

$$4.83 \times 10^{21} T^{3/2} e^{-(0.72 - 0.16)/kT} \cong 10^{16} \text{ m.}^{-3}$$

This density is much smaller than the density of electrons removed from the donors, and therefore our assumption that *all* electrons were provided by donors is an excellent approximation.

TABLE 11-1

Properties of Silicon and Germanium at 300°K

	Silicon	Germanium
Energy gap	1.09 e.V.	0.72 e.V.
μ_n, electrons	0.12 m.^{2}/volt sec	0.39 m.^{2}/volt sec
μ_p, holes	0.05 m.^{2}/volt sec	0.19 m.^{2}/volt sec
Intrinsic σ	1×10^{-3} mho/m.	2 mhos/m.

The Fermi level for this particular specimen of germanium containing donors is therefore above the middle of the forbidden band. There are then many more electrons in the conduction band than holes in the valence band, conduction is almost exclusively by electrons, and this is called n-type germanium because the negative carrier predominates. Electrons are said to be the *majority* carriers in n-type material, and holes are the *minority* carriers. The Fermi level would be closer to E_g if N_d were larger or if T were lower. As T becomes very large or N_d very small, E_0 approaches $E_g/2$ and the crystal behaves like intrinsic germanium. (These statements are verified in prob. 11-17.)

If, as is usual, E_0 lies more than a few kT below E_g, essentially *all the donors are ionized.* The number N_n of electrons in the conduction band is then the same as the number of donors, and

$$\sigma = N_n e \mu_n = N_d e \mu_n \tag{11-7}$$

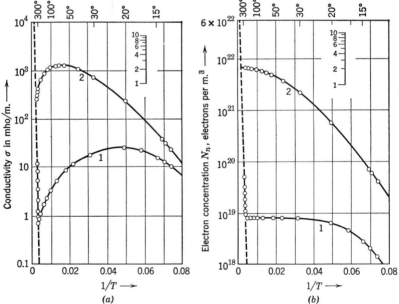

Fig. 11-7. (*a*) log σ vs. $1/T$ for two samples of germanium with different impurity concentrations. (*b*) log N_n vs. $1/T$ from Hall effect experiments for the same samples as in (*a*). The donor and acceptor concentrations are 10^{19} and 10^{18}, respectively, in sample 1, and 7.5×10^{21} and 10^{20} in sample 2. [From P. P. Debye and E. M. Conwell, *Phys. Rev.*, **93**, 693 (1954).]

It should be emphasized that eq. 11-5 is always valid, that the first equality of eq. 11-7 holds whenever $N_n \gg N_p$, and that the second equality holds only if $N_d \gg N_a$ and if all the donors are ionized. Since E_0 lies well below E_g for most practical semiconductors, $f(E_g)$ is usually much less than unity. Therefore relatively few quantum states in the conduction band are occupied, and the conduction process is the process described in the development of eq. 9-19. (An important exception to this statement is the tunnel diode, which will be described at the end of Sec. 11-4.)

At very low temperatures, E_0 is so close to E_g that not all the donors are ionized. The fraction ionized is sensitive to the binding energy of an electron to a donor. The densities N_n at low temperatures have been measured by performing the Hall-effect experiment and using eq. 9-24. In this way the binding energies of electrons to various donors have been measured. Experimental data for the variation of N_n and σ with T are given in Fig. 11-7. The variation of μ_n with T is responsible for the fact that the two curves do not have the same shape. The dashed-line region at the left (high T) is the intrinsic region, where E_0 has fallen to $E_g/2$ and remains at this value.

If acceptors, instead of donors, are added to germanium the conductivity also increases above the intrinsic conductivity. The Fermi level lies below the middle of the forbidden gap; at room temperature all the acceptors are ionized; holes are the majority carriers; the crystal is called p-type (positive carriers); and E_0 and σ can be calculated in the same way as for an n-type crystal. The Hall constant and Hall angle are *positive* for a p-type semiconductor (they are negative for most metals and for an n-type semiconductor). The observation of the Hall effect thus provides an experimental way of distinguishing n-type from p-type crystals.

Representative values of σ for crystals with various amounts of intentional impurities are given in Table 11-2. The concentrations of donors or acceptors are ordinarily only a few parts per million. A great deal of attention to the purification of germanium by repeated crystallizations is necessary in order to reduce the unintentional impurities below such concentrations (purification is also required to remove recombination centers, as will be apparent in Sec. 11-5).

There are always some acceptors even in a crystal with intentionally added donors. As long as there are more donors than acceptors, electrons from the donors fill the vacant states introduced by the acceptors, and the remaining electrons appear in the conduction band. This statement can be verified by carrying out an analysis of the position of E_0 and showing that the number of holes is negligible (probs. 11-13 and 11-14).

TABLE 11-2

Conductivity of *n*- and *p*-Type Germanium

| | | Conductivity at 300°K | |
Chemical Additive	Concentration, atoms/m.^{3}	Type	Magnitude, mho/m.
No intentional impurity		Intrinsic	2
Arsenic	8×10^{19}	n	5
Arsenic	1.5×10^{21}	n	90
Arsenic	5×10^{22}	n	2000
Gallium	9×10^{19}	p	3
Gallium	8×10^{20}	p	30
Gallium	1×10^{22}	p	300

From P. P. Debye and E. M. Conwell. See *Phys. Rev.*, **93**, 693 (1954).

We therefore say (crudely) that donors and acceptors "compensate" each other, and whichever species is in excess determines the conductivity type.

If the Fermi level is at least several kT from both bands, a simple relation connects the equilibrium values of N_n and N_p in the same specimen of any semiconductor:

$$N_n N_p = N_i{}^2 \qquad (11\text{-}8)$$

where N_i is the concentration of either carrier that would have occurred in an intrinsic specimen of this semiconductor at the same temperature. (Proof of this expression is left for a problem, prob. 11-27.) Doping to enhance the concentration of one carrier thus suppresses the concentration of the other. This relation is similar in form and in physical origin to the equations of chemical equilibria, such as the constancy (at a given T) of the product of the concentrations of H^+ and OH^- in water solutions.

The analysis of this section can be applied to any non-metal. Most non-metals have energy gaps and donor and acceptor binding energies much greater than those in germanium. If the binding energy of an electron to a donor is of the order of 1 or 2 e.V., the Fermi level can be shown to lie *between* the donor level and the bottom of the conduction band (for typical temperatures and donor concentrations). Of course in such a semiconductor it is no longer true that all the donors are ionized.

If a solid is to be a good insulator it must have a large enough E_g and be pure enough for the Fermi level to be more than a volt from the nearest band edge.

11-4 *p-n* JUNCTIONS

A *p-n* junction is an internal boundary between *p*-type and *n*-type regions of a single crystal. It is *not* a boundary formed by pressing together a *p*-type and an *n*-type semiconductor. Such a boundary would have too large a concentration of imperfections to have interesting electrical properties. In order to emphasize this point we shall describe briefly some methods of making *p-n* junctions.

One method is to change the *doping* (intentionally added impurity content) during the growing of a single crystal from the melt; the melt first contains a concentration of (for example) a donor impurity sufficient to make the crystal *n*-type, and then a concentration of an acceptor impurity sufficient to make the crystal *p*-type is inserted into the melt. Another method is to place a tiny pellet of an acceptor element (usually indium) on an *n*-type crystal and then to heat the crystal until the acceptor *alloys* with a small surface region of the germanium. As this crystal is slowly cooled, the heavily doped *p*-type germanium crystallizes as an extension of the original *n*-type crystal. A third method is similar to the second, but diffusion instead of alloying is employed; a much higher temperature is generally used, no molten phase need be present, and the junction occurs at the depth below the surface where the diffusing acceptor concentration just equals the original donor concentration. Another method (*epitaxial growth*) consists of evaporating germanium and an acceptor element slowly onto a hot *n*-type germanium crystal. The crystal temperature is high enough for the deposited atoms to move into the perfect lattice positions by surface mobility, thus giving a *p*-doped overgrowth with the same orientation as the original crystal. Of course donors and acceptors can be interchanged in these examples, and more complicated combinations, like the *n-p-n* transistor with two *p-n* junctions, can be produced by an extension of these techniques.

An electron energy diagram of a *p-n* junction in equilibrium is presented in Fig. 11-8. The Fermi level E_0 is constant throughout the circuit of *n*-type and *p*-type germanium, solde ed connections at the ends, and external metal wire. E_0 must be constant by the argument presented at the end of Sec. 9-3. It is instructive to examine closely the way the electron and hole currents across the junction balance in equilibrium. Although this examination will be only a detailed special case of the general argument of Sec. 9-3, it prepares the way for the study of the *p-n* junction when an external voltage is applied.

We shall consider first the electron currents and examine these at the convenient point $x = x_0$, the edge on the p-side of the *transition region* (the region in which the electrostatic potential and energy bands are changing). The current that can flow to the *left* at $x = x_0$ is proportional to the number of electrons in the conduction band of the p-type germanium. This number is small and is proportional to $e^{-E_1/kT}$, where E_1 is defined as $E_g - E_0$ on the p-side of the junction. These electrons experience an accelerating field to the left if they are in the region $x < x_0$, and they flow across the junction. The vector I_{n0} indicates schematically the magnitude of this current of electrons that can move "down the hill." The current I_{n0} is sometimes called the *thermally generated current* since it is a current of electrons that have been thermally excited into the conduction band on the p-side of the junction.

The electron current that can flow to the *right* at $x = x_0$ consists of electrons that have "climbed the hill" from the n-side. The number of electrons in the conduction band on the n-side is far greater than on the

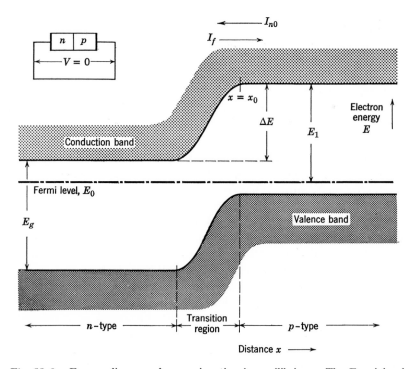

Fig. 11-8. Energy diagram of a p-n junction in equilibrium. The Fermi levels in the two parts are aligned, and the currents of electrons and of holes are the same in each direction.

p-side, but the energy barrier prevents most of them from diffusing into the p-region. The number in the conduction band on the n-side of the junction is proportional to $e^{-(E_g - E_0)/kT}$ (evaluated on the n-side), and the fraction of these that can climb the barrier is $e^{-\Delta E/kT}$. But, according to Fig. 11-8, E_1 is the sum of ΔE and the quantity $(E_g - E_0)$, evaluated on the n-side. The number participating in current to the right is hence proportional to

$$e^{-\Delta E/kT} e^{-(E_g - E_0)/kT} = e^{-E_1/kT}$$

which is the same as the number contributing to I_{n0}. The current vector I_f is therefore drawn equal in length to I_{n0}. The net current of electrons is zero, as it must be if the system is in equilibrium. Precisely similar arguments could be made for holes.

If there were no step ΔE in the potential energy of an electron at the junction, the currents in each direction would not be the same. A net current of electrons to the right (and holes to the left) would charge the n-side positively relative to the p-side until the situation shown in Fig. 11-8 prevailed. The transition region thus has positive charges on its n-side and negative charges on its p-side. Since the germanium atoms are neutral and since the crystal has negligible other impurities, only the ionized donor and acceptor atoms can provide this charge. The width of the junction is determined by the concentrations of these fixed charges as functions of x. If the concentrations are large, a sufficiently strong double layer of charge to produce a given ΔE can be produced with a very narrow region. If the concentrations are smaller the junction region is wider. A typical width is about 1 micron (10^{-6} m.). Outside the transition region electrical neutrality prevails since there are as many free electrons as ionized donors (and free holes as ionized acceptors). Inside the transition region there are very few free carriers, and there is a net charge density almost equal to the product of e and the difference between the concentrations of donors and acceptors. The electrostatic potential as a function of x can be determined by inserting this charge into Poisson's equation, and the width of the transition can be determined from the solution of this equation.

The transition region becomes wider if an externally applied potential increases ΔE. A larger step in the potential requires a larger double layer of charge. Since the charge densities are fixed, the transition region must become wider in order to furnish the required charge. (See prob. 11-34.)

The most important property of a p-n junction is the current I as a function of the applied voltage V. This $I(V)$ characteristic is strongly non-linear and asymmetrical, and therefore a p-n junction is a good recti-

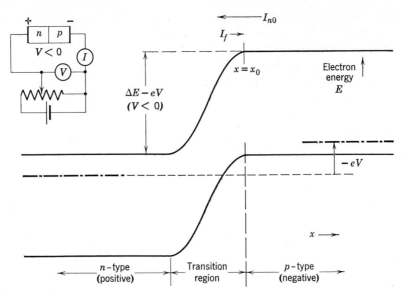

Fig. 11-9. Energy diagram of a *p-n* junction with *reverse* bias. V is less than zero, and the current is small and nearly independent of V. Only the electron currents are shown, but the hole current is also small and nearly independent of V.

fier. Before the theory of $I(V)$ is described it is necessary to specify some assumptions and conventions:

1. We neglect recombination of carriers in the transition region. This is an important physical assumption, which will be examined later in this section. (See Sec. 10-5 for the meaning of "recombination.")

2. We assume that all the applied voltage appears across the transition region. Of course ordinary "IR drops" in the n- and p-sides of the unit may be appreciable, but the correction for these can be made in an obvious way once I is known as a function of the V appearing across the transition region. This assumption also implies that no appreciable voltage drops occur where the n- and p-sides are soldered to the external wires. Negligible voltage drops appear at these connections in practical junction devices.

3. We define $V > 0$ when the p-side is positive. This is called *forward bias*, since it will develop that this is the bias direction for easy current flow. When the n-side is positive, V is less than zero, and this is called *reverse bias*. Forward bias for electrons is also forward bias for holes. There are thus really two rectifiers in parallel. We shall analyze

only the electron flow; the analysis for the hole flow is identical and merely adds a term to the $I(V)$ characteristic that is identical in form to the term from electron flow.

4. We treat the flow of carriers across the junction as a one-dimensional problem. The treatment is therefore only an approximation but a sufficiently good one to demonstrate the physical processes.

Figure 11-9 presents an energy diagram like Fig. 11-8 but with a reverse bias voltage applied ($V < 0$, n-side positive). Since the ordinate is the potential energy of an electron, making the n-side positive lowers the energy levels on that side. Far from the junction on either side the Fermi levels are defined just as before, because the flow of current does not change the thermal equilibrium distribution of electrons appreciably. Near the junction the Fermi level is no longer defined, because here the application of a voltage has profoundly changed the distribution of electrons. The number of electrons participating in the current I_{n0} is the same as in equilibrium since there is no barrier for these electrons; they now "fall down" a somewhat higher "hill," but the rate of arrival at the top of the hill is the same, and this rate determines the current. On the other hand, the current I_f is greatly reduced, since the barrier height has been increased by $-eV$ (with $V < 0$). The increase in barrier

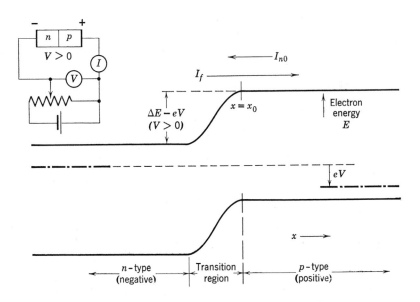

Fig. 11-10. Energy diagram of a p-n junction with *forward* bias. V is greater than zero, and the current is large and varies rapidly with V. Only the electron currents are shown, but the hole current is also large and sensitive to V.

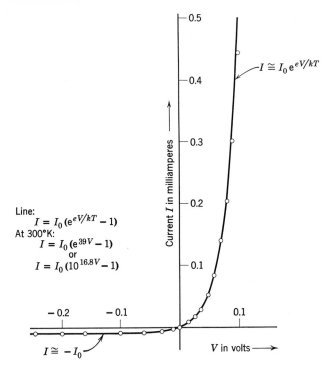

Line:
$$I = I_0(e^{eV/kT} - 1)$$
At 300°K:
$$I = I_0(e^{39V} - 1)$$
or
$$I = I_0(10^{16.8V} - 1)$$

$-I \cong I_0 e^{eV/kT}$

$I \cong -I_0$

V in volts $\longrightarrow$

Current *I* in milliamperes $\longrightarrow$

Fig. 11-11. Current as a function of voltage for a *p-n* junction. $I_0 = 10^{-5}$ amp for this particular junction.

height multiplies the current I_f by a factor $e^{-(-eV)/kT}$, which is much less than 1. In thermal equilibrium I_f was equal to I_{n0}; therefore I_f now equals $I_{n0}e^{eV/kT}$ (with $V < 0$, of course). Thus the net current to the right is

$$I = I_{n0}e^{eV/kT} - I_{n0} = I_{n0}(e^{eV/kT} - 1) \qquad (11\text{-}9)$$

If eV is more negative than $\sim -4kT$, the current is practically constant at the saturation value $-I_{n0}$.

It is unnecessary to analyze the forward bias case separately, since eq. 11-9 applies to it as well, but it is worth illustrating in Fig. 11-10. It should be noted that now I_f is increased from its thermal equilibrium value by the factor $e^{eV/kT}$, which factor is now greater than 1. There is now a larger concentration of electrons on the *n*-side than on the *p*-side at the energy level of the bottom of the conduction band on the *p*-side. This concentration gradient is the driving force that produces a net current of electrons to the right in Fig. 11-10. Thus eq. 11-9 applies to forward bias as well as reverse. Note that we have made use of assumption 1 here, since we have assumed that the electrons in excess of the thermal equilibrium density do not recombine in the transition region but travel on into the p-side.

An expression like eq. 11-9, but with a hole saturation current I_{p0} in place of the electron saturation current I_{n0}, applies to the hole flow. If we write I_0 as the sum of the saturation currents for electrons and holes we have the $I(V)$ characteristic of a p-n junction:

$$I = I_0(e^{eV/kT} - 1) \tag{11-10}$$

This equation is plotted with a linear scale in Fig. 11-11 (see also prob. 11-36). The approximations valid whenever $|eV| > 4kT$ are indicated on the plot. Equation 11-10 has been experimentally confirmed for practical p-n junction rectifiers over a wide range of values of V and I_0, but imperfections in the junction frequently reduce the coefficient of V in the exponential below its value (e/kT) for an ideal junction.

The theory of the dependence of I_{n0} (or I_{p0}) upon the properties of the germanium will now be sketched. The foregoing arguments show that the density of electrons in the conduction band on the p-side is

$$N_n^{(p)} = N_n^{(eq)}e^{eV/kT} \tag{11-11}$$

at the edge of the transition region. $N_n^{(eq)}$ is the equilibrium density of electrons at this point, the density when $V = 0$. The concentration $N_n^{(p)}$ is controlled by the voltage applied to the junction as outlined above, but it is *diffusion* under the driving force of the electron concen-

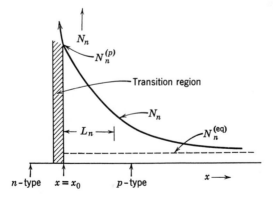

Fig. 11-12. Concentration of electrons on the p side of a p-n junction with forward bias. The concentration of injected minority carriers decreases by $1/e$ in a diffusion length, L_n.

tration gradient (rather than carrier motion induced by the electric field) that is the important process of current flow.

As the distance from the junction increases, N_n decays to the equilibrium value as shown in Fig. 11-12. The excess above the equilibrium density of minority carriers is removed by recombination with majority carriers, as explained in Sec. 10-5. The minority carriers (in our case, electrons) have an average lifetime τ_n before recombining; τ_n decreases as the concentration of recombination centers increases. In this characteristic time τ_n an electron diffuses a characteristic distance $L_n = (D_n\tau_n)^{\frac{1}{2}}$, where L_n is called the *diffusion length* and D_n is the diffusion constant for electrons, and the density of electrons falls off proportionally to e^{-x/L_n} away from the junction (see prob. 11-46). The excess density of minority carriers is therefore

$$N_n - N_n^{(eq)} = N_n^{(eq)}(e^{eV/kT} - 1)e^{-(x-x_0)/L_n} \qquad (11\text{-}12)$$

Equation 11-12 reduces to eq. 11-11 at $x = x_0$, which demonstrates the validity of eq. 11-12 except for the dependence on x. The x dependence is derived from the solution of the one-dimensional diffusion equation (see prob. 11-46). In practical junctions the x dependence is more complicated than eq. 11-12 because of surface recombination and diffusion in three dimensions, but eq. 11-12 is still a fair approximation.

The current of diffusing electrons can be expressed by using the definition of the diffusion constant D_n:

$$J = -D_n(dN_n/dx) \qquad (10\text{-}1)$$

where J is the number of electrons per unit area crossing a plane perpendicular to the x-direction per second, and N_n is (as above) the density of electrons (number per unit volume). The current I is the product of J, e, and the junction area A:

$$I = -AeD_n\left(\frac{dN_n}{dx}\right)_{x=x_0} \qquad (11\text{-}13)$$

This is the current of minority carriers by diffusion (that is, the current produced by a concentration gradient). The current of minority carriers by conduction (produced by the electric field) is much smaller than this, since the electric field is very small in the n- and p-type regions. The field is necessarily small there because of the short-circuiting effect of the large density of *majority* carriers (see prob. 11-45).

When eq. 11-12 is used in eq. 11-13, we obtain

$$I = \frac{AeD_nN_n^{(eq)}}{L_n}(e^{eV/kT} - 1)$$

Therefore

$$I_{n0} = \frac{AeD_n N_n^{(eq)}}{L_n} \qquad (11\text{-}14)$$

and a similar expression holds for I_{p0}. The dependence of I_{n0} on $N_n^{(eq)}$, the equilibrium concentration of electrons on the p side, in this equation has been checked by experiments on actual junctions. The dependence on L_n given by eq. 11-14 is not precisely correct because of the complicating effects of recombination at the surface of the crystal.*

Since $N_n^{(eq)}$ is the number of minority carriers (electrons) on the p side, it is an exponential function of temperature (see prob. 11-28). The saturation reverse current therefore increases very rapidly as the temperature rises. Furthermore, at a sufficiently high temperature the Fermi level on both sides of the junction approaches $E_g/2$, the barrier at the junction disappears, and no rectification occurs. This deterioration of properties with rising temperature constitutes a serious limitation on the applicability of p-n junction rectifiers. Since silicon has a larger energy gap than germanium, $N_n^{(eq)}$ is much smaller in silicon for the same acceptor concentration. Although I_{n0} still varies rapidly with temperature, the variation is less objectionable with silicon junctions because operation at higher temperatures is possible before I_{n0} exceeds the maximum tolerable value.

The value of the diffusion constant D_n for this theory can be most easily obtained by measuring the electron mobility μ_n and then using the *Einstein relation*,

$$D_n = (kT/e)\mu_n \qquad (11\text{-}15)$$

The reason for the existence of such a relation is that the transport of electrons under the driving force of the electric field (conductivity) and the transport under the driving force of a concentration gradient (diffusion) are both limited by the collisions experienced by the electrons. Both D_n and μ_n are therefore proportional to the mean free path $\bar{l}$. (See prob. 10-20.)

If the conductivity of the n region is much greater than that of the p-region, the density of minority carriers in the n-region is much less than in the p-region (see Fig. 11-6 and accompanying discussion, or prob. 11-44). Therefore $N_n^{(eq)}$ in the p-region is much larger than the

---

* It should be noted that this theory of the p-n junction is not the only theory that leads to eq. 11-10. Another theory is based on the calculation of the way the generation of charge carriers in the transition region depends on the applied voltage. The charge generation at deep traps is the reverse of the recombination process at these traps. The charge generation theory leads to a different equation in place of eq. 11-14 but appears to agree well with many experiments, especially on silicon p-n junctions. See L. B. Valdes, *The Physical Theory of Transistors*, McGraw-Hill, New York, 1961, pp. 201–203 and references cited therein.

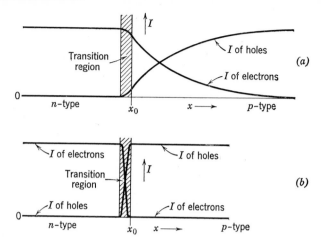

Fig. 11-13. Hole and electron currents as functions of position in a *p-n* junction. (a) Good rectifier, diffusion length $\gg$ length of transition region; $I_{n0} \gg I_{p0}$ for this particular junction. (b) *Ohmic* contact, diffusion length $\ll$ length of transition region.

comparable quantity in I_{p0}, and I_{n0} is much greater than I_{p0}. Thus, if it is desired that almost all the current across the junction be carried by electrons, the junction should be constructed with high-conductivity *n*-type germanium and low-conductivity *p*-type. (This result will be needed in the *n-p-n* transistor discussion.)

For each electron that recombines in the *p*-region, a hole flows into this region from the right. The electron and hole currents are sketched in Fig. 11-13a, where it has been assumed for clarity that $I_{n0} \gg I_{p0}$. Note that the diffusion length is very much greater than the width of the transition region, as required by assumption 1 at the beginning of this analysis. In order to obtain a sufficiently large diffusion length, the density of recombination centers must be very small, which in turn requires extreme purification of the germanium and careful treatment of the surface.

If recombination is very effective in the transition region, a non-rectifying (ohmic) junction is obtained. The electrons from the *n* region recombine in the junction, thermal equilibrium occurs at every point in the junction, and a non-equilibrium density of electrons $N_n^{(p)}$ cannot be created on the *p*-side (Fig. 11-13b). This is the reason why useful *p-n* junctions must be internal boundaries within a nearly perfect, highly purified single crystal. Pressing a crystal of *n*-type germanium against a crystal of *p*-type would produce a junction with a large density of recombination centers introduced by the surface imperfections.

The effect of light on a *p-n* junction is of considerable interest. Each photon absorbed creates a hole-electron pair, as illustrated schematically in Fig. 10-4. The majority carriers thus created are not of primary interest, since practical light intensities do not increase the majority carrier densities appreciably above equilibrium values. But the minority carrier densities produced by light may be many times the small concentrations present in equilibrium. If the light is absorbed within about a diffusion length of the junction, an appreciable additional minority carrier current can flow across the junction.

We shall concentrate on the *electron* currents, as we did in discussing $I(V)$ for a junction in the dark; the arguments can be duplicated exactly for the hole currents. If additional electrons (in addition to the thermally generated, equilibrium concentration) are created by light incident on the *p*-side of the junction, an additional current I_l (in addition to and in the same sense as the current I_{n0}, Fig. 11-8) flows "down the hill" at the junction. If the junction is reverse bia ed (Fig. 11-9) and if the dark current I_{n0} is small, the increment I_l produced by the light can be easily measured, and a *p-n* junction can thus serve as a sensitive photocell or *photodiode*. It can also be used as a detector for X-rays, γ-rays, or energetic charged particles, all of which also create hole-electron pairs.

If instead of applying a reverse bias from an external circuit, the *p-n* junction is open-circuited, a forward-bias voltage (the *photovoltaic effect*) appears at its terminals. The forward bias is just enough that $I_f = I_{n0} + I_l$ (see Fig. 11-10). If the junction is connected to a load resistance R, a voltage V somewhat smaller than the open-circuit photovoltage occurs and a current $I = I_{n0} + I_l - I_f$ flows; V and I are, of course, related by $V = IR$. Thus a *p-n* junction can be used to convert photons or energetic charged particles into electrical power. Large area silicon and gallium arsenide *solar cells* have been constructed with an efficiency of about 15% for converting sunlight into electrical power for such applications as power for the electronic equipment in space vehicles.

We return now to *p-n* junctions without illumination and note that a sharp rise in the reverse-bias current occurs in all junctions when the bias is sufficiently large. This abrupt rise in current is caused by the generation of hole-electron pairs inside the transition region by fast-moving carriers. It is called *avalanche breakdown* and is similar in physical origin to the Townsend avalanche breakdown in gas discharges. As the field in the transition region increases, the velocities attained by the free carriers increase and the probability that these rapidly moving carriers ionize atoms of the solid also increases. Such ionization of course produces additional carriers within the transition region. Eventually (as the bias voltage continues to increase) a field is reached at

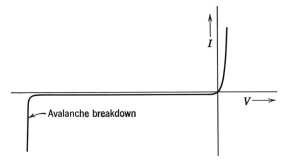

Fig. 11-14. Current as a function of voltage in a *p-n* junction. Unlike Fig. 11-11, the voltage scale here is long enough to exhibit avalanche breakdown, which occurs at voltages from a few volts to a few kilovolts, depending on the extent of doping.

which one additional hole or electron is created for each hole or electron passing through the junction. The current then rises abruptly to a value limited only by the circuit external to the diode, and the voltage corresponding to this field is the breakdown voltage.

If this current is limited by the external circuit to values that prevent excessive heat dissipation, breakdown causes no damage to the *p-n* diode, and such diodes are useful as voltage reference devices in voltage-regulating circuits.* This process, however, does limit the voltage that can usefully be rectified in conventional rectifier circuits. A typical $I(V)$ curve is given in Fig. 11-14; the reverse-bias voltage at which breakdown occurs can be varied widely by varying the width of the junction, which is in turn controlled by the doping.

An interesting quantum effect occurs for extremely narrow junctions; it is easily confused with avalanche breakdown, since $I(V)$ resembles Fig. 11-14 with a sharp increase in current at only a few volts bias, but it arises in quite a different way. As the reverse-bias voltage increases, the electric field in the narrow transition region becomes extremely large, as illustrated schematically in Fig. 11-15. It should be recalled that electrons are not absolutely excluded from a forbidden band; electrons can appear in such a band, but they have exponentially damped (*real* coefficient of x in the exponent) wave functions instead of sinusoidal, traveling-wave functions like those in an allowed band. Electrons coming from the valence band (Fig. 11-15) are thus not reflected precisely at the band edge at the junction, but their wave functions have *ex-*

---

* Such diodes are frequently called *Zener diodes*, a rather unfortunate name since only those with breakdown voltages $< \sim 5$ volts owe their interesting properties to the Zener effect, which will be discussed shortly.

ponential tails extending into the forbidden band on the other side of the junction. If the field becomes large enough, there is an appreciable probability that electrons will cross the junction by this process, which is called *tunneling* (because it relies on the exponential tails of wave functions) or the *Zener effect*.

Clearly this probability, which depends sensitively upon the tunneling distance, is extremely sensitive to the field, to which the tunneling distance is inversely proportional, and hence to the reverse-bias voltage (see eq. 8-5 and prob. 5-28). Thus an abrupt rise in current occurs as the critical voltage is reached. (The Zener current is not observed in the *p-n* diodes commonly used as rectifiers and detectors since the transition regions in these junctions are relatively wide, and avalanche breakdown occurs before the field becomes large enough to exhibit tunneling.)

If extremely large concentrations of donors and acceptors are incorporated into a semiconductor crystal, a *p-n* junction diode can be made that depends for its interesting properties on the tunneling current. Such a crystal is to be sharply distinguished from the semiconductors we have been considering (and that we shall consider in Sec. 11-5). The method of analysis used earlier in this chapter applies, but the heavy doping produces drastic changes in the position of E_0, the sharpness of the impurity levels, and the fraction of the impurity atoms that are ionized. The impurity atoms are so close together that the wave functions of electrons on these atoms begin to overlap, and the impurity levels are therefore broadened into a band. This band is only a few hundredths of a volt wide, but the original donor and acceptor levels were only about 0.01 e.V. from the nearest band; therefore the impurity band extends into and becomes indistinguishable from the bottom of the conduction

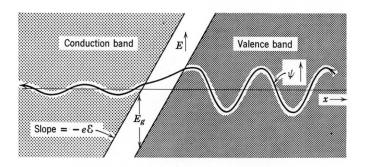

Fig. 11-15. Plot of energy bands with a very strong field ($\sim 10^8$ volts/m.) accelerating electrons toward $-x$. A sketch of a typical wave function is superimposed. This ψ is initially a traveling wave (toward $-x$) in the valence band, then a damped wave in the forbidden band, and finally a traveling wave in the conduction band.

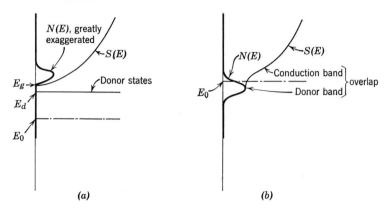

Fig. 11-16. Comparison of an ordinary, nearly pure, n-type semiconductor (a) with a highly doped ("degenerate") n-type semiconductor (b). The $N(E)$ curve in (a) is greatly exaggerated.

band in n-type material (and top of the valence band in p-type), as shown in Fig. 11-16b. The Fermi level E_0 now lies within the conduction band in n-type material and within the valence band in p-type. The position

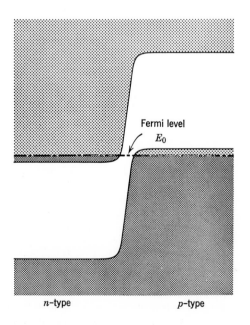

Fig. 11-17. Energy diagram of a tunnel diode in equilibrium. Electrons with energies near E_0 tunnel in both directions, and the net current is zero.

of E_0 is determined, as always, by the supply of electrons and the availability of states; in n-type material, for example, states are filled up to an energy E_0 such that the electron concentration in the combined (impurity + conduction) band equals the concentration of donors.

Figure 11-17 provides the equilibrium (no applied voltage) energy diagram of a p-n junction constructed with both sides consisting of such highly doped material. The transition region is extremely narrow, of the order of 100 Å, since the concentrations of fixed charges (ionized donors and acceptors) are so large and therefore by Poisson's equation the electrostatic potential changes so rapidly (probs. 11-32 and 11-52). Thus electron tunneling across the junction becomes so highly probable that appreciable currents flow back and forth with zero applied field, merely with the built-in field of the junction (of course with zero applied field, the currents to the right and to the left just cancel).

The current I as a function of applied voltage V of such a junction is remarkable for its enormous current density and, especially, for the existence of a region of voltages where dI/dV is <0, that is, where there is a negative resistance $dV/dI < 0$. Figure 11-18 shows how this

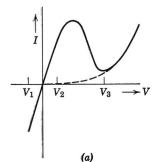

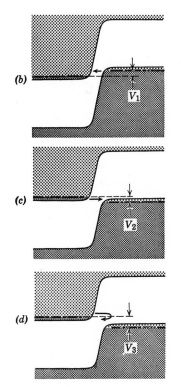

Fig. 11-18. (a) I vs. V for a tunnel diode; the dashed curve is the ordinary diffusion current (eq. 11-10). (b), (c), and (d) Energy band diagrams of a tunnel diode with various bias voltages V. (V_3 is about 0.2 volt for germanium and about 0.5 volt for gallium arsenide diodes.)

behavior arises.   At $V = 0$ large tunneling currents flow in both directions but are equal in magnitude (Fig. 11-17).   Small reverse (V_1) or forward (V_2) biases upset this balance as shown in Figs. 11-18b and 11-18c.   With a further increase of forward bias to V_3 (Fig. 11-18d), however, the electrons can no longer tunnel into the p-side, since there are no allowed states at the same energy as the energy of filled electron states on the n-side (losing appreciable energy in such a short distance is quite improbable, and so the electron cannot descend to the level of the valence band on the p-side).   The tunneling current therefore falls abruptly to zero.   Further increase of forward bias produces an appreciable diffusion current as in eq. 11-10, a current that was negligibly small at V_1 or V_2 compared to the tunneling current.

These *tunnel diodes* or *Esaki diodes* are useful as oscillators, amplifiers, and elements of computing circuits.   The values of V_3 are typically a few tenths of a volt, the value being determined by the energy widths of the occupied states in the conduction band and of the vacant states in the valence band.   The current densities are many amperes per square millimeter, since there is no barrier, and typical devices employ tiny junction areas in order to obtain convenient values of I (milliamperes) and to keep the junction capacitance small.   When such a device is coupled through low-inductance leads to a resonant circuit and biased in the region of negative resistance, the circuit will oscillate if the circuit resistive losses are low enough.   Esaki diode oscillators function at frequencies up to many kilomegacycles, and the upper frequency limit has been set by circuit problems rather than by intrinsic properties of the diode.   These tunnel diodes also function as extremely rapid switches and as information storage elements in computers, with remarkably short access times; operating times of the order of a few tenths of a nanosecond (i.e., a few tenths of 10^{-9} sec) are common.

11-5   JUNCTION TRANSISTORS

An important combination of two p-n junctions within a single semiconductor crystal is the *junction transistor*.   These devices can provide amplification and can perform the other functions performed by a vacuum-tube triode.   Like the p-n junction diode, transistors have frequency and temperature limitations that are different from those of vacuum-tube devices.   Also like the p-n junction diode, transistors can be made in very small sizes and can operate at minute power levels.   They are inherently more rugged and more reliable than vacuum tubes and are especially suitable for computers and for airborne communications and control equipment.   We shall illustrate the physics of tran-

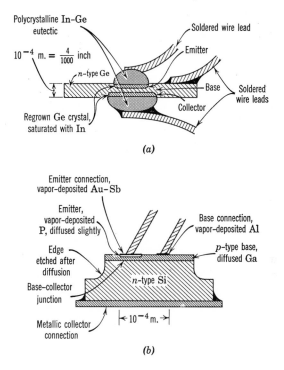

Fig. 11-19. (*a*) A germanium *p-n-p* alloy transistor; some of the germanium dissolved in the molten indium during alloying and recrystallized as *p*-type extensions of the original crystal upon cooling. (*b*) An *n-p-n* silicon mesa transistor. The *p*-type base was made by diffusing gallium to a depth of about 4×10^{-6} m. ($\sim$0.0002 inch). Then deposition from the vapor of a small area of phosphorus (the area being defined by a silica mask) was followed by diffusion to a depth of about 2×10^{-6} m. to form the *n*-type emitter. The vapor deposition of Au-Sb and Al facilitates the connection of Au wires to the emitter and base, respectively.

sistors by a discussion of the *n-p-n* transistor. The *p-n-p* transistor is identical in principle. Other junction transistors have additional junctions or additional connections, but the physics of these devices is similar to that of the *n-p-n* transistor.

An *n-p-n* transistor consists of two *p-n* junctions separated by a very thin *p*-region, of the order of 2×10^{-5} m. (0.001 inch) in thickness. Two methods of making such a transistor are illustrated schematically in Fig. 11-19, but there are many other methods. The sizes indicated on Fig. 11-19 are typical, although much larger units have been made (but still with very thin *p*-regions). The central *p*-region is called the

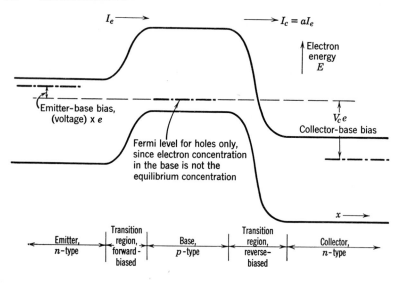

Fig. 11-20. Energy diagram of an *n-p-n* transistor with the appropriate bias voltages. The current vectors are electron currents rather than conventional (positively charged carrier) currents.

base, and the *n* regions are called the *emitter* and the *collector*. Electrical leads are attached to the surfaces of all three regions. The bias voltages are such that the emitter junction is forward-biased and the collector junction is reverse-biased.

An energy band diagram of the *n-p-n* transistor as ordinarily biased is shown in Fig. 11-20. The voltage across the emitter junction is varied by an input signal, and the current through this junction varies in accordance with eq. 11-10. The donor concentration is made much larger in the emitter than the acceptor concentration in the base, and therefore almost all the current in the emitter junction is a current of electrons flowing to the right, rather than holes to the left, as explained in the paragraph following eq. 11-15. The thickness of the base region is much less than the diffusion length L_n. Therefore almost all the electrons injected into the base at the emitter junction flow across the collector junction, where they experience an accelerating field. The collector current I_c is hence only slightly less than the emitter current I_e.*

* The currents used throughout this discussion are electron currents, rather than the conventional, positively charged carrier currents common in electric-circuit theory. Our aim here is to illuminate the process of power gain in a transistor rather than to introduce transistor-circuit theory.

Typical curves of collector current I_c as a function of collector voltage V_c, with emitter current I_e as a parameter, are shown in Fig. 11-21a. When sufficient collector bias is applied the curves have nearly zero slope and the collector current is almost equal to the emitter current, for the reasons given in the preceding paragraph. Figure 11-21b is an enlargement of the region near the origin of Fig. 11-21a. The curves lose their ideal, pentode-like nature only when $|V_c|$ is less than $\sim(4kT/e) \cong 0.1$ volt. This property also follows from the analysis of the preceding paragraph and from the p-n junction theory of Sec. 11-4. Transistors can thus be operated with extremely low voltages and currents, which is one of their most valuable attributes.

If the characteristics of a transistor obeyed this theory perfectly, I_c would exactly equal I_e and the slope $\partial V_c / \partial I_c$ (with I_e constant) at a volt or more collector bias would be millions of megohms. Deviations from ideal behavior of I_c are caused by: (1) The emitter current is not composed exclusively of electrons but contains some holes flowing from the base to the emitter; the holes do not flow to the collector and hence cause a difference between I_c and I_e. (2) There is some recombination in the base, and therefore part of the electron current is lost; an important part of this recombination occurs at the surface of the crystal. Deviations from ideal behavior of $\partial V_c / \partial I_c$ are caused by: (1) Surface conduction paths add a resistance in shunt with the ideal collector junction. (2) The width of the transition region in the collector junction is a function of V_c, as explained early in the discussion of p-n junctions in Sec. 11-4, and this in turn influences I_c in two ways. As an increase in V_c increases the

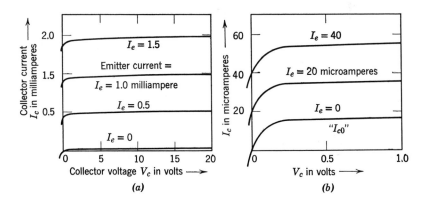

Fig. 11-21. Collector current as a function of collector voltage for an n-p-n transistor. Part (b) is an enlargement of the region near $V_c = 0 = I_c$ of part (a). [From Shockley, Sparks, and Teal, *Phys. Rev.*, **83**, 161 (1951).]

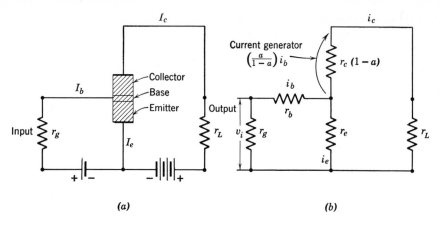

Fig. 11-22. "Common emitter" or "grounded emitter" connection of an *n-p-n* transistor. (*a*) Transistor with labeling of total currents. (*b*) Equivalent circuit and small a-c components of currents.

width of the transition region, the number of thermally generated carriers produced in this region increases and therefore I_c increases. As the transition region becomes wider, the base must become narrower; the loss by recombination of electron current on its way from emitter to collector is thereby reduced, and I_c is slightly increased. To make the performance of a transistor as nearly as possible the ideal performance, extreme care must be exercised in purifying the germanium or silicon and in treating and preserving the crystal surface. The purpose of this care is primarily to minimize recombination.

It is instructive to compute the power gain of a typical transistor in the typical circuit shown in Fig. 11-22*a*. We can readily calculate the power gain of this circuit for small a-c signals if we assume that the collector resistance, $r_c = \partial V_c / \partial I_c$ at constant emitter current, is very * large compared to the load resistance r_L and that the generator resistance r_g is large compared to the input resistance. The input circuit is then sufficiently isolated from the output circuit to make the input power easy to calculate. An equivalent circuit is presented in Fig. 11-22*b*. The resistances employed here are those appropriate to small alternating currents i and voltages v superimposed on the bias currents and voltages (this is the same procedure that is common in vacuum-tube circuit theory).

The resistance r_b is in part the series resistance between the wire

---

* See the precise statement of this requirement in the sentence following eq. 11-17.

soldered to the base region and the actual emitter junction. This is, of course, an average over all parts of the emitter junctions; the resistance would be zero for the part of the junction immediately adjacent to the lead and largest for the most remote part of the junction. This part of r_b is of the order of 100 ohms in typical transistors.

A more consequential part of r_b arises from internal feedback between the collector and emitter junctions. It has already been noted that the increase of collector voltage decreases the base thickness, since the collector transition region becomes wider. In order to maintain a constant diffusion current of electrons from emitter to base, a constant concentration gradient must be preserved (eq. 10-1). When the base becomes thinner, the concentration of electrons injected at the emitter-base junction must therefore decrease in order to keep the gradient dN_n/dx constant. But a decrease in the concentration of injected electrons is the effect that would be produced by a smaller emitter-to-base voltage. Thus a feedback is produced. In the equivalent circuit of Fig. 11-22b this feedback increases r_b, since it reduces the effectiveness of the applied voltage in controlling the emitter-collector current. This contribution to r_b becomes larger if the base width becomes smaller. The total r_b, series resistance plus feedback equivalent resistance, is about 500 ohms in typical transistors.

The resistance r_e is essentially the resistance of a forward-biased p-n junction, but there is a relatively small correction from feedback effects. r_e is of the order of 10 or 20 ohms for a few milliamperes emitter current (and smaller for larger currents; see prob. 11-38).

The a-c input voltage is

$$v_i = i_e r_e + i_b r_b$$

We shall define

$$a = i_c/i_e$$

the ratio of the a-c collector current to the a-c emitter current. Then i_b equals $(1 - a)i_e$, and the input power is

$$v_i i_b = [i_e r_e + (1 - a)i_e r_b](1 - a)i_e \qquad (11\text{-}16)$$

The output power is $i_c^2 r_L = i_e^2 a^2 r_L$, and thus the power gain G is

$$G = \frac{a^2 r_L}{[r_e + (1 - a)r_b](1 - a)} \qquad (11\text{-}17)$$

For a typical transistor $a = 0.98$, $r_e = 20$ ohms, $r_b = 500$ ohms, and the collector impedance r_c is sufficiently large so that an $r_L = 30{,}000$ ohms can be used without violating our assumption about r_c (which can be stated now in terms of a, namely, $r_c(1 - a)$ must be considerably larger than r_L). For these values,

$$G = \frac{(0.98)^2 \times 3 \times 10^4}{[20 + 0.02 \times 500]0.02} = 5 \times 10^4 = 47 \text{ db}$$

In the "common emitter" connection of this illustration, power gain has been produced partly by current gain and partly by the fact that the output impedance is larger than the input impedance.

It should be emphasized that the calculation leading to eq. 11-17 is just a particular example to show how gain is produced and what the important physical parameters are. There is as wide a variety of amplifier, oscillator, detector, and switching circuits for transistors as for vacuum tubes. In addition, the ability to use both n-p-n transistors and p-n-p transistors (which have bias voltages opposite from n-p-n) increases the variety of circuits. It should also be noted that the equivalent circuit that was used in the foregoing example is neither the only nor the most valuable equivalent circuit that can be used to represent a transistor. The "T" circuit used here is probably the simplest circuit, but the performance of a transistor over a wide range of frequencies and temperatures is better described by more complicated equivalent circuits.

The n-p-n transistor can profitably be compared with a triode electron tube. In the triode tube, electrons flow inside the grid wire and change the grid potential as the input signal changes. The grid-cathode potential difference in turn controls the flow of thermionic electrons from cathode to plate. The flow of electrons in the grid wire does not become intermixed with the cathode-anode flow, since the grid is negative (it repels electrons from the cathode) and is cold (electrons cannot be emitted from the grid wire). In the transistor, holes flow in the base region to change the base potential as the input signal changes; in the absence of this hole flow it would not be possible to change the potential difference applied across the base-emitter junction. The base-emitter potential difference in turn controls the flow of electrons from emitter to collector. The flow of holes in the base does not become intermixed with the electron flow since recombination of electrons and holes is slight in the base region. This analogy emphasizes the necessity for two kinds of charge carriers in an n-p-n or a p-n-p transistor.

As the temperature of a transistor increases, the collector current for zero emitter current, usually called I_{c0}, increases exponentially. This current is the current of a reverse-biased p-n junction, and its temperature dependence can be inferred from eq. 11-14. The rapid rise of I_{c0} with temperature limits the applicability of germanium transistors to temperatures less than $\sim 100\,°C$ and of silicon transistors to temperatures less than $\sim 200\,°C$. The current gain a changes slightly with T, but the changes in $(1 - a)$ are relatively larger and are sometimes serious.

The frequency limitations of *n-p-n* transistors are also of interest. There are four important effects that together place an upper limit on the frequency of operation:

1. Some time is required for electrons to diffuse from the emitter to the collector junction. This transit time introduces a phase delay in the output, which is usually not serious, but in addition all electrons leaving the emitter at the same time do not arrive at the collector at the same time. In other words, there is a dispersion of transit times because of the random paths, as illustrated in Fig. 9-14a. If the dispersion is of the same order of magnitude as the period of the signal being amplified, many electrons will arrive at the anode in the wrong phase to give amplification. This dispersion in time equals w^2/D_n, where w is the width of the base region and D_n is the diffusion constant for electrons in the base. It can be reduced by making the base width smaller.

2. Each transition region acts like a capacitance, since a change in voltage across a transition region must be accompanied by a change in the charges on either side of the transition. The emitter-base capacitance is in parallel with a low resistance (a forward-biased junction) and therefore is not very important, but the collector-base capacitance is shunted only by a high resistance and is important because of the limitation it imposes on the output impedance at high frequencies.

3. There is an additional effective capacitance between emitter and base that is much larger than the transition region capacitance listed in 2. This capacitance arises from the necessity for building up an excess concentration of electrons in the base region in order to have a flow of electrons by diffusion from the emitter to the collector. This capacitance is proportional to the product of the emitter current and w^2/D_n.

4. The external base circuit cannot be connected directly to every point of the base-emitter and base-collector junctions but is connected through the series resistance part of the base resistance r_b. If it were not for this resistance, external circuit elements could be directly connected to the junctions and could cancel some of the reactive effects mentioned above.

The combination of items 3 and 4 produces a low-pass RC filter in the input circuit of the transistor, and it is the principal frequency-limiting effect in many transistors. The effects of 1 or 3 could be reduced by making the base width smaller. The effect of 4 could be reduced by making the base width larger. Thus a compromise width must be used, but the compromise value is so small that it is difficult to achieve with alloy junction transistors (Fig. 11-19a). Diffused junction mesa transistors, Fig. 11-19b, can attain not only a small base width but also a small area, thus decreasing capacitances and minimizing the resistance effect 4 above. Mesa transistors have been constructed that operate at frequencies above 1000 megacycles.

All the foregoing analysis applies as well to the p-n-p transistor, with the appropriate changes of sign and of identity of carriers. Furthermore, the basic potentialities of carrier motion in semiconductors are by no means fully exploited by present devices, and much intriguing science and development engineering remains to be done.

11-6 THERMOELECTRICITY

Thermoelectric effects are the interactions between temperature differences and heat flows on the one hand and voltages and currents on the other. The most important effects occur when two dissimilar materials are in contact. All materials, even conducting liquids, exhibit thermoelectric effects, but the effects are much larger in semiconductors than in metals, and it is for this reason that the discussion of thermoelectricity has been reserved for the present chapter. We give only a brief discussion of one of the effects here and confine the discussion to semiconductors. The references at the end of the chapter should be consulted for a serious study of the subject.

If a material a is in contact with a material b and a current I is flowing, heat Q (joules/sec) flows to or away from the contact. This is the *Peltier effect* and is quantitatively described by the equation $Q = \pi_{ab}I$, where π_{ab} is the Peltier coefficient.* Closely related to the Peltier effect is the *Seebeck effect*, the emf developed in a circuit of two materials a and b in which the two junctions ab are at different temperatures. If a combination ab gives a large π_{ab}, it also gives a large Seebeck voltage. In addition to these effects at junctions, a heating or cooling (the *Thomson effect*) occurs within a single material if a current is flowing and a temperature gradient exists. The sign of the Thomson heat flow depends on the material; the flow is proportional to the product of I and the temperature gradient dT/dx and hence reverses when I is reversed. All these effects are in addition to the ordinary I^2R ("joule") heating whenever there is a current I and to the flow of heat by the ordinary process of thermal conductivity whenever there is a temperature gradient.

---

* The sign convention is that $I > 0$ corresponds to *electrons* flowing from b to a (conventional current flowing from a to b), and if $\pi_{ab} > 0$, this electron flow removes heat from the surroundings (i.e., the junction acts as a cooling device). Thus Q is the heat removed from the surroundings.

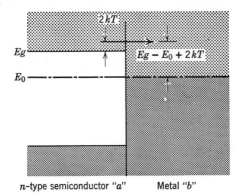

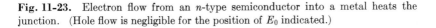

n-type semiconductor "*a*"   Metal "*b*"

Fig. 11-23. Electron flow from an *n*-type semiconductor into a metal heats the junction. (Hole flow is negligible for the position of E_0 indicated.)

Study of Fig. 11-23 will show why a large Peltier effect occurs at a metal-semiconductor junction. We have taken as an example an *n*-type semiconductor, and we consider first electron flow to the right. The Fermi levels in *a* and *b* are aligned, as explained in conjunction with Fig. 11-8 for a *p-n* junction. Thus each electron that flows to the right enters the metal with an energy considerably in excess of the Fermi brim energy. It suffers collisions with the lattice in the metal and in a very short distance becomes just another electron in the metal with an energy E_0. The extra energy has warmed the lattice. Conversely, electrons coming from a metal to an *n*-type semiconductor *cool* the region of contact: Only the most energetic of the electrons in the Fermi distribution within the metal can enter the semiconductor; when the energetic electrons are thus removed from the distribution, heat must be supplied by the lattice to restore the equilibrium distribution, heat is drawn from the surroundings, and a strong cooling effect occurs.

Similar analyses can be made for a contact between a *p*-type semiconductor and a metal and for a contact between *n*- and *p*-type semiconductors. It should be noted that thermoelectric effects do not involve minority carriers or recombination of injected carriers. Therefore the stipulation of physical perfection necessary for a *p-n* junction to have interesting electrical properties does not apply here; soldered contacts or even pressed contacts give the same performance as an internal *p-n* junction within a single crystal.

The magnitude of the Peltier coefficient can be calculated from Fig. 11-23. The energy released per electron is $E_g - E_0$ plus the kinetic energy that the electron had in the conduction band of the semiconductor.   This kinetic energy is, on the average, $2kT$.   Although the average energy of all the electrons in that band is $\frac{3}{2}kT$, the more energetic electrons are more likely to cross a plane normal to the current flow, and the average energy *of those crossing* turns out to be $2kT$ (this is the answer to prob. 2-11).   Thus the heat released per electron is $E_g - E_0 + 2kT$, and since I equals the number of electrons per second flowing from b to a multiplied by e,

$$\pi_{ab} = (1/e)(E_g - E_0 + 2kT) \tag{11-18}$$

which is large if E_0 is well removed from the band edge E_g.[*]   (This analysis does not reveal it, but a much smaller Peltier heat arises from subtler causes at the contact between two different metals.)

The Seebeck voltage and the Thomson heat can be inferred by similar but more involved arguments.   The temperature dependence of the Fermi level, the diffusion flow of carriers from hot to cold regions, and the energy dependence of the probability of carrier scattering by impurities and lattice vibrations enter into these calculations.

Thermoelectric effects in semiconductors may be used to provide compact cooling devices (e.g., for cooling transistor circuits) and may be used to convert heat (e.g., from a nuclear reactor) into electrical power. In either application, the following material parameters enter: (1) The Peltier heat.   (2) The electrical conductivity σ, since I^2R represents an unwanted heating in the refrigerator application and a loss of power in the generator.   (3) The thermal conductivity σ_t, since the flow of heat by this process is an unwanted load on the refrigerator and a loss of input heat in the generator.   In designing semiconductors for thermoelectric applications, heavy doping to raise the Fermi level decreases π but increases σ, and therefore the donor or acceptor concentration has an optimum, compromised value.   The optimum doping is much greater than is typical for transistors, and $E_g - E_0$ in eq. 11-18 is accordingly smaller (it was for this reason that we did not neglect $2kT$ with respect to $E_g - E_0$ in that equation).   The thermal conductivity σ_t must be as small as possible.   Two techniques are used to reduce σ_t by reducing the phonon mean free path: (1) Employing compounds of heavy elements increases the phonon density by lowering the Debye Θ. Although the density of carriers of heat is thus increased, their mean free path is reduced by phonon-phonon collisions much more rapidly than

---

[*] In our example of electrons flowing from n-type semiconductor a to metal b, $I < 0$, $\pi_{ab} > 0$, $Q = \pi_{ab}I$ is <0, and the flow of electrons across the junction delivers heat to the surroundings.   This direction of heat flow agrees with our sign convention, but it is simpler and more instructive to determine the heating or cooling effect directly from the physics (e.g., by a sketch like Fig. 11-23) rather than by trying to remember sign conventions.

their density is increased (since the number of such collisions increases as a high power of this density). (2) Employing mixed crystals the atoms of which are chosen with the same valence (e.g., Se and Te) to minimize electron scattering and thereby to maximize σ, but different masses to maximize phonon scattering.

A typical thermoelectric material is doped $(Bi, Sb)_2(Se, Te)_3$, a single crystal with the ratio 2/3 of metal to non-metal throughout but with either a Bi or an Sb atom at random at each positive ion site, and similarly for Se and Te.* In addition to optimizing the material, it is necessary to optimize the value of the current I. If I is too small, very little cooling or power generation (both proportional to I) occurs, and meanwhile thermal conduction (independent of I) causes an unwanted heat flow. If I is too large, the joule heating (proportional to I^2) causes an unnecessarily large heating in the refrigerator and power loss in the generator.

REFERENCES

General

W. Shockley, *Electrons and Holes in Semiconductors*, Van Nostrand, New York, 1950, Chapters 1, 4, 10, and 12.

N. F. Mott and R. W. Gurney, *Electronic Processes in Ionic Crystals*, Clarendon Press, Oxford, 2nd Ed., 1948, Chapters 3 and 5.

E. Spenke, *Electronic Semiconductors*, McGraw-Hill, New York, 1958.

L. B. Valdes, *The Physical Theory of Transistors*, McGraw-Hill, New York, 1961.

W. C. Dunlap, Jr., *An Introduction to Semiconductors*, Wiley, New York, 1957.

R. A. Greiner, *Semiconductor Devices and Applications*, McGraw-Hill, New York, 1961.

J. N. Shive, *Semiconductor Devices*, Van Nostrand, Princeton, 1959.

Semiconductors, edited by N. B. Hannay, Reinhold, New York, 1959.

Thermoelectricity

R. R. Heikes and R. W. Ure, Jr., *Thermoelectricity: Science and Engineering*, Interscience, New York, 1961.

A. F. Joffe, *Semiconductor Thermoelements and Thermoelectric Cooling*, Infosearch, London, 1957.

J. N. Shive, *Semiconductor Devices*, Van Nostrand, Princeton, 1959, Chapter 23.

PROBLEMS

11-1. The observed energy gap E_g of germanium is $0.75 - 0.0001T$. The principal reason for the variation of E_g with T is the thermal expansion of the lattice.

---

* For such materials with large carrier concentration and small phonon thermal conductivity, the contribution to σ_t by electrons or holes becomes appreciable. This carrier contribution is, of course, the dominant contribution in metals but is negligible with respect to the phonon contribution in the typical semiconductors discussed earlier in this chapter.

Explain how the sign of the temperature dependence of E_g verifies that the bands have crossed (Fig. 9-4) as R is imagined to have decreased from $R = \infty$.

11-2. The observed energy gap in silicon is $E_g = 1.21 - 0.0004T$. Assume that the temperature dependence of E_g arises entirely from the thermal expansion of the lattice. Compute $(dE_g/dR)_{R=R_0}$ and compare with Fig. 9-4. That figure is schematic, and the foregoing assumption is not completely correct; therefore the comparison is not perfect.

11-3.* Let the effective mass m^* for electrons be three times that for holes in an intrinsic semiconductor. Where is E_0 relative to the center of the energy gap?

11-4. It may be shown that for pure germanium near room temperature, $\mu_n + \mu_p \cong AT^{-3/2}$ by using eq. 9-18 and by noting that for scattering by lattice vibrations $\bar{l}$ is proportional to T^{-1} (as for a metal) and $(\overline{1/v})$ is proportional to $T^{-1/2}$ (Maxwell distribution, compare eqs. 9-18 and 9-19). Determine the constant A for germanium from Table 11-1. Use eq. 11-6 to compute the resistivity ρ of intrinsic germanium at $T = 290$, 300, and 310°K and plot ρ vs. T. Compare with the plot for copper (Fig. 9-16) in the same region of T after multiplying all the ρ values for copper by an appropriate power of 10 so that they can be plotted on the same plot as the germanium values.

11-5.* Calculate σ for intrinsic germanium at 300°K from the data of Table 11-1, and compare with σ in Table 11-1. Do the same for silicon.

11-6. The electron mobility μ_n in the "III-V" semiconductor InSb is 6.7 m.²/volt sec at 300°K and the hole mobility is $<0.1\mu_n$. The energy gap is 0.23 e.V. What is σ of intrinsic InSb at 300°K? Compare approximately the temperatures at which InSb and Ge, with equal donor concentrations and no acceptors, become intrinsic.

11-7. Plot on the same diagram $\log_{10} N_n$ and $\log_{10} N_p$ as functions of E_0 for $0 < E_0 < E_g$, where $E_g = 0.72$ e.V. and $T = 300$°K. Make your plot accurate for the region of greatest interest in most semiconductor problems, namely $\sim 4kT < E_0 < E_g - \sim 4kT$, and sketch $\log N_n$ and $\log N_p$ approximately for the remaining values of E_0.

11-8.* Calculate E_0 at 300°K for a germanium crystal containing 5×10^{23} arsenic atoms per cubic meter.

11-9. In prob. 11-8, compute N_n, N_p, and the number of electrons remaining in donor states.

11-10.* Calculate E_0 at 300°K for a germanium crystal containing 5×10^{22} gallium atoms per cubic meter.

11-11.* Calculate E_0 at 100°K for a germanium crystal containing 5×10^{22} arsenic atoms per cubic meter.

11-12. It is sometimes naively thought that if the concentration of donors is (for example) twice the concentration of acceptors, N_n equals $2N_p$. Explain why this is wrong.

11-13.* Calculate E_0 at 300°K for a germanium crystal containing 10^{23} arsenic atoms per cubic meter and 5×10^{22} gallium atoms per cubic meter.

11-14.* In prob. 11-13, compute the conductivity of the specimen. Why is this σ not the sum of the σ of a specimen containing 10^{23} arsenic atoms per m.^{3} and of the σ of a specimen containing 5×10^{22} gallium atoms per m.^{3}?

11-15. Consider that we know μ_n, μ_p, E_g, and other properties of germanium. Consider a specific sample of doped germanium at room temperature. (*a*) σ is measured; can we determine E_0 from this measurement? How or why not? (*b*) The Hall constant R is measured; can we determine E_0 from this measurement? How or why not? (Assume that the value found for R indicates that the concentration of one carrier type far exceeds that of the other.) (*c*) Suppose E_0 is known. Can we determine σ from this information? How or why not?

11-16. The conductivity σ_0 of a sample of germanium is measured at room temperature. The sample is then remelted with sufficient added arsenic so that an arsenic impurity of one part per million is added to the doping in the original crystal. The new crystal has a conductivity σ of 1000 mhos/m. and is n-type. (*a*) What was σ_0? (*b*) Infer as much as you can about the kind and amount of the doping in the original crystal.

11-17.* In Sec. 11-3 the statement was made that as T becomes very large or N_d very small E_0 approaches $E_g/2$. Derive an expression for the value of N_d as a function of temperature such that $E_0 = (E_g/2) + kT$, and use your expression in order to discuss the validity of the statement in question. *Hint:* The number of electrons in the conduction band is equal to N_d plus the number of holes in the valence band; assume that there are no acceptors.

11-18. Curve 1 in Fig. 11-7*b* exhibits distinctly different behavior in the regions $T > {\sim}300°$K, ${\sim}30°$K $< T < {\sim}300°$K, and $T < {\sim}30°$K. Why does N_n have the observed behavior as a function of T inside each region? Estimate from data in the text and in the figure caption the two values of temperature that divide the temperature scale into these three regions.

11-19.* Consider sample 1 of Fig. 11-7*b*. Read the value of T that gives $N_n = \frac{1}{2}N_d - N_a$. Where is the Fermi level E_0 relative to the energy E_d of the donor levels at this T? From N_n and T calculate $E_g - E_0$ at this temperature and hence $E_g - E_d$.

11-20. The binding energy of electrons to phosphorus or antimony donors in *silicon* is about 0.04 e.V. Sketch $\log_{10} N_n$ vs. $1/T$ for silicon with the donor and acceptor concentrations of sample 1 in Fig. 11-7*b* and reproduce curve 1 of that figure on the same plot. *Hint:* Do not calculate the individual points, but estimate from $E_g = 1.09$ e.V. the temperature at which silicon becomes intrinsic and sketch the low temperature region by adjusting curve 1 of Fig. 11-7*b* for the difference in binding energies.

11-21.* Calculate σ for germanium crystals all of which have a concentration of 2×10^{22} indium atoms per cubic meter but each of which has a different concentration of antimony atoms. Compute for 10^{21}, 10^{22}, 10^{23}, and 10^{24} antimony atoms per cubic meter and for two or three other concentrations chosen to permit a complete plot of σ vs. antimony concentration. Draw such a plot on log-log paper. ($T = 300°$K.)

11-22. Compute the Hall constant R for the same specimens described in prob. 11-21 and plot $|R|$ as a function of antimony concentration on log-log paper. ($T = 300°$K.) *Note:* Near the concentration of antimony that makes N_n equal to N_p, the expression for R that is valid when both carriers are present should be used: $R = (N_p\mu_p{}^2 - N_n\mu_n{}^2)/e(N_p\mu_p + N_n\mu_n)^2$.

11-23.* The Hall angle of a specimen of n-type germanium is $5.7°$ at a magnetic induction of 0.3 weber/m.^{2} The conductivity of this specimen is 100 mhos/m. Compute μ_n and the number of conduction electrons per cubic meter. What is the net number of donors per germanium atom? How does the sensitivity of this electrical method of "analysis" compare with that of chemical analysis? (The density of germanium is 5350 kg/m.^{3})

11-24. Calculate μ_n for specimen 1 of germanium from the curves of Fig. 11-7. Plot log μ_n vs. log T for $13 < T < 200°$K. μ_n is proportional to what power of T? Why is μ_n for sample 2 smaller than for sample 1?

11-25.* A silicon resistor for temperature compensation of transistor circuits has a *positive* temperature coefficient of resistance. It has a doping that makes the carrier concentration independent of temperature, and thus σ is proportional to μ. The resistance is observed to increase 0.7% for each $1°$C temperature increase near room temperature. Express $\mu = \mu_0 T^n$ and find n. *Hint: $d\rho/\rho = -d\mu/\mu = -n\, dT/T$.*

11-26.* Calculate E_0 and σ for barium oxide from the following data: $E_g > 4$ e.V.; donor levels at $E_g - 1.5$ e.V.; $N_d = 10^{22}$ m.$^{-3}$; number of acceptors $\ll 10^{22}$ m.$^{-3}$; $T = 1000°$K; μ_n at $1000°$K $= 5 \times 10^{-4}$ m.^{2}/volt sec. (Specify the position of E_0 relative to E_g.)

11-27. Derive eq. 11-8.

11-28.* How does the concentration of holes depend on temperature in an n-type semiconductor? Assume that the temperature and donor concentration are such that: (1) all the donors are ionized; (2) the number of donors is much greater than the number of holes. *Hint:* See eq. 11-8.

11-29. Does the p-type or the n-type material of the particular p-n junction illustrated in Fig. 11-8 have the higher conductivity? Assume that the figure is to scale and that the material is germanium, and estimate the ratio of the net concentration of donors on the n side to the net concentration of acceptors on the p side ($T = 300°$K).

11-30.* For a p-n junction, "inside the transition region there are very few free carriers." Demonstrate the validity of this statement by sketching approximately $N_n + N_p$ vs. distance x from the n- to the p-side of a germanium p-n junction in equilibrium at $T = 300°$K. Let $|N_d - N_a| = 10^{22}$ on each side. What is the minimum value of $N_n + N_p$?

11-31. A germanium p-n junction has a net donor concentration on the n-side of 2×10^{21} m.$^{-3}$ and a net acceptor concentration on the p-side of 2×10^{23} m.$^{-3}$ Compute E_0 on each side and the barrier height at $T = 300°$K.

11-32.* Calculate the width of the transition region in an alloy p-n junction in germanium for the following conditions: In the p region ($x < 0$), the net concentration of acceptors is 10^{24} m.$^{-3}$; in the n region ($x > 0$), the net concentration of donors is 8×10^{21} m.$^{-3}$; the concentrations of donors and acceptors change abruptly at $x = 0$ and are constant elsewhere. *Hint:* Assume that, for $d_p < x < 0$, all the acceptors are ionized and there are no free carriers (negative space charge); assume that, for $0 < x < d_n$, all the donors are ionized and there are no free carriers (positive space charge); solve Poisson's equation in one dimension $[d^2V/dx^2 = -\rho/\kappa_e\epsilon_0$, with ρ the charge density and $\kappa_e = 16$ for germanium] for $V(x)$ subject to the conditions of zero electric field outside of the transition region and for a potential barrier of 0.43 volt; the required width is $d_p + d_n$.

11-33. Show that the potential barrier of 0.43 volt is consistent with the concentrations of donors and acceptors given in prob. 11-32 and with thermal equilibrium at 300°K (no applied voltage).

11-34.* If an external reverse bias voltage much greater than 0.43 volt is applied to the junction of prob. 11-32, how does the transition (space-charge) region width vary with voltage? (That is, $d_p + d_n$ is proportional to what function of V?)

11-35. How can measurements of the capacitance C of a p-n junction be used to determine the width of the transition region? For the junction of prob. 11-32, how does C depend on the bias voltage?

11-36. Plot log I from eq. 11-10 vs. log V for positive V's from 0.01 to 0.3 volt. Plot log $(-I)$ vs. log $(-V)$ for negative V's from 0.01 to 10 volts on the same plot. Assume that $I_0 = 10^{-6}$ amp and $T = 300°$K.

11-37.* What is the ratio ρ of the forward current at 0.5 volt to the reverse current at -0.5 volt for an ideal p-n junction at $T = 300°$K? How does ρ depend on T near 300°K?

11-38.* The low-frequency "resistance" of a non-linear circuit element is defined by writing $r = dV/dI$ and by neglecting any capacitance or inductance. Apply this to eq. 11-10, and obtain r in terms of I for a p-n junction that is forward-biased by a voltage greater than $4kT/e$ [make the approximation to $I(V)$ that is valid under these conditions]. Express r in ohms in terms of the current I in milliamperes for $T = 300°$K.

11-39.* A p-n junction diode is constructed with conductivities of 500 mhos/m. (n side) and 2000 mhos/m. (p-side). It is 2 mm long and $\frac{1}{2} \times 1$ mm in cross section; the junction is at the center. The observed I_0 is 1 microampere and T is 300°K. Compute V from eq. 11-10 for currents of 1 and 10 milliamperes. Correct this V for the IR drops across the p and n halves of the unit for each current.

11-40.* Calculate the diffusion constant D_n for electrons in germanium at room temperature (300°K).

11-41.* Evaluate I_n for the following germanium p-n junction: The junction area is 10^{-6} m.^{2}, the diffusion length of electrons on the p side is 10^{-4} m., the conductivity of the p-type material is 2000 mhos/m., and the temperature is 300°K.

11-42. Make a sketch like Fig. 11-13a, except with $I_p = \frac{1}{2}I_{n0}$ (instead of $I_p = 0$, as in that figure) and with $L_p = L_n$.

11-43. Suppose that some holes were injected into the semiconductor at the contact between the metal lead wire and the n side of a p-n junction. How and why would this process adversely affect the useful characteristics of the device as a rectifier if L_p is about the same as the length of the n-side?

11-44.* Find an expression for the ratio of hole to electron currents across a p-n junction in terms of conductivities and diffusion lengths.

11-45. The purpose of this problem is to verify the statement below eq. 11-13 that diffusion, rather than conduction under the influence of the electric field, is the important process causing the motion of carriers at a p-n junction. Let $N_p = 10^{23}$ m.$^{-3}$ on the p-side of a germanium p-n junction at 300°K. (a) What is $N_n^{(eq)}$? (b) Suppose that the diffusion length for electrons on the p-side is $L_n = 10^{-4}$ m. and suppose that by forward biasing $N_n^{(p)}$ becomes $10N_n^{(eq)}$. Calculate the current density of electrons by diffusion by applying eq. 10-1 at $x = x_0$, Fig. 11-12, with $D = 0.010$ m.^{2}/sec. (c) Estimate an upper limit to the current of electrons by conduction as follows: First place an upper limit on $\mathcal{E}$ by setting the answer to (b) equal to $N_p e \mu_p \mathcal{E}$, i.e., by assuming that hole conduction carries the entire current even at the junction. Then calculate the electron conduction current density, which is $10N_n e \mu_n \mathcal{E}$. ($d$) Compare the answer to (c) with the electron diffusion current from (b).

11-46. It can be shown that if both diffusion and recombination with a lifetime τ are significant processes, and if N_n does not depend on y or z, the differential equation for the variation of $N_n(x,t)$ is

$$\partial N_n/\partial t = D(\partial^2 N_n/\partial x^2) - (N_n - N_n^{(eq)})/\tau$$

(a) Suppose that a non-equilibrium concentration of electrons N_n is created that is initially and remains independent of x. Solve for $N_n(t)$ and note how τ enters in your answer. (b) Suppose that a steady-state occurs (N_n independent of time t) with a concentration N_{n0} at $x = x_0$ and (of course) a concentration $N_n^{(eq)}$ at very large x; find $N_n(x)$ for $x > x_0$ and compare with eq. 11-12.

11-47. Suppose that the light to be detected by a p-n junction photodiode is focused to an extremely small spot with diameter much less than the length of the diode. How near the junction must the light spot be in order to produce nearly the maximum response?

11-48.* Sufficiently intense light shines on a reverse-biased p-n junction photodiode so that $I = 10^8 I_0$. What is the open-circuit photovoltage that this same light intensity would produce in the same photodiode? If this device is to be used as an energy (light to electrical) converter, neither the open-circuit connection ($R = \infty$) nor the short-circuit ($R = 0$) produces power. Sketch $I(V)$ with the light intensity described and estimate graphically the magnitude of the load resistor R such that maximum power is delivered to R.

11-49.* The mean value of the total solar radiation at normal incidence outside the earth's atmosphere is called the *solar constant* and is 2.0 calories/cm^2 minute.

Use the efficiency figure given in the text to calculate the area of a silicon solar cell necessary (*a*) to power a 100-watt lamp, and (*b*) to power a ten-transistor circuit with power levels in each stage typical of the center of Fig. 11-21*b*.

11-50. How does the tunneling current in a tunnel diode depend on *T*? How does the diffusion current depend on *T*? $I(V_3)$ in Fig. 11-18*a* increases sufficiently rapidly with *T* for the negative resistance region to vanish for a germanium diode at $\sim$200°C and for a silicon diode at $\sim$450°C. Why does the negative resistance disappear, and why do the absolute temperatures at which it vanishes in germanium and in silicon have the observed ratio?

11-51.* A typical current density in a tunnel diode at V_3 in Fig. 11-18*a* is 10^8 amp/m.^{2} If I_0 of the diffusion current of this diode is 10^4 amp/m.^{2}, estimate the voltage V_3.

11-52.* Estimate the concentration of donors and acceptors (assume N_d on the *n*-side equals N_a on the *p*-side) in a germanium tunnel diode that is required to produce a transition region 100 Å in width. *Hint:* Use the approach of prob. 11-32; set the barrier height equal to the band gap.

11-53. Estimate the concentration of donors in germanium that will give a strong overlapping of wave functions of electrons on donors and hence an impurity band suitable for a tunnel diode. To do this, use hydrogenic 1*s* wave functions, set $\kappa_e = 16$, and replace *m* in eq. 6-5 by *m*/5, the effective mass. Find r_0 such that $\psi(r_0) = (1/e)\psi(0)$, and then find the volume of a cube $2r_0$ on a side. How many such cubes are there per cubic meter? Compare this number with the concentration calculated in the preceding problem.

11-54.* Estimate the capacitance of a germanium tunnel diode with a transition region width equal to 100 Å and a junction diameter of 0.002 inch. Assume (a fair approximation) that the electric field in the transition region is constant, as in a plane-parallel condenser.

11-55. Consider a *p-n* junction that makes a useful tunnel diode. Compare it with an ordinary, diffusion diode in the following respects: (*a*) Doping concentration. (*b*) Width of transition region. (*c*) Majority carrier density. (*d*) Nature of the $I(V)$ curve for reverse bias. (*e*) Mechanism of charge transport across the junction. (*f*) Type of statistical distribution of electrons in the conduction band on the *n*-side and of holes on the *p*-side.

11-56.* Estimate from w^2/D_n the dispersion in transit times for electrons at room temperature in an *n-p-n* transistor with a base width of 2×10^{-5} m.

11-57. Compare quantitatively the shape of I_{c0} as a function of V_c in Fig. 11-21*b* with that predicted for an ideal *p-n* junction.

11-58. Plot N_n vs. *x* through the base region of an *n-p-n* transistor for two values of the collector voltage, and hence two values of the effective base thickness, and explain the internal feedback part of r_b in terms of your plot (which need be only schematic). This feedback is called the *Early effect*.

11-59. Show that the capacitance *C* described as item 3 of the list of four frequency limitations on an *n-p-n* transistor is proportional to the product of the emitter current *I* and w^2/D_n. To do this, plot N_n vs. distance *x* from $x = 0$

at the beginning (emitter side) of the base to $x = w$ at the collector side of the base. N_n has a value determined by the emitter-base voltage V at $x = 0$, $N_n = 0$ at $x = w$, and $N_n(x)$ is a straight line between these extremes. Next plot another straight line starting at $N_n + dN_n$ at $x = 0$ and ending at $N_n = 0$ at $x = w$. The area between these lines is dQ/e, where dQ is the charge that must flow into the base in order to increase N_n and hence to change V. The upper curve gives a larger current $(I + dI)$ by eq. 10-1. Calculate dQ in terms of dN_n from the geometry of the plot, dN_n in terms of dI from eq. 10-1, and dI/dV expressed in terms of I from eq. 11-10 (dropping the "-1," since the junction is forward biased, as in prob. 11-38). Combine these to get $C = dQ/dV$.

11-60.* Calculate the current and voltage gains of the circuit of Fig. 11-22 by the approximation method and with the values of r_L and of the transistor parameters used in the calculation of the power gain.

11-61. In the *common base* (or *grounded base*) circuit for an *n-p-n* transistor the input is applied between emitter and base and the output is extracted between collector and base. Draw the circuit with the correct biases. Apply the same approximation used in the text $[r_L \ll (1 - a)r_c]$ to compute the general expression for the power gain in terms of a, r_L, r_e, and r_b. Calculate the gain for the values of these parameters used in the text.

11-62.* In the *common collector* (or *grounded collector*) circuit for an *n-p-n* transistor the input is applied between base and collector and the output is extracted between emitter and collector. Perform the operations requested in problem 11-61 but for the common collector circuit. Assume that the generator resistance is small compared to the collector resistance.

11-63. In order to decrease the transit-time dispersion in the base and hence to enhance the high-frequency performance, transistors are constructed with a built-in electric field in the base region directed so as to force carriers from emitter to collector. Show that a gradient of doping concentration will produce such a built-in field; use analysis like that of Sec. 11-3 to locate the band edge relative to the Fermi level, which is constant. Show that the diffusion of gallium from the emitter side during preparation of the mesa transistor of Fig. 11-19*b*, which gives a higher gallium concentration at the emitter side of the base, produces an electric field of the correct direction to hasten electrons through the base.

11-64.* A thin bar of *n*-type germanium about 0.02 m. long has soldered leads on its ends. The right end is made 20 volts negative with respect to the left end. Small indium alloy *p-n* junctions are created on one side of the bar and 0.005 m. from each end (and therefore 0.01 m. apart). The left junction is pulsed positively for 1 microsecond and hence injects a 1-microsecond pulse of holes into the bar. The right junction (collector) is biased negatively and collects holes; the injected pulse of holes arrives at the collector a time t after it was injected. Compute t. Why is t different for this problem from the pulse transmission time that would be observed with a metal or with majority carriers in a semiconductor (see prob. 9-45)? (This experiment provides a valuable way of measuring μ and the rate of recombination.)

11-65. Give a diagram and analysis similar to Fig. 11-23 and associated discussion but for a p-type semiconductor in contact with a metal.

11-66. In Fig. 11-23 it was assumed that the hole flow was negligible. Assume that that figure is drawn to scale and estimate the ratio of hole current to electron current.

11-67. The sign of the Seebeck voltage V_{ab} is the same as the sign of the Peltier heat π_{ab}. Explain how a heated probe wire touched to a semiconductor can be used to tell whether the specimen of semiconductor is n- or p-type.

11-68.* In Bi_2Te_3, $E_g = 0.21$ e.V. and $\mu_n = 0.11$ m.^{2}/volt sec. A representative σ for n-type Bi_2Te_3 for thermoelectric applications is 10^5 mhos/m. For such a specimen at $T = 300°K$, calculate $E_g - E_0$ and π for a contact with a metal. (Let $m^* = m$.)

12

Physical Electronics

12-1 INTRODUCTION

Physical electronics includes three different subjects: (1) The emission and absorption of electrons at the surfaces of solids, which will be considered in Secs. 12-2 to 12-7. (2) The collisions of electrons with atoms or ions in a gas, which will be considered in Sec. 12-8. (3) The trajectories of electrons and ions in electric and magnetic fields, which will be considered in Sec. 12-9. These rather different areas of physics are treated together because a mastery of all three is required to understand physical experiments or practical devices involving electron or ion beams. Now that the quantum physics has been developed in Chapters 5 to 10, we can return in this chapter to some of the phenomena discussed in an introductory manner in Chapter 4 and give a more thorough explanation of the basic physics involved.

The study of the quantum physics of solids in Chapters 8 to 10 provided the basis on which to discuss the emission of electrons from surfaces. The study of the quantum physics of atoms and molecules in Chapters 5 to 7 provided the basis on which to discuss collision processes. The physics of electron and ion trajectories is essentially part of classical physics, but interesting limitations are imposed on the classical theory by quantum physics.

434

12-2 THERMIONIC EMISSION

The emission of electrons by a hot metal or semiconductor is called thermionic emission. It is the principal practical source of electrons in commercial electron tubes and in laboratory experiments. The basic physics of thermionic emission is rather simple. An energy barrier a few electron volts in height at the surface of a solid prevents the emission of most of the electrons in the solid. At any temperature, however, *some* of the electrons have enough energy to surmount this barrier. The current of such electrons is a very sharply increasing function of T. The current from any known solid at room temperature is too small to be practically useful. Therefore practical emitters are always heated, usually to temperatures between 1000° and 2500°K.

In this section we shall first examine the nature of the surface-energy barrier. Then we shall calculate the thermionic-emission current from a metal and examine the effect of an applied electric field on the emission. Finally we shall investigate thermionic emission from semiconductors.

The potential energy of an electron in a metal as a function of distance along a line of atom centers is illustrated in Fig. 12-1. This plot is like Fig. 8-9 except that the present sketch includes the surface, and therefore the regular array is not repeated to the right of the vertical line. There is no sharp definition of the exact position of the surface. Since there are no nuclei to the right of the vertical line, the potential energy curve does not turn downward at the right but approaches a horizontal asymptote, which represents the potential energy of an electron outside the metal.

The conduction band electron energies are also sketched on Fig. 12-1. The allowed energy band extends upward indefinitely, but at 0°K all the energy states with $E > E_0$ are empty, and at *any* temperature *most*

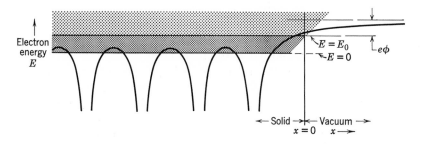

Fig. 12-1. Schematic diagram of the surface of a metal. An electron with energy $E_0 + e\phi$ inside the solid would have zero kinetic energy after emission.

of the states with $E > E_0$ are empty. It should be recalled that the kinetic energy of translation equals zero at the bottom of the conduction band, the point marked $E = 0$ in Fig. 12-1. The energy that must be given to an electron that initially had the energy E_0 in order to remove it from the solid is $e\phi$. This statement defines the work function $e\phi$ and is a refinement of the definition given in Sec. 4-2.

We next calculate the thermionic emission current density j from a metal at a temperature $T°\mathrm{K}$. The electrons emitted are the electrons with sufficient momentum normal to the surface to overcome the surface barrier. A necessary, but not sufficient, condition for emission of an electron is that its energy must be greater than $E_0 + e\phi$. This condition is not sufficient since the electron's momentum may not be directed toward the surface. If the momentum is directed away from the surface, of course the electron will not be emitted. If the momentum is directed generally toward the surface, but not perpendicular to the surface, the electron will be turned back from the surface unless it has an energy considerably greater than $E_0 + e\phi$. This situation is illustrated in Fig. 12-2. The surface force is in the $-x$-direction and reduces the x component of momentum (p_x) of electrons striking the surface. If p_x is initially less than a critical value p_{xc}, the surface force reflects the electron back into the metal, regardless of the value of p_y. The energy $p_y{}^2/2m$ associated with the y component of momentum does not help the electron to surmount the barrier.

The critical value p_{xc} can be calculated by noting that the energy barrier has a height $E_0 + e\phi$, the difference in energy between an electron with zero kinetic energy inside the metal and an electron with zero kinetic energy outside the metal. Therefore the electron in order to escape must have at least the following value of the energy associated with motion in the x direction: $p_{xc}{}^2/2m = \frac{1}{2}mv_x{}^2 = E_0 + e\phi$. In other words,

$$p_{xc} = \sqrt{2m(E_0 + e\phi)} \tag{12-1}$$

Of course p_x must be positive.

The emission current per unit area is the product of the charge e and the number of electrons emitted per second per square meter. The number of electrons emitted is the product of v_x and the number per unit volume with $p_x > p_{xc}$. (This calculation of a *flux* or flow of particles in terms of v_x and the number per unit volume was explained in conjunction with Fig. 2-1; in the present calculation the factor of $\frac{1}{2}$ does not appear, since if $p_x > p_{xc}$ then p_x must be positive, and hence we do not include in the number per unit volume the electrons going toward $-x$.) Therefore

$$j = ev_x \times (\text{Number/m.}^3 \text{ with } p_x > p_{xc}) \tag{12-2}$$

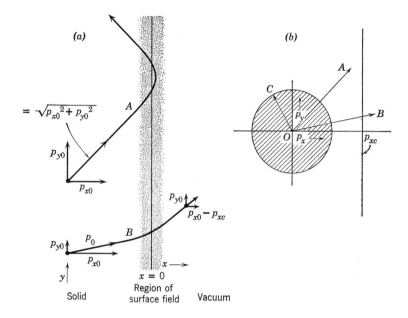

Fig. 12-2. (*a*) Two electron trajectories with the same initial energy ($> E_0 + e\phi$) inside the solid and therefore the same initial magnitude of momentum p_0; *A* is reflected by the surface field, and *B* is emitted. (*b*) The same situation as in (*a*) but plotted in *momentum space* instead of coordinate space. The vector terminating at *C* represents an electron with the energy E_0. The vectors ending at *A* and *B* represent the reflected and emitted electrons, respectively, of (*a*).

We calculate the required density of electrons, as in Sec. 9-4, by multiplying the density of states $S(E)$ by the probability of occupation $f(E)$. Because of the condition on p_x we seek this density in terms of the *momenta*, rather than in terms of the *energy* as in Sec. 9-4. The number of traveling-wave states per cubic meter with p_x between p_x and $p_x + dp_x$, and similarly for p_y and p_z, is

$$(2/h^3)\, dp_x\, dp_y\, dp_z \qquad (12\text{-}3)$$

from eq. G-13 of Appendix G. The fraction of these that are occupied is given by the Fermi function (eq. 9-4). But our calculation need be valid only for $E \gg E_0$, since the work function $e\phi$ is of the order of 1 volt or greater. Therefore we can use the approximate form of the Fermi function, eq. 9-2:

$$f(E) = \mathrm{e}^{-(E-E_0)/kT} = \mathrm{e}^{-(p_x{}^2+p_y{}^2+p_z{}^2-2mE_0)/2mkT} \qquad (12\text{-}4)$$

We can now insert eqs. 12-3 and 12-4 into eq. 12-2 and use the fact that $mv_x = p_x$:

$$j = \frac{2e}{mh^3} \int_{p_y=-\infty}^{\infty} \int_{p_z=-\infty}^{\infty} \int_{p_x=p_{xc}}^{\infty} p_x \, dp_x \, dp_y \, dp_z \, e^{-\left(\frac{p_x^2+p_y^2+p_z^2-2mE_0}{2mkT}\right)}$$

$$= \frac{2e}{mh^3} \int_{-\infty}^{\infty} e^{-\frac{p_y^2}{2mkT}} dp_y \int_{-\infty}^{\infty} e^{-\frac{p_z^2}{2mkT}} dp_z \int_{p_{xc}}^{\infty} e^{-\left(\frac{p_x^2-2mE_0}{2mkT}\right)} p_x \, dp_x \quad (12\text{-}5)$$

The first two integrals can be evaluated by using the fact that $\int_{-\infty}^{\infty} e^{-ax^2} = (\pi/a)^{1/2}$, and each integral has the value $(2\pi mkT)^{1/2}$. The third integral can be evaluated by inserting p_{xc} from eq. 12-1 and substituting u for the function in the exponent:

$$\int_{\sqrt{2m(E_0+e\phi)}}^{\infty} e^{-\left(\frac{p_x^2-2mE_0}{2mkT}\right)} p_x \, dp_x = mkT \int_{e\phi/kT}^{\infty} e^{-u} \, du = mkTe^{-e\phi/kT}$$

If we put these evaluations into eq. 12-5 we obtain

$$j = \left(\frac{4\pi mek^2}{h^3}\right) T^2 e^{-e\phi/kT} = A_0 T^2 e^{-e\phi/kT} \quad \text{amp/m.}^2 \quad (12\text{-}6)$$

This is the Richardson-Dushman thermionic-emission equation. A_0, a universal constant, equals 1.20×10^6 amp/m.^{2} deg^2. We shall defer comparison of this equation with experiment until after the discussion of a modification of it.

The effect on j of the application of an electric field to the surface of the metal will now be considered. In order to do this it is necessary to look more closely at the nature of the force on the emitted electron as a function of its distance x from the surface of the metal. It is fortunate that we need to know the force only some distance away from the surface $(x = 0)$, since near $x = 0$ the situation is complicated by the crystal structure and the atomic electron distributions (see prob. 12-7). The force between the electron and the metal can be computed from ordinary electrostatics when an electron is many lattice constants away from the surface. Under these conditions, the metal surface in Fig. 12-1 can be assumed to be plane and continuous, as shown in Fig. 12-3a.

All the electric field lines must intersect the surface at right angles, since the metal is a conductor. Evidently it would be tedious to calculate the positive charge on the surface as a function of position and then to calculate the force exerted on the electron by this charge distribution. There is a short cut in this electrostatic problem that is based upon the *method of images*. The field to the right of $x = 0$ is identical in Figs.

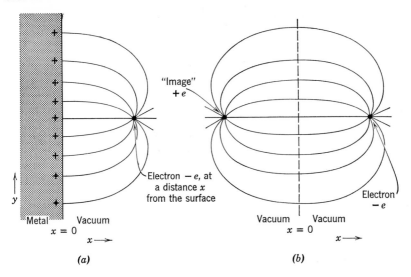

Fig. 12-3. (*a*) Electric field lines for an electron near the surface of a metal. (*b*) Electric field lines for an electron and a charge $+e$ at equal distances either side of $x = 0$; the field for $x > 0$ is identical with the field in (*a*), since the field lines in (*a*) are perpendicular to the metal surface.

12-3*a* and 12-3*b*. This means that the force on the electron at a distance x from the surface is the same as if the metal surface were replaced by a charge $+e$ at $-x$. The force on the electron is therefore $-e^2/(16\pi\epsilon_0 x^2)$, since the distance from the electron to its *image* is $2x$.

The potential energy associated with this force is $-e^2/(16\pi\epsilon_0 x)$, computed on the assumption that $P = 0$ at $x = \infty$. If there is no external electric field applied to the emitting surface, the only contribution to P is the image force contribution, and this term is illustrated in Fig. 12-4*a*. If an accelerating field $\mathcal{E}$ is applied, the situation illustrated in Fig. 12-4*b* prevails. There is now an additional contribution $-e\mathcal{E}x$ to the potential energy of an electron. The electron needs somewhat less ($e\,\Delta\phi$) than an energy $E_0 + e\phi$ in order to be emitted. Therefore more electrons will surmount the barrier, and the resulting increase in j is called the *Schottky effect*.

In order to calculate this increase we first calculate x_0, the position of the maximum of $P(x)$ in Fig. 12-4*b*. Since the total potential energy of the electron is $P = (-e^2/16\pi\epsilon_0 x) - \mathcal{E}ex$, we can locate the position x_0 of the maximum by setting $dP/dx = 0$:

$$\frac{e^2}{16\pi\epsilon_0 x_0{}^2} - \mathcal{E}e = 0 \qquad\qquad x_0 = \left(\frac{e}{16\pi\epsilon_0\mathcal{E}}\right)^{\!\frac{1}{2}} \qquad (12\text{-}7)$$

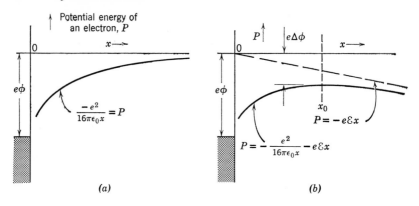

Fig. 12-4. Potential energy P of an electron as a function of its distance x from the surface of a metal. (a) No external field. (b) External field accelerating electrons toward $+x$; a lowering $e\,\Delta\phi$ of the work function has been produced ($\mathcal{E}$ and $\Delta\phi$ are greatly exaggerated).

We can now calculate $e\,\Delta\phi$, since this is the value of P at $x = x_0$:

$$e\,\Delta\phi = \frac{-2e^{3/2}\mathcal{E}^{1/2}}{(16\pi\epsilon_0)^{1/2}} \qquad \Delta\phi = -\left(\frac{e\mathcal{E}}{4\pi\epsilon_0}\right)^{1/2} \qquad (12\text{-}8)$$

This $e\,\Delta\phi$ is the amount by which the work function $e\phi$ is lowered when an accelerating field $\mathcal{E}$ is applied. The thermionic emission equation including the effect of the applied electric field is therefore

$$j = A_0 T^2 e^{-\frac{e}{kT}\left[\phi - \left(\frac{e\mathcal{E}}{4\pi\epsilon_0}\right)^{1/2}\right]} \qquad (12\text{-}9)$$

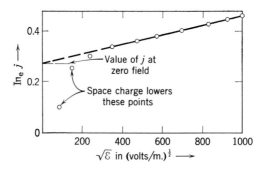

Fig. 12-5. Schottky effect with a tungsten filament at 2100°K. The current at zero field is about 0.8 of the current at 10^6 volts per meter. The units of j and the vertical position of the line (the "Schottky line") are arbitrary.

At a constant temperature, a plot of $\ln_e j$ vs. $\sqrt{\mathcal{E}}$ should be a straight line. This result agrees well with experiment, as illustrated in Fig. 12-5, which is a plot of the logarithm of the cathode current density in a diode as a function of the square root of the electric field at the cathode (proportional to the square root of the anode voltage). The current at very low anode voltages is limited by space charge; much of the emitted current is turned back at the potential minimum (potential energy maximum) outside the cathode created by the space charge. For higher values of $\mathcal{E}$, the Schottky-effect equation (eq. 12-9) is well verified. This agreement enables us to extrapolate the Schottky line (Fig. 12-5) back to $\mathcal{E} = 0$ in order to obtain the *zero field* emission current density, which is the j predicted by eq. 12-6.

Comparison of eq. 12-6 with experiment is shown in Fig. 12-6. The natural logarithm of the zero-field j divided by T^2 is plotted as a function of $10{,}000/T$, and a straight line is obtained. From the slope and

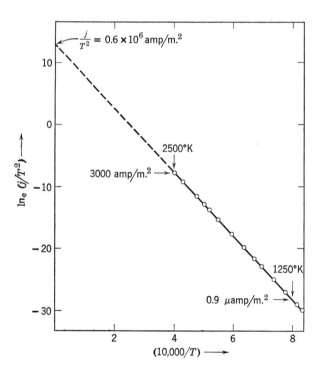

Fig. 12-6. Thermionic emission from tungsten. The ordinate is the logarithm to the base e of the current density (amp/m.²) divided by T^2. The A factor can be determined from the $(10{,}000/T) = 0$ intercept, and the work function can be determined from the slope.

TABLE 12-1

Thermionic Emission Constants

Metal	ϕ, volts	A, amp/m.^{2}	Metal	ϕ, volts	A, amp/m.^{2}
Cr	4.60	0.48×10^6	Pt	5.32	0.32×10^6
Fe	4.48	0.26	Ta	4.19	0.55
Mo	4.20	0.55	W	4.52	0.60
Ni	4.61	0.30	LaB$_6$	2.66	0.29

From C. Herring and M. H. Nichols, *Revs. Mod. Phys.*, **21**, 185–270 (1949), and J. M. Lafferty, *J. Appl. Phys.*, **22**, 299 (1951).

$(10,000/T) = 0$ intercept of such a line we can obtain the work function $e\phi$ and *A-factor* (the experimental value of the quantity A_0 in eq. 12-6) for various materials. Some experimental values are tabulated in Table 12-1. It is very difficult to calculate ϕ values from the quantum theory of metals: it has been done only for the alkali metals. These metals melt at too low temperatures to permit thermionic determinations of ϕ, but the theory is in good agreement with photoelectric determinations of ϕ.

The A values in Table 12-1 can be compared with the theoretical value of 1.20×10^6 amp/m.^{2} The experimental determinations are not accurate because of the large extrapolation of data required (see Fig. 12-6 and prob. 12-6), but the disagreements are larger than the experimental error. There are two principal causes for this disagreement:

1. We have tacitly assumed in analyzing the data of Fig. 12-6 that ϕ is a constant independent of temperature. ϕ should not be exactly constant, since the expansion of the solid with temperature and other, more minor effects change ϕ slowly with T. Because this variation is small compared to ϕ itself, it is possible to use a Maclaurin expansion and to write $\phi = \phi_0 + T(d\phi/dT)_{T=0}$ to a good approximation. If this ϕ is substituted into eq. 12-6, we obtain

$$ j = [A_0 e^{-\frac{e}{k}\left(\frac{d\phi}{dT}\right)_{T=0}}]T^2 e^{-\frac{e\phi_0}{kT}} \tag{12-10} $$

The quantity in brackets is the predicted A factor and can be considerably different from A_0.

2. We have tacitly assumed a perfectly uniform surface of the emitter with the same ϕ at all positions, but actually ϕ is different for different crystal faces of the emitter, and the experiments are performed on poly-

crystalline wires. Suppose that half the emitter area has a work function ϕ_1 and half has ϕ_2, where ϕ_2 is considerably greater than ϕ_1. Practically all the emission is from the patches with the lower work function $e\phi_1$. The emitting area is therefore only half the nominal cathode area, and the measured A will be one-half the theoretical A_0 for a uniform emitter. This *patch effect* also complicates the Schottky effect, since there are strong local electric fields between adjacent patches.

Both these effects are probably important in the measurements summarized in Table 12-1. Another possible effect is probably not of much practical importance but deserves brief mention. This is the partial reflection of the electron waves as they emerge from the surface. The reflection effect was described in Sec. 5-4d; it can be serious for a sharp step in the potential energy of an electron. But the reflection coefficient r is far less than unity if the potential energy changes slowly with distance, as required by the image force theory. The factor $(1 - r)$ really should multiply the right side of eqs. 12-6, 12-9, and 12-10, but is very nearly equal to unity. There are nevertheless small but interesting interference effects between the partial reflections at $x = 0$ and at $x = x_0$ that produce a periodic deviation of experimental data from the Schottky line.

Pure tungsten is the only one of the three kinds of practical thermionic emitters to which the theory just presented is directly applicable. Tungsten is a useful emitter because it has the highest melting point and lowest vaporization rate at high temperatures of any metal. A tungsten filament is ordinarily operated at a temperature of about 2500°K. A higher temperature would give greater emission current density (j), and would even give a greater j per watt of filament heating power, but higher temperatures would reduce the life of the filament because of evaporation. At $T = 2500°K$ the thermionic emission current density from tungsten is about 3000 amp/m.^{2} Because of the low efficiency (low j per watt of heating power), pure tungsten emitters are employed only where other emitters cannot be used. Their application is therefore confined to power tubes (transmitting tubes) and X-ray tubes, where the anode voltage is so high that energetic positive ions can be produced that damage other kinds of emitters.

The second kind of practical emitter is the thoriated-tungsten emitter, which is a filament of tungsten containing a few per cent of ThO_2. The filament must be *activated* in a vacuum by the following treatment: First it is heated to 2800°K for a few minutes to clean the surface by evaporation and to reduce some ThO_2 to Th. Then it is cooled to about 2100°K. At this temperature more thorium diffuses to the surface than evaporates, and thus a thorium coating is created. This coating is

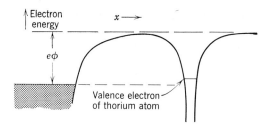

Fig. 12-7. A thorium atom near a tungsten surface. Transfer of the valence electron to the metal will lower the energy of the system and will lower the work function.

probably only one atom thick; that is, it is a *monatomic layer*. The filament is then ready to use and is operated at about 1900°K. The monatomic layer lowers the ϕ of the filament enough so that j is about 10,000 amp/m.^{2} at 1900°K. The emission efficiency is therefore much greater than that of pure tungsten, but the thoriated-tungsten emitter cannot be used in tubes operating at a power level greater than about 1000 watts. Bombardment by the high-energy positive ions present in such tubes would destroy the thorium layer.

The reason that the thorium layer lowers the work function can be understood by studying Fig. 12-7, which shows a neutral thorium atom near a tungsten surface. The work function is about the same as for clean tungsten even though the electron energy vs. x curve is modified by the presence of the atom. This neutral atom is unstable, since the ionization energy of thorium (about 4 e.V.) is less than the work function of tungsten (4.52 e.V.). The outermost electron of the thorium atom transfers to the metal, and this process produces a positively charged layer of thorium ions on the surface. A dipole layer is therefore created, with the positive side outward. Such a dipole layer lowers the work function by increasing the electrostatic potential, and therefore decreasing the potential energy of an electron, just outside the layer.

Another consequence of electron transfer is the binding of the thorium ion to the surface, since a lower energy state has been reached by transfer of the electron. The thorium is said to be strongly *adsorbed* on the surface. Similarly an oxygen atom is strongly adsorbed by accepting an electron into the electron affinity level of the oxygen atom; since the dipole layer in this case has the negative pole away from the surface, the work function is increased.

The third kind of practical emitter is the oxide-coated cathode, which consists of a nickel sleeve with a coating of tiny crystals of a mixture of barium, strontium, and calcium oxides. Since it provides current densities of the order of 5000 amp/m.^{2} at temperatures near 1000°K, its efficiency is much greater than that of tungsten emitters. The relatively low temperature of operation permits the cathode to be indirectly heated, which simplifies circuit applications because a common heater supply circuit can be shared by many tubes. Since the oxides are unstable in air, the nickel sleeve is originally coated with carbonates. During the exhaust of the tube the cathode is heated, and the carbonates decompose to oxides. The further increase of cathode temperature to about 1300°K is necessary in order to activate the cathode, a process that probably involves the creation of oxygen vacancies as described in the next paragraph.

The physics of the oxide-coated cathode is qualitatively the same as that of barium oxide, and therefore we shall discuss the process of thermionic emission from BaO. Although this process is not well understood, the principal physics is probably as follows: Reducing agents (such as silicon) in the nickel sleeve combine with some of the oxygen in the BaO coating, thereby creating oxygen vacancies that diffuse into the

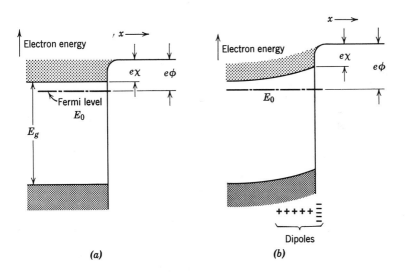

Fig. 12-8. Surface of a semiconductor. (*a*) The highly improbable situation of a neutral surface. (*b*) A negatively charged surface, with the positive charges necessary to make the solid neutral distributed over several hundred angstroms beneath the surface; the Fermi level is constant, since the system is in equilibrium, but an electric field exists near the surface and tilts the energy level of the bottom of the conduction band. $e\chi$ is defined as the energy that must be supplied to an electron initially at rest in the conduction band in order to remove it from the solid.

BaO. Each vacancy has two trapped electrons and serves as an electron donor. Thus the BaO is n-type with a Fermi level E_0 close to the bottom of the conduction band. The electron affinity energy $e\chi$ (the difference in energy of an electron when at rest outside the crystal and at rest in the conduction band inside, Fig. 12-8a) is only about 1.0 e.V., which value has been determined by photoelectric measurements. There is not yet any satisfactory way of calculating χ or even of showing why it should be lower for BaO than for some other crystals. The work function is about 1.1 to 1.5 e.V.

The simple semiconductor model of the preceding paragraph is inadequate to explain the experimental facts about oxide-coated cathodes, such as the fact that j changes rapidly with minute changes in the surface. One such change is the great reduction in j if a little water vapor is present in the residual gas in the vacuum tube containing the cathode. The effect of water vapor is probably to oxidize barium ions near the surface and thereby to produce a net negative charge on the surface (additional O^{--} ions are attached to Ba^{++} ions in the surface). A double layer of charge is therefore produced with the negative pole outward, and ϕ is accordingly increased. Much less than a monatomic layer is required to produce a significant change in ϕ because the length of the dipole is much greater than that produced by a charge layer on a metal. The positive charges on the semiconductor are not at the surface but are distributed over a distance of hundreds of angstroms into the crystal, as illustrated in Fig. 12-8b.* A significant change in ϕ occurs even if only one in every thousand barium atoms at the surface is oxidized. Thus, even though eq. 12-6 should apply to semiconductors as well as to metals, the change of ϕ with temperature and with surface conditions makes it difficult to predict j values for the oxide cathode.

In the practical application of thermionic emission, consideration must almost always be given to the limitation of the electron current by *space charge*. We shall consider such limitation only rather sketchily, since the subject is treated thoroughly in books on electricity and magnetism.

We shall consider as an example a plane-parallel vacuum diode, and

---

* The difference between metal and semiconductor is that great concentrations of charge at the surface of the metal can be produced by a slight change in the conduction electron distribution. No appreciable field can exist inside the metal because of its high conductivity. In the semiconductor, on the other hand, the only way to obtain a net charge in a given region is to ionize all the donors and remove all the conduction electrons. The charge density that can be produced in this way is small because of of the small donor density. Thus the surface dipole region must be thick in order to have one positive charge inside the crystal for each negative charge at the surface. (See prob. 11-32.)

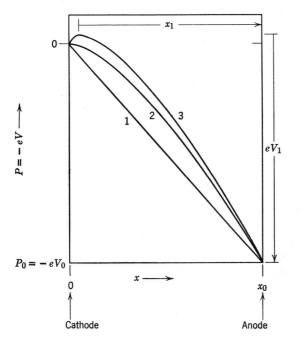

Fig. 12-9. The potential energy of an electron as a function of distance from a plane thermionic emitter (cathode) to a plane collector (anode) parallel to it and a distance x_0 away. The parameters x_1 and eV_1, measured from the potential energy maximum and therefore slightly different from x_0 and eV_0, are introduced for convenience in eq. 12-12 and prob. 12-18. Both the height and the distance from the cathode of the space-charge barrier in curve 3 are exaggerated if V_0 is more than $\sim$10 volts.

we illustrate the potential energy $P = -eV$ of an electron as a function of its distance from the cathode in such a tube in Fig. 12-9. Here we have taken a fixed (accelerating) potential on the anode (collector) and sketched $P(x)$ for three different cathode (emitter) temperatures. Curve 1 corresponds to a low cathode temperature T_c and a tiny current density j; the density of electrons on their way from cathode to anode is so small that there is a negligible space charge, and the electrostatic potential is essentially the linear function of x that it would be in charge-free space. As T_c and therefore the current increase, the space-charge density becomes appreciable. From eq. 12-2 it is $\rho = -j/v$, where v (the velocity) is $(2K/m)^{1/2}$. The electrostatic potential now must be the solution of the one-dimensional Poisson equation

$$d^2V/dx^2 = -\rho/\epsilon_0 \tag{12-11}$$

A curvature thus develops in $V(x)$, and the field at the cathode decreases.

Eventually, with continuing increase of T_c, the depression of V by the negative space charge produces a zero value of the field $\mathcal{E} = -dV/dx$ at the cathode (curve 2 of Fig. 12-9). Further increase in T_c creates a barrier near the cathode (the *potential minimum* or potential energy maximum), which permits only the most energetic electrons to pass (curve 3). Thus the increase of T_c from curve 2 to curve 3 produces a relatively small change in current to the anode, and we say that the electron emission is *space-charge-limited*. In this regime, change in anode voltage produces a large change in j, but change in T_c does not affect j sensitively. Increase in T_c permits increased thermionic emission, but more of the emitted electrons are turned back to the cathode by the increased barrier height. This contrasts with the emission-limited regime, in which a change in anode voltage produces relatively little effect on j and a change in T_c produces a large effect.

It is not difficult to show (prob. 12-18) that the space-charge-limited current is approximately

$$ j = \left(\frac{2^{5/2} \epsilon_0 e^{1/2}}{9 m^{1/2}} \right) \left(\frac{V_1^{3/2}}{x_1^2} \right) \tag{12-12} $$

where V_1 and x_1 are the anode voltage and the anode distance, both measured from the potential energy maximum.* This limitation of current is applied in the vacuum triode by installing a negative grid near the cathode to vary the height of the potential energy barrier and thus to control the current to the anode.

12-3 CONTACT POTENTIAL DIFFERENCE

One corollary of the thermionic-emission theory will be described next. This is the potential difference that exists across the space between two metals of different work functions when the metals are electrically connected, and it is called the *contact potential difference*. The way it originates is illustrated in Fig. 12-10. In Fig. 12-10a two metals with different ϕ's are illustrated. If they are brought into electrical contact at constant temperature, the Fermi levels E_0 must line up. That is, there must be no energy lost or gained in transferring an electron from a state at the Fermi brim in metal M_1 to a state at the Fermi brim in metal M_2. We can see the reason for this by first assuming that it is not true (as in

---

* In the usual electron tube, the potential energy maximum is so small (relative to V_1) and so close to the cathode (relative to x_1) that V_1 is very nearly the anode-cathode potential difference and x_1 is very nearly the anode-cathode spacing.

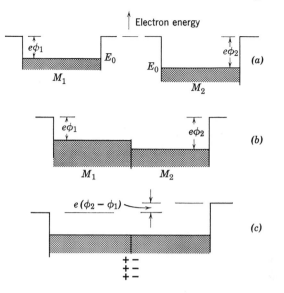

Fig. 12-10. Contact potential difference between two metals with different work functions. (*a*) Isolated metals. (*b*) Situation that would occur on contact if no contact potential difference occurred; this situation could not endure longer than $\sim 10^{-16}$ sec since electrons quickly go from M_1 to M_2. (*c*) Actual situation of metals in contact; the dipole layer at the contact has arisen from electron transfer, and there is an electric field between the external surfaces.

Fig. 12-10*b*). More electrons would go from M_1 to M_2 than would be thermionically emitted from M_2 to M_1, since the energy difference $e(\phi_2 - \phi_1)$ discriminates in favor of the first process. Thus M_1 will rapidly be charged positively with respect to M_2, and this charging will cease only when the difference in Fermi levels has been reduced to zero by the charge double layer at the junction between the two metals. The equilibrium situation is illustrated in Fig. 12-10*c*. (The alignment of the Fermi levels has previously been encountered in Sec. 9-3 and in Sec. 11-4.)

The potential energy of an electron just outside M_2 is greater than that just outside M_1 by an amount $e(\phi_2 - \phi_1)$, in which $(\phi_2 - \phi_1)$ is the contact potential difference. Its effects are important principally when different metals are used in a vacuum tube (e.g., the photoelectric experiment tube illustrated in Fig. 4-2). It can readily be seen that the contact potential difference is determined uniquely by the work functions of the metals in the vacuum. No change is made, for example, if the external circuit is composed of many different materials in series.

(This statement is true, of course, only if all the junctions are at the same temperature, since otherwise thermal emf's occur.) If a battery is included in the circuit, the potential difference produced by the battery adds algebraically to the contact potential difference.

Great caution must be taken in predicting the contact potential difference in a vacuum tube since the work functions of solids depend so sensitively on surface contamination. For example, the work function of clean nickel is 4.61 volts and that of a particular oxide-coated cathode is 1.2 volts. But the contact potential difference between a nickel grid and this cathode will be much less than 3.4 volts, because the grid is coated with a layer of BaO evaporated from the cathode. Since the BaO molecules are bound to the nickel with the Ba^{++} ions outward, a dipole layer is formed that reduces the ϕ far below the clean-surface value.

Contact potential difference is of greatest interest in analyzing the results of physical electronics experiments, like the photoelectric experiment. It is of practical importance in the design and application of receiving-type electron tubes.

12-4 THERMIONIC ENERGY CONVERSION

If two electrodes with different work functions and at *different temperatures* face each other, the thermionic emission of electrons can result

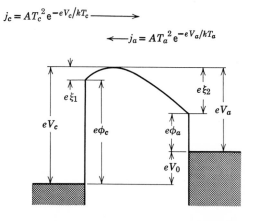

$$j_c = AT_c^2 e^{-eV_c/kT_c} \longrightarrow$$

$$\longleftarrow j_a = AT_a^2 e^{-eV_a/kT_a}$$

Fig. 12-11. Potential energy of an electron as a function of distance in a thermionic energy converter. The cathode at the left is at a temperature T_c much greater than the temperature T_a of the anode. The curved line is the potential energy of an electron; its shape is determined by space charge. The difference in Fermi levels in the two electrodes determines the output voltage V_0.

in the conversion of some of the heat required to maintain the temperature difference into electrical energy. Figure 12-11 is a schematic diagram
of the electron potential energy as a function of distance in one such
thermionic energy converter, in this case two plane-parallel, closely
spaced electrodes in a vacuum tube. The source of heat is applied to the
cathode,* which is thereby maintained at a high enough temperature T_c
to provide appreciable thermionic emission. A very closely spaced
anode at a much lower temperature T_a has a *smaller* work function $e\phi_a$
than the cathode work function $e\phi_c$.

Since the temperature is not the same throughout this tube, we cannot
use the arguments of Secs. 11-4 or 12-3 to assert that the Fermi levels
are aligned, and in general they are not. If the levels were aligned by
connecting the cathode and the anode together through a load resistance
$R = 0$ (i.e., by short-circuiting the device), there would be a net flow
of electrons from cathode to anode, since with $T_c > T_a$ more electrons
can surmount the barrier in the electrode at the higher temperature. If,
on the other hand, R is made infinite (i.e., the device is open-circuited),
the anode will charge negatively until the current density j_a is equal in
magnitude and opposite in direction to j_c. Neither of these choices of R
delivers useful power to the load, of course, and hence there is an optimum value of the load resistance. Figure 12-11 is drawn for a finite
R, appreciable current density $j_c - j_a$, and appreciable output voltage
V_0.

In such a vacuum device, close spacing is vital. If the spacing becomes
as large as a few thousandths of an inch, the space-charge barrier $e\xi_1$
becomes so large that the current density is small, and ξ_2 (the accelerating
voltage necessary to draw the current to the anode) becomes so large that
V_0 is seriously reduced. With such close spacing, transfer of atoms from
one electrode to another can quickly modify artificially prepared surfaces and change their work functions. Thus the two electrodes must
be chosen so that each has a ϕ of the desired value that is stable in the
presence of the other. One way of satisfying this requirement is to provide an identical cathode and anode, each a metal such as platinum
or tungsten coated with alkaline earth oxides. The work function of an
oxide-coated platinum electrode, for example, is 0.5 e.V. higher at 1000°K
than at room temperature, and thus the temperature dependence of ϕ
(rather than the use of different materials) is exploited to make $\phi_a < \phi_c$.

A somewhat different kind of thermionic energy converter uses two
tungsten electrodes in a tube containing cesium vapor. Cesium has an

---

* This electrode is conventionally called the cathode, even though it turns out to
be positive in this device, because it is the principal thermionic emitter. It should
more properly be called the hot electrode.

ionization energy that is smaller than the work function of tungsten, and thus (cf. Fig. 12-7) cesium atoms striking a tungsten surface lose their electrons; if the surface is too hot to retain the cesium, it leaves as ions rather than as neutral atoms. The presence of these ions in the space between the electrodes reduces or completely neutralizes the electronic space charge. Furthermore, the cold anode is coated with cesium ions and hence has a low ϕ_a (cf. again Fig. 12-7). By the proper combination of cesium pressure and cathode temperature so that cesium ions partially cover the cathode surface, ϕ_c can be adjusted to the optimum value (large enough to give a satisfactory V_0 yet small enough to give a satisfactory thermionic emission). Thus the cesium performs three desirable functions that more than offset the undesirable effect of some loss of efficiency by the energy loss of electrons colliding with cesium atoms.

It should be noted that part of the heat supplied to the cathode in a thermionic energy converter is removed by the emitted electrons, each of which removes the energy eV_c. Unfortunately, thermal radiation also removes a large amount of heat at typical temperatures of operation, and this loss seriously reduces the efficiency of such converters.

There is a large variety of thermionic converters of which we have described only two examples, but, like thermoelectric generators, all of them have at present too low an efficiency ($\sim$10 to 15%) to compete with steam turbine generators. Also like thermoelectric generators, however, they are useful in specialized applications where small size, light weight, or other attributes offset their modest efficiencies.

12-5 FIELD EMISSION

If the electric field at the surface of a metal is sufficiently strong and in a direction to accelerate electrons away from the surface, electron emission occurs even though the temperature of the metal is room temperature or below. The field required to produce substantial emission is about 10^9 to 10^{10} volts/m. This process of *field emission* is almost completely independent of temperature. It is a quite different process from thermionic emission as modified by the Schottky effect.

Figure 12-12 is a schematic diagram of the surface of a metal to which a strong electric field is applied in a direction to accelerate electrons away from the surface. This plot is, of course, just like Fig. 12-4*b* except that the field is now very much larger. The conduction band electrons in the metal have traveling-wave wave functions inside the metal. Their Ψ's do not drop exactly to zero at $x = 0$. For $x > 0$, they have the usual exponential-tail wave functions appropriate to an electron in a region

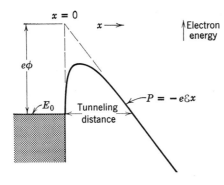

Fig. 12-12. Electron energy as a function of the distance x from the surface of a metal with a very strong accelerating electric field. Electron emission by tunneling through the narrow potential barrier is possible.

where its kinetic energy is negative. The Ψ's are not so simple as those in the square-well problem of Sec. 5-4 because the negative kinetic energy is a function of x, and therefore eq. 5-29 is not quite correct. But the variation of Ψ with x is much more rapid than the variation of $K = E + e\mathcal{E}x$, and therefore Ψ has the form of $e^{-f(x)x}$, where $f(x)$ is a slowly varying function of x.

There is thus a finite probability that electrons in the solid can tunnel through the surface barrier and be emitted. The electrons involved are those with energies near E_0, since there are very few electrons with $E \gg E_0$ and since electrons with $E \ll E_0$ have a much smaller probability of tunneling. The tunneling probability is, of course, very sensitive (exponential dependence) to the tunneling distance, which is inversely proportional to the applied field. Thus the emission current density should be approximately an exponential function of $1/\mathcal{E}$. The theoretical expression (the Fowler-Nordheim equation) is actually

$$j = (1.6 \times 10^{-6}\mathcal{E}^2/\phi)\, e^{-\frac{7\times10^9\phi^{3/2}u}{\mathcal{E}}} \quad \text{amp/m.}^2 \qquad (12\text{-}13)$$

where $u \cong 1 - (14 \times 10^{-10}\mathcal{E}/\phi^2)$.

The numbers 1.6×10^{-6}, 7×10^9, and 14×10^{-10} are approximate and come from combinations of atomic constants, including h. It is worth noting that j for field emission is as sensitive to $\mathcal{E}$ as j for thermionic emission is to T. It should also be noted that field emission is strictly a quantum effect. Its existence and the agreement of eq. 12-13 with experiment reinforce our confidence in quantum mechanics.

Field emission is not widely used as a practical source of electrons.

The required fields are so high that they can be attained only by making the emitter in the form of a point or knife edge and exploiting the concentration of the electric field at the point or edge. Field emission is very useful in studying the electrical properties of surfaces and the adsorption of gases on surfaces, since the experiments can be carried out on tiny areas at ordinary, or even very low (liquid helium), temperatures. Such studies are becoming important in research on catalysis and corrosion. In microwave power tubes the fields at sharp corners of the electrodes sometimes are high enough for field emission to occur and to cause failure of the tube.

12-6 PHOTOELECTRIC EMISSION

We have already considered in Sec. 4-2 the elementary explanation of the photoelectric emission of electrons from solids. We now return to this subject in order to justify and to refine the earlier discussion and in order to describe practical photoemitters.

The discussion in Sec. 4-2 ignored a possible contact potential difference between emitter and collector. Now that this phenomenon has been described it should be apparent that the method followed in Sec. 4-2 (Figs. 4-3, 4-4, and 4-5) in order to determine the maximum energy of emitted electrons requires correction, unless the ϕ_2 of the collector happens to be precisely equal to the ϕ_1 of the emitter. If zero voltage appears on the voltmeter in a photoelectric experiment, a contact potential difference $\phi_2 - \phi_1$ appears across the space between emitter and collector. The correction for contact potential difference merely displaces all the curves of Fig. 4-5 horizontally by the same amount, and therefore the basic conclusions of Sec. 4-2 are unchanged.

It should be recalled that the photoelectric work function $e\phi$ was defined as the energy that must be given to the "most energetic electron" in the solid in order to release it. Of course, correctly speaking, there is no "most energetic electron," as we have noted in Sec. 12-2. The accurate definition of ϕ is identical with the definition used in connection with thermionic emission, namely, that $e\phi$ is the energy difference between E_0 and the energy of a stationary electron outside the metal. The earlier rough definition is an approximation to this, since very few electrons in a metal have energies greater than E_0 by more than $\sim 4kT$, which equals only 0.1 e.V. at room temperature.

The modern theory of the photoelectric effect is similar to the thermionic-emission theory in that the emission current is proportional to the number of electrons with $p_x > p_{xc}$ (see eq. 12-1). An incident photon can be absorbed by producing a transition of one of the conduction band electrons from a state with energy E_1 to a state with energy $E_2 = E_1 + h\nu$. The most favorable situation for emission with the minimum $h\nu$ is that the original momentum be entirely in the x-direction (i.e., $p_x{}^2/2m = E_1$) and also that the acceleration of the electron be entirely in the x-direction (i.e., $p_{x2}{}^2/2m = E_1 + h\nu$). Thus the threshold energy $h\nu_0$ is given by

$$h\nu_0 + E_1 = \frac{p_{xc}{}^2}{2m} = E_0 + e\phi$$

If the maximum initial energy E_1 of the electrons were E_0, this would give the simple Einstein threshold equation (eq. 4-2 with $V_0 = 0$). The electron distribution (eq. 9-1) is actually such that the density of electrons decreases from 99% of its maximum value to 1% of this maximum as E_1 varies from $E_0 - 4kT$ to $E_0 + 4kT$. Thus the threshold is not sharply defined, but the photocurrent drops rapidly toward zero as $h\nu$ is decreased below $e\phi$.

The known form of the Fermi distribution permits the calculation of the photoelectric yield as a function of $h\nu - e\phi$ for ν values near the threshold. (The *yield* is the number of emitted electrons per incident photon and is always $\ll 1$.) We shall not perform the calculation, but merely note that the resulting theory (called the *Fowler theory*) is in excellent agreement with experiment over a wide range of temperatures. Figure 12-13 compares experimental values of the photoelectric yield with the theory. The curve plotted is the universal theoretical curve for all metals and all temperatures. All the data points for silver were then plotted as a plot of $\ln_e I/T^2$ vs. $h\nu$, where I is the photocurrent observed when a constant light intensity was incident. This whole group of points was then moved horizontally and vertically until the best agreement with the line was obtained. Since the universal curve has the abscissa $h\nu - e\phi$ and the data have the abscissa $h\nu$, the relative displacement of the zeros of abscissa gave $e\phi$. The displacement for the silver data gave $e\phi = 4.74$ e.V. The values of ϕ determined in this way agree closely with thermionic ϕ values for the metals on which both types of experiments have been performed.

Photoelectric emission from a semiconductor is quite different from that from a metal because in a semiconductor there are usually no electrons with the energy E_0. There are relatively small numbers of electrons in the conduction band and at the donor levels. Photoelectric currents with small yield values can be obtained whenever $h\nu$ is large enough to eject these electrons. Substantial photocurrents do not begin until $h\nu$ becomes large enough to eject electrons from the filled band. The Fowler theory is not applicable to semiconductors.

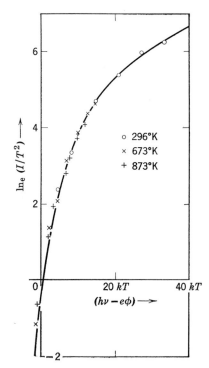

Fig. 12-13. "Fowler plot" for silver from data by R. P. Winch. The logarithm to the base e of the photocurrent divided by T^2 is plotted as a function of the difference in energy between $h\nu$ and $e\phi$, in units of kT. The number of photons incident per second is constant. (From A. L. Hughes and L. A. DuBridge, *Photoelectric Phenomena*, McGraw-Hill, New York, 1932.)

Metals are not useful as practical photoemitters because of their very low yield values. Part of the reason for low yields is the fact that most of the incident light is reflected by the metal. Another part is the necessity for the conservation of momentum and energy in the absorption of a photon. A photon cannot excite a perfectly free electron because of the impossibility of satisfying the conservation of both energy and momentum, as we shall now demonstrate. The initial energy is the energy E_1 of the electron plus the energy $h\nu$ of the photon. For simplicity we assume that the initial momentum of the electron $\sqrt{2mE_1}$ is in the same direction as the momentum $h\nu/c$ of the photon, and therefore the initial momentum is $\sqrt{2mE_1} + h\nu/c$. The final energy is entirely the energy $E_1 + h\nu$ of the electron, and the final momentum is the momentum $\sqrt{2m(E_1 + h\nu)}$ of the electron. But $\sqrt{2m(E_1 + h\nu)}$ does not equal

$\sqrt{2mE_1} + h\nu/c$. If the electron is bound to an atom, the other electrons in the atom can provide the necessary momentum increment. The conduction electrons in a metal are relatively free, however, and are therefore not excited readily by incident photons.

Practical photoemitting surfaces are complicated semiconductors characterized by covalent and ionic binding. As explained in the preceding paragraph, such binding permits higher yields than are possible in metals, but there is as yet no detailed physical theory of the high yield. A common surface (S-1) is made by evaporating cesium in a vacuum onto an oxidized silver surface. The spectral response of this surface is illustrated in Fig. 12-14. The sensitivity at long wavelengths is probably related to the low work function (1.9 e.V.) of cesium, but metallic cesium gives no photocurrent for $\lambda > 6500$ Å. The relatively high yield (about $\frac{1}{300}$ electron per photon at $\lambda = 8500$ Å) is probably a consequence of the fact that the electrons in the $Ag\text{-}Cs_2O\text{-}Cs$ semiconductor are less free than in a metal.

Another practical emitter is the semiconducting compound Cs_3Sb. The spectral response of this surface (S-5) is also illustrated in Fig. 12-14. The yield is the highest of any known surface; it is 0.25 electron per photon at $\lambda = 3600$ Å.

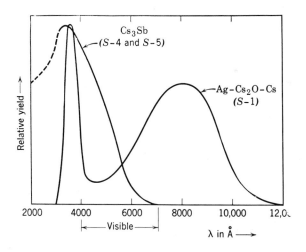

Fig. 12-14. Relative photoelectric yield as a function of wavelength (spectral sensitivity) of two photosurfaces. The scales of ordinates are different for the two surfaces. The Cs_3Sb surfaces S-4 and S-5 are the same, but S-5 is the designation of the surface when it is mounted in a tube with ultraviolet-transmitting glass; when it is in a lime-glass tube it is called S-4 and has zero sensitivity for $\lambda < 3000$ Å. (From V. K. Zworykin and E. G. Ramberg, *Photoelectricity and Its Application*, Wiley, New York, 1949.)

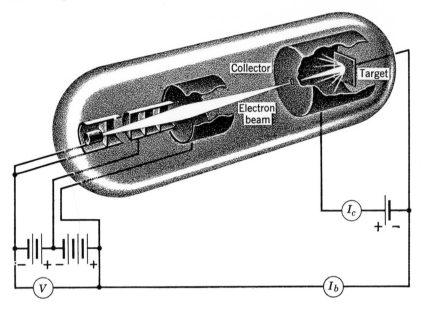

Fig. 12-15. Vacuum tube for measuring the secondary emission yield δ of a solid target. The electron gun at the left sends a focused beam of electrons through the hole in the collector. The collector current I_c consists exclusively of secondary electrons emitted by the target.

12-7 SECONDARY EMISSION

When energetic electrons bombard a solid surface, electrons are ejected from the solid target. The incident electrons are called *primary* electrons, and the ejected electrons are called *secondary* electrons. The ratio of the secondary electron current to the primary current is called the secondary emission *yield* and is given the symbol δ. δ is a function of the energy of the primary electrons and differs for different solids or for the same solid with different surface conditions. δ is independent of primary current density and practically independent of the target temperature.

A vacuum tube for studying the secondary-emission yield of a metal target is illustrated in Fig. 12-15. The cathode and focusing electrodes provide a narrow beam of electrons, all of which go through a small hole in the collector electrode. These electrons strike the target and produce secondaries. If the collector is sufficiently positive with respect to the target, all the secondaries go to the collector. The secondary

TABLE 12-2

Secondary Emitting Properties of Representative Solids

Metals	δ_{max}	K_{max}	Non-Metals	δ_{max}	K_{max}	K_p for $\delta = 1$
Ag	1.5	800	Ge	1.1	400	
C	1.0	300	NaCl	6	600	1400
Cu	1.3	600	MgO	2.4	1500	
Fe	1.3	350	Al_2O_3	4.8	1300	
Li	0.5	85	Oxide cathode	10	1400	
Mo	1.25	375	Mica	2.4	380	3300
Ni	1.3	550	Hard glass	2.3	400	2400
Pt	1.6	800	$Ag-Cs_2O-Cs$	8.8	550	$>20,000$
W	1.4	600	Phosphors			3000–40,000

Most of the values are from K. G. McKay, "Secondary Electron Emission" in *Advances in Electronics*, Vol. I, edited by L. Marton, Academic Press, New York, 1948.

emission yield δ is therefore the ratio of the collector current I_c to the beam current I_b. The beam voltage V can be varied, and thus δ can be studied as a function of the kinetic energy K_p of the primary electrons.

The observed dependence of δ of nickel upon the kinetic energy K_p of the primary electrons is illustrated in Fig. 12-16. Curves practically

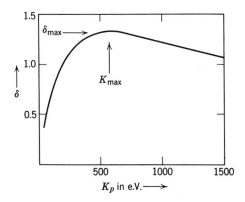

Fig. 12-16. Secondary emission yield δ as a function of the kinetic energy K_p of the primary electrons, for nickel.

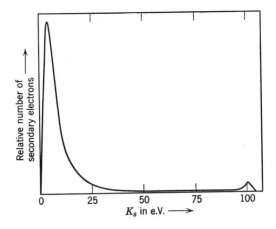

Fig. 12-17. Distribution of energies of secondary electrons from a molybdenum target. Each primary electron had a kinetic energy of 100 e.V. [From G. A. Harrower, *Phys. Rev.*, **104**, 52 (1956).]

identical in shape are obtained for all metals, but with different values for δ_{max} and K_{max}. Values of δ_{max} and K_{max} are given in Table 12-2. δ generally increases if the metal surface becomes oxidized or otherwise contaminated.

Yield curves for non-metals have the same form as Fig. 12-16. Some non-metals, like silicon and germanium, even have the same order of magnitudes for δ_{max} and K_{max} that metals have. For most non-metals, however, δ_{max} and K_{max} are much higher than for metals. Representative values are given in Table 12-2. Another parameter of interest for non-metals is also given in the table. This is the *second cross-over energy*, the larger of the two values of K_p for which $\delta = 1$.

The energy distribution of the secondary electrons from a molybdenum target is shown in Fig. 12-17. The distribution of the energies K_s of secondaries was determined by sending the secondary electrons through a magnetic momentum selector, like the mass spectrometer of Sec. 3-3. (In the present case we know the mass of the electron but seek its K_s, which can be calculated from the measured momentum.) It is evident that there are a few emergent electrons with an energy K_s practically equal to the primary energy K_p. These are elastically scattered electrons. The electrons with lower energies are called the *true secondary electrons*.

It is worth digressing at this point to recall that in the Davisson-Germer experiment (Sec. 4-10) a single crystal was bombarded with electrons. The angular distribution of the elastically scattered electrons was studied. The detector was biased so that only electrons with $K_s \cong K_p$ could enter. We now see that the purpose of this bias was to exclude the great majority of secondary electrons. The angular distribution of the true secondaries is proportional to $\cos \theta$, where θ is the angle the secondary makes with the normal to the surface; this distribution is independent of crystal orientation and lattice constant.

The theoretical understanding of secondary emission is still incomplete. It is not possible to calculate $\delta_{\max}$ for solids with an interesting precision or to answer such questions as: Why is $\delta_{\max}$ for silver greater than $\delta_{\max}$ for copper? But it is possible to show by a simple qualitative theory why the yield curve (Fig. 12-16) has the observed shape. This theory also provides information on such problems as the variation of δ with angle of incidence and the comparison of $\delta_{\max}$ of insulators with $\delta_{\max}$ of metals, and therefore it is worth while to discuss it.

We must first examine the way in which a primary electron loses energy. The principal process is the excitation of the atomic electrons. The primary electron interacts with the electrons in the solid by the Coulomb force. If the encounter is sufficiently close, the atomic electron is excited to a higher, vacant energy state. In each such encounter the primary electron loses a fraction of its energy. (The situation here is quite different from the encounter between a photon and an electron, in which the photon disappears.) On the average, the primary loses about 30 e.V. energy for each electron excited. If the primary electron has an energy of many thousands of electron volts, it can excite *any* electron (including K-shell electrons); but if it has an energy of about 100 e.V., it can excite only conduction band electrons and possibly a few lower-lying electrons. In either case, an electron of the solid is excited to a vacant state above the Fermi level E_0; this is an *internal secondary electron*.

The internal secondary electron may or may not become an *emitted* secondary electron. If it is produced near the surface and has a momentum directed toward the surface, it may arrive at the surface with enough momentum remaining to surmount the surface barrier. But it suffers collisions with the conduction band electrons and loses energy, and if it is produced at too great a depth below the surface it will not be emitted.

The kinetic energy K of a primary electron as a function of the distance x it has penetrated into a solid is plotted schematically in Fig. 12-18a. For K_p in the range 10^4 to 10^5 e.V., there is good theoretical evidence for the relation

$$K = \sqrt{K_p{}^2 - \alpha \rho x}$$

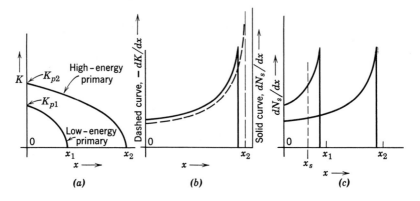

Fig. 12-18. Production of internal secondary electrons. (*a*) Kinetic energy of primary electron as a function of depth of penetration x into the solid; the primary with lower initial energy (K_{p1}) has a smaller penetration depth. (*b*) Rate of production of internal secondaries (dN_s/dx) per unit distance, which is proportional to the rate of loss of energy of the primary electron, $-dK/dx$. (*c*) Comparison of production rates for the two primary electrons of (*a*); although primary 2 produces more internal secondaries, primary 1 produces more in the region $x < x_s$, and only these electrons can be emitted as secondary electrons.

where ρ is the density (kg/m.^{3}) and α is a constant. If K and K_p are in electron volts, α is about 4×10^{10} (e.V.)^2 m.^{2}/kg for the energy range 10^4 to 10^5 e.V. For the lower energies of interest in secondary emission, the calculations and the experiments become very difficult, but we expect the same general shape at K_p values of the order of 10^2 to 10^4 e.V. as at higher values. This shape occurs because a slower primary electron spends more time near an atomic electron and therefore has a greater probability of exciting that electron. Thus the rate of energy loss should always increase toward the end of the range of the primary.

The negative of the derivative of $K(x)$ with respect to x is plotted as the dashed curve in Fig. 12-18*b*. Since one internal secondary is produced (on the average) for each 30 e.V. of energy lost by the primary, the rate of production of internal secondaries is proportional to $-dK/dx$. Of course this proportionality ceases when K becomes $<\sim 30$ e.V., since after that energy is reached no more secondaries can be produced. The solid curve in Fig. 12-18*b* is the rate of production of secondaries plotted with an arbitrary vertical scale but with the same abscissa as the other curves.

Figure 12-18*c* compares the rate of production of internal secondaries by primaries of two different energies. The higher-energy primary

(K_{p2}) produces more internal secondaries, but the lower-energy primary (K_{p1}) produces more internal secondaries near the surface. The only internal secondaries that can escape are those produced at $x < x_s$. (Some measurements of δ as a function of angle of incidence of the primaries have shown that $x_s \cong 30$ Å.) Thus for the case pictured δ is larger for the lower value of K_p. This explains why δ decreases as K_p increases beyond K_{max}.

The decrease of δ as K_p decreases below K_{max} is more obvious. Here the range x_1 of the primary is of the same order of size as or smaller than x_s, and as K_p approaches zero the total number of internal secondaries approaches zero. At very small K_p values ($K_p < 30$ e.V.) there are probably no true secondaries at all, merely elastically scattered primaries.

Thus the shape of the yield curve (Fig. 12-16) can be explained. In summary, δ decreases for small K_p because of a decrease in the total number of internal secondaries produced. δ decreases for large K_p because of a decrease in the number of internal secondaries produced in the region near the surface, even though the total number is increasing.

The fact that δ is practically independent of temperature is consistent with the above analysis, since no thermal excitation was required. The fact that δ is a slowly varying function of the work function of the surface is also consistent with this analysis. Measurements of δ have been made on a tungsten surface as its work function was being changed by adding a layer of atoms, like the layer of thorium atoms discussed in Sec. 12-2. As ϕ was lowered by about a factor of 2, δ increased by a factor of 1.6. This is, of course, a far smaller effect than if a process like thermionic emission were occurring. The effect is caused by the fact that, with a lower work function, a larger fraction of the internal secondaries can escape. Since the average energy of the emitted secondaries is 2 to 5 e.V., the probability of escape is not drastically changed by a change in ϕ.

Secondary emission can cause erratic behavior and poor characteristics of electron tubes unless they are properly designed. The plate of a pentode electron tube can be negative with respect to the screen grid. Secondary electrons from the plate would then be collected by the screen grid. This would reduce or even change the sign of the plate current. To avoid this, a suppressor grid is placed close to the plate and is maintained at cathode potential. Secondary electrons from the plate are thus turned back to the plate, and the operation is just as if $\delta = 0$.

A second effect of secondary emission that must usually be avoided is the charging of insulators. This charging can occur if the δ of the insulator is >1 for values of K_p encountered in the tube. Whenever one electron strikes the insulator more than one electron leaves. The insulator charges positively and attracts more electrons. A mica insulator or glass tube wall can thus be charged and deflect an electron beam from its proper course. Proper design keeps insulators well away from the paths of electrons or makes sure that there is no positive electrode to collect the electrons from the insulator. In the latter case, secondary electrons return to the insulator and no positive charging occurs. This charging effect is used constructively in many cathode-ray tubes. The phosphor screen is an insulator with $\delta > 1$ for the electron energies employed. The screen therefore charges positively until its potential reaches the potential of the graphite coating on the inside of the tube, which is the collector for secondary electrons from the screen. Thus a positive screen potential is maintained even though the screen is a good insulator.

One interesting application of secondary emission is the *multiplier phototube* illustrated in Fig. 12-19. Photons produce photoelectrons from the Ag-Cs$_2$O-Cs surface. These electrons are accelerated toward the first dynode by a potential difference of about 100 volts. When they strike this surface, secondary electrons are emitted. These secondaries are accelerated to the second dynode, and so on. δ of each

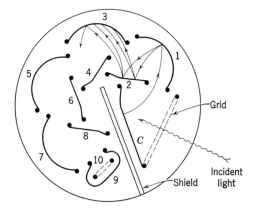

Fig. 12-19. Multiplier phototube. The photoelectrons from the photocathode C are focused onto dynode 1. Secondary electrons from this dynode are focused onto dynode 2, and so on. Each of the dynodes 1 to 9 is at a successively higher electrostatic potential. The anode 10, which is connected to the external output circuit, collects the amplified electron current. [From R. W. Engstrom, *J. Opt. Soc. Am.*, **37**, 420 (1947).]

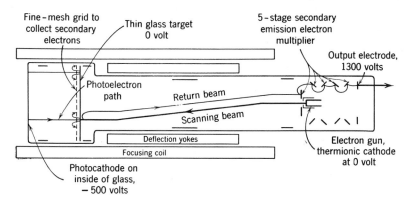

Fig. 12-20. Image orthicon television camera tube. The four unlabeled cylindrical electrodes fix the electrostatic potential in the different parts of the tube. The first secondary emission multiplier dynode is a disc; the other four dynodes are in the form of spoked wheels, and the spokes are arranged like Venetian blinds in order to permit secondary electrons from one dynode to travel to the next dynode. [See Rose, Weimer, and Law, *Proc. I.R.E.*, **34,** 424 (1946).]

stage is about 4.6, and therefore the overall gain of the nine-stage secondary emission multiplier is $(4.6)^9 \cong 1,000,000$.

Another application of secondary emission is to the television camera tube called the *image orthicon*. This tube is much like the vidicon described in Sec. 10-5 and is illustrated in Fig. 12-20. The image of the scene is focused onto a semitransparent photocathode. The emitted electrons are accelerated toward a thin glass target plate. Because of the focusing action of the strong magnetic field, each photoelectron strikes the glass target at the same relative position on that target as it originated from on the photocathode. The secondary electrons from the glass are collected by a fine screen. Thus a spot on the glass charges positively (since $\delta > 1$) whenever light strikes the corresponding spot on the photocathode. The scanning electron beam tests the target for charge. Wherever a spot is positively charged, some of the beam strikes the target. Where a spot has not been charged, the beam is completely reflected. The return beam strikes the first dynode of a five-stage secondary emission multiplier. Thus an amplified signal proportional to the relative darkness of the target is transmitted in synchronism with the motion of the scanning beam. The brief description here of this complicated tube shows how several forms of electron emission can be useful in the same device: The image orthicon employs thermionic emission (for the scanning beam), photoelectric emission, and secondary emission.

12-8   COLLISIONS BETWEEN ELECTRONS AND ATOMS

Collisions between electrons and atoms in a gas have been considered briefly in Secs. 2-5 and 4-5; ionizing collisions have been mentioned in many other places. We now return to this subject and present a systematic description and an elementary quantum-physics explanation of the phenomena. All the processes considered in this section are of importance in gas-discharge devices such as mercury-arc rectifiers, thyratrons, and Geiger counters, and in the investigations leading toward the production of power from nuclear fusion.

The mathematical description of collisions between two particles was presented in Sec. 2-5. In that discussion the assumption was made that the particles had definite radii r_1 and r_2 and that a collision occurred whenever the centers of particles were separated by a distance $r_1 + r_2$. In order to apply that description of collisions to electron-atom collisions we first note that the size of the electron is negligible. We therefore set $r_1 = 0$, and the attenuation dJ of a beam of J electrons per unit area can be obtained from the development preceding eq. 2-19:

$$dJ = -JN(\pi r_2^2)\,dx$$

$$J = J_0 e^{-N\pi r_2^2 x} \text{ electrons/m.}^2 \text{ sec}$$

Here J is the number of electrons per second per square meter, and the electrical current density $j = eJ$. N is the number of gas atoms per cubic meter, and x is the distance in the direction of motion of the beam.

We realize from the quantum physics of atoms and molecules that the description of an atom or molecule as a hard sphere of radius r_2 is completely inadequate. We therefore replace the term πr_2^2 by an equivalent area σ, which is called the *collision cross section*. The fraction of electrons removed from the beam by one process (e.g., elastic collisions) can be completely different from that by another process (e.g., ionizing collisions). Thus σ is different for each process considered. σ is also a function of the kinetic energy of the electrons. The σ of an inelastic process like ionization is, of course, zero if the electron energy is less than the threshold energy required for this process.

We now consider a single process by which electrons are removed from the beam. Such a process might be the excitation of gas atoms from the ground state to the first excited state. The number of electrons removed from the beam per unit area of beam per meter traversed by the beam is

$$dJ = -JN\sigma\,dx \quad \text{electrons/m.}^2 \text{ sec} \tag{12-14}$$

as explained in conjunction with Fig. 2-9; this equation effectively defines σ. $JN\sigma\,dx$ is also the rate at which the process considered (e.g., excitation) occurs in the plate-like element of volume consisting of unit area (1 m.^{2}) of the beam and a thickness dx. The rate per unit volume is $JN\sigma$ processes/m.^{3} The attenuation of the beam is expressed by

$$J = J_0 e^{-N\sigma x} \quad \text{electrons/m.}^2 \text{ sec}$$

or

$$j = eJ = j_0 e^{-N\sigma x} \quad \text{amp/m.}^2 \tag{12-15}$$

where J_0 and j_0 are the values at $x = 0$ (compare eq. 2-19).

Measurements of the cross section σ for a particular process are based either on eq. 12-14 or on eq. 12-15. For example, measurement of the cross section σ_{el} for elastic scattering of low-energy electrons can be made by observing the decrease in j with the distance x traversed. (The apparatus is just like that shown in Fig. 2-8 except that the detector is simply an electrode and the current to this electrode is meas-

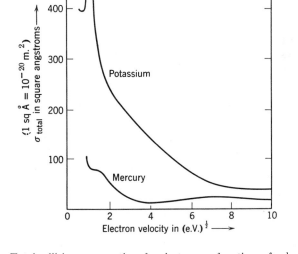

Fig. 12-21. Total collision cross sections for electrons as functions of velocity for two gases. Elastic collisions are primarily responsible for removal of electrons from a beam at the energy values considered here. [From R. B. Brode, *Revs. Mod. Phys.*, **5**, 257 (1933).]

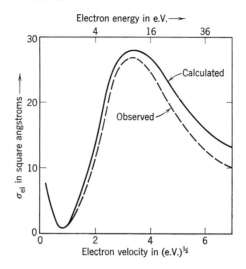

Fig. 12-22. Comparison of quantum theory with experiment for the Ramsauer effect in krypton. (From H. S. W. Massey and E. H. S. Burhop, *Electronic and Ionic Impact Phenomena*, Clarendon Press, Oxford, 1952.)

ured.) Since the electron energy is very low, the elastic collision process is the only process removing electrons from the beam. In order to measure the cross section σ_{ion} for ionization, on the other hand, it is necessary to measure the rate at which ions are formed, since electrons with energy large enough to make ionizing collisions are lost from the beam by other processes in addition to ionization. This measurement is therefore based on eq. 12-14, and σ_{ion} is computed from the measured rate of production of ions.

(a) Elastic collisions

Experimental data for the cross section σ_{el} for elastic collisions are illustrated in Figs. 12-21 and 12-22. As explained in the preceding paragraph, this particular measurement is based on eq. 12-15, and *any* process that removed electrons from a beam would contribute to the measured cross section, but at the energies considered the elastic scattering process dominates all others. The curves for cadmium and zinc are similar to the curve shown for mercury; the curves for sodium and cesium are similar to the one shown for potassium. Three generalizations can be made from these facts: (1) σ_{el} always increases as the electron velocity approaches zero. (2) Superimposed on this general trend is a fine structure characteristic of the atom of the gas. (3) σ_{el} as a function of the electron kinetic energy K is approximately the same for atoms with the same outer electronic structure (there are exceptions to this last statement, however).

The quantum theory of elastic collisions is complicated. We shall give a brief quantum description of a collision without attempting any quantitative work. The incident electron is guided by a packet of plane waves, each of the form $e^{2\pi i\left(\frac{x}{\lambda}-\frac{Et}{h}\right)}$. As these strike the region of varying potential energy in the atom, they are partially reflected. The interference among the partial reflections from different parts of the atom determines the direction of the outgoing wave packet. The fine structure in the curves of Fig. 12-21 is caused by this interference, which is of course a function of λ and hence of the momentum of the electron wave. The general rise toward $K = 0$ occurs because low-energy electrons are relatively more affected than high-energy electrons by the change in potential energy that occurs inside an atom. The similarity between the curves of σ_{el} as a function of electron velocity for atoms with the same outer electronic structure occurs because the potential energy as a function of r of such atoms is similar for r values in the outer shell. Since the outer shell provides most of the volume of an atom, similar $P(r)$ in this region produces similar σ_{el} vs. K.

A striking success of the quantum theory of collisions is its ability to explain the Ramsauer effect. This effect was described in Sec. 5-4 and the data were presented in Fig. 5-9. It is an extreme example of the interference mentioned in the preceding paragraph. Figure 12-22 compares the quantum-mechanical theory with experiment for krypton.

(b) Excitation collisions

The experimental method described in Sec. 4-5 provides information on the first excitation energy of an atom but cannot give information on the excitation cross section σ_{ex} as a function of K. Of the many ways of measuring σ_{ex} we shall mention only two. One of them consists of measuring the number of photons emitted when the excited atoms return to their ground states. Since a single photon (or at most a few photons with distinguishable wavelengths) is produced per excitation process, we can thus count the number of such processes and use eq. 12-14 to compute σ_{ex}. Another method is to analyze (perhaps by a $180°$ magnetic field) the distribution of energies of an initially constant-K beam of electrons after it has passed through a gas. This analysis permits the computation of the number of electrons that have made collisions in which an energy E_1, E_2, $\cdots$, or E_n e.V. was lost. Thus σ_{ex} for the various excitation processes can be computed.

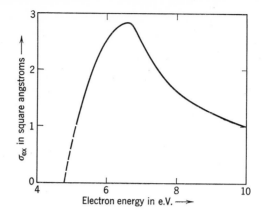

Fig. 12-?3. Excitation cross section for the production of the excited state that gives the $\lambda = 2536$ Å emission line in mercury. The vertical scale is only approximately correct. (From H. S. W. Massey and E. H. S. Burhop, *Electronic and Ionic Impact Phenomena*, Clarendon Press, Oxford, 1952.)

Figure 12-23 presents σ_{ex} vs. K for the excitation of mercury atoms from the ground state to the first excited state. The excited atom returns to the ground state by emitting a photon with $\lambda = 2536$ Å unless the atom first collides with another atom or a wall. Curves similar to Fig. 12-23 occur for all excitation processes, each with a different minimum energy. The shape of such curves is determined by the existence of a threshold energy (the energy of excitation) and the fact that the cross section decreases at large K for the same reason that σ_{el} decreases: A fast electron spends less time near an atom, and therefore is less likely than a slower electron to induce a transition from one electronic state of the atom to another.

The quantum description of excitation by electrons is somewhat different from the excitation by absorption of light (Sec. 5-8). Let us assume for definiteness that the gas atom is sodium, which has only one $3s$ electron. The incident electron produces a strong and unsymmetrical electric field, which is changing with time because of the electron's motion and which accelerates the $3s$ electron. This valence electron will now have a quite different probability of being at various r, θ, and φ positions because of the repulsion from the incident electron. Thus its wave function, instead of being a pure $3s$ wave, will be a mixture of $3s$, $3p$, $3d$, $4s$, $\cdots$, wave functions. This mixture changes with time and has a $|\psi|^2$ that puts the valence electron, on the average, farther away from the incident electron than it would be if its ψ were a pure $3s$ wave function.

The electron may be viewed as spending part of the time in a $3s$ state, part in each of the various $3p$ states, etc. Meanwhile the incident electron has been slowed by the repulsion of the valence electron. As the incident electron recedes, the valence electron may be left in the $3s$ state, in which event the collision has been elastic. The valence electron may be left in a $3p$ state, in which event an excitation collision has occurred. The K of the outgoing electron is less than its initial K by the difference in energy between the $3s$ and the $3p$ states. The difference in momentum is also the difference between the $3s$ and $3p$ state momenta. Excitation to higher states than the $3p$ occurs in similar fashion.

It is worth noting that there are no selection rules in this excitation process, whereas such rules occurred for excitation by photons (Sec. 5-8). This result occurs because the incident electron, unlike the photon, has a wavelength of the same order of magnitude as or smaller than the size of the atom. The electron is therefore fairly well localized and produces an electric field much like the field of a point charge. This field is much different from the nearly uniform field of the photon, whose λ is very much greater than the size of an atom. The non-uniform field of the electron can induce transitions from one state to another regardless of the symmetry of their charge distributions.

Although there are no selection rules for electron excitation, there are selection rules for radiative return to the ground state, and therefore the atom may be excited to a state from which it cannot return to the ground state by an allowed transition. If the transition is allowed, the lifetime of the excited state is $\sim 10^{-8}$ sec. If the transition is forbidden, the lifetime is very much longer (> 1 sec). (The lifetime is not infinite, because the electric field of the photon is not exactly uniform over the the atom.) Excited atoms with such long lifetimes are called *metastable atoms*, and they play an important role in gas discharges. Some of the effects that they can produce are: (1) A metastable atom can give up its energy of excitation to an electron at the surface of an electrode, thereby releasing the electron from the solid if the excitation energy of the metastable is greater than $e\phi$ of the solid. (2) A metastable atom can collide with another atom, which can be either in its ground state or in an excited state, and transfer its energy to the latter; if the total energy of excitation is greater than the ionization energy of the latter atom, the atom is ionized. (3) A metastable atom may be further excited or ionized by another incident electron. These processes are important because they can release electrons from the cathode or produce ionization in a gas discharge without requiring that the incident electrons ever have an energy as great as the ionization potential of the gas atoms.

(c) *Ionization collisions*

These collisions are like excitation collisions except that the excited state is a free state of the atom, and a valence electron leaves the atom. The ionization energy (e times the ionization potential) is the binding energy in its ground state of the most easily removed electron. The experimental determination of ionization cross sections is easier than of excitation cross sections, since positive ions are easy to detect. But there is one complication, namely, the possible production of multiply charged ions. If the kinetic energy of the incident electron is sufficiently large, this electron can remove two or more electrons from the atom. In order to separate the processes of single, double, etc., ionization, the positive ions formed must be sorted according to charge-to-mass ratios. Figure 12-24 shows the results of such an experiment on mercury. The threshold energies for production of the various ions are 10.4 volts for Hg^+, 30 volts for Hg^{++}, 71 volts for Hg^{+++}, and 120 volts for Hg^{++++}.

The theoretical explanation of ionization by electron collision closely parallels that of excitation. It should be noted that ionization can also occur by photons of sufficient energy ($h\nu$ greater than the ionization energy). The processes of collision ionization and photoionization are important in gas discharges, but most ions in practical gas discharges

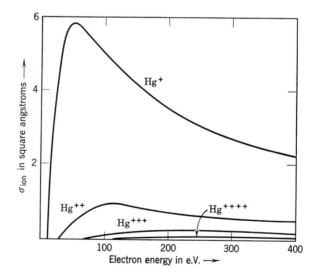

Fig. 12-24. Ionization cross sections of mercury as functions of electron kinetic energy. [From W. Bleakney, *Phys. Rev.*, **35,** 139 (1930).]

are produced indirectly, by processes such as those discussed in connection with excitation collisions.

(d) Excitation- and charge-transfer collisions

An excited atom carries with it the excitation energy. If it collides with an atom in its ground state, the excited atom may emerge in its ground state and the other atom may be excited. The details of this process are closely similar to the excitation process described in (b) above. This process is not usually of interest if both atoms are of the same element, but when different kinds of atoms are present it can be exploited to adjust the populations of excited states for maser or other applications. For example, excited helium atoms (in the lowest excited state of He, about 20 e.V. above the ground state) can transfer their excitation to neon atoms in a gas mixture; the neon atoms are left in a high excited state ($\sim$20 e.V. above the ground state) and they decay by successive radiative transitions to lower excited states at $\sim$19 e.V. and $\sim$17 e.V. The density of atoms in the $\sim$19-e.V. state can be made much larger than the density in the $\sim$17-e.V., since the lifetime of the $\sim$17-e.V. state is very short. This *inverted population* of neon atoms (larger concentration of atoms in the upper energy state) can then be used in a gas maser to produce stimulated emission of $\sim$2-e.V. photons (see the last few pages of Sec. 10-6).

If an ion collides with a neutral atom, its charge can be transferred in a similar manner. This process is of interest in gas discharges even if all the atoms are of the same element, since ions with large kinetic energies thus become neutral atoms and preserve their kinetic energies, whereas slow neutral atoms with approximately kT kinetic energy become ions. The charge-transfer process thus profoundly affects the kinetic energy distributions of both ions and atoms.

Neither of the two types of collisions considered in this subsection is an *electron*-atom collision, but these atom-atom collisions are included in the present list because they invariably accompany inelastic electron-atom collisions and substantially modify the products of such collisions.

(e) Recombination collisions

An electron and a positive ion are strongly attracted to each other. An electron follows an orbit around the positive ion, like the Rutherford scattering orbits of Sec. 3-2 except that it is concave toward the ion instead of concave away from the ion. Unless the electron-ion pair can lose energy while the electron is close to the ion, the electron will continue on its way. The only method it has of losing energy is radiation, and an excited state takes $\sim$$10^{-8}$ sec to decay to a lower state. Since the electron is near the ion for a time $\ll 10^{-8}$ sec, it is very unlikely to be caught and to *recombine*. Thus recombination cross sections are very much smaller than the geometrical cross section of the atom. This process is so unlikely that it can be neglected except in very high-pressure arcs. Recombination of ions and electrons occurs with appreciable probability only at the walls of the discharge tube, where the electrons in the solid can take up the energy. This situation should be compared with electron-hole recombination in semiconductors.

(ƒ) Electron attachment collisions

The process of adding an electron to a neutral gas atom is as unlikely as the recombination process, and for the same reasons. We shall therefore not discuss it further.

The situation is quite different if the gas is composed of molecules rather than atoms. The cross sections for the production of negative ions by electron attachment to molecules can be of the same order as ionization cross sections. The processes generally involve dissociation and therefore will be discussed below. Electron attachment collisions are an important way of removing electrons from a gas discharge.

(g) Dissociation collisions

When an electron strikes a molecule it cannot give an appreciable kinetic energy directly to one of the atoms in the molecule because the electron mass is so small relative to the atomic mass. The process of dissociation by electron impact actually occurs, but in a quite different manner. This process consists of excitation of the *electrons* of the molecule to a state in which the atoms are no longer bound together. Thus the atoms move apart, and dissociation has occurred. The motion of the ions is so slow compared to that of the incident electron that they can be regarded as stationary during the collision. Only after the incident electron has left do the atoms slowly move apart.

The process of dissociation by electron impact can be understood by studying the example of the hydrogen molecule. Figure 12-25 is a plot of the total energy of a hydrogen molecule as a function of the distance R between nuclei. The two lowest curves, (a) and (b), are identical with the curves in Fig. 7-20. Curve (a) is the ground state, and the molecule is ordinarily in this state and is vibrating with the zero-point energy about the position of stable equilibrium. The range of R values for the molecule in the ground state is shaded in Fig. 12-25. Curves (e) and (f) are curves for the hydrogen molecule ion $H_2{}^+$ and are identical with the curves of Fig. 7-19 except that they have been raised 13.6 e.V. This elevation is in order to keep the same definition of $E = 0$ as for the

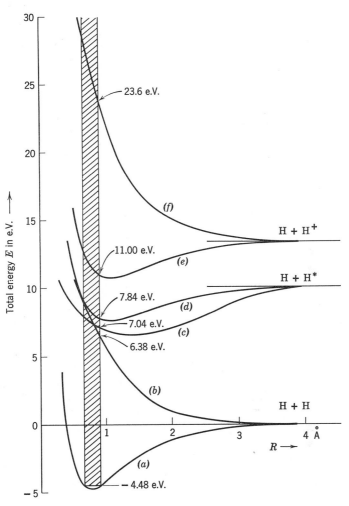

Fig. 12-25. Total energy of a hydrogen molecule in various electronic states as a function of the internuclear separation R. The shaded region is the region of R values in the vibration of the normal hydrogen molecule at room temperature or below. [From H. D. Smyth, *Revs. Mod. Phys.*, **3**, 347 (1931).]

lower curves, namely, the energy of two neutral H atoms at infinite separation. The curves (e) and (f) approach an energy of 13.6 e.V. at infinite separation, the energy of one H atom (in the ground state) and one H^+ ion. Similarly the curves (c) and (d) are for states of electronic excitation that at infinite R reduce to one H atom in the ground state and one excited H atom (written H*). There are still other, higher states not shown.

If an incident electron has an energy greater than 4.48 + 6.38 = 10.86 e.V. it can excite the molecule to the state (b). This is the repulsive state described in Sec. 7-5, and the atoms move apart. Dissociation occurs, and the two atoms share equally (because of their equal masses) the 6.4 to 9 e.V. of kinetic energy that is obtained from the repulsion force. This is the difference in energy between the state (b) in the shaded region and the state (b) at $R = \infty$.

If an incident electron has an energy greater than 11.5 volts, it can excite the molecule to the (c) state, and if it has an energy greater than 12.3 volts it can excite to the (d) state. The molecule is left in a high vibrational state about the new position of equilibrium, 1.1 Å for (d) or 1.4 Å for (c). It will not dissociate, since (regardless of the incident electron energy) the total energy of the molecule is only from 11.5 to 13.5 e.V. above −4.48 e.V., and at least another electron volt would be required to permit dissociation. Similarly, excitation to state (e) produces the hydrogen molecule ion H_2^+ in a vibrating, but not dissociating, state. Excitation by electrons with 28.1 e.V. or greater energy produces neutral H atoms and H^+ ions with appreciable kinetic energies.

Similar and even more complicated behavior occurs with all other molecules. Sometimes negative ions are among the products of dissociation.

Experiments on dissociation are best carried out in the gaseous-ion source of a mass spectrometer. A diagram like Fig. 12-25 can be constructed from three kinds of experiments: (1) Mass and charge determinations of the products of dissociation by the usual mass-spectrometer technique. (2) Measurement of *appearance potential* or *appearance kinetic energy*, the minimum energy of the electrons in the ion source that suffices to produce a particular ionic or atomic species (e.g., 28.1 e.V. for H^+ ions and 15.5 e.V. for H_2^+ ions). (3) Measurement of the initial kinetic energy of ions or atoms formed [e.g., $\frac{1}{2}$(10 to 15 e.V.) for H^+ ions, curve (f)]. Experiments such as these have been a powerful tool in determining molecular structure.

12-9   ELECTRON OPTICS

At many places in our work we have discussed the motion of free electrons and ions in electric and magnetic fields. The trajectories of charged particles in electric and magnetic fields have been exploited to provide indispensable information about the particles. A large variety of directions of $\mathcal{E}$ and $\mathcal{B}$ fields, of electrodes, and of initial conditions were used in such experiments as the determination of e/m of electrons, the cyclotron, the synchrotron, and mass spectrometers. This variety is nevertheless only a tiny fraction of all the useful arrangements of charged-particle deflections in conjunction with electron emission and collision processes.

The systematic study of charged-particle motion is called *electron optics* or *electron ballistics*. The principal underlying equation is the force **F** on a charge q moving with velocity **v**:

$$\mathbf{F} = q\boldsymbol{\varepsilon} + q\mathbf{v} \times \boldsymbol{\mathfrak{B}} \tag{12-16}$$

In many important applications, especially grid-controlled vacuum tubes like ordinary radio tubes, the free charges themselves produce appreciable fields. This effect is called *space charge*. The fields produced by space charge are most conveniently computed from Poisson's equation (eq. 12-11), one of the Maxwell equations that determine $\boldsymbol{\varepsilon}$ and $\boldsymbol{\mathfrak{B}}$ in terms of potentials on electrodes, currents in conductors, and free charges in space. Electron optics is thus founded on four bases: (1) Newton's second law of motion, force equals the product of mass and acceleration; (2) eq. 12-16; (3) Maxwell's equations; (4) information about initial energies and directions of emission of electrons from surfaces.

We shall not attempt a systematic study of electron optics but shall confine our attention to an introduction to one small area of the field: electron lenses and the electron microscope. A configuration of electric or magnetic fields that concentrates an electron beam is called an *electron lens*. The terminology ("lens," "microscope") is borrowed from geometrical optics and is useful because the imaging properties of electron lenses and the properties of combinations of thin lenses are just the same as in light optics.

The focusing action of a lens occurs because of the force expressed in eq. 12-16, and it can be computed and described without any knowledge or use of the wave properties of electrons. In short, it is a geometrical-optics property, not a physical-optics property. But it is worth noting that the same results for focal length and other properties are obtained if we compute the most probable trajectory of a packet of waves (instead of the trajectory of a classical charged particle). This statement is true only if the $\boldsymbol{\varepsilon}$ and $\boldsymbol{\mathfrak{B}}$ fields do not vary appreciably over the length of the wave packet. Since the size of an electron wave packet with an appropriate momentum spread is of the order of 10^{-8} m., this requirement is always met for laboratory-scale fields. Thus the Correspondence Principle is obeyed, and it provides the connection between particle ballistics and wave-packet motion. We can use either point of view, since both give the same answer for laboratory-scale fields. It is easier to compute the focal length of a lens by particle ballistics, but some calculations of aberrations are easier by wave theory.

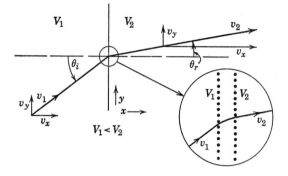

Fig. 12-26. Refraction of an electron trajectory upon passing from a region of potential V_1 to a region of higher potential V_2. The inset shows the bending of the trajectory in the region between the closely spaced grids.

The basic effect of an electric field on the motion of an electron is illustrated by the idealized situation shown in Fig. 12-26. Regions of constant electrostatic potential V_1 and V_2 are separated by two very fine grids, the left at potential V_1 and the right at potential V_2. The electron travels in a straight line except in the region between the grids. In this region, v_y is unchanged since there is no y component of $\mathbf{\mathcal{E}}$, but v_x is increased. Let us assume that the electrostatic potential was set equal to zero at the point where the $K = \frac{1}{2}mv^2$ of the electron equals zero. Then

$$\tfrac{1}{2}mv_1{}^2 = eV_1 \qquad \tfrac{1}{2}mv_2{}^2 = eV_2$$

$$\frac{v_2}{v_1} = \left(\frac{V_2}{V_1}\right)^{\!\frac{1}{2}}$$

In terms of the angle θ_i of incidence and θ_r of refraction:

$$\sin \theta_i = v_y/v_1 \qquad \sin \theta_r = v_y/v_2$$

$$\frac{\sin \theta_i}{\sin \theta_r} = \left(\frac{V_2}{V_1}\right)^{\!\frac{1}{2}} \tag{12-17}$$

Thus refraction at a change in electrostatic potential occurs and obeys the same type of expression as Snell's law of optical refraction at the interface between two media. The electron path bends toward the normal to the equipotential surfaces if V is increasing. If we let $V = 0$ where K of the electron equals zero, we have the ordinary optical law with the index of refraction of the medium replaced by $\sqrt{V}$. The same result is obtained, of course, if we treat the electrons as if guided by waves with $\lambda = h/mv$ (see prob. 12-62).

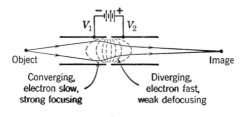

Fig. 12-27. Electrostatic lens composed of two coaxial cylinders. The electron paths bend toward the normals to the equipotential surfaces (these surfaces are indicated by the dashed lines).

Of course the idealized situation of Fig. 12-26 is not common, but the principle of bending toward the normal to the equipotential surface when V is increasing enables us to show how focusing in electrostatic lenses occurs. Furthermore, eq. 12-17 is the basis of numerical methods of ray tracing in a region of continuously varying V.

A practically useful electrostatic lens is illustrated in Fig. 12-27. Two coaxial circular cylinders are at potentials V_1 and V_2. The equipotentials are symmetrical about the plane between the cylinders. The refracting of the electron trajectories is less in the right side than in the left, because the average V is higher there, and therefore a given ΔV between two equipotentials V' and V'' corresponds to a smaller ratio $\sqrt{V''/V'}$. Another way of looking at this is to note that an electron's energy (and hence momentum) is higher in the right side of Fig. 12-27, and therefore the electron is deflected less by the same field. Thus the two-cylinder lens has the effect of a strong positive lens plus a weak negative lens, and therefore it brings divergent rays back to a focus. This

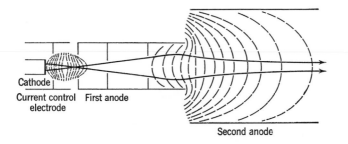

Fig. 12-28. An electron gun. The equipotentials are the dashed lines. All the electrodes are cylindrically symmetrical. (From Zworykin, Morton, Ramberg, Hillier, and Vance, *Electron Optics and the Electron Microscope*, Wiley, New York, 1945.)

result occurs for electrons starting from any point in the object plane, but the image quality deteriorates if the object size is more than a small fraction of the diameter of the cylinders.

A more complicated electrostatic lens system is illustrated in Fig. 12-28. This is an *electron gun* designed to produce a concentrated electron beam on a small spot remote from the electron emitter, as illustrated in Fig. 12-15, for example. Electrons from the oxide-coated cathode are focused as nearly to a point as their initial velocities will permit, the crossover point near the cathode. The lens action between the first and second anodes produces an image of this crossover at a distant point. Variation of the potential of the control electrode varies the total current in the beam in the same way that a grid in a triode varies the current (namely, by varying the depth of the space-charge potential minimum near the cathode). Electron guns are useful in television picture tubes and in many other devices and experiments (Figs. 10-19 and 12-20, for example).

A thin magnetic lens is illustrated in Fig. 12-29. The cylindrical coil

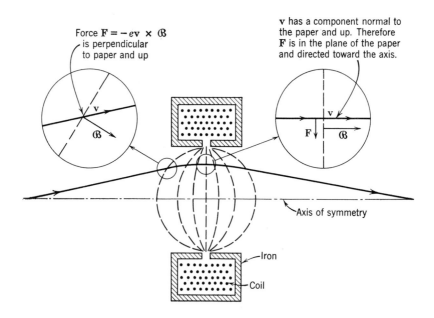

Fig. 12-29. The principle of the operation of a short magnetic lens. Incoming electrons are accelerated in the azimuthal direction. The azimuthal component of velocity (rotation about the axis of symmetry) then produces a force toward the axis. Since this force is proportional to the distance of the electron from the axis, focusing is achieved. (The dashed lines are magnetostatic equipotentials; the magnetic field lines are perpendicular to the equipotentials.)

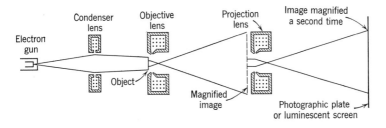

Fig. 12-30. Magnetic electron microscope. Each of the three lenses is similar to that shown in Fig. 12-29. The specimen to be studied is the "object" of the objective lens.

is almost completely surrounded by iron in order to confine appreciable magnitudes of ℬ to a short distance along the axis. As the electron enters the lens it experiences a force that accelerates the particular electron illustrated outward, as shown in the first inset. For *any* incident electron the force is such as to rotate the electron in a clockwise sense (when looking toward the right in Fig. 12-29). Thus the electron acquires an azimuthal velocity component. This component is at right angles to ℬ, and therefore a force is produced toward the axis of the lens, as shown in the second inset. Thus focusing is achieved. One feature of all magnetic lenses is the rotation of the image relative to the object; it is caused by the azimuthal velocity component mentioned. The short magnetic lens described here is only one example. A long solenoid also acts as a lens; solenoidal lenses are used in tubes like Fig. 12-20 and in spectrometers for measuring the energy distributions of β rays from radioactive nuclei.

An important application of electron optics is to the electron microscope illustrated in Fig. 12-30. A narrow beam of electrons with kinetic energies of the order of 100,000 e.V. is incident upon the specimen. The specimen is a very thin (200 to 1000 Å) section of the material to be studied or a *replica* of a thick specimen such as a metal surface. (A replica is made by depositing a thin layer of a plastic or other film on the surface to be studied and then stripping off this thin "negative" of the surface for use in the microscope.) A larger fraction of the incident electrons are scattered out of the beam by regions of the specimen that are either denser or thicker. The electron beam leaving the object therefore has regions of differing intensities, like the light beam leaving the specimen of a light microscope. The object is magnified in two stages, and the highly magnified image is recorded on a photographic plate.

The resolution of the electron microscope is much greater than that of the light microscope. Both are limited by diffraction as mentioned in Sec. 4-12.   The wavelength of a 100,000-e.V. electron is much less than the wavelength of visible light, and therefore much smaller details can be resolved in the electron microscope.   On the other hand, the numerical aperture (sin θ in eq. 4-23) cannot be so large with magnetic lenses as with light optics.   The reason is that it is difficult to correct a magnetic lens for spherical aberration, and therefore very narrow beams must be used in order to minimize this aberration.   Furthermore, since contrast among different parts of the image is obtained by scattering of electrons (instead of by absorption, as in the light microscope), increasing the numerical aperture would permit scattered electrons to join in the image and thereby to reduce the contrast.   Unless the speci-

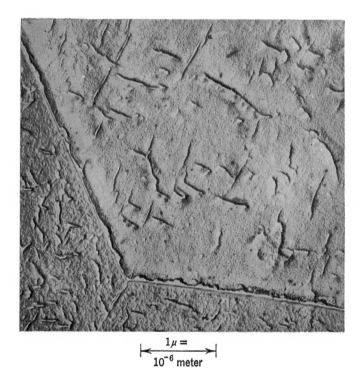

$$1\mu = $$
$$\longleftarrow \qquad \longrightarrow$$
$$10^{-6} \text{ meter}$$

Fig. 12-31. Photograph made with an electron microscope, illustrating precipitates of $Fe_{16}N_2$ in pure iron.   These precipitates are responsible for the magnetic coercive force of this specimen.   Grain boundaries are visible at the lower left.   The surface of the iron was electropolished, and carbon was evaporated onto the surface to form a replica 150 Å thick.   The replica film was stripped, and gold and chromium were evaporated onto it at a 20° angle in order to increase the contrast of the film, which was then used as the specimen in the electron microscope.   (Photograph by C. Wert, J. Kerr, T. Noggle, and A. Vatter, University of Illinois.)

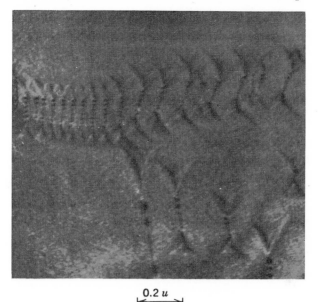

0.2 *u*
|←————→|
2000 Å

Fig. 12-32. Transmission electron micrograph of an array of dislocations in a film of niobium a fraction of a micron in thickness. The connection between etch pits (Fig. 10-9 and Frontispiece) and dislocations is demonstrated here, since each of the approximately 23 dislocations runs from an etch pit (a tiny "V") on the upper surface to an etch pit (an inverted "V") on the lower surface of the thin film. Each dislocation line appears in the photograph as a series of 3 to 6 dots, resulting from electron diffraction effects produced by the extra half-plane of atoms at a dislocation. (From A. Berghezan and A. Fourdeux, *Propriétés des joints de grains*, Presses Universitaires de France, Paris, 1961, pp. 127–147; photograph courtesy of Dr. R. H. Gillette, European Research Associates.)

men is very thin, chromatic aberration because of varying energy losses of electrons in the specimen and multiple scattering of electrons in the specimen may also limit the resolution.

The design of an electron microscope represents a compromise between diffraction (too low a numerical aperture) and spherical aberration and loss of contrast (too high a numerical aperture). The numerical aperture is usually between 0.01 and 0.001 for electron microscopes. Resolutions of 6 Å have been obtained with the most suitable specimens, and resolutions of 20 Å can be obtained even on replicas. Typical electron micrographs are shown in Figs. 12-31 and 12-32. The electron microscope is an important tool in chemical and metallurgical engineering, biological and medical research, and solid-state physics.

REFERENCES

Thermionic Emission

C. Herring and M. H. Nichols, *Revs. Mod. Phys.*, **21**, 185–270 (1949).

A. S. Eisenstein, "Oxide-Coated Cathodes" in *Advances in Electronics*, Vol. I, edited by L. Marton, Academic Press, New York, 1948, pp. 1–64.

V. L. Stout, "Thermionic Emission and Thermionic Power Generation," in *Thermoelectricity*, edited by P. H. Egli, Wiley, New York, 1960.

Photoelectric Emission

A. L. Hughes and L. A. DuBridge, *Photoelectric Phenomena*, McGraw-Hill, New York, 1932.

V. K. Zworykin and E. G. Ramberg, *Photoelectricity and Its Application*, Wiley, New York, 1949.

Secondary Emission

H. Bruining, *Physics and Applications of Secondary Electron Emission*, Pergamon Press, London, 1954.

H. S. W. Massey and E. H. S. Burhop, *Electronic and Ionic Impact Phenomena*, Clarendon Press, Oxford, 1952, Chapter V.

Collisions

H. S. W. Massey and E. H. S. Burhop, *Electronic and Ionic Impact Phenomena*, Clarendon Press, Oxford, 1952.

S. C. Brown, *Basic Data of Plasma Physics*, Wiley and Technology Press, New York, 1959.

D. J. Rose and M. Clark, Jr., *Plasmas and Controlled Fusion*, Wiley and Technology Press, New York, 1961.

Electron Optics

V. E. Cosslett, *Introduction to Electron Optics*, Clarendon Press, Oxford, 2nd Ed., 1950.

C. E. Hall, *Introduction to Electron Microscopy*, McGraw-Hill, New York, 1953.

PROBLEMS

12-1. Why does the free-electron mass m rather than the effective mass m^* (which is different for different solids) appear in the Richardson-Dushman equation (eq. 12-6)? *Hint:* The sorting of electrons into the group that is emitted and the group that is turned back by the surface field occurs at about what region of x values in Fig. 12-1? How does the $P(x)$ to be inserted into the Schrödinger equation to determine the electron wave function vary with x in this region?

12-2. * Use $j(T)$ from eq. 12-6 and $W(T)$ from eq. 4-17 (with the correction for the emissivity ϵ) to calculate the ratio of thermionic current to radiated power. Compute the temperature T_m that maximizes this ratio in terms of ϕ. Show that for observed magnitudes of ϕ, T_m is impractically large, and hence that practical thermionic emitters operate at the highest T that gives a tolerable emitter life.

12-3.* Calculate the thermionic emission current density j (zero electric field) for tungsten at $2500°K$.

12-4. Make the same calculation as in prob. 12-3 but for LaB_6 at a temperature of $1800°K$.

12-5.* Calculate ϕ in volts from the data of Fig. 12-6.

12-6.* Refer to Fig. 12-6, and assume that the value of j at $2000°K$ is perfectly accurate. Assume that the experimental uncertainty in the slope (uncertainty in ϕ) is $\pm 1\%$. Calculate the ratio of A_{max} to A_{min}, where A_{max} is the A value calculated with $\phi = 1.01\phi_0$ and A_{min} is the A value calculated with $\phi = 0.99\phi_0$. (This calculation shows why it is so difficult to obtain good experimental values to compare with A_0.)

12-7.* An anode is a plane parallel to a plane metal cathode and at a distance 0.01 m. from the cathode. A potential difference of 1000 volts is applied between these electrodes. What is the field at the cathode? What is the magnitude of the distance x_0 in angstroms (eq. 12-7)? What is the amount $\Delta\phi$ by which the work function ϕ is lowered by Schottky effect? If the cathode temperature is $1700°K$, what is the ratio of j with this field to the zero field value of j? Assume no space charge in any part of this problem.

12-8. An anode is a circular cylinder of diameter 0.01 m. concentric with a circular cylinder filamentary cathode of diameter 10^{-4} m. A potential difference of 1000 volts is applied between these electrodes. Answer the same questions as for prob. 12-7. (Assume no space charge in any part of this problem; see also prob. 12-9.)

12-9. In prob. 12-8 it is a great convenience to assume that the strength of the applied field at the cathode surface is the same as at the position x_0. This permits us to assume that the dashed line of Fig. 12-4b is approximately a straight line over the region of interest, and hence that the Schottky theory for a constant field can be applied to this case. Calculate $\mathcal{E}$ at x_0, and compare with $\mathcal{E}$ at $x = 0$ (surface of filament). Is the use of the usual Schottky theory justified here?

12-10. Compare the values of x_0 of probs. 12-7 and 12-8 with the order of magnitude of the spacing between atoms in a solid. Is x_0 large enough to cause the atomic structure of the solid to be small compared to x_0? That is, is the *plane* approximation to the surface (Fig. 12-3) justified?

12-11. The Fermi energy E_0 in a metal is a function of the number of electrons per unit volume (see Sec. 9-4). Assume that the only change in E_0 with T is caused by the expansion of the solid. Calculate $d\phi/dT$ for copper by assuming that the only change in ϕ is caused by the change in E_0. Insert the $d\phi/dT$ you calculate from the room-temperature expansion coefficient into eq. 12-10, and calculate A/A_0. Is this deviation from A_0 in the correct direction for metals generally? Is it of the correct order of magnitude?

12-12.* Give a crude estimate of the change $e\,\Delta\phi$ in the work function of a tungsten surface produced by covering it with a monatomic layer of thorium positive ions (Fig. 12-7). Assume that the thorium ions are in a cubic array with nearest neighbor distance equal to 5 Å and that the average charge of each is 0.2e. Assume that the positive and negative charges are in thin parallel sheets (negative on W, positive on Th) a distance 2 Å apart. *Hint:* Use Gauss' law.

12-13. Show that the fraction of thermionically emitted electrons having kinetic energies between E and $E + dE$ is proportional to $e^{-E/kT}$. *Hint:* Sketch $f(E)$ from eq. 12-4 in the region $E > E_0 + e\phi$.

12-14. Devise an experimental tube to measure the energy distribution dj/dE_x of thermionic electrons. What function of current to what electrode is plotted as a function of the voltage of that electrode in order to obtain a straight line? What should be the slope of this line?

12-15. Show that the fraction of thermionically emitted electrons with y-associated energies between E_y and $E_y + dE_y$ is proportional to $e^{-E_y/kT}$. *Hint:* Consider eq. 12-5, integrate over dp_z and dp_x but not over dp_y, and differentiate both sides. This procedure gives dj/dp_y, which can then be converted to dj/dE_y.

12-16.* The total emissivity (see Sec. 4-8) of tungsten at 2500°K is about 0.31, and that of an oxide-coated cathode at 1000°K is about 0.35. Use the j values quoted in Sec. 12-2 to compute the electron-emission efficiency of both types of emitter in amperes of electron emission per watt of heating power. Assume all the heating power to be dissipated by radiation.

12-17. Sketch schematically on a single plot curves of $j(V)$ for a thermionic emitter with two different values of T. Assume T and V values that make both space-charge-limited and emission-limited regimes of $j(V)$ apparent at each T.

12-18. Show from Poisson's equation (eq. 12-11) and the density of electronic space charge $\rho = -j/(2K/m)^{1/2}$, where K is the kinetic energy of an electron, that the space-charge-limited current density j of a plane-parallel diode is given approximately by eq. 12-12. *Procedure:* Choose new variables x and V with slightly different origins from those of Fig. 12-9 by letting $x = 0$ and $V = 0$ at the position of the potential energy maximum; let $x = x_1$ and $V = V_1$ at the anode. Use the conservation of energy to obtain K in terms of $V = -P/e$ and hence to obtain eq. 12-11 in a form where the only variables are V and x. Set $K = 0$ at $x = 0$, usually a good approximation since ordinarily $V_1 \gg kT$. Multiply both sides by $2(dV/dx)$ and note that one side can now be easily expressed in terms of the derivative of $(dV/dx)^2$ and the other in terms of the derivative of $V^{1/2}$. Thus integration is obvious by removing the d/dx signs, and the arbitrary constant can be evaluated by setting $dV/dx = 0$ at $x = 0$. Integrate the resulting equation for dV/dx by separating the variables.

12-19.* If the anode-cathode potential difference of a space-charge-limited thermionic diode with cathode at 1000°K changes from 100 volts to 195 volts, how much does the current change? How much does the height of the space-charge barrier change? (Note that your result justifies the approximation procedure in the preceding problem of assuming that the barrier height is fixed and hence that it can serve as an origin for measuring V.)

12-20. Sketch schematically the potential energy of an electron as a function of its distance x from the cathode in a space-charge-limited vacuum triode with negative control grid. Sketch on the same plot (with a dashed line) the new potential energy when the control grid takes on a new, more negative voltage, and show how the anode current is reduced by this change in grid voltage.

12-21. ˙Draw Fig. 4-3 as it would be if there were a contact potential difference between the collector and the emitter. Assume that the collector is clean nickel and the emitter is clean barium ($\phi = 2.50$ volts for barium). Draw and label an energy diagram like Fig. 12-10a for this situation when the applied voltage is just enough to stop the most energetic electrons ($V = -V_0$) for a particular $h\nu$.

12-22. Draw Fig. 4-24 as it would be if the cathode work function is 1.2 volts and the grid and plate are both contaminated nickel with a work function of 3 volts.

12-23.* Consider a precision measurement of h/e by the Duane-Hunt limit experiment (Sec. 4-6) in which the cathode work function is 4.5 e.V. (pure tungsten) and the anode work function is 9.1 e.V. (oxidized tungsten). Sketch the potential energy of an electron as a function of distance from inside the cathode to inside the anode, with a measured anode-cathode voltage (difference in Fermi levels) of V_0 volts. What is the maximum energy that an electron can lose inside the anode (which equals $h\nu_{max}$)?

12-24. Consider a vacuum tube consisting of a plane platinum electrode and a plane tungsten electrode parallel to it and connected to it by a platinum wire, all at constant temperature. We might at first sight think that the electric field in such a tube could do work and increase the kinetic energy of an electron. This would violate the first law of thermodynamics, since the system has no energy sources. Show that no work can be performed by the field in this tube in each of the following cases: (a) An electron is emitted by the platinum plate and absorbed by the tungsten plate. (b) An electron enters the region between the two plates from an external source, is deflected, and proceeds to an external collector.

12-25.* Webster and Beggs have constructed a thermionic energy converter with both the hot and cold electrodes of platinum coated with a mixture of BaO and SrO. What is the A factor of such an electrode if its $\phi = 1.1$ volts at $T = 330°K$ and $\phi = 1.6$ volts at $T = 1000°K$?

12-26.* Assume that Fig. 12-11 is drawn to scale and ignore the (important) thermal radiation from hot to cold electrodes. What is the efficiency of conversion of heat to electricity in terms of the parameters of the figure? What is the approximate numerical value of the efficiency? What is the maximum efficiency permitted by thermodynamical reasoning if the temperatures are those used in prob. 12-25?

12-27.* As in the preceding problem, calculate the energy output of a device with an area of 10^{-4} m.^{2} if Fig. 12-11 is drawn to scale, if $eV_c = 1.6$ e.V., and if the current is 2 amp for this area. Calculate the efficiency, including the loss by radiation from the cathode at $1000°K$ (eq. 4-17 with $\epsilon = 0.30$). Assume that the cool electrode neither emits nor reflects radiation.

12-28. Sketch the replacement for Fig. 12-11 if the device is open-circuited ($j = 0$). Where is the potential energy maximum if T_a and T_c are related in such a way that $e\xi_1 = e\xi_2$?

12-29.* Verify the necessity for close spacing in a vacuum thermionic energy converter as follows: Assume that $e\xi_2$ in Fig. 12-11 is 0.4 e.V. and use eq. 12-12 (only a fair approximation here because kT is not negligibly small with respect to V_1). Calculate the spacing required if j is to be 2×10^4 amp/m.^{2}

12-30.* What is the current density j from a tungsten point subjected to an electric field of 3×10^9 volts/m.? If the area of the point is 10^{-12} m.^{2}, what is the total current?

12-31. In the field-emission electron microscope, variations in the work function from point to point on a tiny hemispherical emitter tip (radius = 6×10^{-7} m.) of a tungsten wire are revealed by variations in the field-emission current. A spherical fluorescent screen of radius 0.05 m. concentric with the tip is maintained at a high positive potential and indicates the current variation. Sketch the tube, compute the screen potential necessary to give the current density of prob. 12-30, and compute the magnification.

12-32.* Consider an electron, initially at rest, that is excited by a photon with $h\nu = 3$ e.V. The photon, of course, gives up all its energy to the electron. What is the electron's momentum? How does this compare with the original photon momentum $h\nu/c$? (This shows that such an excitation of a *free* electron is not possible, since energy and momentum cannot both be conserved; if the electron is bound to a solid or a surface, the additional momentum can come from other electrons or ions.)

12-33.* If 1 microwatt of light of $\lambda = 3600$ Å is falling on a Cs_3Sb photocell surface, what is the photoelectric current?

12-34. Sketch the logarithm of the photoelectric yield expected for n-type germanium as a function of $h\nu$ of the incident photons (constant intensity, room temperature).

12-35. Assume that the target is clean nickel, the beam current is constant, and the collector voltage is sufficiently positive to collect all the secondary electrons in the experiment of Fig. 12-15. Assume that the beam current is 10 milliamperes. Plot the target current as a function of beam voltage V.

12-36. Assume that the δ vs. K_p curve of platinum has the same shape as Fig. 12-16. Estimate the first cross-over energy of platinum, the smaller of the two values of K_p that give $\delta = 1$.

12-37. Devise a vacuum tube to permit the measurement of the energy distribution of secondary electrons. Sketch the tube, and explain its operation.

12-38. What effect do secondary electrons emitted at the edges of the electrodes have on the e/m experiments of Figs. 1-3 and 1-4? How can this effect be minimized?

12-39. How would Fig. 12-16 be altered if the primary electrons, instead of arriving at normal incidence, made a large angle ($\sim$75°–85°) with the normal to the nickel surface?

12-40.* Use the fact that the mean depth of origin of secondary electrons is ~30 Å to estimate the time delay between bombardment (primary enters the target surface) and emission (secondary leaves this surface). (This time has not been measured, but experiments have proved that it is $<10^{-10}$ sec.)

12-41. All experiments on secondary emission show that δ is independent of the current density j. What does this fact reveal about the process of secondary emission? *Hint:* Suppose excitation by two or more primary electrons were required to produce an internal secondary.

12-42. Consider a vacuum tetrode with cathode at zero potential, negative control grid, positive screen grid at 300 volts, and clean nickel anode at V_a volts. Sketch schematically the current to the anode as a function of V_a, including the effect of the emission of secondary electrons from the anode but neglecting any such emission from the screen grid. (In a pentode tube, a suppressor grid at cathode potential is placed close to the anode to turn back secondaries emitted from the anode.)

12-43.* Calculate the number N of atoms per m.^{3} in a gas if the pressure is p mm of mercury (760 mm = 1 atmosphere) and the temperature is 300°K.

12-44.* If a 1-milliampere beam of electrons of 3-e.V. energy enters a region of mercury gas at a pressure of 0.002 mm of mercury, what is the beam intensity after the beam has passed through 0.1 m. of the gas? ($T = 300$°K.)

12-45.* The diameter of the argon atom is about 3.6 Å. Suppose that an electron with the critical kinetic energy for Ramsauer transmission (0.5 e.V. in argon) encounters a one-dimensional square well with a width of 3.6 Å (this is a very crude model of the argon atom). What must be the depth of the well for 100% transmission of the electron wave?

12-46.* If a beam of 100 amp/m.^{2} of 10-e.V. electrons traverses mercury vapor at a pressure of 0.002 mm of mercury, how many photons with $\lambda = 2536$ Å are produced per unit volume per second? ($T = 300$°K.)

12-47. Why is an alkali-metal gas an especially good gas for a gas-discharge rectifier tube in which a low-voltage drop across the tube is desired?

12-48. Sketch an experimental tube for measuring the ionization cross section of mercury for electron energies between 10 and 30 e.V. What would be a suitable mercury pressure in the tube? What difficulties would be experienced if the mercury pressure were too high? What would be a convenient way of obtaining and maintaining the desired pressure in a tube (*not* attached to a vacuum system)?

12-49.* Calculate the *total* positive-ion current per meter of electron path per milliampere of electron current for 200-volt electrons incident on mercury vapor at a pressure of 10^{-4} mm of mercury. (Assume that all ions formed are collected by the same electrode; let $T = 300$°K.)

12-50. In an experiment like prob. 12-48, the electron energy must be 79 e.V. or greater in order to produce He^{++} ions. On the other hand, it was explained in Sec. 6-4 that the second ionization potential of He was just four times the ionization potential (13.6 volts) of hydrogen. Explain.

12-51. * From the threshold voltages for the production of Hg^+, Hg^{++}, etc., ions given in the text, calculate the energy required to remove the first, second, etc., electron from Hg. Why does this energy increase with the ordinal number of the electron to be removed?

12-52. In a gas maser like the helium-neon maser described in Sec. 12-8*d*, why do we not use transitions to the *ground* state of the emitting atom (neon in the example cited)?

12-53. * Estimate the recombination cross section for electrons incident on positive ions as follows: Take the lifetime of an excited electronic state as 10^{-8} sec, the diameter of the ion as 5 Å, and the energy of the electron as 1 e.V. Estimate the fraction of the encounters between electron and ion that give recombination. Multiply this by the geometrical cross section of the ion to give the recombination cross section. (Experimental data are meager, but the observed order of magnitude of σ_{rec} is 10^{-25} m.²)

12-54. If a 12-e.V. electron is incident on an H_2 molecule and the electron is moving initially in the *x*-direction, in what directions would the two dissociated H atoms be emitted, or would all directions be equally likely?

12-55. Precisely (from theory) how many electron volts is the curve for H + H* at $R = \infty$ above zero in Fig. 12-25?

12-56. Suppose that the curve (*e*) in Fig. 12-25 were moved 0.2 Å to the right. Dissociation by excitation to this state could then occur. How would the distribution of kinetic energies of the resulting H^+ ions differ from those from the (*f*) state? (Sketch distributions as functions of ion kinetic energy.)

12-57. * Estimate the time required to excite a hydrogen molecule to the state (*b*) in Fig. 12-25 by electron impact. Assume a molecular diameter of 3 Å and an incident electron kinetic energy of 100 e.V. Estimate the time required for the atoms of the excited molecule to become separated by an amount $R = 4$ Å by approximating the curve (*b*) in Fig. 12-25 by straight line segments through the points (13 e.V., 0 Å), (0 e.V., 2 Å), and (0 e.V., ∞ Å).

12-58. Consider the structure of methane given in Fig. 7-9 and discuss the dissociation products to be expected if methane is bombarded by electrons in a mass-spectrometer ion source. Specifically, are CH_3, H, or H_2 likely products? Suppose the structure were a linear molecule with the carbon atom in the center, and answer the same questions.

12-59. In the text it was mentioned that it was convenient to treat the electron as a wave packet in electron-optics aberration calculations. Should Planck's constant *h* enter into the results (for spherical aberration, for example)? Why?

12-60. * Derive the equivalent of eq. 12-17, but assume that the kinetic energy of the electron equals K_0 at the place where V is set equal to zero. If $K_0 > 0$, is this electron refracted more or less (by the same electrostatic lens) than an electron with $K_0 = 0$?

12-61. Show that if $\mathcal{B} = 0$ and if $\mathcal{E}$ is an arbitrary function of x, y, and z but independent of time, the trajectories of all charged particles are the same, regardless of their e/M values, if they all have zero velocities at the same point. This result means that through path measurements alone (i.e., without measuring transit times), it is impossible to construct a mass spectrometer unless either a magnetic field or a time-varying electric field is employed. *Hint:* Set eq. 12-16 equal to the product of mass and acceleration. Transform the time variable by writing $t' = at$, where a is a constant, to eliminate e/M from the equation.

12-62. Prove eq. 12-17 by considering the refraction of de Broglie waves.

12-63. Consider the change in refracting power (reciprocal of focal length) of the lens of Fig. 12-29 as the current I in the coil is changed (assume no saturation in the iron). How should the refracting power depend on I (i.e., proportional to I, I^2, $1/I$, etc.)? How should the angle through which the image is rotated depend on I?

12-64.* What is the de Broglie wavelength λ of a 100,000-e.V. electron? Does relativity make a change of more than 1% in your answer? Use this λ, and calculate the limit to resolution imposed by diffraction if $\sin \theta = 0.003$ in an electron microscope. Express your result in angstroms.

12-65. X-rays cannot be focused by lenses, since the index of refraction for all materials is very nearly unity. Explain how an X-ray microscope can nevertheless be constructed by taking advantage of the use of electron lenses to focus an electron beam to a very fine spot ($\ll 1\mu$).

13

Nuclear Physics

13-1 INTRODUCTION

In this chapter we return to the study of elementary
nuclear physics begun in Chapter 3. This study was interrupted to develop wave mechanics, which is indispensable
in understanding nuclear processes. The present chapter
could have followed Chapter 6, but the study of molecules
and solids has been undertaken first because that study
followed more directly from the study of atoms and because it is frequently helpful in the treatment of nuclei
in the present chapter and in Chapter 14. Only a brief
introduction to nuclear physics is provided by Chapters 1,
3, 13, and 14. Nuclear physics is a large area of modern
physics that is being actively investigated.

The phenomena and the quantum-physics explanation
of natural radioactivity will be described in Sec. 13-2.
The proof that the nucleus cannot contain electrons, but
must consist of protons and neutrons, will be given in
Sec. 13-3, which also presents the experimental facts about
the size of nuclei. The emission of β-particles from nuclei
and the evidence for the existence of the neutrino will be
discussed in Sec. 13-4, and γ emission will be considered in
Sec. 13-5. In Sec. 13-6 the basic quantum physics underlying the precise masses of nuclei will be explained, in
preparation for the important question: Why are certain
nuclear species stable and others not? Artificially produced nuclear reactions will be discussed in Sec. 13-7. The

492

particles produced by high-energy accelerators and discovered in cosmic rays will be considered briefly in Sec. 13-8, and the probable role of one such particle (the pion) in nuclear forces will be examined.

Since the nature of nuclear forces is not thoroughly understood, it is not possible to present a theory of nuclear behavior like the theory of the hydrogen atom. The theory of nuclei is like a quantum theory of the hydrogen atom without the use of Coulomb's law: We predict discrete energy levels for bound states, but we are unable to calculate the energy levels. We predict tunneling of charged particles through narrow potential barriers, but we are unable to give all the details of these barriers. Nevertheless, the combination of quantum theory, which tells the qualitative nature of the phenomena, and experimental data, which fill in the quantitative aspects, permits a working understanding of nuclear physics.

13-2 RADIOACTIVITY; α EMISSION

All nuclides (that is, all nuclear species) with $Z > 83$ are radioactive; they disintegrate spontaneously to give energetic particles or γ-rays plus other nuclides. For example, $_{88}Ra^{226}$ (radium) disintegrates into $_2He^4$ (an α-particle) and $_{86}Em^{222}$ (the $A = 222$ isotope of the element emanation is called "radon" because it was observed as a gaseous decay product of radium). This reaction is written in the notation introduced in Chapter 3 (compare eq. 3-10):

$$_{88}Ra^{226} \rightarrow {}_{86}Em^{222} + {}_2He^4 \tag{13-1}$$

The α-particle energies can be measured by their deflections in a magnetic field and by other methods that will be described in Sec. 14-5. Most of the energy arising from the difference in mass between Ra^{226} and $(Em^{222} + He^4)$ appears as kinetic energy of the α-particle, and the rest appears as kinetic energy of the radon (see prob. 13-1). The number of disintegrations per unit time is usually specified in *curies*. One curie is 3.7×10^{10} disintegrations/sec.

The number of disintegrations per unit time in a pure sample of a radioactive element is found experimentally to be an exponentially decreasing function of time. Figure 13-1 gives data for the observed disintegration rate of radon, which disintegrates by α-particle emission. Let n be the number of radioactive nuclei in the sample. Disintegrations decrease the number of radioactive nuclei, and $|dn/dt|$ is the rate of loss of radioactive nuclei of this species. It is also the observed rate of α-particle emission. dn/dt is, of course, a *negative* number. The natural logarithm

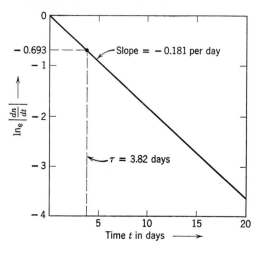

Fig. 13-1. The logarithm to the base e of the number of disintegrations per unit time in a sample of radon ($_{86}Em^{222}$) plotted as a function of time. At $t = \tau$, one-half of the radioactive nuclei have decayed and the disintegration rate is one-half its initial value.

$\ln |dn/dt|$ is shown by Fig. 13-1 to be related to the time t (in days) by

$$\ln \left| \frac{dn}{dt} \right| = C - 0.181t$$

$$\frac{dn}{dt} = -e^{C-0.181t} = -e^C e^{-0.181t}$$

where C is a constant.

In order to determine n as a function of time we must integrate this equation and use the fact that $n = 0$ at $t = \infty$ (all the nuclei will have disintegrated by the time $t = \infty$):

$$\int_n^0 dn = -e^C \int_t^\infty e^{-0.181t} \, dt$$

$$n = \frac{e^C}{0.181} e^{-0.181t} = n_0 e^{-0.181t} \tag{13-2}$$

Here we have let n_0 equal the value of n at time $t = 0$. Thus the number of radioactive nuclei of a single nuclide in a sample is an exponentially decreasing function of time. Similar expressions hold for all radioactive nuclei with a different constant in the exponent for each radioactive nuclide.

The usual way of specifying how fast disintegrations occur is by specifying the *half-life* τ. This τ is defined as the time it takes for one-half
of the initial number of nuclei in a sample to disintegrate. In terms of τ,
the expression for n for *any* radioactive nucleus is

$$n = n_0 e^{-0.693t/\tau} \tag{13-3}$$

We can show that this is consistent with the definition of τ by solving
eq. 13-3 for n at the time $t = \tau$:

$$n = n_0 e^{-0.693} = n_0/2$$

Radioactive elements present in nature exhibit values of τ from as long
times as can be measured ($>10^{10}$ years) to as short times as can be
measured ($<10^{-8}$ sec). (An alternate way of specifying the disintegration rate is by specifying the value of the *disintegration constant* λ, which
is simply the coefficient $0.693/\tau$ of $-t$ in the exponent of eq. 13-3.)

The value of τ for nuclei of a particular nuclide (e.g., radium, $\tau =$
1620 years) cannot be changed appreciably by varying such laboratory
conditions as temperature and pressure. The energies involved in nuclear reactions are very large (several million electron volts), as noted in
Sec. 3-4. The change in the kT energy that can be produced by varying
the temperature is only a few tenths of an electron volt and hence is far
too small at ordinary temperatures to affect the course of nuclear reactions.

The striking point about the exponential decay law of eq. 13-3 is
that the number of nuclei disintegrating per second, $-dn/dt$, is proportional to n, the number of radioactive nuclei still remaining, which can be
verified by differentiating eq. 13-3:

$$\frac{dn}{dt} = -\left(\frac{0.693}{\tau}\right)n \tag{13-4}$$

Since only the number remaining, and not some function of the time
since the radioactive nuclei were created, appears here, a particular
nucleus does not "remember" how long a time has elapsed since it was
created. It has just as large a chance of disintegrating in unit time now
as it has in unit time tomorrow, provided, of course, that it has not disintegrated in the time between these two observations. Whether or not a
particular nucleus will disintegrate in a certain time interval is entirely
a question of probability.

Another experimental fact learned by observations on α-particle
emitters is that there is a relation between the kinetic energy K of the
α-particle and the half-life τ of the various nuclides. This relation is

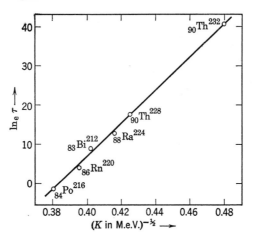

Fig. 13-2. Dependence of the logarithm to the base e of the half-life τ (in seconds) on the kinetic energy K of emitted α-particles for a series of radioactive nuclei.

illustrated in Fig. 13-2. The natural logarithm of the half-life τ is plotted against $K^{-\frac{1}{2}}$ for a series of six α-emitters. Evidently

$$\ln \tau = AK^{-\frac{1}{2}} - B \qquad (13\text{-}5)$$

where A and B are constants. The range of τ values plotted is more than a factor of 10^{18}, ranging from 0.16 sec to 1.4×10^{10} years.*

Both eq. 13-5 and eq. 13-4 can be satisfactorily explained by quantum mechanics; we shall now outline this explanation. Figure 13-3 illustrates the potential energy of an α-particle in the neighborhood of a heavy nucleus. In order to construct this curve we have used two facts: (1) At large distances r from the center of the nucleus, the nucleus repels the α-particle according to the usual Coulomb law. This fact was demonstrated by experiments described in Sec. 3-2. The charge of the α-particle is $+2e$, and the charge of the nucleus after emitting the α-particle is $+(Z - 2)e$ (its original charge was $+Ze$). (2) At short distances ($r < r_0$) the nuclear attractive forces dominate. These forces are strong but of *short range*. Therefore a very steep-sided well of attractive energy results. There are abundant experiments that demonstrate this property of nuclear forces. For example, we know from the Rutherford scattering

* An expression similar to eq. 13-5, but with a different function of K, was proposed very early by Geiger and Nuttall and is called the Geiger-Nuttall law. Since the range of K values of available data is so small, their function gave as good a fit to the data as eq. 13-5.

experiments of Sec. 3-2 that the Coulomb force is valid without modification for $r > 2 \times 10^{-14}$ m.; therefore the nuclear forces must be negligible at such distances. We know that there must be a deep attractive well because of the observed binding energies of nuclei.

The nucleus of Fig. 13-3 contains many protons and neutrons. Occasionally a group of two protons and two neutrons are together near a boundary of the attractive well and have momentum directed toward this wall. In other words, an α-particle is moving toward the wall. We henceforth treat this problem as if the α-particle exists inside the nucleus and is vibrating inside the well in a quantum state of energy E. The fact that the α-particle has such a large binding energy compared to other combinations of n's and p's (see Fig. 3-14 or 13-13) makes this approximation a good one and makes it more reasonable that an α-particle (rather than, for example, a combination of one proton and one neutron) is emitted. We shall assume that the α-particles emitted had a unique energy E inside the nuclear wells from which they came. We have noted above that the nuclear force must create an attractive potential well, and we know from the analysis of Chapter 5 that bound states in such a well have discrete energy values.

If the energy E is less than zero (with V set equal to zero at $r = \infty$), the nucleus is stable with respect to α disintegration, since energy would have to be supplied to remove the α-particle from the nucleus. If E is greater than zero and less than $P(r_0)$, the nucleus is metastable. No net

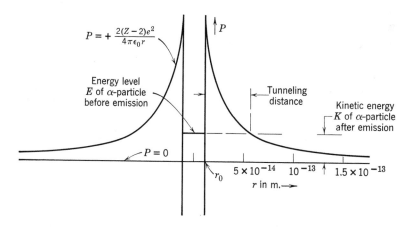

Fig. 13-3. The potential energy of an α-particle near a heavy nucleus ($Z \cong 90$). The particular nucleus shown is radioactive by α-particle emission, since the energy E is above the $P = 0$ line.

energy is required to remove the α-particle, but the potential barrier restrains it. The α-particle clearly cannot go over the top of the barrier; both calculations (prob. 13-7) and the fact that 4 to 8 M.e.V. α particles are scattered as if only the Coulomb force is present indicate that the barrier in Fig. 13-3 is many M.e.V. larger than the kinetic energy of the α-particle. But the α-particle *can tunnel* through the barrier, since we have here the type of situation described in Sec. 5-4c and Fig. 5-6. Because of the exponential tail of the wave function that describes the motion of the α-particle, there is a non-zero probability that the α-particle will appear outside the nucleus. This tunneling process is a prediction of quantum physics; classical physics provides no reasonable explanation of α-particle emission.

The rate of emission by the tunnel effect is the product of the number of times per second that the α-particle strikes the barrier and the probability of tunneling through the barrier. Since the α-particle with an energy of a few million electron volts crosses the nucleus in a time of the order of 10^{-21} sec, the tunneling probability must be very small. This probability is the same each time the α-particle approaches the barrier. The emission rate is hence independent of the age of the nucleus, and the disintegration law (eq. 13-4) can be understood.

The tunneling probability is a very sensitive function of E. A large E, as in Fig. 13-4a, means that the height $(P-E)$ and the width of the barrier are small. A small E means that height and width are large. We learned in Sec. 5-4c that the tunneling probability is an exponential function of the product of the width and the square root of the height of the square-topped barrier of Fig. 5-6. Our present problem is not so simple, but we still expect the tunneling probability to decrease very rapidly as E decreases. E cannot be measured inside the nucleus, but

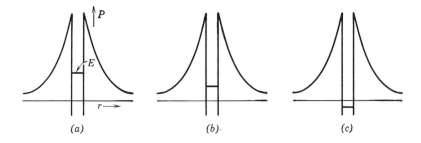

Fig. 13-4. Three nuclei, each with $Z \cong 90$, with three different energy levels for α-particles. The nucleus in (a) is radioactive with a very short half-life. The nucleus in (b) is radioactive with a very long half-life. The nucleus in (c) is stable.

the kinetic energy K of the α-particle after emission can be measured and is approximately equal to E (see prob. 13-1). Thus we expect the lifetimes of radioactive nuclei to be very long if the K of the emitted α-particle is small, and very short if K is large. The complete quantum-mechanical solution of this problem predicts eq. 13-5, including the values of the constants. This agreement between quantum theory and experiment is quite striking since the nuclei considered have τ values from 0.16 sec to 1.4×10^{10} years.

Most radioactive nuclei, both those that occur in nature and those that are artificially produced, decay by β or γ emission (Secs. 13-4 and 13-5), and some decay by spontaneous fission (Sec. 14-2). We have concentrated in this section on α emission in order to demonstrate that quantum mechanics can be fruitfully applied to nuclear processes and to learn through this example how the disintegration law (eq. 13-3 or 13-4) arises from one physical process of disintegration.

13-3 SIZES AND CONSTITUENTS OF NUCLEI

The sizes of nuclei are of interest from several different points of view. In Chapter 3 we used the Rutherford scattering experiment to set an upper limit (a radius $r_0 < 2 \times 10^{-14}$ m. for silver) to the nuclear size in order to proceed in atomic physics with the assumption that the nucleus was a charged, massive *point*. In the present section we shall need size information in order to distinguish among various possible constituents of nuclei. In Secs. 13-6 and 14-2 nuclear size will be an important ingredient in the problem of nuclear stability. We shall therefore explore the question of size here and note several methods of determining it experimentally.

Nuclei can be considered to be spherical in shape with a fairly sharp outer boundary, to an excellent degree of approximation for most nuclei. Of course, some very light nuclei (e.g., $_1\text{H}^2$) cannot be spherical, and most nuclei fail to be spherically symmetric in all their properties in that they possess spin and a magnetic moment. Nevertheless, as we shall learn in Sec. 13-6, the surface energy of nuclei assures a nearly spherical shape for all but the lightest nuclei.

In the α-emission theory outlined above, the height and width of the barrier are sensitive to the nuclear radius r_0, and the tunneling probability is in turn sensitive to this height and width. Thus comparison of the theory (which contains r_0 as a parameter) and experiments provides a method of determining r_0. Values found in this way for heavy nuclides are plotted as crosses in Fig. 13-5.

Another method of determining r_0 is from measurements of the scat-

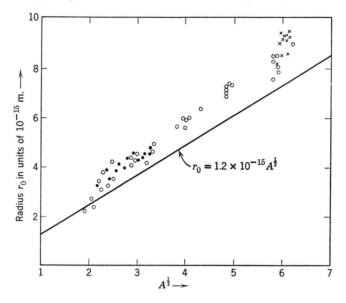

Fig. 13-5. Nuclear radius as a function of the cube root of the mass number. The crosses are from α-emission studies. The circles are from fast neutron scattering experiments (Sec. 13-7). The dots are from the comparison of the binding energies of mirror nuclei. The line is a fit to data from electron scattering and μ-mesic atom experiments.

tering of fast neutrons by nuclei; the probability of scattering can be measured and reveals the cross-sectional area of the target nucleus, as will be described in Sec. 13-7. Data from this method are plotted as the circles in Fig. 13-5. The scattering of high-energy electrons ($\lambda < r_0$) can also be used to explore the nuclear size by determining the variation of nuclear charge with distance r from the center of the nucleus. These experiments show that the charge drops rapidly to zero as r becomes greater than $r_0 = 1.2 \times 10^{-15} A^{1/3}$. Since the radius of a sphere of volume ι is proportional to $v^{1/3}$, this relation shows that the volume of a nucleus is approximately proportional to its A, and that the volume per nucleon is nearly constant.

The electrostatic repulsion of the protons in a nucleus is greater if more of them are packed tightly together. The nuclear-force binding in a nucleus (Z, A) is the same as in a nucleus $(Z + 1, A)$ since the total number of nuclear-force interactions is the same, but the electrostatic repulsion energy of the additional proton lowers the total binding energy of the nucleus an amount that depends on r_0. Comparison of the binding energies of pairs of such *mirror nuclei* thus gives a measure of r_0 (dots in Fig. 13-5).

The finite size of the nucleus makes a minute change in the energies of *s* states of atomic electrons (only *s* states have a non-zero $|\psi|^2$ at $r = 0$). The potential energy is *not* $P = -Ze^2/r$ when $r < r_0$, and thus the size of the nucleus slightly affects the atomic energy levels. Under the most favorable conditions this *isotope shift* can be measured in atomic spectra, but it is a tiny effect. This effect gives an r_0 in approximate agreement with other methods and thereby reveals the lack of a strong interaction between the electron and a nucleus. The electron interacts only by the Coulomb force and by the (weaker) interaction between its magnetic dipole moment and that of the nucleus; if there were a strong interaction (e.g., like a nucleon-nucleon force), the isotope shift would be greatly enhanced.

An ingenious method similar to that of the preceding paragraph employs *μ-mesic atoms*. These are atoms in which a negative *μ-meson* (or *muon*) occupies a quantum state in the field of a nucleus. This particle has a charge $-e$, a mass $207m_0$, and (like the electron) an extremely weak interaction with nuclei. Its larger mass reduces the radial extent of its wave function in bound states (see eqs. 6-3 and 6-5) relative to that of the electron, and thus the effect of nuclear size on energy levels is enhanced in μ-mesic atoms relative to ordinary atoms by a factor of the order of $(207)^3$. Investigations of the X-ray line spectra of such atoms give $r_0 = 1.2 \times 10^{-15}A^{\frac{1}{3}}$.

All these methods give approximately the same values for nuclear sizes, but the values observed for $r_0(A)$ by the different methods differ by more than the probable errors of the experiments. Such differences are to be expected, since the radius of a nucleus is defined differently by each type of experiment. For example, for α emission the radius is the value of r at which the maximum in the potential energy occurs. For neutron scattering, on the other hand, the radius is the value of r that makes πr^2 equal to the scattering cross section, which is not quite the same because the nucleus is not a homogeneous, hard sphere with a sharply defined boundary. The near agreement among the various methods attests to the usefulness of the concept of a radius, however, and with these values of r_0 in mind we can proceed with the study of nuclear structure.

We turn next to the question of the *constituents* of nuclei and recall that it was asserted in Sec. 3-4 that the nucleus consists of protons and neutrons. Neutrons, protons, and electrons are the elementary particles emitted by nuclei. It is probable, therefore, that the nucleus consists of at least some of these particles. We shall prove that the nucleus cannot contain electrons. Although this does not constitute a proof that the nucleus contains protons and neutrons, it disposes of the chief rival proproposal for nuclear constitution, namely, that the nucleus consists of protons and electrons.

The first argument that electrons cannot be in the nucleus is that there is not "room" for them. Consider a nucleus with $A = 20$ which has, according to Fig. 13-5, a radius $r_0 = 3 \times 10^{-15}$ m. If an electron were bound in this nucleus its ground-state wave function would presumably be like Fig. 5-5*b*. The diameter of the nucleus (6×10^{-15} m.) would be of the order of $\lambda/2$, where λ is the de Broglie wavelength of the electron. The wavelength $\lambda = 1.2 \times 10^{-14}$ m. thus computed is so short that the electron's velocity must be nearly equal to c, and this circumstance permits (through eqs. 1-13 and 1-17) an easy estimation of its kinetic energy K:

$$\lambda = 1.2 \times 10^{-14} = \frac{h}{mv} \cong \frac{h}{mc} \cong \frac{hc}{K}$$

$$K \cong 100 \text{ M.e.V.} \tag{13-6}$$

This energy is far greater than nuclear binding energies per nucleon, which are about 8 M.e.V. (Fig. 3-14). It therefore seems very unreasonable for an electron in the nucleus to have so large a K. If it has a K of only a few million electron volts, its λ is much too long (its wave function extends to much too large radii) to agree with the observed nuclear size.

The second argument is concerned with the *spins* of nuclei. *Hyperfine structure* of atomic spectral lines is produced by the relative orientations of the nuclear magnetic moments and the electron spin and orbital magnetic moments. The measured quantities are the positions and intensities of the very closely spaced lines that differ only in the energy of the relative spin orientations. From such measurements and other experiments the spin of a nucleus can be determined. The spin of the electron is $\frac{1}{2}$, and the spin of the proton is $\frac{1}{2}$. If $_1\text{H}^2$ were composed of two protons and one electron, the individual spins of the three particles could be added (with either a plus or minus sign, depending on the orientation of each particle) to give either $\frac{3}{2}$, $\frac{1}{2}$, $-\frac{1}{2}$, or $-\frac{3}{2}$. The observed spin of $_1\text{H}^2$ is 1, which does not agree with this suggested composition. The correct spin is predicted if $_1\text{H}^2$ is assumed to be composed of one proton (spin $\frac{1}{2}$) and one neutron (spin $\frac{1}{2}$). The spins of all other nuclei are also consistent with neutrons and protons as the nuclear constituents.

The fact that high-energy electrons (β^--particles) are emitted by some radioactive nuclei might seem to be an argument for electrons as a constituent of nuclei, but actually it is the third argument *against* the elec tron as a nuclear constituent. If there were electrons in the nucleus the energy E of an electron would either be less than zero (stable nucleus, no emission) or greater than zero. If it were greater than zero, the electron would be emitted immediately, since there is no potential barrier like Fig. 13-3 for electrons. The electron's potential energy as a function of distance from a positive charge is an attractive curve; at large distances it is like Fig. 6-1 for the potential energy caused by one proton. It is illustrated schematically in Fig. 13-6. The existence of long-lived radioactive β^--emitters is therefore incompatible with a nucleus containing electrons.

None of the above arguments applies as an argument against protons or neutrons as constituents of the nucleus, and thus we infer that a nuclide (Z, A) contains Z protons and $N = A - Z$ neutrons, where N is called the *neutron number*. We have already (Sec. 3-4) referred to nuclides with the same Z as isotopes; nuclides with the same A are called *isobars*.

It must not be thought that any property of a particular nucleus is simply the sum of the comparable properties of its neutrons and protons. We have already seen (Sec. 3-4) that the mass of a nucleus is less than that of its constituents, the difference representing the binding energy. A nucleus is therefore more complex than a mere assembly of neutrons and protons, each preserving its own properties.

Another property that is not simply the sum of neutron and proton contributions is the nuclear magnetic moment, which can be measured with great precision by nuclear magnetic resonance (NMR) experiments

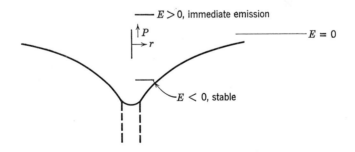

Fig. 13-6. Potential-energy curve for an electron near a nucleus. If electrons were in the nucleus they would have either $E < 0$ (stable, no β^- emission) or $E > 0$ (immediate emission, no β^- emission with measurable half-life). The dashed line (incorrect) is the way $P(r)$ would behave for $r < r_0$ if the electron interacted strongly with the nucleus.

analogous to the electron spin resonance described in Sec. 6-5. For example, the magnetic moment of the proton is $+2.79255 \, \mathfrak{M}_N$, of the neutron is * $-1.91280 \, \mathfrak{M}_N$, and of the deuteron is $+0.85735 \, \mathfrak{M}_N$, each written in terms of the nuclear magneton

$$\mathfrak{M}_N = \frac{eh}{4\pi M_p} \tag{13-7}$$

where M_p is the mass of the proton (cf. eq. 6-14). Although the spins $\frac{1}{2}$ of the proton and $\frac{1}{2}$ of the neutron add to give a spin of 1 for the deuteron, the magnetic moments associated with these spins do not exactly add. It turns out that the observed deuteron moment can be understood if the nucleons are described by a mixture of s-state and d-state wave functions, the latter contributing some orbital angular momentum. Thus not only do the neutrons and protons in a nucleus interact, but they also interact in a way that makes the ground state more complicated than an s-state; since a spherically symmetrical potential energy (like the hydrogen atom) produces a pure s-state as the lowest energy state, the interaction energy must not be spherically symmetrical. This and similar analyses emphasize that the properties of nuclei are not simply the sum of the properties of their constituent protons and neutrons.

We shall return to the question of nuclear structure in Sec. 13-6, but meanwhile we shall complete the survey of processes by which an unstable nucleus can decay.

13-4   β EMISSION AND ELECTRON CAPTURE

Many nuclei decay into other isobaric nuclei (i.e., nuclei with the same A but different Z) by the emission of electrons, which are called β^--particles when they originate in this way. Since electrons are not constituents of nuclei, the β^--particle must somehow be created during the process of emission; † since there is no barrier to restrain electrons that are energetic enough to be emitted (Fig. 13-6), the long life of many β^--emitters must originate in the low probability of this creation process. The conversion of some of the charge and mass of a nucleus into the charge, rest mass, and kinetic energy of an electron is a complicated process for which we shall not attempt to provide much illumination here, but we shall describe some features of the process.

---

* The negative sign means that the magnetic moment is directed opposite to the spin.

† The nucleus is not unique in emitting a particle that it did not contain: The hydrogen atom can emit photons, particles with much too long wavelengths to have existed inside the atom.

Perhaps the simplest β decay is the disintegration of a free neutron into a proton and an electron,

$$n \rightarrow p + \beta^- + \bar{\nu} \tag{13-8}$$

with a half-life τ for the free neutron of 13 minutes. In addition to the three familiar particles involved, the *antineutrino* $\bar{\nu}$ appears. This elusive particle (and its relative the neutrino ν) has no charge, no rest mass, spin $\frac{1}{2}$, and almost no interaction with matter. A neutrino or anti-neutrino (the distinction between the two will not concern us until Sec. 13-8) has such an insignificant interaction with nuclei if it has an energy of a few M.e.V. or less that it can traverse the entire sun with only a tiny probability of being captured or of causing a reaction. How, then, can we know that it is emitted in a process like eq. 13-8?

One item of evidence for the emission of neutrinos comes from measuring the distribution of kinetic energies of emitted β^--particles. For example, the disintegration ($\tau = 5.0$ days)

$$_{83}\text{Bi}^{210} \rightarrow {}_{84}\text{Po}^{210} + \beta^- + \bar{\nu} \tag{13-9}$$

releases an energy of 1.17 M.e.V., which is known from mass determinations of the isobars Bi^{210} and Po^{210}. But the energy distribution of the electrons (Fig. 13-7) shows that almost all of the β^--particles have less than 1.17-M.e.V. kinetic energy. Thus to conserve energy we must postulate that some other particle, though undetectable by ordinary means, has carried away a share of the energy. This particle (the anti-

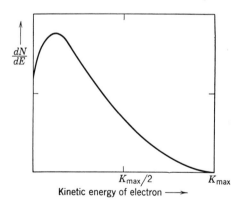

Fig. 13-7. Distribution of kinetic energies of electrons from $_{83}\text{Bi}^{210}$, a β^--emitter. The maximum kinetic energy $K_{\max}$ equals the Q value of the reaction; almost all the energy not accounted for in the β^- kinetic energy goes into the kinetic energy of the antineutrino, but a slight amount goes into the recoil energy of the polonium nucleus.

neutrino) cannot have a rest mass, however, since occasional electrons *do* have $K = 1.17$ M.e.V., leaving no energy available for the rest energy Mc^2 of the undetected particle.

There is other evidence for the neutrino in the conservation of momentum. If the neutron was initially at rest, the linear momentum on the right side of eq. 13-8 must be zero; but the measured proton and β^- momenta do not add to zero. Furthermore, the spin angular momenta of the neutron, proton, and electron are each $\pm \frac{1}{2}$; only by the emission of a fourth particle, also with half-integer spin, can the angular momentum be conserved.

Finally, nuclear reactors emit such copious supplies of antineutrinos and elaborate detecting equipment can be made so sensitive that energetic antineutrinos can be detected by the reaction

$$\bar{\nu} + p \rightarrow n + \beta^+ \tag{13-10}$$

Thus under the most favorable conditions a direct confirmation of the existence of these particles can be obtained.

The reaction of eq. 13-10 produces a *positron*, a positively charged particle that is the antiparticle of the electron; it has the same mass, magnitude of charge, spin, and magnitude of magnetic moment as the electron. The reaction eq. 13-10 is not a practical positron source, because of its minute cross section. Many artificially produced positron emitters are known, although none occurs in nature. β^+ emission from one such emitter occurs as follows:

$$_{11}Na^{22} \rightarrow {_{10}Ne^{22}} + \beta^+ + \nu \tag{13-11}$$

with a $\tau = 2.6$ years. The positron can readily be identified by measuring its e/m.

Nuclei that are unstable with respect to β^+ decay can nearly always also decay by electron capture. In this process, one of the inner (usually K-shell) atomic electrons is captured by the nucleus, and a neutrino is emitted. For example,

$$_{11}Na^{22} + \beta^- \rightarrow {_{10}Ne^{22}} + \nu \tag{13-12}$$

After the nuclear reaction has occurred, the inner atomic electron shell is filled by another atomic electron with the emission of a characteristic X-ray. This capture process occurs almost exclusively for s electrons because the wave functions of all other electrons are zero at the nucleus ($r \cong 0$). The capture probability is highest for K-shell ($1s$) electrons, since their ψ's are larger at $r = 0$ than the ψ's with $n > 1$; this process is therefore often referred to as *K-capture*. There is also an appreciable, but smaller, probability of L-capture, involving the $2s$ electrons of the L-shell.

Some of the nuclei in a sample of $_{11}\mathrm{Na}^{22}$ (for example) will decay by the process of eq. 13-11 and some by eq. 13-12. We say that there are two *branches* of the decay process. The branching ratio, the fraction of decays that goes by a given branch, can be calculated quite exactly by quantum mechanics. The calculation of the probability of emission of β^{+}-particles or of the capture of β^{-}-particles involves the product of the wave function of the nucleons and of the β^{+}- or β^{-}-particle at the nucleus. The nucleon wave function is not accurately known, but the free-particle ψ for the β^{+}-particle in eq. 13-11 and the atomic ψ for the β^{-}-particle in eq. 13-12 *can* be accurately calculated. Since the nucleon wave function is the same for the two branches, the branching ratio depends only on well-known wave functions. Quantum mechanics can frequently be successfully applied in such a fashion to nuclear physics problems, even though the wave functions for particles inside the nucleus cannot yet be readily calculated.

A nuclear neutron thus can transform into a proton with β^{-} emission, or a nuclear proton can transform into a neutron with β^{+} emission or electron capture. Evidently some nuclei have too little and some too much positive charge for stability. The lifetimes and energies of such processes vary widely, corresponding to different degrees of instability and to differences in other nuclear properties.

The technique of calculating the energy released (Q) in nuclear reactions from tables of atomic masses (see Appendix C) was introduced in Sec. 3-4 by reference to a particular reaction. We shall now continue the discussion of such calculations by considering a general β^{-}-emitter with atomic mass $_{Z}M^{A}$, the mass of the nucleus and Z electrons. The general reaction is schematically

$$_{Z}M^{A} \rightarrow {}_{Z+1}M^{A} + \beta^{-} + \bar{\nu} \tag{13-13}$$

The actual nuclear mass on the left is $(_{Z}M^{A} - Zm_0)$, since Z electrons were counted in the atomic mass. On the right, the sum of the actual masses is $[_{Z+1}M^{A} - (Z + 1)m_0]$ plus the mass m_0 of the β^{-}-particle. Thus it happens that the Q value, the energy released, is simply $(_{Z}M^{A} - {}_{Z+1}M^{A})c^2$; this is the kinetic energy shared among the three products (but only a tiny fraction of it goes to the recoil nucleus, since it is so heavy compared to the β^{-}-particle). The use of atomic masses thus automatically provides the mass m_0 for the emitted β^{-}-particle. The general agreement to tabulate *atomic* masses was, in large part, because of the convenience in just such β^{-} calculations. (When actual kinetic energies are to be calculated, it should be recalled that 1 amu = 931.5 M.e.V.)

The situation is quite different for β^+ decay, a process unknown at the time that tabulation of atomic masses became standard. The general β^+ decay is

$$_ZM^A \rightarrow {}_{Z-1}M^A + \beta^+ + \nu \qquad (13\text{-}14)$$

and the actual mass sum on the right is $[_{Z-1}M^A - (Z - 1)m_0]$ plus the mass m_0 of the β^+-particle. Thus the Q value in positron emission is

$$Q = (_ZM^A - {}_{Z-1}M^A - 2m_0)c^2 \qquad (13\text{-}15)$$

Therefore the use of atomic masses requires a correction of $2m_0c^2$ (that is, 1.022 M.e.V.) in the computed energy in β^+ decay.

13-5 γ EMISSION AND INTERNAL CONVERSION

We have already described considerable evidence that the behavior of the nucleus is governed by the laws of quantum mechanics. Therefore we expect discrete energy levels for bound states as in any other quantum physics problem, regardless of the detailed character of the nuclear force. Data to support this conclusion were available even before the development of quantum mechanics. For example, some α-emitters emit several groups of α-particles, each group with a different energy. When an α-particle in one of the lower energy groups is emitted, the new nucleus is left in an *excited state* and decays to the ground state by the emission of one or more γ-rays, high-energy photons. This process is similar in principle to the emission of a photon by an atom when it makes a transition from an excited state to a lower excited state or to the ground state.

Nuclei are frequently left in excited states after β emission, and therefore γ-rays often accompany β-particles. Most such γ-rays are emitted so quickly after β emission that the delay is unmeasurable, but hundreds of instances are known of much longer, measurable lifetimes of excited states, ranging from 10^{-10} sec to many years. Pairs of these long-lived, metastable nuclei and the ground-state nuclei that have the same A and Z are called *isomers;* the isomeric nuclide that is the excited state is given a superscript m, as in Br^{80m}.

The long life of the excited state (i.e., the existence of isomers) arises from selection rules for photon emission. In photon radiation by atoms (Sec. 5-8), the coupling of the photon's electric field (which is nearly constant over atomic dimensions of the order of a_0, since $\lambda \gg a_0$) to the atomic energy states is weak, and magnetic interactions are negligible. In photon radiation by nuclei these simplifying conditions are not met, and the selection rules are more complicated because (in addition to the electric dipole transition characteristic of atomic spectra) *quadrupole* electric, dipole and quadrupole magnetic, and even more involved transitions are possible. But the simple dipole transition, when allowed, is much more probable than these more complicated transitions that involve large changes in nuclear spin. When the change in spin (angular momentum of the nucleus) is greater than 1, however, the electric dipole transition is not allowed; when the change in spin is greater than 2, even the quadrupole transition is not allowed. Thus radiative transitions become increasingly improbable as the difference in spin between initial and final states increases.

Excited states and isomerism add considerable complexity to the decay schemes of radioactive nuclei. The simplest possibility is for $_Z(M)^{Am}$ to emit a γ-ray and to become the other member of the isomeric pair, $_Z(M)^A$. But the excited nucleus can frequently β-decay to another nuclide, in addition to the γ decay, and complex situations like that in Fig. 13-8 (where some of the nuclei making up a sample of a nuclide decay by one route and some by another) are by no means rare.

A process related to γ emission, since it can also reduce the state of excitation of a nuclide without changing A or Z, is *internal conversion*, the transfer of excitation energy from a nucleus to one of the atomic elec-

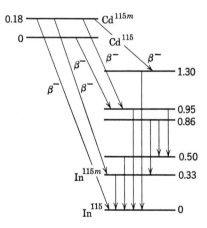

Fig. 13-8. The decay scheme of the isomers Cd^{115m} and Cd^{115}. The ordinate is approximately proportional to the energy; the number on each state is the energy of that state in M.e.V. The transition $Cd^{115m} \rightarrow Cd^{115}$ is so highly forbidden by the large change in angular momentum that $β^-$ decay to an excited state of In^{115} is more probable. The 5 excited states of In^{115} include the isomeric state In^{115m} that has a long (4.5 hours) half-life because the transition from that state to the ground state requires a large change in angular momentum. (From Hollander, Perlman, and Seaborg, *Rev. Mod. Phys.*, **25**, 469, 1953.)

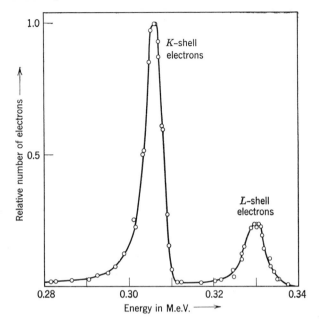

Fig. 13-9. Spectrum of kinetic energies of internal conversion electrons emitted by In[115]. (From Graves, Langer, and Moffat, *Phys. Rev.*, **88**, 344, 1952.)

trons, usually a K or L electron, which is thereupon ejected from the atom. If the energy difference between the excited nuclear state and the lower state to which the nucleus goes is E, the kinetic energy of the internal conversion electron is E minus the binding energy of that electron. Thus the energy calculation proceeds just as if a γ-ray of $E = h\nu$ were emitted and subsequently absorbed by ejection of an atomic electron, as in the photoelectric absorption of X-rays (Fig. 6-14). But this is *not* the physical process, since internal conversion electrons frequently occur without the emission of any γ-rays at all. The physical process is in fact similar to electron capture and consists of a direct interaction between the atomic electron, whose $|\psi|^2$ is appreciable at the nucleus, and the nucleus. The emitted electrons have a sharp line energy spectrum (Fig. 13-9), since *discrete* states of both the nucleus and the atomic electron are involved. Such line spectra are frequently superimposed on the continuous energy distribution of electrons from β decay.

We shall now return to γ emission and consider the widths of γ-ray emission lines. If no other process widens a line, the width will be the *natural line width* given simply by the $\Delta E \, \Delta t \geqslant h$ indeterminacy (eq.

4-22). If the half-life of the excited state of a γ emitter is, for example, 10^{-16} sec, the spread ΔE in energies of the emitted photons must be about 40 e.V.

A change in the energy of the emitted photon that is frequently larger than the width of the line is the Doppler shift entailed by the recoil of the emitting nucleus. The emission of a γ-ray with momentum $h\nu/c$ implies that the emitting nucleus is given a momentum $Mv = h\nu/c$ in the opposite direction. The sum of the kinetic energy $\frac{1}{2}Mv^2$ of the recoil and the energy $h\nu$ of the γ-ray equals the difference in energy levels between the two energy states of the emitting nucleus. For example, the recoil energy of a $_{29}Cu^{64}$ nucleus when it emits a 1.35-M.e.V. γ-ray is about 15 e.V., and thus the photon energy is about 1 part in 100,000 lower than if recoil did not occur.

Isomeric transitions with long half-lives, like the 0.0144 M.e.V. transition in $_{26}Fe^{57}$ with a $\tau = 10^{-7}$ sec, have extremely small natural line widths ($\sim 10^{-8}$ e.V. for this example). For such emitters, the Doppler shift is more consequential; it is about 0.002 e.V. for the $_{26}Fe^{57}$ transition cited. An important consequence of this shift to lower energy is that a γ-ray from $_{26}Fe^{57m}$ does not have enough energy to excite a $_{26}Fe^{57}$ nucleus (in its ground state) to the excited state; in other words, there is no resonance absorption.

Now if the emitting nucleus is bound in a solid, it can recoil only if it emits or absorbs one or more *phonons*. Furthermore, the phonons involved must be those of minimum wavelength (eq. 9-13 and Fig. 9-10c and g), since the emitting nucleus starts to move while all the other nuclei in the solid remain fixed; that is, the other nuclei are not initially moving along with it in correlated motion, as they are for long-wavelength phonons. We can calculate the energy of these phonons in an iron crystal from eq. 9-13 and find $h\nu_m = 0.036$ e.V. Since this energy is greater than the recoil energy, there is considerable probability that no phonons will be emitted; if the temperature is not too high, there is a considerable probability that no phonons will be absorbed. Thus in an appreciable fraction of the γ-emission acts, no independent nuclear recoil will take place, the recoil momentum being taken up by the crystal as a whole.

The emission or absorption of such extremely narrow γ-ray lines without Doppler shift is called the *Mössbauer* effect. Since the full energy difference between nuclear energy levels goes into the $h\nu$ energy of the photon (no energy removed by the recoil), this photon can be absorbed in another nucleus of the same species (again without Doppler shift and recoil) and excite the same level. The phenomenal sharpness of this resonant absorption permits intriguing experiments. For example, the fractional line width of the Fe^{57} Mössbauer line is about 10^{-8} e.V. $\div$ 14,400 e.V., or about 10^{-12}. With such a narrow line, the Zeeman effect in the γ-ray emission line can be measured and used to probe the magnetic field at the nucleus. Even investigations of some aspects of the general theory of relativity, involving the variation in time-measuring devices as they move from place to place in a gravitational field, have been made with Mössbauer emitters and absorbers serving (through comparison of their frequencies) as the "clocks." In most experiments it is advantageous to be able to vary the γ-ray energy slightly, for example, in order to sweep through the various Zeeman components. This modulation is easily accomplished by moving the emitter or absorber at a relative speed v. The Doppler shift in the energy E of the line is then $E(v/c)$, as shown in prob. 13-32. Since the usefulness of the sharp Mössbauer lines lies in their ability to reveal extremely small energy or frequency differences, typical experiments do not involve much modulation, and moving the emitter or absorber at speeds of the order of 0.01 to 100 m./sec is common practice.

13-6 NUCLEAR STABILITY

Now that we have studied the principal modes of decay of unstable nuclides, we are in a position to consider the general questions of nuclear stability: Will a particular combination of neutrons and protons form a stable nuclide? If it is unstable, how much energy and what kind of particles will be released when it decays? How swiftly is this decay likely to occur? In this section we shall combine quantum theory and experimental data on nuclide masses to supply the main features of the answers to these questions.

The relative number of protons (Z) and of neutrons $(N = A - Z)$ in all the *stable* nuclei are illustrated in Fig. 13-10. By *stable* we mean here that these nuclei have not been observed to disintegrate.* It is evident from Fig. 13-10 that for small Z, Z is approximately equal to N, but for large Z the number of neutrons becomes considerably greater than the number of protons. Thus $_8O^{16}$ has 8 protons and 8 neutrons, but $_{82}Pb^{208}$ has 82 protons and 126 neutrons.

This behavior of the relation between Z and N can be understood by using three facts: (1) The nuclear force between two neutrons is the same as that between two protons or between a proton and a neutron.

---

* In addition to these stable nuclei, about a dozen other nuclei are found in nature that are radioactive but with extremely long half-lives ($\sim10^9$ to $\sim10^{15}$ years). For example, $_{19}K^{40}$ constitutes about 0.01% of natural potassium; it has a $\tau = 1.3 \times 10^9$ years.

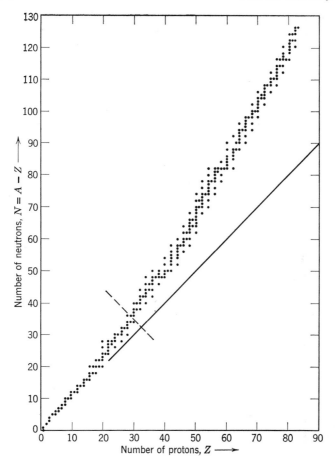

Fig. 13-10. The number of neutrons plotted as a function of the number of protons for all the stable nuclides. The solid line corresponds to $N = Z$. The dashed line indicates the section at which Fig. 13-14 applies.

(2) The Exclusion Principle. (3) Coulomb's law. The Exclusion Principle was introduced in Sec. 6-3 for electrons, but it applies as well to neutrons or protons. Two neutrons, for example, cannot occupy the same quantum state. But a neutron can occupy any quantum state not filled with a neutron, independent of the protons and their quantum states.

As more neutrons are added to a nucleus they must occupy higher-energy quantum states, since the lower states are filled. On the other hand, adding more neutrons means a greater binding (hence lower energy) because of the increased number of nuclear-force bonds. In $_8O^{16}$, there are 16 particles and therefore the nuclear binding associated with 16 particles. On the other hand, only 8 quantum states for neutrons and 8 for protons are filled. This is a lower energy than the combination of 7 protons and 9 neutrons, which would form the very unstable nucleus $_7N^{16}$. (It emits a β^--particle and becomes the stable nuclide $_8O^{16}$.) If the Coulomb repulsion is neglected (a good approximation for nuclei with small Z), the energy levels for protons and neutrons with the same quantum number are the same. The ninth quantum state for neutrons is therefore at a higher energy than the eighth quantum state for protons.

These same arguments apply for large Z, but the Coulomb repulsion between protons becomes more and more significant since it increases approximately as Z^2. Thus the neutron quantum states are somewhat lower in energy than the proton states, and a lower energy for the nucleus is obtained by using fewer protons relative to the number of neutrons. This same repulsion effect is the reason that all nuclei with $Z > 83$ are radioactive.

The binding energy B of a nuclide was defined in Sec. 3-4 as the energy that would have to be supplied to break up a nucleus of that nuclear species into its constituent nucleons. This definition can be expressed in the form of the following equation:

$$B = Z(_1H^1) + N(_0n^1) - _ZM^A \qquad \text{amu} \qquad (13\text{-}16)$$

where $_ZM^A$ is the mass of the atom with atomic number Z and mass number A. B is frequently expressed in M.e.V. by multiplying the right side of this equation by 931.5 M.e.V. per amu.

We shall now apply experimental data and quantum ideas to understand the nature and Z- and A-dependence of the various contributions to the binding energy; this will be a refinement of the introductory analysis just carried out in conjunction with Fig. 13-10. In this way, a formula for B can be constructed in which each term represents the effect of one kind of contribution, but each term has a coefficient that cannot be calculated exactly for lack of detailed knowledge of the nuclear forces. By fitting this formula to empirical data for atomic masses, we can obtain the coefficients of the terms. The resulting *semi-empirical mass formula* is of far-reaching importance in nuclear stability problems involving nuclides with $A > \sim 10$ or 15. We shall proceed term-by-term:

1. Every nucleon occupies about the same volume in the nucleus, as shown by the fact that the volume $\frac{4}{3}\pi r_0^3$ of the nucleus is proportional to the number A of nucleons (Fig. 13-5). Each nucleon has a strong attraction to its immediate neighbors, but this attraction (like the covalent

Approximate range
of nuclear force

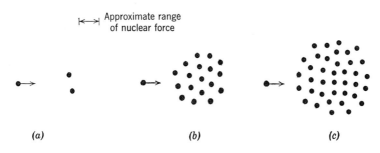

(a)   *(b)*   *(c)*

Fig. 13-11. Schematic diagram of the binding energy increment when one nucleon is added to a nucleus. (*a*) The nucleon adds the binding to two neighbors; this binding is less than it could add if it were to acquire more neighbors, and eq. 13-17 does not apply. (*b*) The nucleon adds the binding to as many neighbors as it can have; the binding now has the *saturation* value, and eq. 13-17 begins to be valid. (*c*) A is much larger here than in (*b*), but the added nucleon still has the same number of neighbors and adds only the same, saturation binding energy as in (*b*).

binding in molecules) rapidly saturates. That is, each nucleon added to the nucleus establishes bonds with its near neighbors, giving a nearly constant binding contribution per nucleon (rather than interacting with *all* the nucleons as would be the case, for example, for the Coulomb force between charged particles). This situation is illustrated schematically in Fig. 13-11. This *volume* binding energy term, the most important single term in the formula, is thus simply proportional to A, and the fit to mass data indicates that it is approximately $15.7A$ M.e.V.

2. Clearly a nucleon near the surface of the nucleus, like an atom near the surface of a liquid, cannot provide the full saturation amount of binding. The number of such nucleons is proportional to the nuclear surface area $4\pi r_0^2$, which in turn is proportional to $A^{2/3}$. Thus a negative *surface* term proportional to $A^{2/3}$ appears in the formula, negative because the full saturation binding was already included in the first term and thus the binding of the surface nucleons has been overstated. This term is not much smaller in magnitude than the volume term. The geometrical shape that corresponds to the minimum surface-to-volume ratio is the sphere, and therefore nuclei are nearly spherical in shape. The fit to data gives the value of the surface term as $-17.8A^{2/3}$ M.e.V.

3. Protons introduce a Coulomb repulsion as well as the nuclear force attraction, and we have seen in connection with Fig. 13-10 how this repulsion energy discriminates against protons in heavy nuclei. This term in the binding energy can be calculated from ordinary electrostatics if it is assumed that the positive charge is spherically symmetrical and if the coefficient in $r_0 = (\text{constant})A^{1/3}$ is taken from nuclear data.

The result is a negative *electrostatic* term (opposing binding) in the binding energy equal to $-0.71Z(Z-1)A^{-\frac{1}{3}}$.

4. We also saw in Fig. 13-10 that stability, and therefore large binding energy, is associated with $N = Z$ for light nuclei such as $_8O^{16}$ in which the Coulomb effect is insignificant. We might guess that a negative term proportional to an even power of $(N-Z)$ should occur in B, since either a neutron or a proton excess decreases B. Figure 13-12 gives a crude approximation to the nucleon energy levels near the highest filled level; in this approximation, all the neutron and proton levels are separated by the same energy Δ. Keeping A constant, we remove a proton from level 3 and add a neutron to level 4; $(N-Z)$ now equals 2, and the energy has increased (B has decreased) an amount Δ. Transferring two nucleons in this way $(N-Z=4)$ increases the energy an amount Δ (for the first) plus 3Δ (for the second) or a total of 4Δ. We thus find as a generalization of this process that the transfer of $(N-Z)/2$ nucleons decreases the binding energy by an amount $-\Delta(N-Z)^2/4$. Our argument tells us nothing about the magnitude or possible A-dependence of the level spacing, but it is not surprising that Δ decreases as A increases (see also Table 13-1 in Sec. 13-7). This argument is thus scarcely a complete theory of this *asymmetry* term, but it does show the origin of the important quadratic dependence on $(N-Z)$. The fit to the data gives for this term the value

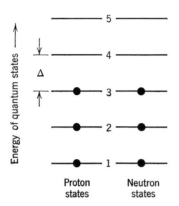

Fig. 13-12. Schematic diagram of nuclear energy levels near the highest filled levels. Many additional levels lie above and below this small section, and in an actual nucleus the levels are not equally spaced.

$$-23.6A^{-1}(N-Z)^2 = -23.6A^{-1}(A-2Z)^2 \text{ M.e.V.}$$

5. In the preceding paragraph, we really should have considered the protons in pairs and the neutrons in pairs, since two protons (one with $+\frac{1}{2}$ spin and one with $-\frac{1}{2}$ spin) can occupy each spatial quantum state, as in the structure of atoms. Thus there is an energy advantage (lower energy, higher B) if there is an *even* number of protons or an even number of neutrons or both. If we imagine filling the nuclear energy levels (as we did in the Fermi distribution of electron energies in a metal), starting with the lowest level and continuing until we have used all of an even

number of nucleons, we can get a lower energy if both N and Z are even. Suppose, for example, that the highest filled proton state is above the highest filled neutron state. If both N and Z are *odd*, we can profitably (i.e., with an increase in B) remove one proton from the nucleus and add one neutron; A remains the same but now N and Z are both even. Similarly, removing a neutron and adding a protron would be profitable if the highest filled proton state is below the highest filled neutron state. In either event, a binding energy advantage accrues to populating levels in such a way that all proton and all neutron levels are either completely full (each with two nucleons) or completely empty, and such populating obviously requires an *even* number of each kind of nucleon. Because of this *pairing* effect,* when A is even there is a larger B when both N and Z are even than when they are both odd. When A is odd, either N or Z must be odd and the other even, and this pairing effect is of no consequence. Thus a term in B appears that is zero if A is odd; for even N and even Z it is $+132A^{-1}$ M.e.V. (from the fit to mass data), and for odd N and odd Z it is $-132A^{-1}$ M.e.V.

The sum of all these terms is

$$B = 15.7A - 17.8A^{2/3} - 0.71Z(Z-1)A^{-1/3}$$
$$- 23.6A^{-1}(A-2Z)^2 \pm 132A^{-1} \quad \text{M.e.V.} \quad (13\text{-}17)$$

where the plus in the last term applies when N and Z are both even, the minus applies when they are both odd, and the last term is omitted altogether when A is odd. This formula is an excellent fit to nuclear mass data for $A > \sim 15$, as shown in Fig. 13-13. Because of the surface and volume terms in this expression, this way of treating nuclear binding energies is usually called the *liquid drop* model of the nucleus, but the reasoning underlying it is based on experimental nuclear physics and quantum theory, rather than on reasoning by analogy with a liquid drop.

The deviations from eq. 13-17 exhibited in Fig. 13-13 are also of interest, but we shall not analyze them in detail. Filled shells of quantum states occur in nuclei as in atoms, but the up-and-down structure in the binding energy curve for nuclei is far less than in the plot of ionization potential as a function of Z for atoms. The smaller effect of filling shells

---

* An *exchange* energy of the same sign enhances the pairing effect. Since the nuclear force is attractive, there is a lower energy if two protons (or two neutrons) occupy the same spatial quantum state, since they therefore move in the same way and stay closer to one another. This is similar in principle to the exchange energy described in Sec. 9-8, but there having electrons with the *same spin* lowered the energy because it kept electrons (which repel each other) farther apart by requiring that they be in *different spatial* states.

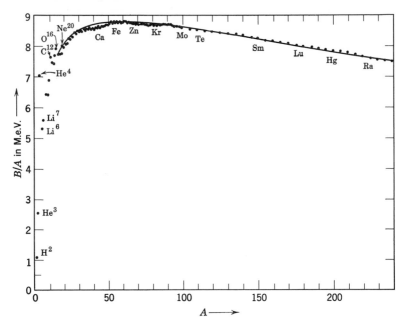

Fig. 13-13. Binding energy per nucleon plotted as a function of the mass number A. The solid curve is the approximate formula eq. 13-17 without the last term of that equation. Many nuclides, especially those with large A, are not plotted. Note the relatively large B/A of the "even-even" nuclei He^4, C^{12}, O^{16}, and Ne^{20}, and the especially large B/A relative to its neighbors of the "magic number" nucleus He^4. (Modified from R. B. Leighton, *Principles of Modern Physics*, McGraw-Hill, New York, 1959; used by permission.)

in nuclei occurs because the energy of the nucleons' interaction with each other is much greater than the difference in energy between, say, two angular momentum states. Thus when a shell is filled by having either the number of neutrons or the number of protons equal to one of the numbers 2, 8, 20, 50, 82, or 126, a small local maximum in B as a function of A occurs. These integers, colloquially referred to as *magic numbers*, originate in the same way that the numbers 2, 6, 10, and 14 originate in atomic structure but are different in nuclei because of the profound interaction between nucleon spin and orbital angular momentum. The *shell model*, which is essentially a refinement of the liquid drop model that includes consideration of nuclear quantum numbers and filled shells, has had considerable success in predicting the spins, magnetic moments, and relative abundances in nature of nuclides.

The semi-empirical mass formula can be applied to a host of problems, but we shall illustrate its use by a single application, namely the application to the understanding of how many nuclides of a given A can be stable (in the region $A > \sim 15$, where eq. 13-17 applies). Let us consider first an odd A, and we thereby eliminate the last term in eq. 13-17. Let us take $A = 65$ as an example; the nuclear mass $_ZM^A$ in M.e.V. from eqs. 13-16 and 13-17 is plotted for $A = 65$ in Fig. 13-14. The parabola, which owes its shape to the quadratic dependence of the asymmetry term upon $(N - Z)$, agrees reasonably well with the B and mass values for the four $A = 65$ nuclides for which data are available (crosses in the figure). The nuclide $_{29}Cu^{65}$ that is closest to the minimum mass at Z_{min} is evidently stable. All other isobars decay into this nuclide either

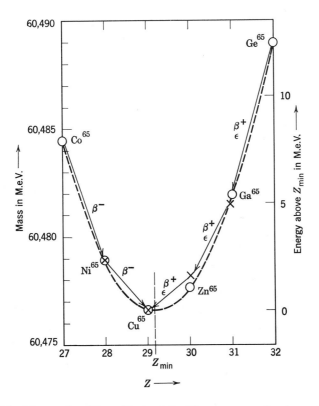

Fig. 13-14. Masses of nuclides with $A = 65$. The crosses are the observed masses, and the dashed curve and the circles are masses calculated from eqs. 13-16 and 13-17. Both ordinate scales are in M.e.V., and the scale at the right is mass-energy relative to the minimum in the parabola at $Z = Z_{min}$. Only the single nuclide $_{29}Cu^{65}$ is stable. "ϵ" means electron capture. (Modified from R. B. Leighton, *Principles of Modern Physics*, McGraw-Hill, New York, 1959; used by permission.)

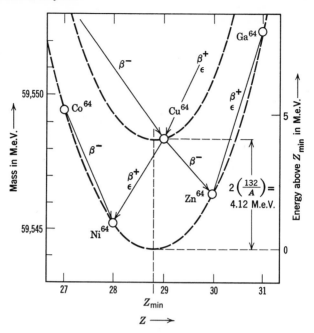

Fig. 13-15. Calculated masses of nuclides with $A = 64$. The upper parabola applies for Z and N both odd, and the lower for Z and N both even. Other details are the same as in Fig. 13-14, but now *two* nuclides are stable. (Modified from R. B. Leighton, *Principles of Modern Physics*, McGraw-Hill, New York, 1959; used by permission.)

by β^- emission (if Z is too small for stability) or by β^+ emission and electron capture (if Z is too large). Thus we predict that for odd A there will be only a single stable isobar. This is a rather remarkable prediction, since the variety and abundance of stable nuclides is so great (Fig. 13-10), but it is almost always fulfilled. The only exceptions are $A = 113$, 115, and 123, in all of which cases Z_{min} comes almost midway between two integer Z values.*

The situation for even A is more complicated because the last term in eq. 13-17 must be included; the masses for $A = 64$ are sketched in Fig. 13-15. There are now *two* parabolas, the upper for Z and N both odd (high energy and mass, low binding energy, negative sign of last

---

* Also in all of these cases, the β transition from one isobar to another is very highly forbidden, since the neighboring isobars differ by many units in angular momentum. The combination of a low transition energy and a highly forbidden type of transition evidently produces an unmeasurably long half-life in the higher mass member of each isobaric pair.

term in eq. 13-17) and the lower for Z and N both even (low energy and mass, high binding energy, positive sign of this term). The separation in energy of these two parabolas is $2 \times 132A^{-1} = 4.12$ M.e.V. Now *two* isobars can be stable when Z_{min} falls near an odd integer, with Z's even and differing by two. If Z_{min} falls near an even integer, two odd-Z nuclides can decay to that integer and only a single, even-Z isobar (or rarely three isobars) is stable. These predictions are also in remarkable

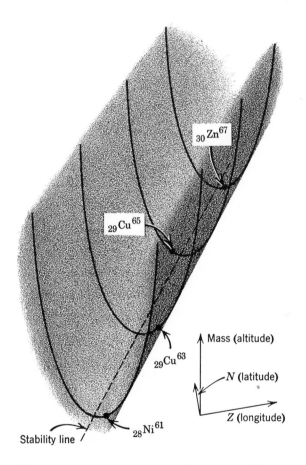

Fig. 13-16. Schematic representation of the stability valley. The mass (or energy) of each nuclide is represented as the altitude above a plane in which Z and N are the Cartesian coordinates. Only odd-A nuclides are shown in order to avoid drawing three separate valleys (odd-A; odd-Z, odd-N; even-Z, even-N). Stable nuclides are plotted as the black dots. The valley generally follows the course of the nuclides in Fig. 13-10. Compare also Fig. 13-17, in which the curved line is the bottom of the valley, for data on nuclides in this region of A values.

agreement with the observed stable nuclides for $A > 14$: All even-A nuclides have either one or two isobars (except for four A's, for each of which there are three isobars), and both Z and N are even for nearly all * even-A nuclides.

Similar analysis can be made for the number of stable isotopes, and there are many other applications throughout nuclear physics of the semiempirical mass formula.

Energy plots such as Figs. 13-14 and 13-15 suggest visualizing the energies of nuclei (or their masses) as altitudes above a plane in which Z and N are the Cartesian coordinates. The two-dimensional plot of Fig. 13-10 is thus replaced by a surface consisting of a deep trough, the *stability valley*. The bottom of the valley follows the locus of Z_{min} at each A, and the cross section along a line of constant A for odd-A nuclides is the parabolic shape of Fig. 13-14. The narrowness of the valley suggests that "canyon" would be a more descriptive label, and clearly complications ensue if even-A nuclides are included; but the stability valley concept is nevertheless a fruitful visualization (Fig. 13-16).

The principal observed data on nuclides, both stable and unstable, are conveniently and conventionally summarized on the *nuclide chart*, a section of which is reproduced in Fig. 13-17. Study of the half-lives on such a chart reveals that unstable nuclei that decay with large energy release (emitting particles with large energies) have short half-lives. Since the walls of the stability valley become steeper the farther we go from the bottom (the curved line on Fig. 13-17), decay energies become larger and half-lives shorter for nuclides farther from the center line of the chart. The blank hexagons thus represent nuclides that could exist for only very short times and that therefore have not been observed.

The correlation of half-lives and decay energies from the nuclide chart suggests the following generalization: The higher the energy released in a nuclear process, the shorter is the half-life of the parent nucleus. In β^- emission, for example, high decay energy means that there are many more momentum states available for both the electron and the neutrino; for large energies, the half-life τ of β^--emitters is related to the maximum energy K_{max} of the emission spectrum approximately by $\tau \sim (K_{max} + m_0c^2)^{-5}$. The emission of γ-rays also occurs faster for higher excitation states: An increase of a factor of 10 in $h\nu$ for electric dipole transitions decreases τ by a factor of 1000, and the dependence is even more sensitive for forbidden (isomeric) transitions. We have already seen in Sec.

---

* The only exceptions (odd-odd nuclides) occurring in nature are $_{23}V^{50}$ and $_{73}Ta^{180}$; they are present in small relative abundances and may be unstable.

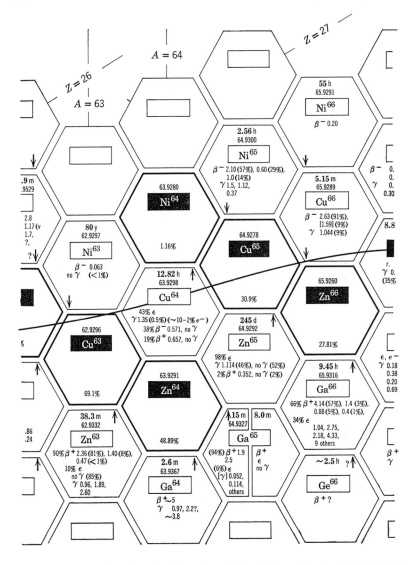

Fig. 13-17. A small section of a chart of nuclides. A heavy border indicates that a nuclide occurs in nature; atomic masses in amu and percentage abundances are given for these nuclides. Half-lives are given for unstable nuclides (y = years, d = days, h = hours, m = minutes). Decay processes of unstable nuclides are shown with the percentage of each type of emission. There is a wide variety of such charts, and most include additional information on spins, neutron cross sections, magnetic moments, and other properties of nuclides. (Courtesy of W. H. Sullivan, Oak Ridge.)

13-2 the extremely sensitive dependence of τ for α emission on the Q value of the disintegration, for reasons understood by the analysis of the tunneling process.

The radioactive emission of *nucleons* from nuclides with measurable lifetimes does not occur.* An emitting nucleus would have to have an energy of at least $\sim$6 to 8 M.e.V. above the energy of the product nucleus if it were to emit a proton or a neutron, since Fig. 13-13 shows that the binding energy per nucleon is about this amount. Of course the nucleon that could be emitted most easily might not have the average binding energy, but we can compute from the mass formula that the *neutron separation energy*, which is the binding energy of the most easily removed neutron and is the analogue of the first ionization potential energy for atoms, has about this same value. But typically nuclides with energies of even only 1 M.e.V. above the product nuclide can quickly β-decay to nuclides that are closer to the stability line. Thus nucleon emission competes unfavorably with β decay because of the large magnitude ($\sim$8 M.e.V.) of the minimum energy required for nucleon emission. It was for this reason that we could ignore nucleon emission in the foregoing discussion of the stability of isobars.

α emission, on the other hand, is a more successful competitor because of the large binding energy (28.2 M.e.V.) of the α-particle, about 7 M.e.V. per nucleon. Thus for heavy nuclei (where B/A is about 7.6 M.e.V., as shown in Fig. 13-13), an excitation energy of only about 0.6 M.e.V. per nucleon or $\sim$2 M.e.V. per α-particle above the energy of the product nucleus is required as the minimum energy for emission. α emission is, however, restrained by the potential barrier (Fig. 13-3), which permits the decay with appreciable probability only for considerably higher energies (see prob. 13-43).

This last observation emphasizes that in our consideration of nuclear stability we must not confine our attention exclusively to the mass formula: Barriers are also important. Many heavy nuclei that are counted as stable by our definition could exothermically emit α-particles, but the decay energy of these is so small that the decay probability is too small to permit disintegrations at an observable rate. Furthermore, all nuclei with $A > \sim$110 could release energy by fissioning into two nuclei with nearly equal masses. This process does occur for $_{92}U^{238}$, but with a very long half-life ($\sim$10^{16} years). α emission and spontaneous fission are responsible for the termination at the upper (large A) end of the array of stable nuclides. For example, the artificially produced nu-

---

* An example that at first sight looks like an exception is the delayed neutron emission to be studied in Sec. 14-2. This is not in fact an exception, however, since these neutrons are emitted essentially instantaneously after the nuclide that emits them is formed; the delay arises in the formation of the parent nuclide from *its* ancestors.

clide $_{100}\text{Fm}^{256}$ has a half-life for spontaneous fission of only 3 hours, even though its neutron-proton ratio is optimum for stability. We shall consider fission in some detail in Sec. 14-2.

13-7 NUCLEAR REACTIONS

We turn now from consideration of disintegration processes, in which a single nucleus spontaneously decays into a different nucleus, to nuclear reactions, processes in which an incoming particle produces a transformation inside the nucleus with which it collides. In this section we shall first describe how the probability of such reactions is specified by specifying the *reaction cross section*, and then we shall consider the physics of reactions.

The concept of a cross section, which is a quantitative measure of the probability of a given reaction, was introduced in Secs. 2-5 and 12-8 in conjunction with collisions in gases, but it will now be described again in the language appropriate to nuclear processes. For definiteness, we shall consider the example of the deuteron bombardment of lithium and specifically the reaction

$$_1\text{H}^2 + {}_3\text{Li}^6 \rightarrow {}_2\text{He}^4 + {}_2\text{He}^4 \quad (13\text{-}18)$$

The lithium nuclei are initially at rest in a metal foil, and the deuterons are incident at right angles to the foil with a known energy. In some of the encounters between a deuteron and a lithium nucleus this particular reaction will occur. Suppose that there are N lithium nuclei per cubic meter and that the sample thickness dx is so small that almost all the deuterons penetrate this thickness without reacting. Let $d\mathcal{P}$ be the probability that a particular incident deuteron will produce this reaction. The cross section is then defined by writing

$$d\mathcal{P} = \sigma N \, dx \quad (13\text{-}19)$$

The geometrical interpretation of this equation is presented in Fig. 13-18. A unit area of the layer of thickness dx con-

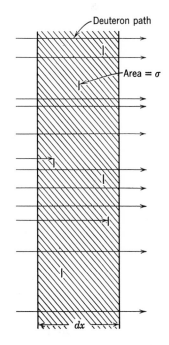

Fig. 13-18. Schematic diagram to illustrate the definition of σ. About one-sixth of the deuterons make reactions in this dx since the sum $N\sigma$ of the "target areas" is about one-sixth of the total area of the sample.

tains $N\,dx$ nuclei. Suppose that each of these nuclei has this property: if a deuteron strikes within an area σ centered on the nucleus, the desired reaction takes place. Then a fraction $\sigma N\,dx$ of incident deuterons produces reactions.

It must be emphasized that σ is not simply related to the actual geometrical cross section of the "target" nucleus, although σ and the geometrical cross section are frequently of the same order of magnitude. *If* every incident particle which struck the target nucleus (i.e., hit within a circle of radius r_0 equal to the nuclear radius) caused a reaction and no other particles caused reactions, then σ would equal $\pi r_0{}^2$. But this is not the description of an actual encounter. The probability of the occurrence of the reaction once the particle is near the target depends on the details of the reaction and upon the kind and energy of the incident particle. σ thus has in general no simple connection with the nuclear size,* but the σ's for most nuclear reactions are of the same order of magnitude as the geometrical cross section of nuclei.

If a beam of J particles/m.^{2} sec is incident on the target foil, since the probability that each one will react is given by eq. 13-19, there are $JN\sigma\,dx$ reactions per unit area normal to the beam per second. Measuring the number of reactions (in our example by counting the number of emergent α-particles) thereby gives a method of measuring σ. If the reaction described by σ is the only reaction that is removing particles from the incident beam, the attenuation dJ of the beam is simply

$$dJ = -JN\sigma\,dx$$

If there are many such layers, each of infinitesimal thickness dx,

$$J = J_0 e^{-N\sigma x} \tag{13-20}$$

where J_0 is the intensity of the beam at $x = 0$. This equation is precisely comparable to eq. 2-19 for molecular beams and to eq. 12-15 for electron beams. The definitions and arguments associated with eqs. 13-19 and 13-20 closely parallel the treatment of electron-atom collisions in Sec. 12-8.

Frequently the particles producing the reaction are not in a collimated beam but are moving in all different directions. For the analysis of such situations we define the *flux* of particles as

$$\text{Flux} = (\text{Particles/m.}^3) \times (\text{Average speed}) = n\bar{v} \tag{13-21}$$

where n and $\bar{v}$ are the quantities in the parentheses. If a flux $n\bar{v}$ of particles is present in a region where there are N target nuclei per cubic meter,

---

* An important exception to this statement is the high-energy neutron scattering cross section to be discussed later in this section.

and if the cross section for the reaction is σ, the number of reactions per second per cubic meter is $n\bar{v}\sigma N$. This statement follows from the fact that the total path length of 1 m.^{3} of the particles (regardless of direction) is $n\bar{v}$ per second. From eq. 13-19, the probability of a reaction is σN per unit path length. Hence the reaction rate is $n\bar{v}\sigma N$ per m.^{3} sec.

Cross sections of nuclear reactions are usually measured in *barns;* 1 barn is 10^{-28} m.^{2} Some reactions produced by slow neutrons have σ's as large as 10^5 barns for neutrons of just the right kinetic energy, but most σ's are of the order of magnitude of 0.1 to 10 barns. The σ of an endothermic nuclear reaction (one with a $Q < 0$) is zero unless the incident particle has sufficient kinetic energy to produce the reaction.

We shall next consider the physical processes in nuclear reactions. In the following paragraphs we shall provide at least qualitative answers to such questions as: What happens inside the nucleus during a reaction like that of eq. 13-18? How does the cross section σ for this reaction depend on the deuteron's kinetic energy?

In Sec. 3-4 we analyzed a relatively simple collision process, in which α-particles were scattered by the Coulomb field of a nucleus. One of the reasons for the relative simplicity was that the α-particle had too little energy to come within the range of nuclear forces; the large repulsive barrier (Fig. 13-3) kept α-particles of modest (a few M.e.V.) energies safely within the region of space where Coulomb's law exactly describes the potential energy. This same barrier produced a related simplification: No excitation or transformation occurred in the struck nucleus; that is, the collision was perfectly elastic. The third reason for the relative simplicity was that classical mechanics and quantum mechanics happen to give the same predictions for scattering by an inverse-square-law field. Of course the details of the descriptions are quite different at small r values during the actual scattering, but the predicted angular distributions (necessarily measured at large distances from the nucleus) are the same. In the language of cross sections, we should say that the *differential cross section* $\sigma_\theta\, d\theta$ for scattering into an angle between θ and $\theta + d\theta$ was the same on the two bases of calculation.

As we turn to collisions in which the incoming particles reach closer to the nucleus, these simplifications disappear. We consider first collisions that are still elastic but have a large enough kinetic energy of incoming particle and a small enough impact parameter for the nuclear force interaction to become important. Such collisions can be handled by quantum mechanics, which treats the striking particle as a wave encountering a variation in potential energy with distance. Measurements of the differential scattering cross section σ_θ as a function of the kinetic energy K of the incoming particle and as a function of the outgoing angle

θ can provide considerable information about the size and depth of the nuclear potential well. A particularly interesting special case is the scattering of 15 to 50 M.e.V. neutrons. The energy of these neutrons is such that they have wavelengths less than the radius r_0 of the nucleus and are not affected by the fine structure of nuclear energy levels. Whenever such a neutron hits within a distance r_0 of the center of a nucleus, it is elastically scattered because of the strong interaction between neutrons and other nucleons. The cross section for this process for these projectiles is hence $\sigma = \pi r_0^2$, and this is one of the methods of measuring r_0.

If an incoming particle is composed of more than one nucleon, it can lose a nucleon to the nucleus with which it collides. A typical example of such a *stripping reaction* is the loss of its neutron by a deuteron when it encounters a nucleus:

$$_1\text{H}^2 + {_3}\text{Li}^6 \rightarrow {_3}\text{Li}^7 + {_1}\text{H}^1 \tag{13-22}$$

A shorthand label for this kind of reaction is "a (d,p) reaction," meaning that a deuteron was incident and a proton emergent; to describe this particular reaction in this shorthand, we should say that it is the reaction $\text{Li}^6(d,p)\text{Li}^7$.

The inverses of stripping reactions such as the (d,p) and (d,n) reactions are the *pick-up* reactions (p,d) and (n,d). In these, as in stripping reactions, it appears that to a good approximation only a single nucleon is involved. That is, no significant energy is transferred to the nucleus as a whole or to nucleons other than the one being exchanged.

The most important nuclear reactions are those in which a *compound nucleus* is formed. When a neutron or proton hits and remains in a nucleus, it necessarily excites the nucleus violently. The energy of excitation is the kinetic energy of the incoming nucleon plus the binding energy (typically 6 to 8 M.e.V.) that each such particle adds when it joins a nucleus, the nucleon separation energy described in Sec. 13-6. The nucleon is accelerated into the nucleus by the nuclear forces and acquires this additional separation energy in the process. A deuteron, α-particle, or even a heavier nucleus can similarly play the role of the incoming projectile and add its nucleons and energy to the target nucleus. The added energy at first is concentrated in the incoming nucleon or group of nucleons and in a relatively few struck nucleons, but after only a few transits of the nucleus (10^{-22} to 10^{-21} sec) the energy is effectively shared among all the nucleons. A compound nucleus has been formed, in a highly excited state. Such a compound nucleus has a definite excitation energy but has "forgotten" what the incoming particle was. For example, either proton bombardment of $_3\text{Li}^7$ or deuteron bombardment of $_3\text{Li}^6$ leads to the compound nucleus $_4\text{Be}^8$; once formed at a certain excitation energy, the subsequent decay of $_4\mathrm{Be}^8$ is independent of how this compound nucleus was formed.

The compound nucleus typically survives for about $10^{\sim -16}$ sec. In this time, $10^{\sim 6}$ transits of a nucleon across the nucleus occur, and there is an appreciable probability that a fluctuation in the statistical distribution of the excitation energy among the nucleons will provide one nucleon (or bound group of nucleons, such as an α-particle) with enough energy to be emitted. It typically happens, especially if the incoming particle had a large kinetic energy, that the compound nucleus can decay in several different ways. For example, the compound nucleus N^{14} can dissociate into B^{10} and an α-particle, into C^{12} and a deuteron, into C^{13} and a proton, or into N^{13} and a neutron. These possible products of the reaction compete with one another, and the fraction of reactions that produces each product depends on the N^{14} excitation energy but not on the particular way in which it was formed.

An example of a reaction in which a compound nucleus is formed is

$$_2\mathrm{He}^4 + {}_4\mathrm{Be}^9 \rightarrow {}_6\mathrm{C}^{12} + {}_0 n^1 \tag{13-23}$$

The compound nucleus $_6\mathrm{C}^{13}$ is formed in this reaction but is customarily omitted from the equation. Another example is

$$_1\mathrm{H}^2 + {}_1\mathrm{H}^3 \rightarrow {}_2\mathrm{He}^4 + {}_0 n^1 \tag{13-24}$$

The Q value of this reaction is easily computed to be 17.6 M.e.V.; the sum of the kinetic energies of the neutron and the outgoing $_2\mathrm{He}^4$ nucleus equals the sum of this Q and the kinetic energy of the incident proton (we assume that the H^3, the hydrogen isotope called *tritium*, was initially at rest).

The incident particles in eqs. 13-23 and 13-24 are both positively charged, and therefore they must have considerable kinetic energy in order to enter the nucleus. The Coulomb barrier (Fig. 13-3) is many M.e.V. in height for α-particles incident on heavy nuclei. It is much less for smaller Z, such as in eq. 13-24, and it is, of course, a factor of 2 less for protons than for α-particles.* But for all except the very lightest target nuclei, the incident charged particles must have kinetic energies at least of the order of a few M.e.V. in order to produce reactions like eqs. 13-23 and 13-24 with appreciable cross sections. The original im-

---

* Also, some tunneling can occur, but if the tunneling probability is less than $\sim 10^{-3}$, σ is so small that the reaction may not be observable. We cannot make use of the extremely small tunneling probabilities characteristic of α emission from radioactive nuclei since we cannot provide beams of particles externally that make as many attempts per second ($\sim 10^{22}$) as the α-particles striking the barrier from the inside.

petus for constructing cyclotrons and other particle accelerators was to provide energetic charged particles for the study of nuclei by measurements of cross sections and other details of such reactions.

When the neutron is the incident particle, of course no energy barrier impedes its access to the nucleus. For example, the $B^{10}(n,\alpha)Li^7$ reaction

$$_0n^1 + {_5}B^{10} \rightarrow {_3}Li^7 + {_2}He^4 \tag{13-25}$$

can be produced with large cross section by neutrons with nearly zero kinetic energies. In fact, the reaction cross section is larger for neutrons that have been slowed by elastic collisions with nuclei until they have only *thermal energies* ($\sim kT$, which equals $\frac{1}{40}$ e.V. at room temperature) than for more energetic neutrons. In general, the σ of neutron-induced reactions is proportional to the reciprocal of the neutron velocity v at low velocities, although there frequently is some fine structure in $\sigma(v)$ superimposed on $\sigma \sim 1/v$. The $1/v$ relation comes about because the probability of capture of the neutron by the nucleus is proportional to the time the neutron spends near the nucleus, which is in turn proportional to the reciprocal of the neutron velocity. Because of the absence of a potential barrier, the neutron is especially versatile as an incident particle for the production of nuclear reactions.

It should be emphasized that the excitation energy of the compound nucleus formed when a thermal neutron has joined a target nucleus is *not* just the $\sim\frac{1}{40}$ e.V. of kinetic energy (which could produce no nuclear effects) but is $\frac{1}{40}$ e.V. *plus* the neutron separation energy, which is typically 6 to 8 M.e.V. (and can produce drastic effects). A neutron with just the right kinetic energy to excite one of the energy levels of the compound nucleus has a very large cross section for the production of this compound nucleus. Suppose, for example, that the compound nucleus has energy levels at energies $E_1, E_2, \cdots$ above its ground state and that the separation energy of the added neutron is E_0. Then the cross section for producing the compound nucleus will have peaks (*resonances*) at neutron kinetic energies $(E_1 - E_0), (E_2 - E_0), \cdots$. These resonances are similar in principle to the large increase in cross section for inelastic collisions of electrons with atoms whenever the kinetic energy of the electron is just sufficient to excite the struck atom to one of its excited electronic states (Franck and Hertz experiment, Sec. 4-5). Thus neutron cross sections as functions of neutron kinetic energies can give information about the excited states of nuclei, but only in the range of excitation above the ground state exceeding the separation energy of a neutron.

The typical spacing of these excited states is given in Table 13-1, but

TABLE 13-1

Typical Spacing between Nuclear Energy Levels

	$A = 20$	$A = 200$
Near ground state	$\sim 10^6$	$\sim 10^5$
6 to 8 M.e.V. above ground state	$\sim 10^4$	~ 10

An order-of-magnitude approximation to the spacing between energy levels is given, with energies in e.V.

of course in a particular nucleus in a particular energy range there may be many times as many levels as would be indicated by this table, or there may be no levels at all. The larger density of levels, and hence the prevalence of resonances in the slow neutron cross sections, in heavy nuclei results from the many possible arrangements of nucleons in a nucleus that contains so many of them. The complexity and small level spacing of large-A nuclei compared to small-A nuclei are reminiscent of the complexity and small level spacing in the atomic spectrum of iron, for example, compared to that in the spectrum of hydrogen.

Another way of using nuclear reactions to study nuclear energy levels is to study the energy distributions of the outgoing particles, such as the protons emitted in the reaction $Al^{27}(d,p)Al^{28}$ when monoenergetic deuterons are incident. Such distributions consist of groups of particles, each group with a definite kinetic energy K as illustrated in Fig. 13-19. When protons in the group with maximum kinetic energy K_{max} are emitted, the product nuclei are left in their ground states; a group with K less than K_{max} by an amount K_1 reveals an excited state of the product nucleus at an energy K_1 above the ground state.

It may happen that the compound nucleus has too small an energy to emit a nucleon or group of nucleons. It can then lose its excitation energy by emitting a γ-ray, as in the prevalent (n,γ) reactions typified by the reaction

$$_0n^1 + {}_{92}U^{238} \rightarrow {}_{92}U^{239} + \gamma \qquad (13\text{-}26)$$

Such reactions are usually called *radiative capture* reactions.

The two preceding reactions (with products Al^{28} and U^{239}) introduce the subject of *artificial radioactivity*. The reactions studied earlier in this section were all *artificial disintegrations*, reactions in which an incoming particle stimulated the decay of a stable nucleus to form another stable nucleus. In artificial radioactivity reactions, on the other hand, the

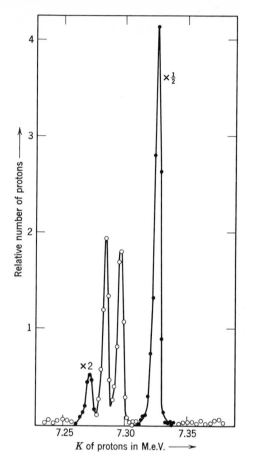

Fig. 13-19. Distribution of proton kinetic energies from the reaction $Al^{27}(d,p)Al^{28}$ when 7.01 M.e.V. deuterons are incident on aluminum. The distribution was measured by an instrument identical in principle to a mass spectrometer. Shown here is only a tiny part of the whole energy spectrum, which contains $\sim$100 peaks in an 8 M.e.V. energy range. (From Buechner, Browne, Mazari, and Sperduto, *Phys. Rev.*, **101**, 188, 1956, and *Rev. Sci. Inst.*, **27**, 899, 1956.)

compound nucleus decays to form a product nucleus that is unstable with a measurable half-life. Some typical reactions of this kind (in addition to the two cited above) are

$$\left. \begin{array}{l} {}_0n^1 + {}_7N^{14} \rightarrow {}_6C^{14} + {}_1H^1 \quad \text{ followed by} \\ {}_6C^{14} \rightarrow {}_7N^{14} + \beta^- + \bar{\nu} \quad \tau = 5580 \text{ years} \end{array} \right\} \quad (13\text{-}27)$$

$$_2He^4 + _{11}Na^{23} \rightarrow _{13}Al^{26} + _0n^1 \qquad \text{followed by}$$
$$_{13}Al^{26} \rightarrow _{12}Mg^{26} + \beta^+ + \nu \qquad \tau = 7 \text{ sec} \qquad (13\text{-}28)$$

$$_0n^1 + _{27}Co^{59} \rightarrow _{27}Co^{60} + \gamma \qquad \text{followed by}$$
$$_{27}Co^{60} \rightarrow _{28}Ni^{60} + \beta^- + \bar{\nu} \qquad \tau = 5.3 \text{ years} \qquad (13\text{-}29)$$

It should be emphasized that all the measurable half-lives in these reactions are associated with the long life of the β-emitters created as decay products of the compound nuclei. Each compound nucleus above has more than 7 M.e.V. excitation energy, and any nucleus with this high degree of excitation will lose most of its excitation by some process (usually the emission of a nucleon or a γ-ray) in an immeasurably short time.

In all such nuclear reactions there are several conservation laws that are always obeyed. Conservation of mass-energy is, of course, one of these, and calculations of Q values from measured kinetic energies (or vice versa) are based on this principle; Q values so obtained are at least as useful as precision mass spectrometry in establishing a table of atomic masses. Conservation of linear momentum is a second law; and if the momenta of the initial particle and target nucleus are known, this principle permits calculation of the way the energy of the reaction will be distributed between the kinetic energies of the two products. Conservation of nucleons is a third; no reactions have been observed in which the sum of the A values on the left of the equation did not equal the sum of the A values on the right. Conservation of charge is a fourth. There are also several other, more involved, conservation laws involving spin, statistics, parity, and isotopic spin, attributes of nuclei that we have not defined and shall not study here.

13-8 PARTICLES AND NUCLEAR FORCES

Particles, the ingredients of modern physics, have played a central role in the development of quantum physics. The laws of quantum physics were discovered through the analysis of simple combinations of protons and electrons. But as physical experimentation and theory applied to these particles developed an understanding of atoms and nuclei, the physicists' attitude toward these particles also underwent development. Early in the twentieth century, we should have said that the list of known particles consisted of electrons, protons, and the α-, β-, and γ-particles emitted by radioactive substances. Electrons and β-particles were soon shown to be identical. The α-particle was found to be a composite nucleus, and thus it no longer qualified as a member of this list.

The early quantum physics experiments revealed that the photon, whether of the few e.V. characteristic of visible light or of the few M.e.V. characteristic of γ-rays, was an important and stable particle. Later, the neutron was needed to explain nuclear structure, since it became increasingly difficult to believe in a nucleus composed of protons and electrons, and then the neutron was discovered as a free particle. The neutrino was postulated as a necessary companion of the electron in β decay and was later detected directly.

Another development of the concept of particles stemmed from the prediction by Dirac that the electron, proton, and most other particles should each have an *antiparticle* associated with it. The discovery of the positron, also called the β^+-particle or the anti-electron, followed soon thereafter. Much later the antiproton and the antineutron were discovered as products of the bombardment of nuclei by protons with energies of thousands of M.e.V. The antiproton $\bar{p}$ has the other properties of a proton but is negatively charged.

The most significant distinction between a particle and its antiparticle stems from the fact that they can annihilate one another. Thus the stability of a particle depends on whether it is surrounded by particles like itself or by its antiparticles. For example, when a positron passes through matter (which of course contains many electrons), it annihilates with an electron, producing two or three γ-rays; the sum of the energies of these γ-rays equals the kinetic energy of the two β-particles plus $2m_0c^2$. The inverse of this process is *pair production*, the creation of an electron-positron pair by a γ-ray with $h\nu > 2m_0c^2$ when it strikes a nucleus.

Similarly, a proton and an antiproton annihilate each other, but the products that carry away the energy in this process are pions (formerly called π-mesons). We shall consider pions later in this section, but meanwhile we note that they are also produced by n–$\bar{n}$ annihilation and thus appear to interact strongly with all nucleons.

In summary, there are the following stable particles: electron, proton, photon, and neutrino. The positron β^+ and the antiproton $\bar{p}$ are stable in the literal sense in that they do not spontaneously disintegrate; but they are unstable in the practical sense in that, since matter is full of electrons and protons, they are promptly annihilated upon striking any kind of matter. The neutron, though unstable, has a half-life (of 13 minutes) that in most experiments and in applications in nuclear reactors is long enough for its instability to be of no consequence.

If the list of particles stopped at this point, we should see some role in the structure of nuclear matter (some "application," as it were) for all the particles. But in experiments with the high-energy particle accelerators, experiments designed initially to further the understanding of the nucleus, many new particles were found. Many of these had previously been discovered, under more difficult experimental conditions, in studies of the cosmic rays, the energetic (up to $\sim 10^{18}$ e.V.) particles incident upon the earth from outside its atmosphere. All of these additional particles are unstable and decay into other particles on the list until they eventually decay to the stable particles. Of course, in order to produce these new particles in the laboratory, it is necessary to supply by a high-energy accelerator at least the rest energy Mc^2 of each particle.

The masses of all these particles are tabulated in Table 13-2. Not

TABLE 13-2

The Particles of Nuclear and High-energy Physics

		Mass*	Particle		Anti-particle*
Photons		0		$h\nu$	
Leptons	Neutrino †	0	ν		$\bar{\nu}$
	Electron	1	β^-		β^+
	Muon	207	μ^-		μ^+
Mesons	Pion	273	π^+	π^0	π^-
	Charged K meson	966	K^+		K^-
	Neutral K meson	966	K^0		$\overline{K^0}$
Nucleons and Hyperons	Proton	1836	p		$\bar{p}$
	Neutron	1838	n		$\bar{n}$
	Lambda hyperon	2181	Λ^0		$\overline{\Lambda^0}$
	Sigma hyperons	2328	Σ^+		$\overline{\Sigma^+}$
		2335	Σ^0		$\overline{\Sigma^0}$
		2343	Σ^-		$\overline{\Sigma^-}$
	Xi hyperons	2584	Ξ^0		$\overline{\Xi^0}$
		2584	Ξ^-		$\overline{\Xi^-}$

* The antiparticle always has a sign opposite to that of the particle. Thus $\bar{p}$ and $\overline{\Sigma^+}$ are negatively charged. The photon and the neutral pion do not have antiparticles. Masses are expressed in units of the electron mass m_0.

† There appear to be two kinds of neutrinos: $\bar{\nu}_e$ and ν_e accompany the electrons and positrons emitted in β decay; ν_μ and $\bar{\nu}_\mu$ accompany the muons emitted in the decay of pions.

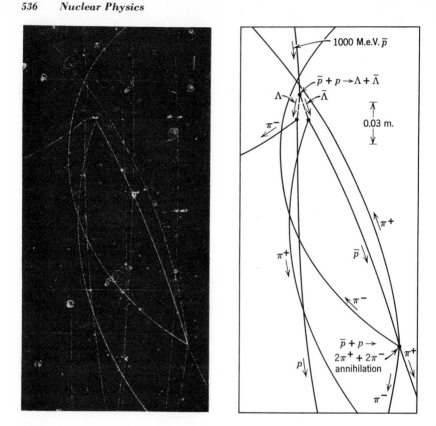

Fig. 13-20. Production of a Λ-hyperon pair by $\bar{p}$–p annihilation when an energetic antiproton strikes a proton in a hydrogen bubble chamber. The $\bar{\Lambda}$ subsequently decays to another antiproton, and it in turn annihilates with another proton to produce 4 pions. The experimental method will be described in Sec. 14-6. The curvature of the tracks was caused by a strong magnetic field normal to the plane of the photograph. (Courtesy of the 72-inch bubble chamber group of the Lawrence Radiation Laboratory and Jay Orear.)

tabulated is the body of additional information now known about them, such as decay schemes, lifetimes, and spins. In the table the labels photons and nucleons are familiar. *Leptons* are particles that interact very weakly with nuclei; the muon (formerly called the μ-meson) was first detected in cosmic rays and is commonly produced as a decay product of the pion. *Mesons* are particles that have masses intermediate between electrons and nucleons and interact strongly with nuclei; the pion appears to play the central role in nuclear forces. Nucleons are familiar particles. *Hyperons* are highly unstable particles with masses greater than nucleon masses. The creation and decay of Λ hyperons are illustrated in Fig. 13-20.

An over-all theory of this list of particles is lacking. That is, we do not know why the particles have the observed masses and other properties, although some rather powerful generalizations about the properties and modes of decay have been discovered. Furthermore, since many of these particles are of recent discovery, we cannot be confident that the list will not be increased by further discovery. The production, interactions, and properties of particles are being actively studied. This field of science is called *high-energy physics* because particle accelerators or cosmic rays with energies of thousands of M.e.V. are required in order to produce particles heavier than pions.

We now return to the pion and examine briefly its probable role in nuclear forces. First it should be noted that an alternate to the concept of the electromagnetic field to describe the ordinary Coulomb force between two charged particles is the concept of exchange of photons. In this approach, one particle is considered to emit and the other to absorb the photon, and this exchange leads to Coulomb's law by a theory that will have to remain mysterious here. Yukawa conceived the idea of treating the nuclear force in this way and specifically of the exchange of a particle with a non-zero rest mass. We shall not be able to present this theory here, but we shall describe briefly some of its successful predictions, especially the way in which the exchange of particles with a non-zero rest mass leads to a finite range of the nuclear force.

If a proton emits a pion with rest mass M_0, conservation of energy appears to be challenged. But if the pion is absorbed again in a short time Δt so that the complete reaction is

$$p \rightarrow p + \pi^0 \rightarrow p \qquad (13\text{-}30)$$

we can know the energy of the system only within an indeterminacy

$$\Delta E \cong h/\Delta t \qquad (13\text{-}31)$$

Thus emission and re-absorption of a particle with a rest mass $M_0 c^2 = \Delta E$ is consistent with the conservation of energy if the particle encounters the same or another nucleon within a time $\Delta t \cong h/M_0 c^2$. If we assume that the pion travels at about one-third the speed of light and if we insert the pion mass $273 m_0$, we obtain for the critical distance r_c between nucleons *

$$r_c \cong \frac{c\,\Delta t}{3} \cong \frac{h}{3 \times 273 m_0 c} \cong 3 \times 10^{-15} \text{ m.} \qquad (13\text{-}32)$$

* If we assume that the pion travels faster, we should have to include appreciable kinetic energy in ΔE along with $M_0 c^2$. If we assume that it travels slower, we obtain a shorter range.

Thus we estimate roughly that if the process of emission and absorption of pions is somehow responsible for the nuclear force, the range of this force (the internucleon distance within which the force is strong) should not be greater than $\sim 3 \times 10^{-15}$ m. This result agrees well with the observed range of the nuclear force inferred from nucleon-nucleon scattering experiments and from the typical nucleon-nucleon spacing in nuclei. Thus the process of emission and absorption of particles with the mass of pions leads to a finite range of interaction with a magnitude of this range typical of the observed range of the nuclear force.

The pion explanation of the nuclear force is also consistent with the *charge independence* of this force. When we were discussing the first term (nuclear force binding) in the binding energy formula eq. 13-17, we ignored the distinction between protons and neutrons. The agreement of this formula with experiment thus gives us evidence that the n-n, n-p, and p-p forces are the same (except of course for the Coulomb p-p force, which is treated separately), a conclusion that has been verified by experiments in which one of these particles was scattered by another. The pion interaction explanation of this charge independence is that there is no fundamental difference between the two kinds of nucleons. Each nucleon is continually emitting and absorbing pions, either by the scheme of eq. 13-30 or by one of the following schemes:

$$n \to n + \pi^0; \quad p \to n + \pi^+; \quad \text{or} \quad n \to p + \pi^- \qquad (13\text{-}33)$$

These are all fundamentally the same process, with pions of the same mass forming a kind of cloud around each nucleon. Thus the pion exchange concept of the nuclear force does not distinguish among n-n, n-p, and p-p forces.

A third success of the pion-exchange idea is its explanation of the saturation of the nuclear force. This property develops naturally from the pion-exchange concept since only nearest neighbors are involved in the exchange. Adding a nucleon at some distance from a pair of nucleons does not affect the pion-exchange interaction between these two.

A fourth success of this pion explanation is that it illuminates the question of the magnetic moments of nucleons. The proton's moment is nearly three times the value that would be expected if the proton's mass were substituted in eq. 6-14; the proton's moment is $2.793\mathfrak{M}_N$, where $\mathfrak{M}_N$ is the nuclear magneton (eq. 13-7). The relation between spin and magnetic moment is thus more complicated for the proton than for the electron, which result is consistent with the view of the proton as a particle surrounded by a pion cloud composed of particles at least some of which are charged. Furthermore, the fact that a neutron has a magnetic moment at all is also consistent with this view. We should not expect a simple neutral particle to have a magnetic moment when it is spinning, but a particle with a positive core surrounded by a negatively charged pion cloud should exhibit a magnetic moment, while remaining neutral in over-all charge.

We have thus seen how some of the simplest features of the nuclear force are probably related to the nucleon-pion interaction. The foregoing presentation certainly does not deserve to be called a theory, but complicated mathematical theories of the nuclear force have been constructed along these lines that are rather successful in dealing quantitatively with the deuteron binding problem and even with the binding of heavy nuclei.

REFERENCES

R. B. Leighton, *Principles of Modern Physics*, McGraw-Hill, New York, 1959.

R. D. Evans, *The Atomic Nucleus*, McGraw-Hill, New York, 1955.

R. M. Eisberg, *Fundamentals of Modern Physics*, Wiley, New York, 1961.

D. Halliday, *Introductory Nuclear Physics*, Wiley, New York, 2nd Ed., 1955.

H. A. Bethe and P. Morrison, *Elementary Nuclear Theory*, Wiley, New York, 2nd Ed., 1956.

F. K. Richtmyer, E. H. Kennard, and T. Lauritsen, *Introduction to Modern Physics*, McGraw-Hill, New York, 5th Ed., 1955.

PROBLEMS

13-1.* The Q value of the reaction of eq. 13-1 is 4.88 M.e.V. The radium nucleus is originally at rest, and therefore the initial momentum is zero. When this nucleus disintegrates, the radon and the α-particle share 4.88 M.e.V. of kinetic energy, and the sum of their momenta is, of course, zero (since momentum is conserved). What is the kinetic energy of each disintegration product?

13-2. Show that 1 curie = 3.7×10^{10} disintegrations per sec is the activity of 0.001 kg of $_{88}Ra^{226}$. This nuclide has a half-life τ equal to 1620 years.

13-3.* What is the *average* life of a radioactive nucleus with half-life τ?

13-4.* It frequently happens that a *parent* nuclide decays with disintegration constant λ_p to produce a *daughter* nuclide that is itself radioactive with disintegration constant λ_d. The number of nuclei of the two species is then governed by $dn_p/dt = -\lambda_p n_p$ and $dn_d/dt = \lambda_p n_p - \lambda_d n_d$. Find n_p and n_d as functions of t if at $t = 0$, $n_p = n_0$ and $n_d = 0$. *Hint:* Solve the first differential equation; assume a solution for the second of the form $n_d = C(e^{-\lambda_p t} - e^{-\lambda_d t})$, and find the constant C.

13-5. Show from the result of the preceding problem that if $\lambda_p \ll \lambda_d$ (i.e., $\tau_p \gg \tau_d$), the disintegration rate ($= -\lambda_d n_d$) of the daughter nuclei is the same as that of the parent nuclei after a time t such that $\lambda_d t \gg 1$.

13-6. Neptunium $_{93}Np^{237}$ with $\tau = 2.2 \times 10^6$ years is not present in nature, but there is good reason to believe that it was as abundant as uranium at the time the elements were formed. $_{19}K^{40}$ with $\tau = 1.3 \times 10^9$ years is present in a small concentration (0.01%) in natural potassium. What can you infer from these facts about the age of the elements?

13-7.* Calculate the height in million electron volts of the nuclear barrier for α-particles (Fig. 13-3) if $r_0 = 9 \times 10^{-15}$ m. and $Z = 88$. Assume that the rounding at the top of the barrier is negligible.

13-8.* Calculate the thickness of the nuclear barrier for α-particles (the tunneling distance in Fig. 13-3) if $r_0 = 9 \times 10^{-15}$ m., $Z = 88$, and $K = 4.88$ M.e.V.

13-9. $_{60}Nd^{144}$ is an α-emitter with a half-life of 2×10^{15} years. Make a sketch like Fig. 13-3 with curves for both $Z = 87$ and $Z = 60$. Estimate the α-particle energy corresponding to $\tau = 2 \times 10^{15}$ years for $Z = 87$ by inspecting Fig. 13-2. Estimate the α-particle energy corresponding to this τ and $Z = 60$ by recalling that the tunneling probability for a square well is inversely proportional to the product of the tunneling distance and the negative kinetic energy. (The observed energy is 1.9 M.e.V.)

13-10.* Calculate the density of nuclei in kilograms per cubic meter by using the relation between r_0 and K that is given by the line in Fig. 13-5.

13-11.* In order to appreciate the small size of the isotope shift in atomic spectra, find the fraction of the time that a $1s$ electron is inside the Li^7 nucleus. Repeat for the U^{238} nucleus. Comment on the relative probability that a $2p$ electron will be inside the nucleus.

13-12. Repeat the lithium $1s$ part of prob. 13-11 but for a μ-mesic lithium atom. What complication occurs if the same calculation is attempted for the μ-mesic uranium atom? Calculate the energy levels for the μ-mesic hydrogen atom.

13-13. Use the approximation technique employed in connection with eq. 13-6 to calculate the λ of a 1000-M.e.V. electron. Discuss the suitability of such electrons as incident particles in scattering experiments designed to determine the way in which the electrostatic potential varies with r inside the nucleus.

13-14.* Calculate the K of a proton or neutron with the same λ as that of the electron in the analysis leading to eq. 13-6. Does the value of K make it seem reasonable that protons and neutrons are constituents of nuclei?

13-15. Discuss the possibility that electrons are in the nucleus with an argument based on the Indeterminacy Principle. This argument is similar to which of the three arguments of Sec. 13-3?

13-16.* Use eq. 6-15 and the data given in the text for the magnetic moment of the proton to calculate the frequency at which radiation will be absorbed in a proton magnetic resonance experiment (in terms of $\mathfrak{B}$). In what region of the electromagnetic wave spectrum is this frequency if $\mathfrak{B} = 1$ weber/m.²?

13-17.* Compute the Q value of eq. 13-8.

13-18. Consider a particular neutron decay event (prob. 13-17) in which the neutrino carries away no energy. If the neutron was at rest before the decay, what are the kinetic energies of the proton and the emitted electron?

13-19. Sketch the energy distribution of the neutrinos from the β disintegration for which the β^- spectrum is given in Fig. 13-7.

13-20.* In order to learn the approximate size of β-ray spectrometers, compute the radius of curvature of the path of a 1-M.e.V. β^--particle in a constant magnetic induction of 0.1 weber/m.^{2}

13-21. If the neutrino has zero mass, what should be the ratio of the missing energy to the missing momentum in β decay? (This ratio is actually observed.)

13-22.* A positron can become bound for a short time to an electron when the positron is slowed in matter. The positron and the electron move about their common center of mass in a kind of atom called *positronium*. What are the atomic energy levels of positronium in electron volts? *Hint:* Apply the concept of eq. 7-9 to eq. 6-2.

13-23. Find a general expression for Q in the electron capture process, in terms of atomic masses, by starting with the process in terms of nuclear masses (as in the analysis leading to eq. 13-15).

13-24. Compare the result of prob. 13-23 with eq. 13-15. Why is the number of nuclides that are unstable with respect to electron capture greater than the number unstable with respect to β^+ emission?

13-25.* Consider K-electron capture in a nuclide with $Z = 50$ and $Q = 2.0$ M.e.V. What fraction of the energy Q is carried away by the neutrino?

13-26.* Calculate the wavelength of a 1-M.e.V. γ-ray. Does it make sense to say that a γ-ray existed as a particle within a nucleus before emission?

13-27. Consider the magnitude of the γ-ray wavelength calculated in prob. 13-26 and discuss the possibility of measuring γ-ray energies by crystal diffraction, as used for X-rays.

13-28. Sources of both β^-- and β^+-particles with continuous energy distributions are available because of the process of β decay. Many line sources (discrete energy distributions) of β^--particles are known. Why are there no line sources of β^+-particles?

13-29. Predict approximately and compare with the experimental data in Fig. 13-9 the difference in energy between the two peaks. *Hint:* Adjust the data for molybdenum in Fig. 6-13 by the method described in connection with eq. 6-17.

13-30. Verify the statements made in Sec. 13-5 of the text about the recoil energies of Cu^{64} and Fe^{57} when emitting γ-rays.

13-31.* What is the minimum number of atoms in an Fe^{57} crystal in order that the recoil of the crystal as a whole does not displace the Mössbauer line by an energy equal to the natural line width?

13-32. Show that if a Mössbauer γ-ray emitter is moving relative to an absorber with a velocity $v \ll c$, the emission and absorption lines are displaced in energy by an amount $\Delta E = E(v/c)$. *Hint:* For such small velocities, the simple relation $\nu' = \nu + \Delta\nu$ as in the acoustic Doppler effect is valid. [The relativistic relation valid for all velocities is $\nu' = \nu(1 - v/c)(1 - v^2/c^2)^{-\frac{1}{2}}$.]

13-33.* Use the conclusion of the preceding problem to calculate the relative velocity of an Fe^{57} emitter and an Fe^{57} absorber that will shift the energy of the Mössbauer line by 10^{-8} e.V., approximately one line width. (This calculation shows the remarkably small order of magnitude of velocities required to modulate the γ-ray lines in typical Mössbauer experiments.)

13-34. Calculate the binding energy per nucleon (B/A) from eq. 13-16 and tabulated masses for He^4, Li^6, and Ne^{20} and compare with Fig. 13-13.

13-35. What terms in eq. 13-17 are analogous to terms in the binding energy of atoms in a liquid drop? Explain the physics of the comparable terms in the liquid drop binding energy formula, especially the way saturation of interatomic forces occurs.

13-36. Estimate the magnitude of the Coulomb repulsion energy in a nucleus by calculating the potential energy P_{max} of a proton in M.e.V. at a distance $r_0 = 6 \times 10^{-15}$ m. from the center of a spherically symmetrical charge distribution of 49 protons. P_{max} is the energy of the fiftieth proton; the first brought to the nucleus has zero energy, and the average electrostatic energy per proton is of the order of $P_{max}/2$. Compare $50P_{max}/2$ with the third term of eq. 13-17. (Let $P = 0$ at $r = \infty$; see Fig. 13-5 for the reason this value of r_0 is chosen for $Z \cong 50$, $A \cong 115$.)

13-37.* Combine eqs. 13-16 and 13-17 to provide a formula for $_zM^A$.

13-38.* Calculate the value Z_{min} of Z that minimizes the mass for isobars with $A = 65$. Use the result of prob. 13-37.

13-39. Use the equation of prob. 13-37 to find the mass difference between $_{z+2}M^{A+4}$ and $_zM^A$, and thus to find a general expression for the Q value in the α-decay process $_{z+2}M^{A+4} \rightarrow {}_zM^A + {}_2He^4$.

13-40. Use the expression of prob. 13-39 to compute the Q value of eq. 13-1; the observed value is 4.88 M.e.V. What term of eq. 13-17 makes the largest positive contribution to Q?

13-41. Show that the principal effects responsible for the maximum in B/A vs. A and for the general shape of the curve of Fig. 13-13 are the surface energy term and the Coulomb energy term in eq. 13-17. This can be done with very little calculation as follows: Let $Z \cong 0.42A$ (a good approximation for not too light nuclides), let $Z - 1 \cong Z$, and drop the odd-even term in eq. 13-17. Show that $B/A \cong 15.1 - 17.8A^{-1/3} - 0.125A^{2/3}$. Calculate B/A from this expression for $A = 10, 20, 50, 100,$ and 200, plot B/A vs. A, and compare with Fig. 13-13.

13-42. For many even values of A, only a single isobar is stable. Sketch the replacement for Fig. 13-15 appropriate to this situation. (Do not calculate; start with the conclusion that only a single isobar is stable.)

13-43.* Calculate from tabulated atomic masses the Q value of the α decay of $_{80}Hg^{202}$. Why is $_{80}Hg^{202}$ found in nature? (This problem shows the importance of barriers in the analysis of stability.)

13-44. Both the reaction eq. 13-18 and the reaction eq. 13-22 can occur when deuterons bombard Li^6, each reaction being described by a certain value of the cross section. Find the generalization of eq. 13-20 that is appropriate under such circumstances.

13-45.* The total cross section (absorption plus scattering) of nickel for 1-M.e.V. neutrons is 3.5 barns. What is the fractional attenuation dJ/J of a beam of such neutrons on passing through a sheet of nickel 10^{-4} m. in thickness? (The density of nickel is 8900 kg/m.^{3}; the σ quoted refers to the naturally occurring distribution of isotopes.)

13-46.* Estimate the scattering cross section in barns of nickel for 40-M.e.V. neutrons.

13-47. The nuclear energy levels in the natural isotopic mixture of cadmium nuclides are such that the total neutron cross section is >2000 barns for neutron energies <0.2 e.V. How thick a layer of cadmium is needed to prevent 99% of incident thermal neutrons from arriving at a detector? (The density of cadmium is 8640 kg/m.^{3})

13-48. Why are stripping reaction cross sections smaller for α-particles than for deuterons incident upon the same nuclei with the same kinetic energies?

13-49.* Neutrino-nucleus reaction cross sections for $K = 1$ M.e.V. are of the order of 10^{-48} m.^{2} What is the probability that a 1-M.e.V. neutrino will be absorbed in passing through the earth along a diameter? (The diameter of the earth is 1.3×10^7 meters, and its mean density is 5500 kg/m.^{3}; assume a mean atomic weight of 38.)

13-50. Compute the energy of excitation of the product nucleus when a thermal neutron is captured by $_{29}Cu^{63}$; see Fig. 13-17 for the relevant masses. (This energy is also the *neutron separation energy* or *neutron binding energy* of Cu^{64}.)

13-51.* The levels revealed by the proton kinetic energy peaks in Fig. 13-19 are how far above the ground state of Al^{28}? (Neglect the kinetic energy of the Al^{28} nucleus; the mass of Al^{28} is 27.98191.)

13-52. Compare schematically the way σ depends on the kinetic energy of the incoming particle for n-induced and p-induced reactions. (Assume that there are no resonances in the energy range considered.)

13-53.* Calculate the minimum kinetic energy of an α-particle required in order that the α-particle can surmount the potential barrier of the $_4Be^9$ nucleus. (This is the energy required to produce the reaction of eq. 13-23 with a large cross section. α-particles with somewhat less energy than this can produce the reaction by tunneling, but the tunneling probability is small for very low-energy α-particles or for target nuclei with larger Z.)

13-54. Ne^{20} has energy levels at 1.63, 4.36, 5.4, 6.74, 7.18, 7.22, 7.45, 7.85 M.e.V. and many more at higher levels above the ground state. P^{31} has energy levels at 1.26, 2.23, (15 levels omitted), 7.78, 7.90, 7.95, 8.03, 8.04, 8.11, 8.22 M.e.V. and many more at higher levels above the ground state. Compare the $A = 20$ level spacings and the trend in level spacing from $A = 20$ to $A = 31$ with Table 13-1.

13-55.* Why must at least two γ-rays (instead of only one) be produced by β^+-β^- annihilation? What is the angle between the directions of emission of these two γ-rays if the positron and electron were both initially at rest?

13-56. A positron loses energy in a solid rapidly enough for its energy usually to be reduced to thermal energy before it is annihilated. Why is it more likely to interact with (and annihilate) a conduction-band electron in a metal than to interact with an inner-shell electron?

13-57.* Assume that a positron with zero kinetic energy annihilates a conduction-band electron in copper with kinetic energy equal to E_0. The two γ-rays can be emitted in any direction, but consider the particular pair of γ-rays that make equal angles with the initial velocity of the electron involved. Draw a vector diagram of the initial electron momentum and the momenta of the two γ-rays. What is the angle between the directions of emission of the two γ-rays?

13-58. Do you expect the antineutron to decay, and if so into what products and with what half-life?

13-59.* A pion at rest decays by the following scheme: $\pi^- \rightarrow \mu^- + \bar{\nu}_\mu$. Calculate the energy of the muon. (*Hints:* The momentum of a particle with kinetic energy K and zero rest mass, like the photon, is K/c. Consider whether the classical expressions for the energy and momentum of the muon are appropriate.)

13-60. Estimate from Fig. 13-5 the approximate distance between nearest-neighbor nucleons in nuclei. Compare with eq. 13-32.

14

Experimental and Applied Nuclear Physics

14-1 INTRODUCTION

In this chapter we shall discuss the most important terrestrial applications of nuclear physics and some of the experimental methods used to detect nuclear particles and to study nuclear reactions. Clearly we cannot pursue all the lines of study originating from the basic nuclear phenomena described in Chapter 13. We shall be forced to omit many intriguing topics, such as solar and stellar energy, mechanisms for the origin of the elements and analysis of their relative abundances, the age of the earth, and the phenomena and origin of cosmic rays.

Section 14-2 will give the basic physics of nuclear fission, the nuclear reaction that at present has the most prominent applications. Fission provides the energy in nuclear reactors and some of the energy in atomic weapons. Nuclear reactors, which apply the fission reaction to the production of useful power and of new nuclides, will be discussed in Sec. 14-3. Another kind of potentially important nuclear reaction, the fusion reaction, will be described briefly in Sec. 14-4.

The energetic charged particles and photons emerging from nuclear reactions interact strongly with matter. This interaction, described in Sec. 14-5, is of interest in itself and is the basis for most methods of detecting and measuring the properties of nuclear particles. After discussing the fission reaction (which is a useful means of

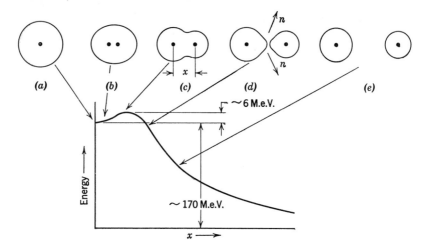

Fig. 14-1. Schematic diagram of the fission of a heavy nucleus. The 6-M.e.V. barrier height and the 170-M.e.V. kinetic energy of fission fragments are appropriate values for U^{235}. The parameter x measures the separation of the two fragments and is poorly defined until the stage d is reached.

detecting neutrons) and after considering the interaction between charged particles and matter, we shall have all the information we need to explain the operation of nuclear particle detectors. This explanation will be given in Sec. 14-6. Finally, some of the applications of radioactive nuclides, most of which are by-products of nuclear reactors, will be described in Sec. 14-7.

14-2   NUCLEAR FISSION

Nuclear fission, the splitting of a nucleus into two nearly equal parts, is illustrated schematically in Fig. 14-1. The nucleus at (a) is nearly spherical in shape, but if this nucleus is excited (e.g., by particle bombardment) it can execute violent vibrations. It should be recalled that the list of stable nuclides terminates at the high Z end because of the electrostatic repulsion of the protons.* In a fissionable nucleus this repulsion is very powerful when a configuration like (b) or (c) is reached.

* And because of the asymmetry energy. At large Z, the electrostatic repulsion energy discriminates against protons and thereby prevents Z from equaling $N = A - Z$. Whenever $Z \neq N$, the asymmetry energy lowers the binding energy. Thus both the third (Coulomb repulsion) and fourth (asymmetry) terms on the right in eq. 13-17 give large negative contributions to the binding energy at large Z.

The two parts are restrained from flying apart, however, by the surface energy (the second term on the right in eq. 13-17), since the surface-to-volume ratio of the shape (*b*) or (*c*) is larger than that of the sphere (*a*). The surface energy thereby provides a barrier, and it is this barrier that is responsible for the existence in nature of nuclides with $A > \sim 110$.

Nuclides with A of the order of 250 have such a small barrier that the spontaneous oscillations of the nucleus can surmount it with a short half-life, but this spontaneous fission process becomes unmeasurably slow for $A < 238$. A much more important fission process is induced fission, fission excited by an external particle. Suppose, for example, that a thermal neutron is incident upon $_{92}U^{235}$. It adds its kinetic energy (negligible) and its binding energy (6.3 M.e.V.) to the nucleus, and the compound nucleus $_{92}U^{236}$ then has enough energy to surmount the barrier. Once past the configuration (*c*), the surface energy rapidly decreases by "necking," as in the break-up of a large water drop. The nearly spherical fragments in (*d*) fly apart, accelerated by the Coulomb

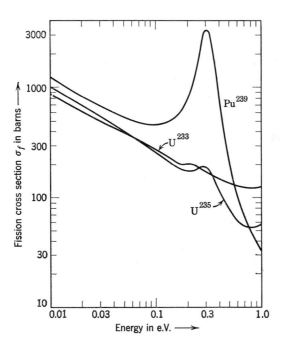

Fig. 14-2. Cross sections for neutron-induced fission in three nuclides that are fissionable by slow neutrons. Note the resonances superimposed on the general rise in σ_f as the neutron kinetic energy decreases.

repulsion. Two or three neutrons are typically ejected, presumably during the $(c) \rightarrow (d)$ stage of the process.

Many nuclides are fissionable by fast particles, which add their kinetic energies as well as their binding energies to nuclei, but of the naturally occurring nuclides, only U^{235} undergoes fission by slow neutrons. This is the only natural nuclide with a barrier low enough to be overcome by the neutron binding energy alone. Two other nuclides in which fission can be induced by slow neutrons are $_{92}U^{233}$ and $_{94}Pu^{239}$, which can be produced in nuclear reactors by neutron capture in $_{90}Th^{232}$ and $_{92}U^{238}$, respectively, and subsequent β decay. The fission cross sections for these three nuclides are given in Fig. 14-2. (See also prob. 14-6.)

It should be clear from the foregoing sketch of the fission process that there is *not* a unique reaction equation for a given fissionable nuclide and a given incoming particle. The fission fragments vary in A and Z, with a considerable preference for a somewhat unequal distribution of mass between them. A typical reaction is

$$_{0}n^{1} + _{92}U^{235} \rightarrow _{39}Y^{95} + _{53}I^{139} + 2(_{0}n^{1}) \tag{14-1}$$

The N/Z ratio near $A = 235$ is much larger than near $A = 95$ or $A = 139$. The ejection of the two neutrons during the fission process helps provide the proper N/Z for the fragments, but the ratio is still much too large for stability (see Fig. 13-10, and imagine a line from $Z = 0$, $N = 0$ to $Z = 92$, $N = 141$; fission fragments lie near this line). Therefore the fission fragments are always radioactive.

The particular fragments of eq. 14-1 decay according to the following schemes:

$$_{39}Y^{95} \xrightarrow[10 \text{ min}]{} _{40}Zr^{95} \xrightarrow[63 \text{ days}]{} _{41}Nb^{95} \xrightarrow[35 \text{ days}]{} _{42}Mo^{95} \tag{14-2}$$

$$_{53}I^{139} \xrightarrow[2.7 \text{ sec}]{} _{54}Xe^{139} \xrightarrow[41 \text{ sec}]{} _{55}Cs^{139} \xrightarrow[9.5 \text{ min}]{} _{56}Ba^{139} \xrightarrow[85 \text{ min}]{} _{57}La^{139}$$

$$\tag{14-3}$$

The stable nuclides that terminate the chain (such as Mo^{95} and La^{139} in this example) are called *fission products*, but this term is also applied to long-lived radioactive decay products in such decay chains. A β^{-}-particle is emitted in each stage of these chains; the value of the half-life τ for each disintegration is written under the arrows in the equations above. Most of the β^{-}-particles are accompanied by γ-rays, and all are accompanied by antineutrinos.

Figure 14-3 presents the observed distribution of fission products, corresponding to the distribution of reaction equations of the general type eq. 14-1, from the slow-neutron fission of U^{235}. The average number of neutrons ejected in the fission process in this nuclide is 2.43.

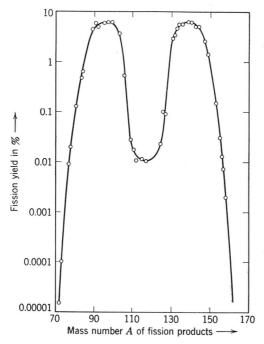

Fig. 14-3. Distribution of fission products from the thermal-neutron fission of U^{235}. The per cent of fissions that produce a product with mass number A is plotted as a function of A. [From J. M. Siegel, C. D. Coryell, and others, *Revs. Mod. Phys.*, **18**, 513 (1946).]

Although almost all the decay processes in chains like eqs. 14-2 and 14-3 are β^-- and γ-emission processes, occasionally a nuclide has enough excitation energy to emit a neutron. An example of such a chain is given in Fig. 14-4. The neutron separation energy of Kr^{87} is 5.1 M.e.V., which is less than the excitation energy (5.4 M.e.V.) with which this nuclide is formed when Br^{87} decays by β^- emission ($\tau = 56$ sec) to the upper state of Kr^{87}. This excited state rapidly decays to the ground state, in 97% of the cases by γ emission and in 3% by the emission of 0.3-M.e.V. neutrons. Thus the neutron emission in such decay chains is rare, because of the rarity of excited states with as much as 5 to 8 M.e.V. of energy and because of the competition of energetic γ emission. Nevertheless, we shall find in the next section that this neutron emission is important in nuclear reactors because it is *delayed* an appreciable time after the fission event (in the example, by the $\tau = 56$ sec of the β^- decay

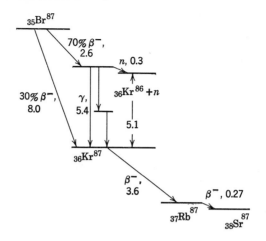

Fig. 14-4. Delayed neutron emission. The numbers are energy differences in M.e.V. In 70% of the β^- decays of $_{35}Br^{87}$, $_{36}Kr^{87}$ is formed in an excited state with an energy 5.4 M.e.V. that is greater than the neutron separation energy 5.1 M.e.V. Either neutron emission or γ decay of this excited state occurs so rapidly that the possible 9.0-M.e.V. β^- decay to $_{37}Rb^{87}$ is not observed. (Modified from R. D. Evans, *The Atomic Nucleus*, McGraw-Hill, New York, 1955, used by permission.)

of Br^{87}). About 0.7% of the neutrons produced in fission are delayed, and the average time of delay is about 9 sec.

A very important feature of the fission process is the large amount of energy released. This energy can be estimated by referring to the nuclear mass data that were summarized in Fig. 13-13. For stable isotopes in the range $70 < A < 160$ (the range of fission fragments), the binding energy per nucleon is about 8.5 M.e.V. For $_{92}U^{235}$, the binding energy per nucleon is about 7.6 M.e.V. The total energy released when one $_{92}U^{235}$ undergoes fission and the fragments decay to stable nuclei is therefore about $235(8.5 - 7.6) \cong 200$ M.e.V. Measurements of the sum of the kinetic energies of the two fission fragments show that this energy averages about 165 M.e.V. The remaining energy is released in the kinetic energies of the neutrons and the energies of the various β^--particles, antineutrinos, and γ-rays from the disintegrations of the fission fragments.

A more accurate and informative calculation of the energy released in the fission process can be obtained through the use of the semi-empirical binding energy formula eq. 13-17. We shall consider only odd-A examples (which permits us to drop the last term of eq. 13-17), and we can conveniently combine eqs. 13-16 and 13-17 to obtain a formula for the odd-A nuclide mass in M.e.V.:

$$_ZM^A = 923.8A - 0.8Z + 17.8A^{2/3}$$
$$+ 0.71Z(Z - 1)A^{-1/3} + 23.6A^{-1}(A - 2Z)^2 \quad (14\text{-}4)$$

The energies involved in the particular example eq. 14-1 of a fission reaction are tabulated in Table 14-1. The first column is the energy arising from the neutron mass, proton mass, and neutron binding energy (the first two terms of eq. 14-4). The remaining columns are the successive terms of eq. 14-4. The first row gives the energy differences in M.e.V. between the left and right sides of eq. 14-1; this is the energy shared among the fission fragments Y^{95} and I^{139} and the promptly emitted neutrons. The second row gives the energy term-by-term in the over-all energy release, after all fission fragments have decayed to stable fission products. The $\sim$30-M.e.V. energy difference between the totals of the two rows is the energy of the β^--particles, γ-rays, and antineutrinos emitted in the radioactive decay of the fission fragments.

As expected, Table 14-1 shows that the Coulomb energy is the important driving force in fission, whereas the surface energy is the important term inhibiting fission. The asymmetry term is responsible for the energy released in the decay of the fission fragments, all of which have N/Z ratios too large for stability. It should be emphasized again that eq. 14-1 is only one of many ways in which neutron-induced fission of U^{235} can occur, but the energies calculated for other ways compare closely with those in Table 14-1.

The reason for the practical importance of the fission reaction is that the products of the reaction include the neutrons necessary to produce

TABLE 14-1

Energy Released in Nuclear Fission

	Nucleon Masses and Volume Binding	Surface	Coulomb	Asymmetry	Total
Immediate (fission fragments)	−15.7	−170.4	354.9	4.5	173.3
Ultimate (stable products)	−10.1	−170.4	257.8	125.0	202.3

The various terms in the energy released in fission are tabulated in M.e.V., using the particular fission reaction eq. 14-1 as an example.

the reaction. A *chain reaction* is hence possible. This reaction is analogous to the burning of fuel gas in a flame: The reaction of one molecule of the fuel with oxygen produces heat that raises the kinetic energy of adjacent molecules sufficiently so that they in turn react. No nuclear reaction other than the fission reaction gives the right kind of products to keep the chain of reactions operating. Since there is an average of 2.43 neutrons emitted in the slow-neutron fission of U^{235} and only one is required to continue the chain, there is some margin for loss of neutrons by escape from the region containing the U^{235} and for loss by absorption in other nuclei.

14-3 NUCLEAR REACTORS

The nuclear fission reactor is the principal application of nuclear fission. There is a large variety of nuclear reactors, differing in the fissionable material (*fuel*), power level, neutron flux, and purpose. We shall discuss briefly three examples to illustrate the principles and the problems, but many variations of these types and other types have been constructed. The first of these examples is a large but low-power reactor, which uses natural uranium and was designed for experimental purposes. The second is a small, powerful reactor, which employs *enriched* uranium (uranium containing a much larger fraction of U^{235} than natural uranium) and generates useful electrical power. The third is a *breeder* reactor, which produces more fissionable material than it consumes and also provides usable power.

In all these reactors we shall be interested in the neutron absorption processes, since neutrons are the precious commodity in a nuclear reactor. The essential feature of a chain reactor is that the loss of neutrons by nuclear reactions (other than fission) and by escape from the reactor must be small enough so that there is one neutron from each fission left over to create a new fission. Radiative capture is the principal nuclear reaction competing with fission. In U^{235} it is

$$_0n^1 + {}_{92}U^{235} \rightarrow {}_{92}U^{236} + \gamma \tag{14-5}$$

with a cross section of 107 barns for thermal neutrons. Natural uranium consists largely of U^{238}, and only about 1 part in 140 is U^{235}. Radiative capture in U^{238} is

$$_0n^1 + {}_{92}U^{238} \rightarrow {}_{92}U^{239} + \gamma \tag{14-6}$$

followed by

$$_{92}U^{239} \rightarrow {}_{93}Np^{239} + \beta^- + \bar{\nu} \qquad (\tau = 23 \text{ min})$$

and

$$_{93}Np^{239} \rightarrow {}_{94}Pu^{239} + \beta^- + \bar{\nu} \qquad (\tau = 2.3 \text{ days})$$

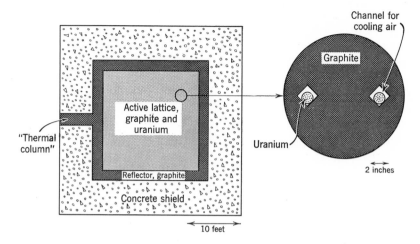

Fig. 14-5. Cross section of Oak Ridge graphite reactor. The inset shows the natural uranium fuel rods in place in square holes in the graphite moderator. The thermal column provides neutrons with thermal energies for experiments like the neutron diffraction experiments of Sec. 4-11. Not shown are a large number of holes in which experiments can be performed, solids can be irradiated in order to study the effects of neutrons on them, or stable nuclides can be irradiated in order to produce radioactive isotopes.

The fissionable Pu^{239} (plutonium) thereby produced is α-radioactive with a half-life of 24,000 years.

The Oak Ridge graphite reactor is illustrated schematically in Fig. 14-5. The *active volume* or *core* is composed of a lattice of 1-inch-diameter natural uranium rods on 8-inch centers in a regular array and embedded in very pure graphite. The purpose of this structure is to conserve neutrons. The neutrons from fission have an initial energy of the order of 2 M.e.V., and the cross section for fission by such fast neutrons is very small. The cross sections for scattering and for radiative capture in the U^{238} (eq. 14-6) by fast neutrons are also small, and therefore most of the fast neutrons pass out of the uranium rods into the graphite. They are slowed by successive collisions with carbon nuclei in the graphite, which is therefore called the *moderator*. When a neutron has been scattered many times in the graphite it has only thermal energy, about $\frac{1}{40}$ e.V. Thereafter it neither gains nor loses energy (on the average) by further collisions with the carbon nuclei, which also have thermal energy. Thermal neutrons are thus incident upon the uranium rods and produce fissions with high probability, since σ for fission by thermal neutrons in

U^{235} is large (580 barns). Slowing the neutrons in graphite, rather than in uranium, has permitted them to avoid capture by U^{238} as much as possible. The cross section for the capture process in U^{238} (eq. 14-6) has six large and several smaller resonances at neutron energies between 5 and 200 e.V., and similar resonances occur in U^{235}. The neutrons usually have an energy greater than 200 e.V. when they leave the uranium rod in which they were released and have less than 5 e.V. before they again encounter a uranium rod.

Each 41 fissions produce about 100 neutrons. On the average about 33 of these are lost by capture in U^{238}, despite the use of the graphite moderator to reduce this number. About 8 are lost by radiative capture in U^{235} (eq. 14-5). About 11 are lost by reactions with nuclei in the moderator; carbon has a very small capture cross section, but impurities with large cross sections also remove neutrons. About 2 neutrons are absorbed by the structure and the aluminum cladding that protects the uranium fuel cylinders from oxidation. About 4 neutrons escape from the active volume or are used in experiments. If the reactor were smaller, the number of escaping neutrons per fission would be larger. There is a *minimum size* that must be exceeded if the reactor is to operate at all; if the size is smaller than this, the surface-to-volume ratio is so large that escape of neutrons at the surface makes it impossible to balance the neutron budget.

All these processes subtract about 58 neutrons from the original group of 100. *Control rods* are inserted into or withdrawn from the core of the reactor to absorb neutrons and thereby to control the power level. Their absorption (about one neutron), when added to the approximately 58 absorbed by other processes, is adjusted so that of the original group of 100 neutrons, *exactly* 41 remain to produce new fissions and to continue the chain reaction. In the reactor described, these rods are made of steel containing 1.5% boron; natural boron has a large thermal neutron cross section, about 750 barns. Suppose, for example, that the rods are withdrawn 10%, and therefore they absorb only 0.9 neutron per 100 initial neutrons. Then, each time the cycle of fission, moderation, and fission occurs, the number of neutrons is increased 0.1%. If *all* the neutrons from fission were emitted at the time of the fission, this cycle would take about 10^{-2} sec (see prob. 14-18). Thus the power level, which is proportional to the number of fissions per second, would increase at the rate of 10% in 1 sec (neglecting the effect of delayed neutrons). This rapid rise in power level illustrates how precisely the neutron cycle must balance in order to keep the reactor operating safely at a steady power level.

The fact that 0.7% of the fission neutrons are delayed by an average time of 9 sec greatly aids in the control of a reactor. It is instructive to consider again the example of the previous paragraph and to include now the fact that although most of the neutrons are prompt, some are delayed. At the time the control rod is withdrawn 10%, 0.7% of the neutrons being released are delayed, and their number is determined not by the fission rate at this time, but by what the fission rate was some time previously (an average of 9 sec previously). Without the delayed neutrons, the neutron cycle does not balance: The withdrawal of the control rod decreased the neutron consumption by 0.1%, but 0.7% of the neutrons are delayed. Thus the reactor power level cannot increase until $\frac{6}{7}$ of the delayed neutrons have been emitted. The effect of the delayed neutrons is hence to retard the rapid rise of power level when there is some fluctuation in the neutron production or absorption, and the time for changing the reactor power appreciably is of the order of several seconds or even minutes. This effect makes manual control possible, although automatic control based on power-level-indicating instruments in the reactor is customary.

Safety rods of boron steel automatically drop into the reactor (and shut it down) in the event of electrical power failure or if signals from instruments indicate that the operation of the reactor is not according to plan.

It was noted above that a reactor must be larger than a minimum size. If it is smaller than this size, too many neutrons leave the active volume without producing fissions. In order to turn back as many neutrons as possible, there is a region of graphite, called the *reflector*, on the border of the active volume (Fig. 14-5). This reflector scatters a large proportion of the neutrons incident upon it back into the active volume.

The shielding surrounding the reflector adds greatly to the size of a reactor. This shielding is usually many feet of concrete, and its purpose is to prevent substantial amounts of γ-rays or fast neutrons from penetrating into the regions where operators must work.

The 200 M.e.V. of energy per fission in this reactor produces about 4000 kilowatts of power. Most of the power comes from the kinetic energy of the fission fragments. These nuclei move only very short distances (about 10^{-6} m.) since they are multiply charged and therefore lose their energy rapidly by ionizing the material through which they are moving. Their kinetic energy is dissipated in the fuel cylinders. The rest of the 200 M.e.V. per fission (β, γ, and neutron energies) is largely dissipated in the moderator, but of course the antineutrinos carry their energy away ($\sim$10 M.e.V. per fission). Large fans cool the reactor by pulling air through it and exhausting the air through a tall stack. The temperature rise of the air is too small to permit the efficient production of electrical power.

The neutrons captured by U^{238} produce the fissionable element plutonium, $_{94}Pu^{239}$, as explained in eq. 14-6. The plutonium can be chemically separated from the uranium in the fuel cylinders after a reactor has operated for a long time. Thus a concentrated form of fissionable material can be obtained.*

The use of natural uranium in this graphite-moderated reactor is responsible for its large size and small power density. Neutron capture in U^{238} is an unwelcome parasite on the fission reaction in U^{235} and forces the designer of a natural uranium reactor to adopt a large size and a *heterogeneous* geometry (uranium and moderator in different regions of space) to conserve neutrons.

In choosing a particular element as a moderator in a reactor we must consider its mass number A and its neutron scattering and absorption cross sections. To be most effective in slowing neutrons, a moderator nucleus should have a large scattering cross section, which causes a large number of neutron-moderator collisions per unit neutron path length. It should also have a small A, since the fraction of a neutron's energy that it loses at each collision is greater the nearer the neutron's mass is to the mass of the moderator nucleus. To be most efficient, a moderator nucleus should also have as small a capture cross section as possible, to reduce the waste of neutrons in absorption reactions.

The availability of concentrated fissionable material, such as uranium enriched in U^{235} or pure U^{235}, Pu^{239}, or U^{233}, makes possible much greater flexibility in reactor design. In particular, much smaller reactors with enormous power densities suitable for the production of electrical power are made possible. Water or another hydrogenous substance usually replaces graphite as a moderator in such reactors. By this change the number of neutron-moderator collisions required to reduce the neutron energies to thermal energy is decreased, because the thermal neutron scattering cross sections for H^1 and H^2 are greater than those of carbon and the mass numbers A are much less. Thus the neutron path length for the moderation process is decreased, and the scale of the reactor can be reduced accordingly. Ordinary water ($H_2^1O^{16}$) is a very

---

* Another way of producing concentrated fissionable material is to separate the U^{235} (0.7%) from the U^{238} (99.3%) in natural uranium. Since the chemical properties of these two isotopes are identical, they must be separated by some process that depends on their different masses. The most useful process exploits the different rates of diffusion of gaseous $U^{238}F_6$ and $U^{235}F_6$. The molecules of the latter move slightly faster (on the average) than those of the former (see prob. 14-24). If a mixture of these gases is permitted to diffuse at low pressure through a porous barrier, the transmitted gas is about 0.3% richer in U^{235} than the starting mixture. The process must therefore be repeated thousands of times in order to provide substantially pure U^{235}, and a very large plant is required to effect this enrichment of natural uranium.

effective moderator, but some neutrons are captured by the reaction

$$_0n^1 + {}_1H^1 \rightarrow {}_1H^2 + \gamma \qquad (14\text{-}7)$$

Heavy water (H_2O^{16}) does not slow neutrons quite so quickly, but, since no capture reactions of any consequence occur with deuterium (H^2), heavy water is a more efficient moderator than ordinary water in that it wastes fewer neutrons in capture reactions.

The second example of reactor to be described is a small *homogeneous* reactor employing a water solution of an enriched uranium (or plutonium) salt as fuel and moderator. The fissionable material and moderator can be intimately mixed in this reactor since there is very little U^{238} to capture neutrons of intermediate energies (5 to 200 e.V.). Although radiative capture in U^{235} (or Pu^{239}) can still occur, the parasitic reactions are not nearly so important as in the first example above. The power density is of the order of megawatts per kilogram of active volume and is limited primarily by the effectiveness of heat transfer and by the high-temperature and radiation resistance of materials. Useful power is obtained, as indicated in the schematic diagram of Fig. 14-6.

The fuel-moderator solution is contained in a stainless-steel sphere

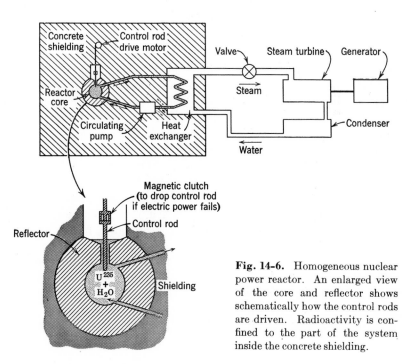

Fig. 14-6. Homogeneous nuclear power reactor. An enlarged view of the core and reflector shows schematically how the control rods are driven. Radioactivity is confined to the part of the system inside the concrete shielding.

with stainless-steel pipes connected to it. Only in this sphere is there an appreciable rate of fissioning. In all the other parts of the system, neutrons produced by fissions most probably travel out of the pipes instead of producing more fissions. In other words, the size of the pipes and other parts is small compared to the critical size required to maintain the chain reaction. The system is kept under pressure so that it can operate at a temperature above the normal boiling point of water.

An interesting feature of this reactor is the way it can operate in a stable manner without external control. The solution is circulated through a heat exchanger, which heats and vaporizes water in a conventional steam-turbine system. (The water in this system does not become radioactive, since it is subjected only to β and γ radiation from fission products, not to substantial fluxes of neutrons; electrons and photons can produce some nuclear reactions, but these reactions rarely have radioactive products.) Automatic controls open the steam valve as required in order to maintain constant generator speed. Suppose that this valve is suddenly opened somewhat wider, in response to the sudden connection of a load to the generator. The flow through the valve increases, and the temperature in the heat exchanger decreases. This in turn causes the temperature in the reactor to decrease. The contraction of the solution as the temperature drops increases the density of U^{235} nuclei in the active sphere because some of the solution is drawn through the neck from the inactive region. The fission rate increases, and therefore the temperature rises until the reactor returns to its initial temperature. The reactor is thus stable and controls itself. Control rods are provided, however, to assist in starting up the reactor and to operate as safety devices. They are supported by electromagnets. In the event instruments indicate leaks, too high neutron flux, too rapid rise of flux, or other dangerous conditions, current to the electromagnets is cut off, the control rods fall into the reactor, and the flux is quickly reduced.

A third example of reactor is a *fast breeder*. The purpose of this reactor is to produce useful power and to produce more fissionable material than it consumes. The active volume is only a few liters and consists entirely of enriched uranium and a coolant, which is a liquid solution of sodium and potassium. The coolant is circulated through a heat exchanger that produces high-pressure steam, as in the homogeneous reactor described above. The use of liquid metal as the heat-transfer fluid permits the attainment of high temperatures (and therefore high thermodynamic efficiency) without the necessity for operating the reactor under pressure. Liquid metal coolant also simplifies the problem of a leak-free circulating pump for the radioactive fluid. Since the fluid is a good electrical conductor, a force on the fluid can be obtained by passing a strong electrical current I at right angles to the direction of flow and by providing a d-c magnetic field at right angles to I and to the flow direction. The *electromagnetic pump* has neither packing nor moving parts and is therefore valuable for circulating radioactive materials at high temperatures.

No moderator is used in this reactor, and the fissions are produced by the fast neutrons direct from previous fissions. Since there is very little U^{238} in the reactor, a moderator need not be employed to prevent capture of 5–200 e.V. neutrons by this nucleus. The neutrons produced in fission have a spectrum of energies extending from zero to about 8 M.e.V., with an average energy of about 2 M.e.V. The average cross section for fission by such neutrons is much less than for fission by thermal neutrons. But the cross section for radiative capture in U^{235} (eq. 14-5) is reduced even more for fast neutrons compared to thermal neutrons For thermal neutrons, 85% of the neutrons produce fission and 15% produce U^{236}, which is of no value (the cross sections for these processes are 580 and 107 barns, respectively). For fast neutrons, relatively more fissions are produced. Thus this *fast* reactor wastes fewer neutrons than *thermal* reactors like the first two described.

A *blanket* of natural uranium surrounds the active volume of this reactor. Any neutrons escaping from the core are almost certainly absorbed in the blanket and produce the reactions of eq. 14-6. After a long period of operation, the blanket is removed and the Pu^{239} is chemically separated from the uranium. The reactor has therefore produced useful fissionable material in a concentrated form.

Each 40 fissions produce about 100 neutrons in this reactor. Forty of these neutrons are required to produce new fissions and thus keep the chain reaction going. About 10 neutrons are lost by absorption in the reaction of eq. 14-5, absorption in the NaK alloy, container, and control rods, and by escape. Approximately 50 are left to be absorbed in the blanket, producing 50 fissionable Pu^{239} nuclei by the reaction of eq. 14-6 Thus by the consumption of 40 fissionable nuclei the breeder reactor has produced about 50 fissionable nuclei from U^{238}. Since there is 140 times as much of this isotope in nature as there is U^{235}, the breeder reactor is capable of increasing the supply of fissionable materials, and the reactor obtains the designation *breeder* from this fact. Furthermore, thorium is much more abundant than uranium, and a blanket of $_{90}Th^{232}$ (100% of natural thorium is this isotope) in a breeder produces the fissionable isotope U^{233} by a reaction similar to eq. 14-6. Breeder reactors thus are capable of enlarging the supply of fissionable material while simultaneously producing electrical power.

Only three of a host of types of nuclear reactors have been described here, and in this brief study of nuclear reactors we have considered only a few of the many scientific and technological problems involved. Many of these are nuclear-physics problems, but many are solid-state physics, chemistry, chemical-engineering, and mechanical-engineering problems. For example, the corrosion of pipes and containers is a very serious problem for reactors at high temperatures and with large densities of ionization. The operating temperatures of reactors are therefore limited, and low thermodynamic efficiency results. Another problem is the effect of fast neutrons on solids. Such neutrons knock atoms out of their ordinary positions in the crystal lattice and thus create vacancies and interstitial atoms. These imperfections alter the mechanical properties of solids in ways that are not yet well understood. Active research is in progress on corrosion in the presence of radiation and on the effects of neutrons on solids.

It should be recognized that nuclear reactors have other useful products as well as electrical power. We have already noted that they can produce additional fissionable fuel. The ionizing radiation in reactors can induce chemical reactions such as nitrogen fixation or cross-linking in polymers. Radioactive fission products and radioactive nuclides produced by neutron capture in elements placed in reactors are of enormous value in science, industry, agriculture, and medicine. These and other applications in research and industry assure the continued importance of the fission reactor.

14-4 NUCLEAR FUSION

Nuclei with small A can release energy by joining together to form heavier nuclei. Inspection of Fig. 13-13 shows that nuclides with A up to about 25 can participate in such exothermic fusion reactions, just as large-A nuclides can undergo exothermic fission. We shall study briefly in this section the fusion reactions that are most promising for the controlled production of useful power and outline the problems encountered in exploiting these reactions.

The most interesting fusion reactions appear to be the following:

$$_1\text{H}^2 + {_1}\text{H}^2 \rightarrow {_1}\text{H}^1 + {_1}\text{H}^3 \qquad (Q = 4.04 \text{ M.e.V.}) \qquad (14\text{-}8)$$

$$_1\text{H}^2 + {_1}\text{H}^2 \rightarrow {_0}n^1 + {_2}\text{He}^3 \qquad (Q = 3.27 \text{ M.e.V.}) \qquad (14\text{-}9)$$

$$_1\text{H}^2 + {_1}\text{H}^3 \rightarrow {_0}n^1 + {_2}\text{He}^4 \qquad (Q = 17.6 \text{ M.e.V.}) \qquad (14\text{-}10)$$

$$_1\text{H}^2 + {_2}\text{He}^3 \rightarrow {_1}\text{H}^1 + {_2}\text{He}^4 \qquad (Q = 18.3 \text{ M.e.V.}) \qquad (14\text{-}11)$$

The first two reactions occur with essentially equal probability when deuterons collide with deuterons; deuterium (heavy hydrogen) accompanies ordinary hydrogen in nature in a concentration of 0.015%. The third reaction requires tritium, a radioactive isotope of hydrogen present only in a minute concentration in nature; it is a β^--emitter with an energy of 0.018 M.e.V. and a half-life $\tau = 12.4$ years. It can be produced from lithium by the reaction

$$_0n^1 + {}_3Li^6 \rightarrow {}_1H^3 + {}_2He^4 \tag{14-12}$$

and is produced in nature by cosmic-ray neutron bombardment of atmospheric nitrogen by the reaction $_7N^{14}(n,{}_1H^3)_6C^{12}$. The fourth reaction requires He^3, which is present in natural helium in a concentration of a few parts per million.

All of these reactions release energy, and the third and fourth release far more energy per nucleon (and hence per kilogram) than the fission reaction. But there are profound problems in harvesting this power in a fusion reactor. The principal origin of these problems is the circumstance that all the reactants in promising fusion reactions are *charged*. Thus they must have considerable kinetic energies in order to surmount the Coulomb barrier and to approach one another within the range of the nuclear force. Figure 14-7 shows how the reaction cross sections for the above reactions fall rapidly to zero at low deuteron kinetic energies because of this barrier. The problem, then, is to utilize the energy of

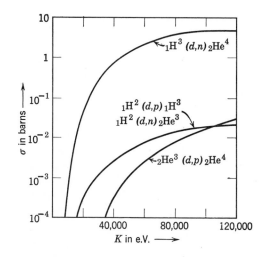

Fig. 14-7. Fusion reaction cross sections as functions of the kinetic energy of the deuteron. The other reactant for each reaction is initially at rest.

each fusion collision to provide at least one deuteron with sufficient kinetic energy to produce a second fusion collision. The seriousness of this problem becomes apparent if we convert the required deuteron energy (about 10^4 e.V.) into an equivalent temperature; the required temperature is about 10^8 °K!

The nuclei produced in a fusion reaction will inevitably collide and share their energy with the deuterium nuclei constituting part of the fuel, and thus a high-temperature *plasma* (a gas with equal densities of positively and negatively charged particles) will be formed as soon as fusion begins. But the energetic deuterons must be *contained* to produce new reactions; they must not leave the active volume of the reactor or be degraded in energy below $\sim 10^4$ e.V. Clearly there is no wall material that will sustain anything like the required temperature, and if the wall is cooler than the plasma it will reduce the kinetic energy of the deuterons that collide with it. For masses as large as the sun or the stars, the gravitational field holds the reactants together. The small stellar surface-to-volume ratio also reduces the ratio of radiation loss (surface) to energy production (volume), and the sun and stars obtain their energy from fusion reactions (often more complicated than the reactions in the above equations). For very short times, a reacting mass remains together because a finite acceleration cannot eject all the reactants instantaneously from the reacting region. This transient, uncontrolled fusion reaction is the energy source in the *hydrogen bomb*, in which the reactants are heated initially by a fission explosion (*atomic bomb*). Neither of these methods is helpful for containing a terrestrial, steady-state fusion reaction.

Thus electric and magnetic fields appear to be the only promising means of confining the hot plasma, and many ingenious proposals have been devised for magnetic containment. The use of strong magnetic fields is not a simple solution of the containment problem, however, because of the instabilities that occur when large densities of currents of both electrons and ions in the plasma can add their own fields to the externally applied field. Furthermore, in order to minimize radiation and the loss of the kinetic energies of any particles leaking through the magnetic field, the device must be large, and the simultaneous requirements of large magnetic field and large volume necessitate the consumption of considerable energy in the coils producing the field.

At the temperature of operation, the fusion reaction cross section must be large enough to permit fusion collisions to occur sufficiently often to offset the energy losses by competing processes, notably radiation. Thus the choice of reaction and the temperature attained must result in a substantial cross section; the σ required is estimated to be at least 10^{-3} barn.

Large-Z nuclei must be excluded from the active volume. Such nuclei retain some of their electrons even at high temperatures, and atoms that are only partially stripped of their electrons can effectively radiate ordinary atomic radiation. Even if high-Z nuclei have no electrons attached, they increase the radiation loss because of the X-rays produced by the acceleration of charged particles as they approach and are deflected by such nuclei. Thus impurities must not be released from the tank containing the fuel and the active volume. As Z increases above 1 or 2, this soft X-ray energy loss increases, and the fusion cross section decreases because of the increasing Coulomb barrier. Thus even lithium is not an attractive competitor to deuterium as a fusion fuel, and heavier elements are not promising at all.

Ordinary hydrogen ($_1H^1$) is also unsuitable as a fuel. Reactions such as $He^3(p,\gamma)He^4$ have very small cross sections, since the compound nucleus decays to $He^3 + p$ in a time short compared to the characteristic time for γ emission. Thus most $p - He^3$ collisions result only in the scattering of the proton. Furthermore, most of the energy of those (p,γ) processes that do occur goes into γ-rays, and it is difficult to convert this energy into the kinetic energy of the protons required to produce additional fusion. Thus of all possible fusion reactions, only eqs. 14-8 to 14-11 seem likely candidates for a successful fusion reactor.

As yet no self-sustaining controlled fusion reaction has been achieved on earth. Many experimental devices have been built, and each has met with some partial success. From experience with these experiments, it would appear that a fusion reactor *might* consist of the following parts: (1) A deuterium-tritium plasma at a temperature of $\sim 10^8$ °K. (2) A strong magnetic field to confine the plasma, perhaps produced by a superconducting coil. (3) A tank to contain the fuel gases, which diffuse into the active volume of plasma. (4) An initial source of energy, such as a radio-frequency electric field or an external ion beam, to heat the plasma to initiate fusion. (5) A blanket of Li^6 surrounding the active volume to generate tritium by the reaction of eq. 14-12. (6) A *neutron multiplier*, perhaps by the reaction $_4Be^9(n,2n)2_2He^4$, to compensate for the loss of some of the neutrons from eq. 14-10 and to make sure that one $_1H^3$ nucleus is created for each one consumed. (7) A heat exchanger, turbine, and generator system to harvest the fusion energy (most of which would appear as neutron heating of the blanket) and to produce electrical power. It should be emphasized that intriguing problems, especially the problems of stable magnetic containment and of rapid initial heating, remain to be solved before such a reactor can be successfully designed and constructed.

14-5 INTERACTION BETWEEN ENERGETIC PARTICLES AND MATTER

In this section we shall consider the processes that occur when an energetic particle, such as might be emergent from a nuclear reaction, is incident upon matter. We shall first concentrate on the energy loss of the incident particle, and then we shall turn our attention to the effects produced in the absorber.

(a) Heavy charged particles

A heavy charged particle (like a proton, α-particle, or fission fragment) has a fairly definite *range* in a gas, liquid, or solid. The particle loses energy primarily by the excitation and ionization of atoms in its path, but it loses some energy by elastic collisions with nuclei. The energy loss occurs in a large number of small increments. The primary particle has such a large momentum that its direction is usually not seriously changed during the slowing process. Eventually it loses all its energy and comes to rest. The distance traversed—called the range—is a function of the charge, mass, and energy of the primary particle, the number of atoms per unit volume in the material traversed, and the atomic number and average ionization potential of the atoms composing this material.

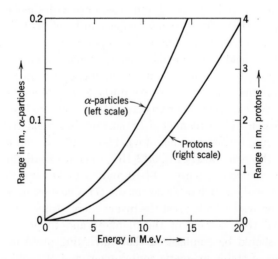

Fig. 14-8. Ranges of α-particles and protons in air at 15°C and 760 mm pressure. (From *Experimental Nuclear Physics*, Vol. I, edited by E. Segrè, Wiley, New York, 1953.)

Figure 14-8 shows the ranges of α-particles and protons in air.   An α-particle has a much shorter range than a proton with the same energy because the α-particle is slower and more heavily charged.   A slow particle loses more energy by ionizing atoms than a fast particle, since the slower particle spends a longer time in an atom, and there is a greater probability that an electronic transition will occur in the atom.   The energy loss per unit distance $-dK/dx$ is approximately proportional to the reciprocal of the square of the velocity of the particle.   This effect can be observed in the ionization along the path of a single heavy particle; the number of ions produced per unit distance is small at the beginning of the path, rises to a maximum at the end of the path, and then falls sharply to zero where the particle becomes too slow to ionize at all (the end point of the range).   The measurement of the range of a heavy charged particle is a common way of measuring its energy.

The energy loss rate $-dK/dx$ (also called the *stopping power*) varies from material to material because the atomic energy levels in solids and gases vary, and hence the average energy loss per excitation or ionization act varies.   But the density of the stopping material is far more important, since the number of electrons per unit volume is approximately proportional to density and since it is primarily the excitation of electrons that causes the energy loss.   Thus if we express range in terms of kilograms per square meter of matter traversed (or replace $-dK/dx$ by $-(1/\rho)(dK/dx)$, where ρ is the density), the stopping powers of various substances become comparable.   For example, the range of 10-M.e.V. protons is about 1.4 kg/m.^{2} in air, 1.7 kg/m.^{2} in aluminum, and 2.1 kg/m.^{2} in copper.

The ranges of fission fragments with energies of the order of 80 M.e.V. are only ∼0.02 m. of air (∼0.03 kg/m.^{2}), which is about one thousandth of the range of 80-M.e.V. protons.   The reason for the short range is, of course, the large charge of the fission fragments; $-dK/dx$ is proportional to the square of the charge of the moving particle.   A fission fragment does not simply have a charge Ze, however, since it acquires electrons as it loses energy.   Energy loss by fission fragments is further complicated by the fact that toward the end of their ranges they are moving too slowly to excite atomic electrons, and they lose energy only by elastic collisions with atoms.

(b) *Electrons*

The interaction of electrons with matter has been discussed briefly in Sec. 12-7, and an idealized plot of the rate of ionization by an electron was given in Fig. 12-18*b*.   The slowing-down of electrons with energies less than about 1 M.e.V. is primarily accomplished by the same ionization and excitation collision processes that slow heavy particles, but there are two reasons why the ranges of electrons are not so well defined as those of protons or other heavy particles. First, the path lengths of electrons with the same energy have much larger fluctuations about a mean value than those of protons; only a few large-deflection collisions suffice to stop an electron, and hence the statistical fluctuations in these are more significant than the fluctuations in the large number of small-deflection collisions characteristic of protons. Second, the fact that the electron is much lighter than a proton means that the momentum of an electron is much less than that of a proton with the same energy. Therefore an electron does not preserve its direction so well as a proton. Two electrons incident in the same direction on a solid may follow quite different paths, and even if the lengths of these paths were the same, the total distance traveled into the solid may be quite different in the two cases. Thus energetic electrons do not possess so sharply defined a range in matter as do heavy particles, and therefore the energies of electrons are not accurately determined from range measurements.

Very energetic electrons ($K > 1$ M.e.V.) lose an appreciable fraction of their energies by producing continuous X-rays (also called *Bremsstrahlung*), and the cross section for this process increases with increasing K. The rate of energy loss for electrons thus has a minimum (at $K \cong 1$ M.e.V. in large-Z absorbers) and rises toward higher energies because of the radiation loss. (Even protons create continuous radiation, but a proton must have an energy of the order of 1000 M.e.V. before this process becomes important.)

(c) Neutrons

Neutrons have only very weak interactions with matter. Slow neutrons are scattered by the weak magnetic interaction between their magnetic moments and the magnetic moments of atoms. Slow and fast neutrons can be scattered by the nuclear-force interaction and can produce nuclear reactions; the cross sections for slow neutron reactions can be quite large because of resonances (see Fig. 14-2 and prob. 13-47). Fast neutrons can be slowed by elastic collisions with nuclei, and if the nuclei are light (like protons) a neutron can lose an appreciable fraction of its energy in a single collision.

(d) Photons

A high-energy photon beam is not nearly so strongly attenuated by matter as a beam of charged particles, since photons are uncharged and hence are not so effective as electrons or protons in producing excitation or ionization. A photon is not slowed as it passes through matter, since

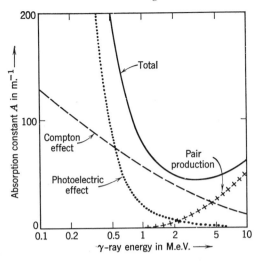

Fig. 14-9. Absorption of γ-rays in lead. The contributions to the total absorption constant by the three important processes are added to give the "Total" curve. The relative contributions are different for different elements as absorbers. (From W. Heitler, *The Quantum Theory of Radiation*, Clarendon Press, Oxford, 3rd Ed., 1954.)

it always travels with the speed of light. It can be scattered out of a beam, or it can be absorbed; in either event it is lost from the beam. Thus it does not lose energy in a large number of small steps, and therefore it does not have a definite range. The attenuation of a monochromatic photon beam follows the equation

$$I = I_0 e^{-Ax} \tag{6-16}$$

which was introduced in connection with X-ray absorption in solids.

The absorption constant A for γ-rays (photons) of various energies is illustrated in Fig. 14-9. Also illustrated in the figure are the three principal processes that remove photons from a beam. These processes are: (1) The photoelectric effect, which is identical with the process discussed in Sec. 6-6 in conjunction with X-rays and is the most important process for X-rays and γ-rays with energies less than 0.5 M.e.V. (2) The Compton effect, which was described in Sec. 4-7. (3) Pair production, which is the creation of an electron-positron pair by a photon passing near a nucleus. The rest energy $2m_0c^2$ of the pair is twice the rest energy 0.51 M.e.V. of an electron, and therefore the γ-ray must have at least an energy of 1.02 M.e.V. in order to produce a pair. If the energy of the incident γ-ray was 2.02 M.e.V., for example, the electron and positron would share 1.00 M.e.V. of kinetic energy. The measurement of their kinetic energies can serve as a method of measuring the γ-ray energy.

In the foregoing brief survey of the principal processes stopping or absorbing energetic particles in matter we have concentrated on the effects produced on the incoming particle. It is also of interest to examine the effects on the medium performing the stopping or absorbing. Fast charged particles ionize atoms along their paths in gases, liquids, and solids; and some of the electrons thereby produced have sufficient energy to create additional ionization. In gases, one electron-ion pair is formed for each ∼30 e.V. of energy lost by the charged particle. Most detecting devices for nuclear particles make use of this ionization.

Heavy charged particles can also give considerable kinetic energy to nuclei that scatter them by the Rutherford scattering process. The nuclei so accelerated are of minor consequence in gases or liquids. But such a nucleus in a solid can leave its position in the perfect crystal lattice, leaving a vacancy behind it. Fast neutrons (with kinetic energies of a few M.e.V. or more) can also create such displacements when they are scattered by nuclei in a crystal. Atoms displaced in this way often have enough energy to displace their neighbors, and one or more vacancy-interstitial pairs can be formed for each collision of the primary particle with a nucleus of the solid. The vacancies and interstitials created in this manner modify the properties of the solid; since the modification frequently degrades the useful properties of solids, the process is called *radiation damage*. Both fission-fragment and fast-neutron damage adversely affect the properties of nuclear reactor materials, as has been noted in Sec. 14-3. Semiconductors and other electronic materials are even more sensitive to radiation damage because of their sensitivity to imperfections.

The effects of electrons and photons on solids are generally of little consequence because both types of particles carry little momentum and therefore cannot efficiently displace heavy atoms of a crystal. Nevertheless, energetic electrons (kinetic energies greater than about 0.5 M.e.V.) create some displacements, and even photons can produce a few, by first creating energetic electrons through the Compton effect or photoelectric absorption of γ-rays.

14-6 DETECTORS FOR NUCLEAR PARTICLES

We can describe here only a few of the many varieties of detectors for nuclear particles that are in current use. In the following, we shall tacitly assume a particle energy of about 0.1 to 10 M.e.V. unless a higher energy is specified.

Most methods of detecting nuclear particles are based on the ionization produced in matter by the particles. If the incident charged particle is a proton, α-particle, or other energetic charged heavy particle, it loses energy rapidly by ionization in a gas or solid, and such particles usually lose *all* their energy inside a detector. Electrons, positrons, and beams of γ-rays generally lose only a fraction of their initial energies in passing through a detector employing a gas. The γ-ray is especially difficult to detect in a gas because of its high penetrating power, and beams of γ-rays lose only a small fraction of their initial photons in such detectors.

(a) *Proportional counter*

This detector consists of a metal chamber filled with gas, with a thin wire (a few thousandths of an inch in diameter) in the center. The wire is connected to a source of potential that is positive with respect to the walls of the counter. A thin aluminum or mica window at the end of the counter permits particles to enter if their energies are not too low. (Another form of this counter admits particles through a thin-walled cylinder instead of through an end window.) The voltage applied to the central wire is a few hundred volts and is not large enough to cause a discharge. When a particle creates ionization inside this chamber, the electrons produced flow toward the central wire. Although the electrons acquire energy from the electric field, the field in most of the chamber is too small to give an electron more than a few tenths of an electron volt between collisions. But the field is very large in the region near the fine wire, and an electron can there acquire enough energy between collisions to enable it to ionize gas atoms. Thus additional electrons and positive ions are produced. The *gas amplification* achieved thereby can

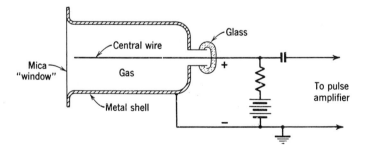

Fig. 14-10. "End window" proportional counter. Proportional counters are also constructed with thin walls and no end windows, and with other geometrical arrangements. Geiger counters have the same parts and geometrical arrangements but different gases and voltages.

be as large as a factor of 10^3 or 10^4. The end window form of the proportional counter is shown in Fig. 14-10.

The pulse of current is proportional to the number of initial ion pairs produced by the primary particle. If the primary particle gives up all its energy in the counter and is brought to rest, the pulse magnitude is proportional to the energy of the primary particle. The current pulse produces a voltage pulse by the IR drop across the resistor (Fig. 14-10), and this pulse is transmitted to a pulse amplifier. Primary particles usually arrive in too rapid succession to permit connecting the output of the amplifier directly to a mechanical register. Electronic computing circuits are therefore interposed between the amplifier and the register.

If photons (γ-rays) are being counted, the principal source of ionizing particles is the emission of electrons from the counter walls, since the energetic, penetrating photon has a small probability of producing ionization directly in the gas.

If fast neutrons are to be counted, a gas containing hydrogen can be used. Elastic collisions of the neutrons with protons produce a few energetic protons, which are then detected as usual. If slow neutrons are to be counted, the counter can be filled with the gas BF_3, and the reaction of eq. 13-25 provides energetic charged particles (the $_3Li^7$ and $_2He^4$ atoms move so rapidly that their electrons are stripped off as they pass through a gas). Another way of detecting slow neutrons is to coat the inside of the chamber with $_{92}U^{235}$ and to employ the fission reaction (eq. 14-1) to provide charged particles.

(b) Geiger counter

This counter is similar in construction to the proportional counter of Fig. 14-10 but the voltage applied to the Geiger counter is large enough for the ionization produced by the primary particle to create a discharge that spreads along the length of the wire. Even a single ion pair, produced by a single primary particle, can thus produce a discharge. The incident particle *triggers* an incipient discharge, and the pulse of current in this discharge is the same whether the initial ionization by the incident particle was large or small. The discharge spreads primarily by the photons produced by electron-atom excitation collisions. These photons provide photoelectric emission of electrons from gas molecules, which in turn produce ionization and additional photons.

As in the proportional counter, most of the ionization is produced very close to the central wire. Electrons are rapidly swept out of this region by the high electric field; there remains a positive space charge produced by the positive ions, which move much less rapidly than electrons. The discharge stops when the positive ion space charge near the central wire becomes so large that the electric field near the wire is too small to sustain the discharge. By this chain of events a pulse is transmitted to the scaling circuit for each particle that enters the counter. The pulse height depends on the voltage applied to the counter but does not depend on how much ionization was produced by the primary particle.

After the current pulse has ended, the Geiger counter is not immediately sensitive to further ionizing radiation. The positive space charge prevents any current pulse at all from being generated until sufficient time (the *dead time*) has elapsed to permit most of the space charge to be swept away. Even after the end of the dead time, the pulses are small until all the space charge has disappeared (the *recovery time*). Dead time and recovery time are of the order of 100 microseconds.

A typical Geiger counter contains about 10% alcohol and 90% argon. The alcohol in this combination prevents spurious pulses from following real ones; the spurious pulses arise from photoelectric and secondary electron emission from the walls. A typical pressure is about 100 mm of mercury, and a typical voltage is 1000 volts.

The valuable feature of the Geiger counter is its large and constant output pulse, which permits processing of the pulse by relatively simple electronic circuits. Its disadvantages are its relatively long recovery time and its lack of information about the energy of the particle counted.

(c) Scintillation counter

In this detector, which is illustrated in Fig. 14-11, the primary particle produces internal secondary electrons in a luminescent crystal or luminescent plastic. A typical crystal is sodium iodide activated with thallium (see Sec. 10-6). The crystal is coated with a highly reflecting paint on all sides but one, and the uncoated side faces the photocathode of a multiplier phototube (see Sec. 12-7). The phototube has an amplification of about 10^6 and a photocathode efficiency of about $\frac{1}{20}$ electron per photon.* For each photon produced by luminescence in the crystal, therefore, about 50,000 electrons flow in the output circuit of the tube. The pulse is proportional to the number of photons produced by the primary particle in the crystal. There are many more atoms in a crystal a few inches on a side than there are in the gas of a proportional counter or Geiger counter of reasonable size. Therefore a primary particle, such as a γ-ray or high-energy electron, that would lose only a small fraction

---

* This is, of course, an *average* efficiency, and fluctuations about such an average value can be very important, especially when only a few photons are produced by the incoming particle. Twenty photons produce, on the average, 1 photoelectron, but they may produce 0, 2, or even more photoelectrons.

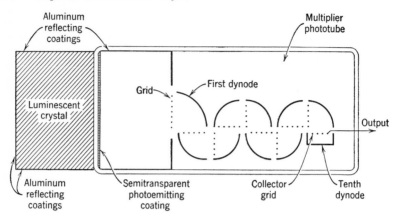

Fig. 14-11. Scintillation counter. The energetic incident particle excites luminescence in the phosphor, and this light produces photoelectrons from the coating on the inside of the end of the multiplier phototube. The secondary-emission multiplier is the same in principle as the tube of Fig. 12-19.

of its energy in a proportional counter may lose all its energy in a scintillation counter. The pulse amplitude from a scintillation counter is usually measured, and this measurement permits the determination of the energy of a particle stopping in the counter if its nature is known, or its nature if its energy or momentum is known.

Another advantage of scintillation counters is their speed of response. Sodium iodide activated with thallium has a decay time of only 0.3 microsecond, and other crystals have even shorter times. The organic compound *trans*-stilbene has a decay time of only 0.006 microsecond, and organic liquid scintillators, such as a 1% solution of *p*-terphenyl in toluene, not only have an even faster response but also can be conveniently constructed in large sizes (several cubic meters). Solid solutions of luminescent organic molecules in plastics, such as terphenyl in polystyrene, are even more popular as fast scintillators.

(d) Semiconductor detector

A flat *p-n* junction just beneath the surface of a sheet of a semiconductor such as silicon serves as a nuclear particle detector. The geometry and principle of operation are similar to those of the silicon or gallium arsenide solar cells described in Sec. 11-4. Electron-hole pairs created by the incident particle in the transition region of a reverse-biased *p-n* junction cause a current to flow. In silicon, one electron-hole pair is created for each 3.5 e.V. lost by the incident particle. If the particle stops in the transition region, the current pulse height is an accurate measure of the energy of the particle. These detectors can be made extremely small (e.g., for insertion into living animals) and very rugged (e.g., for measuring the radiation incident upon earth satellites). An interesting feature of these devices is the possibility of varying the length of the active region merely by varying the magnitude of the reverse bias (see probs. 11-32 and 11-34).

(e) Photographic detectors

Photographic film can be used as a detector of nuclear particles in the same way as it is used in electron and X-ray diffraction experiments. The blackening of the plate or film is proportional to the ionization produced by the primary particles.

A somewhat different way of using the photographic effect is illustrated in Fig. 14-12. This figure is a developed section of a thick photo-

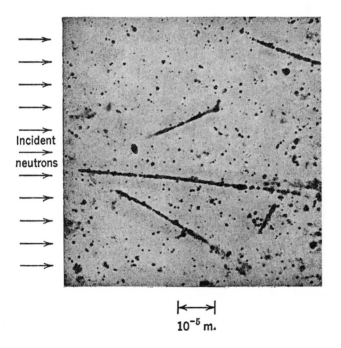

10^{-5} m.

Fig. 14-12. Proton tracks in a photographic emulsion. The proton tracks were caused by the collisions between energetic neutrons (all with the same energy) and protons in the emulsion. Only a part of the track at the upper right is contained within this section of the emulsion. (From C. F. Powell and G. P. S. Occhialini, *Nuclear Physics in Photographs*, Clarendon Press, Oxford, 1947.)

graphic emulsion that was exposed to fast neutrons. Occasionally a neutron collided with a proton, since there are many hydrogen atoms in the organic compounds in the gelatin of an emulsion. The protons received many electron volts or even million electron volts of energy and ionized strongly along their paths. The individual developed grains can be counted and the length of the proton's path measured. The proton's energy can be determined from its path length, and the grain count permits identification as a proton. Of course the neutrons are not visible at all in Fig. 14-12, but nevertheless the photographic emulsion has served as a neutron detector.

Photographic emulsions are especially applicable to high-energy physics, since the stopping power of the dense emulsion is high and since the actual trajectories of particles can be seen. Thus complicated collision processes involving meson or hyperon production can be analyzed by studying the directions of the emergent tracks and the density of silver in the tracks. The developed-grain density along a track is approximately proportional to the energy loss $-dK/dx$ by ionization, and thus information can be obtained from the track density about the mass or energy of the particle producing the track.

(f) Bubble chamber

If a liquid is heated under pressure nearly to its boiling point at that pressure, and then if the pressure is suddenly released, the liquid will begin to boil. Bubble formation in this boiling ordinarily begins at sharp points on the walls of the container or at dirt particles that serve as nuclei. But if an ionizing particle traverses the liquid within a few milliseconds after the pressure is released, the ionization produces a local heating that initiates bubble formation along the track of the particle. The bubble chamber is a volume of liquid that is rendered sensitive in this way. After the bubbles have developed for a few milliseconds, the tracks are photographed, and a record such as Fig. 13-20 is thereby obtained.

The bubble chamber is especially suited to high-energy physics experiments utilizing particle accelerators. The chamber can be made sensitive each time a pulse of particles emerges from the accelerator, and the large volume and large density (relative to gas-filled devices) of the chamber permit appreciable numbers of interesting collision processes even when the incident particles have energies of thousands of M.e.V. A magnetic field is usually provided at the chamber in order to distinguish the sign of the charged particles and to measure their momenta. Liquid hydrogen, although dangerous and expensive, is an attractive liquid for such experiments, since the interpretation of processes such as those shown in Fig. 13-20 is simplified if protons are the only nucleons in the chamber.

(g) Spark chamber

The spark chamber utilizes an incipient electrical discharge in a gas, like that used in a Geiger counter, to give track-geometry information, like that provided by a bubble chamber. The spark chamber typically consists of a series of thin parallel plates spaced a few millimeters apart. The first, third, fifth, etc., plates are grounded, and the second, fourth, sixth, etc., plates are connected to a voltage source. This voltage has a steady component of the order of 100 volts that provides a *clearing* field, and a large, pulsed component that is almost, but not quite, sufficient to produce a discharge between adjacent plates. A charged particle traversing this arrangement leaves a trail of ionization. If the voltage pulse is applied in the few tenths of a microsecond that the electrons in this trail remain free (before becoming attached to molecules in the gas or swept out of the gas), a small, localized spark discharge between each pair of plates will occur along the particle's path. A photograph of all the individual sparks reveals the sections of the trajectory of the particle between the plates, and collision events or deflections in a magnetic field can be observed as they are in the bubble chamber. The voltage pulse is quickly removed, and the small steady field clears away the ions to make the chamber ready for the next pulse.

The spark chamber has an important advantage over the bubble chamber in that it can be rendered sensitive for only a very short time (tenths of a microsecond). Suppose, for example, that in an intense beam from a high-energy particle accelerator a rather rare particle is produced. Scintillation or Čerenkov counters followed by appropriate electronic circuits can signal the production of this kind of particle, and the pulse voltage is then applied to the spark chamber. The track of this particular particle can thus be determined, and there is little likelihood that one of the much more numerous *background* particles will cause another spark track in the short time that the chamber is sensitive. Thus rare events and processes can be studied without the confusing or even misleading background of commoner events.

(h) Čerenkov counter

When a charged particle moves through a medium with a velocity greater than the velocity of light in that medium, weak electromagnetic radiation called Čerenkov radiation is emitted. Čerenkov photons emerge at an angle θ with the vector velocity of the particle, where

$$\cos \theta = v_l/v \qquad (14\text{-}13)$$

Here $v_l = c/n$ is the velocity of light in the medium with index of refraction n, and v is the velocity of the particle.

Although the intensity of light produced by this process is only a few percent of that produced under comparable conditions in a scintillator, the Čerenkov effect is the basis for an important detector in high-energy physics because of the *selectivity* of the Čerenkov counter. No Čerenkov radiation occurs if $v < v_l$, and this circumstance permits suppression of the background of low-velocity particles always present in high-energy experiments. Furthermore, when v is greater than v_l, the radiation emerging at a selected angle θ can be conducted to a multiplier phototube by an optical system, and thus a pulse in the phototube will be produced only when a particle with velocity computed from eq. 14-13 passes through the counter. Thus an unambiguous measurement of particle velocity can be made, and this measurement can be combined with energy or momentum measurements to determine the particle mass.

14-7 APPLICATIONS OF RADIOACTIVE NUCLIDES

The nuclear reactor is the source of a wide variety of relatively inexpensive radioactive nuclides (frequently called *radioactive isotopes* or *radioisotopes*). The fission products are the most abundant such nuclides (e.g., the Zr^{95} of eq. 14-2), but reactors can also produce radioactive isotopes by neutron irradiation of elements placed in the cores of the reactors (e.g., the Co^{60} of eq. 13-29). In addition, radioactive nuclides can be produced by bombardment of stable nuclides by heavy particles from accelerators such as the cyclotron.

Applications of these radionuclides are of two kinds: (1) The radionuclide is a *source* of γ or β radiation. (2) The radionuclide is a *tracer* to reveal, through the easily detectable radiation accompanying its decay, the position of a chemical element in a process or an experiment.

The application of radionuclides as sources is convenient because they are inexpensive and small compared to other sources such as X-ray machines or electron accelerators. A typical application is the use of Co^{60} to inspect welds or castings. The radioactive source, so small that it is essentially a point, is placed on one side of such a structure, and a photographic film is placed on the other side. The exposed and developed film reveals any flaws present in the part, as explained for X-ray radiographic examination (Sec. 4-4).

Another typical application as a source is in the thickness gauge illustrated in Fig. 14-13. Sheet metal or plastic is being reduced in thickness by rolling. A β^- source is placed under the rolled sheet and a detector

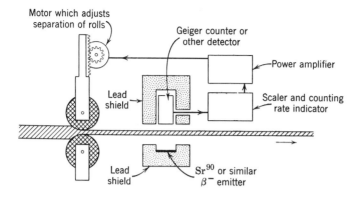

Fig. 14-13. Schematic diagram of a method of using a β^--emitter in an automatic thickness control for a rolling operation.

above it. The output of the detector is one count for each particle transmitted through the sheet. The source provides β^--particles of all energies from zero to a maximum value (Sec. 13-4). The lower-energy β^--particles are absorbed in the sheet, and most of the higher-energy β^--particles are transmitted and detected. The thicker the sheet, the fewer are the particles that are detected. For example, a steel sheet 0.025 inch thick stops 90% of the particles with energies less than about 1.50 M.e.V. If the sheet decreases to 0.024 inch, 90% of the particles with energies less than 1.45 M.e.V. are stopped, and therefore the number entering the detector increases. In a typical application the detector output is about 2% greater for 0.024-inch than for 0.025-inch sheet. Hence the detector output can be "fed back" to control the rolls (Fig. 14-13), and a constant sheet thickness can be maintained.

Radioactive sources are now widely used in medicine. In many cases, the application takes advantage of the fact that the body concentrates certain chemical species in a particular region of the body. If the radioactive nuclei are a part of such a chemical compound they can be concentrated in the region of the body to be irradiated.

Many other applications of radioactive sources have been, or are being, developed. Possible large-scale uses are the sterilization of foods, the polymerization of plastics, the production of light (through the use of a phosphor to convert high-energy β^- radiation to visible light), and the production of small amounts of electrical power (through the use of thermionic or thermoelectric converters).

The applications of radioactive nuclides as tracers are wide and varied.

A small amount of a radionuclide can be added to the petroleum in a long pipeline when the petroleum pumped into the line changes from one form to another (e.g., crude oil from one well to crude oil from another). At the other end of the line a detector indicates when the tracer appears and hence when to turn a valve to send the different oils into different tanks. Another tracer application is the location of leaks in buried pipes. A small sample of a radionuclide is placed in the fluid in the pipe, the area around the pipe is surveyed with a detector, and a leak is indicated by a high counting rate. A third application is in the study of wear on bearings and sliding surfaces. For example, piston rings for an internal-combustion engine can be made radioactive by alloying radioactive elements during manufacture or by the neutron irradiation of the finished piston rings. The rings are installed, and the radioactivity of the circulating lubricating oil and of the cylinder walls is studied. Thus the amount of metal worn away or transferred to the cylinder walls can be determined for various oils and conditions of operation.

Applications of tracers to experimental research in metallurgy, physics, chemistry, and biology are widespread. The evaporation rate of materials that evaporate very slowly can be sensitively measured with tracers. For example, the rate of barium evaporation from an oxide-coated cathode can readily be measured by mixing some Ba^{140} with the cathode coating. After a known time at a known temperature, the counting rate of the material deposited on a collector surrounding the cathode is measured. The amount of barium transported can be computed from this counting rate. Tracers are practically essential in studying the diffusion of impurities in metals or self-diffusion of the metal atoms themselves. Tracers can be used to follow the history of atoms in chemical reactions. They have been widely used in studying biological processes. For example, some radioactive nuclide can be incorporated in a particular kind of protein molecule, which is fed to an animal. The metabolism of the protein can then be studied by finding the fraction of the radionuclide in various organs of the animal as a function of time after the feeding time. Medical diagnosis, oceanography, and a host of other fields profit from the use of tracers.

In all applications of radioactive nuclides the experimenters must be careful to protect themselves and others from too large doses of radioactivity. A well-designed experiment or process provides shielding where needed. Heavy, expensive shielding is required primarily for γ-rays, since β radiation is easily stopped by thin shields. Only a few radionuclides (e.g., Sr^{90}) emit β^--particles but no γ-rays. Since nuclear-particle detectors are so sensitive, only minute quantities of radionuclides (microcuries or millicuries) are necessary for tracer applications, and this fact simplifies the protection of personnel. The quantities of radio-nuclides needed for sources of radiation are usually much larger (curies or thousands of curies), and therefore thick lead shields are required if the sources emit γ-rays.

REFERENCES

Nuclear Fission and Nuclear Reactors

R. L. Murray, *Introduction to Nuclear Engineering*, Prentice-Hall, Englewood Cliffs, N.J., 2nd Ed., 1961.
S. E. Liverhant, *Nuclear Reactor Physics*, Wiley, New York, 1960.
Nuclear Engineering, edited by C. F. Bonilla, McGraw-Hill, New York, 1957.
G. Murphy, *Elements of Nuclear Engineering*, Wiley, New York, 1961.

Nuclear Fusion

D. J. Rose and M. Clark, Jr., *Plasmas and Controlled Fusion*, Wiley, New York, 1961.
S. Glasstone and R. H. Loveberg, *Controlled Thermonuclear Reactions*, Van Nostrand, Princeton, 1960.
A. Simon, *An Introduction to Thermonuclear Research*, Pergamon, New York, 1959.

Interaction between Energetic Particles and Matter

H. A. Bethe and J. Ashkin in *Experimental Nuclear Physics*, Vol. 1, edited by E. Segrè, Wiley, New York, 1953, pp. 166–357.
R. B. Leighton, *Principles of Modern Physics*, McGraw-Hill, New York, 1959, Chapter 14.
D. Halliday, *Introductory Nuclear Physics*, Wiley, New York, 2nd Ed., 1955, Chapter 7.
D. S. Billington and J. H. Crawford, Jr., *Radiation Damage in Solids*, Princeton University, Princeton, 1961.
G. J. Dienes and G. H. Vineyard, *Radiation Effects in Solids*, Interscience, New York, 1957.

Detectors and Applications

Methods of Experimental Physics, Vol. 5, *Nuclear Physics*, edited by L. C. L. Yuan and C-S. Wu, Academic Press, New York, 1961.
G. Friedlander and J. W. Kennedy, *Nuclear and Radiochemistry*, Wiley, New York, 1955.
W. J. Price, *Nuclear Radiation Detection*, McGraw-Hill, New York, 1958.
"Spark Chamber Symposium," *Rev. Sci. Inst.*, **32**, 479–531 (1961).

PROBLEMS

14-1.* What is the shape (x-dependence of E) of the curve in Fig. 14-1 for large values of x? At about what value of x should this shape begin?

14-2.* What is the ratio of the total surface area of two spheres of equal volume to the area of one sphere with a volume equal to the total volume of the two? Estimate the magnitude of the surface energy term (eq. 13-17 or eq. 14-4) for U^{235}, and apply the ratio just calculated to this magnitude to estimate the surface energy change in fission. This is the energy inhibiting fission.

14-3. Sketch a qualitative comparison between $E(x)$ in Fig. 14-1 for U^{235} and for a nuclide with $A \cong 200$. Comment on the possibility of fission when $A \cong 200$ after considering the dependence of the probability of competing processes upon the nuclear excitation energy (last paragraphs of Sec. 13-6).

14-4. A slow neutron captured by U^{235} adds more energy than a slow neutron captured by U^{238} (and hence U^{238} requires fast neutrons to produce fission). Why? *Hint:* Consider the last term of eq. 13-17.

14-5. Compare the fission cross section $\sigma_f(K)$ for U^{235} and $K < 0.1$ e.V. with the statement in Sec. 13-7 that neutron cross sections approach a $1/v$ law at low velocities v.

14-6. Use the data of Fig. 14-2 to sketch $\sigma_f(K)$ for the slow-neutron fission of U^{235} on a *linear* σ and K scale. Sketch the Maxwellian distribution of neutron energies to the same abscissa scale. Estimate roughly the total fission cross section for thermal neutrons with $T = 298°$K. (The value observed is about 580 barns.)

14-7.* Thermal neutrons are incident on pure U^{235} metal. What is the distance L along the path of a single neutron such that there is a probability of $\frac{1}{2}$ that the reaction of eq. 14-5 will take place in a distance less than or equal to L? (The density of natural uranium is 18,700 kg/m.^{3}; do *not* set $\frac{1}{2} = \sigma NL$.)

14-8. After studying Fig. 14-3, write two common fission reaction equations other than eq. 14-1. If you have access to a nuclide chart or table of nuclides, give also the fission-fragment-to-fission-product decay chain, like eq. 14-2 or eq. 14-3, for one of these reaction equations.

14-9. Why is the mass spectrum of fission products (Fig. 14-3) symmetrical?

14-10. The average τ of delayed neutrons from U^{235} is 9 sec, but the average τ of all the β^- emitters in the fission-product decay chains is much longer. Why? *Hint:* Consider the minimum energy required for neutron emission and the correlation between decay energy and τ (Sec. 13-6).

14-11. There are a few delayed neutrons in fission. Why are there no delayed protons? Why are there no positrons?

14-12. Verify the calculations in the first row of Table 14-1.

14-13.* Repeat prob. 14-12 but for the (rare) situation in which the two fragments are nearly equal in mass. (Let $Z_1 = 46 = Z_2$ and let $A_1 = 117 = A_2$ to avoid the necessity of including the odd-even term.)

14-14. On a sketch of the stability valley (Fig. 13-16), draw the lines corresponding to fission and to the subsequent decay of fission fragments to fission products. Discuss the changes in Coulomb energy and asymmetry energy (Table 14-1) along each line. (Note that the valley floor rises as A increases.)

14-15. Calculate the energy in kilowatt hours that is released by the fission of one kilogram of U^{235}. Compare with the energy released in the combustion of one kilogram of coal. (The heat of combustion of coal is about 13,000 Btu per pound; 1 Btu = 2.93×10^{-4} kilowatt hour; 1 pound = 0.454 kg.)

14-16.* Suppose that a neutron with non-relativistic kinetic energy K hits head-on a C^{12} nucleus at rest. What fraction of its energy does it lose? (Use only the conservation of energy and momentum.) Answer the same question for a neutron that strikes a H^2 and a H^1 nucleus.

14-17. * The preceding problem assumed head-on collisions and thus gave the maximum possible energy loss per collision. Assume that the actual average energy loss per collision is one-half the value calculated for a head-on collision. How many collisions are required in a graphite moderator to reduce the energy of a neutron, which was initially 2 M.e.V., to thermal energy (0.025 e.V.)? Estimate the time required to reduce the energy from 2 M.e.V. to 0.025 e.V. The density of the particular graphite is 1650 kg/m.^{3}, and the scattering cross section of carbon is nearly constant at about 4 barns over this energy range.

14-18. The thermal neutron absorption cross section σ_a of hydrogen is 0.33 barn, σ_a of deuterium is 0.00046 barn, and of oxygen is zero. What is the thermal neutron mean free path λ_a for absorption in ordinary water and in heavy water? What is the lifetime $\tau = \lambda_a/v$ of thermal neutrons in ordinary water and in heavy water? Let v, the average neutron speed, be 2200 m./sec. The typical lifetime in a reactor is considerably less than these values because of absorption by the fuel.

14-19. * There are about 50,000 kg of natural uranium in a reactor similar to the one illustrated in Fig. 14-5. The average thermal neutron flux is about 3×10^{15} neutrons/m.^{2} sec. How many fissions per second are produced? Assume that 200 M.e.V. of energy is released per fission, and calculate the power produced in kilowatts. Count each fission as the equivalent of about ten disintegrations of radioactive nuclei, and calculate the number of curies that is equivalent in radioactivity to the reactor. (The thermal neutron fission cross section in U^{235} is 580 barns.)

14-20. * Use the data of the preceding problem to calculate the fraction of U^{235} "burned up" in a year of continuous operation of the reactor described in that problem.

14-21. What is the power density (watts/kg) in U^{235} in a reactor with a slow-neutron flux of 10^{18} neutrons/m.^{2} sec? (This calculation shows why heat transfer considerations are so important in reactor design.)

14-22. * A fast breeder reactor has a flux of 10^{18} fast neutrons per m.^{2} sec and produces about 2×10^8 watts/m.^{3} Assume that the average density of U^{235} in the reactor core is about 6,000 kg/m.^{3} (about one-third the density of metallic uranium). Estimate from these data the fission cross section of U^{235} for fast neutrons.

14-23. Highly purified (de-ionized) water is used in the steam-water loop of Fig. 14-6. Why? *Hint:* What effect do delayed neutrons in the fuel have on H_2O? On H_2O containing impurities?

14-24. * Calculate the ratio of the rms speed of $U^{238}F_6$ molecules to the rms speed of $U^{235}F_6$ molecules in a gas in thermal equilibrium.

14-25. For use in the gaseous diffusion separation of uranium isotopes, the compound UF_6 is chosen in preference to other gaseous uranium compounds, for both physical and chemical reasons. What physical reason makes fluorine particularly attractive for this purpose? *Hint:* Consult Appendix C, and compare fluorine with chlorine.

14-26. * The cross section σ for production of Co^{60} by the reaction of eq. 13-29 is 20 barns for thermal neutrons. A thin sheet of Co^{59} weighing 0.010 kg is placed in a reactor in which the thermal neutron flux is 2×10^{18} neutrons/m.^{2} sec, and it is left for 100 hours. How many Co^{60} nuclei are produced? What is the activity in curies of this sheet? Describe qualitatively how your method of calculation would have to be altered if the cobalt were left in the reactor for 1 year. (The density of cobalt is 8900 kg/m.^{3})

14-27. Consider the radioactive nuclides in Fig. 13-17. Which ones can be produced by irradiation of the appropriate stable nuclides in a nuclear reactor? Which ones could in principle be produced from the decay of fission fragments? (The relative abundance of $A \cong 65$ nuclides in the fission product spectrum is essentially zero, as shown in Fig. 14-3.)

14-28. * Calculate the energy released per amu in the reaction eq. 14-10, and compare with the fission reaction.

14-29. * The combustion of gasoline releases about 5×10^7 joules/kg. What mass of gasoline is equivalent in energy content to 1 kg of water, *if* the water could be "burned" by the (d,d) reaction of eq. 14-8?

14-30. Calculate the Q value of eq. 14-10 and the kinetic energies of the two products.

14-31. Why is the use of neutrons as reactants in a fusion reactor not an attractive possibility?

14-32. * Estimate the height in M.e.V. of the Coulomb barrier in the reaction of eq. 14-10 by letting the nuclear radius be $r_0 = 1.2 \times 10^{-15} A^{1/3}$ (not an accurate expression for very light nuclides). Convert this energy to an equivalent temperature. (This procedure gives only a rough estimate of the temperature required for fusion reactions because of the inaccurate r_0 and because of tunneling, but it does show that a high temperature is required.)

14-33. The negative muon μ^- can serve in place of an electron to bind two protons (or deuterons) together in a combination that is identical (except for the scale of size) with the hydrogen molecule ion (Sec. 7-3). What is the internuclear spacing in this ion? [The spacing is small enough to permit the (d,d) reaction in heavy hydrogen (eq. 14-8) to occur with appreciable probability, but the lifetime of the muon (2.2 μsec) is evidently too short to base a fusion reactor on the muon as a "catalyst."]

14-34. * Calculate the thickness of an aluminum plate required to stop 10-M.e.V. protons incident at right angles to the plate; the density of aluminum is 2700 kg/m.^{2} (Protons of this energy, and of higher energies, are present in the Van Allen radiation belts $\sim$2000 to 10,000 miles above the earth, and instruments and men require protection in space vehicles at these altitudes.)

14-35. As a fission fragment loses energy it eventually reaches a velocity comparable to the velocities of the outer electrons of the atoms through which it is passing, and then it can no longer ionize. Estimate the energy of a fission fragment of $A = 100$ at this stage by equating its velocity to that of a 10-e.V. electron.

14-36.* A beam of γ-rays consists of equal numbers of 0.5-M.e.V. and of 1.0-M.e.V. photons. After this beam passes through 0.01 m. of lead, what is the ratio of the 1.0-M.e.V. to the 0.5-M.e.V. component?

14-37. A beam of γ-rays has equal numbers of photons at all energies up to a maximum energy of 1 M.e.V. How is this spectrum modified after the beam has passed through a thickness of lead of 0.001 m.? of 0.05 m.?

14-38. The photoelectric contribution to the absorption constant A for 0.5-M.e.V. γ-rays is only 1/10,000 as much in aluminum as in lead. Why is it so much less in aluminum? The total absorption constant A for 0.5-M.e.V. γ-rays in aluminum is 23 m.$^{-1}$ What do you infer about the dominant absorption process at 0.5 M.e.V. in aluminum?

14-39.* Consider the Compton effect for 1.0-M.e.V. γ-rays in a heavy element like lead (Fig. 14-9). What is the energy of the Compton-scattered photon at $\varphi = 90°$? What is the probable fate of this photon?

14-40. Show that pair production cannot take place in free space. *Hint:* Energy and momentum must be conserved.

14-41.* Estimate the number of positive-ion electron pairs produced in a proportional counter by a 10-M.e.V. proton if the counter size and pressure are large enough for all the proton's energy to be absorbed. If the gas amplification factor is 10^3, how many coulombs flow in the counter when this particle is absorbed? If the pulse of current flows for about 0.001 sec, and if the resistor (Fig. 14-10) is 10^4 ohms, estimate the height of the voltage pulse delivered to the amplifier.

14-42. The thermal neutron cross section in B^{11} is <0.05 barn and in B^{10} is 4000 barns. What is the average cross section in natural boron? How many electrons are produced by ionization of the gas in a proportional counter by the (n,α) reaction in B^{10} (eq. 13-25)? (BF$_3$ neutron counters are frequently filled with isotopically enriched $B^{10}F_3$.)

14-43. The luminescence process in solids or liquids is characterized by a *Stokes shift*, a difference between the wavelength of the emitted light and the wavelength of stimulating light. It might be thought that this fact does not concern us in the scintillation counter application, where stimulation does not involve visible light, but the Stokes shift is vital to the functioning of the counter. Explain.

14-44. Either scintillation or semiconductor detectors can be used in the following way to determine the mass spectrum of a beam of heavy charged particles: The particles first traverse a *thin* detector in which they lose only a small fraction of their energies. The current pulse for each particle is proportional to $-dK/dx$, which is in turn proportional to $1/v^2$, where v is the velocity of the particle. The particles then pass into a *thick* detector and lose all their energy; the current pulse for each particle is proportional to K. Explain how to combine these data to obtain the particle masses.

14-45. How and why does the effective length of a semiconductor nuclear particle detector depend on bias voltage? If we wish a large effective length, why must the device be constructed of a very pure, high-resistivity semiconductor?

14-46. A beam consisting of two kinds of particles (e.g., α-particles and protons), one kind with energy K_1 and the other with energy K_2, is incident upon a semiconductor nuclear particle detector. Sketch the number of pulses as a function of pulse current, first when the bias is such that the ranges of both kinds of particles are less than the thickness of the p-n transition region and then for successively smaller biases.

14-47. How can you tell that not all the track in the upper right-hand corner of Fig. 14-12 appears in the illustration? *Hint:* The incident neutrons were monoenergetic.

14-48. A much older device than the bubble chamber is the *cloud chamber*, in which the ions produced by charged particles serve as nuclei for condensation in a supersaturated vapor. Although this device gives the same type of track information as the bubble chamber, the latter is much more popular in high-energy physics. Why?

14-49.* What are the two limiting directions as $v \to v_l$ and as $v \to c$ for visible Čerenkov radiation in water ($n = 1.33$)?

14-50. How would you construct an electron detector that would count *only* electrons with energies greater than 500 M.e.V.? (Be as quantitative as you can; find the necessary data in a handbook.)

14-51.* Radionuclides are useful sources of small amounts of power in space vehicles, remote communications stations, and similar applications. Calculate the power in watts per kilogram of Ce^{144}, a β^--emitter with an average energy of 0.10 M.e.V. and $\tau = 285$ days, and of Po^{210}, an α-emitter with an energy of 5.30 M.e.V. and $\tau = 138$ days.

14-52. Sketch the geometrical arrangement of a Co^{60} γ-ray source, a welded steel tank, and a photographic film assembled in order to make a radiograph of one of the welds. What considerations affect the choice of source-to-weld distance and of weld-to-film distance?

14-53.* The evaporation of barium from an oxide-coated cathode is being studied. The collector surrounding the cathode collects all the evaporated material. We require that the total number of disintegrations ("counts") in the collected material be 10,000 in order that statistical fluctuations in this number be only of the order of 1%. We can count for 10 minutes. How many curies of Ba^{140} must be deposited? If the total amount of barium (radioactive plus nonradioactive) evaporated is 10 micrograms, what is the *specific activity* (curies per kilogram) of radioactive barium that must be used? What is the concentration of radioactive barium (number of radioactive nuclei per normal nucleus)? (τ for Ba^{140} is 12.8 days.)

14-54. In tracer experiments like that of the preceding problem it is sometimes possible to choose one of several different radioisotopes with which the experiment can be performed. Is it better to choose a short ($\ll 1$ month), medium (~ 1 month), or long ($\gg 1$ month) half-life isotope? Why?

14-55. Cosmic-ray neutrons create C^{14} in the upper atmosphere by the reaction $N^{14}(n,p)C^{14}$. The equilibrium concentration of this radionuclide in growing matter is the same as in the CO_2 in the air and in the soil and gives approximately 260 disintegrations per second per kilogram of carbon. When growth stops, the equilibrium no longer obtains; no new C^{14} is produced, and the C^{14} in the plant or animal decays with $\tau = 5568$ years. Explain how radioactivity measurements can be used to compute the age of wood, cloth, or skin that is a few thousand years old.

Appendix A
Physical Constants

FUNDAMENTAL CONSTANTS *

Planck's constant	$h = 6.6257 \times 10^{-34}$ joule sec
Velocity of light	$c = 2.99793 \times 10^{8}$ m./sec
Charge of the electron (magnitude)	$e = 1.6021 \times 10^{-19}$ coulomb
Mass of the electron (rest mass, m_0)	$m = 9.1091 \times 10^{-31}$ kg
Ratio of proton mass to m	$M_p/m = 1836.1$
Boltzmann constant	$k = 1.3806 \times 10^{-23}$ joule/°K
Avogadro's number	$N_0 = 6.0224 \times 10^{26}$ molecules per kg molecular weight

OTHER USEFUL CONSTANTS AND RELATIONS

Unit mass, $C^{12} = 12.000000$ scale $(1/N_0)$	amu $= 1.6605 \times 10^{-27}$ kg
Energy equivalent of 1 amu	$= 931.49$ M.e.V.
Energy equivalent of electron rest mass	$= 0.5110$ M.e.V.
Charge-to-mass ratio of electron	$e/m = 1.7588 \times 10^{11}$ coul/kg
Wavelength of photon with energy E e.V.	$\lambda = 12{,}398/E$ Å
Wavelength of electron with non-relativistic kinetic energy K e.V.	$\lambda = 12.26/K^{1/2}$ Å
Energy levels in hydrogen-like atoms	$E_n = -13.60Z^2/n^2$ e.V.
Characteristic length in hydrogen-like atoms	$\rho = 0.529/Z$ Å
Potential energy of electron in field of charge $+Ze$, with r in Å	$P = -14.40Z/r$ e.V.
Bohr magneton	$\mathfrak{M}_B = 0.9273 \times 10^{-23}$ joule m.²/weber
Gas constant	$R = 8.314 \times 10^{3}$ joules/kilomole °K
Thermal energy per degree K	$= 8.617 \times 10^{-5}$ e.V./°K
Thermal energy at 290°K = 17°C	$= 1/40$ e.V.
mksa units	$\epsilon_0 = 8.8542 \times 10^{-12}$ farad/m.
	$\mu_0 = 4\pi \times 10^{-7}$ henry/m.

* From J. A. Bearden and J. S. Thomsen, Supplement No. 2 to Vol. **5,** Series X of *Nuovo Cimento*, pp. 267–360 (1957), with later modifications by J. A. Bearden. See also E. R. Cohen, J. W. M. DuMond, T. W. Layton, and J. S. Rollett, *Rev. Mod. Phys.*, **27,** pp. 363–380 (1955).

Appendix B
Periodic System
of the Elements

I	0
H	He
1	2

I	II	III	IV	V	VI	VII	0
Li	Be	B	C	N	O	F	Ne
3	4	5	6	7	8	9	10
Na	Mg	Al	Si	P	S	Cl	Ar
11	12	13	14	15	16	17	18

I	II	III	IVa	Va	VIa	VIIa	VIII			Ia	IIa	IIIa	IV	V	VI	VII	0
K	Ca	Sc	Ti	V	Cr	Mn	Fe	Co	Ni	Cu	Zn	Ga	Ge	As	Se	Br	Kr
19	20	21	22	23	24	25	26	27	28	29	30	31	32	33	34	35	36
Rb	Sr	Y	Zr	Nb	Mo	Tc	Ru	Rh	Pd	Ag	Cd	In	Sn	Sb	Te	I	Xe
37	38	39	40	41	42	43	44	45	46	47	48	49	50	51	52	53	54
Cs	Ba	*	Hf	Ta	W	Re	Os	Ir	Pt	Au	Hg	Tl	Pb	Bi	Po	At	Em
55	56		72	73	74	75	76	77	78	79	80	81	82	83	84	85	86
Fr	Ra	†															
87	88																

* *Rare earths, lanthanide series*: $_{57}$La, $_{58}$Ce, $_{59}$Pr, $_{60}$Nd, $_{61}$Pm, $_{62}$Sm, $_{63}$Eu, $_{64}$Gd, $_{65}$Tb, $_{66}$Dy, $_{67}$Ho, $_{68}$Er, $_{69}$Tm, $_{70}$Yb, $_{71}$Lu.

† *Actinide series*: $_{89}$Ac, $_{90}$Th, $_{91}$Pa, $_{92}$U, $_{93}$Np, $_{94}$Pu, $_{95}$Am, $_{96}$Cm, $_{97}$Bk, $_{98}$Cf, $_{99}$Es, $_{100}$Fm, $_{101}$Md, $_{102}$No, $_{103}$Lw.

Appendix C
Atomic Masses
and Atomic Weights

Each mass $_ZM^A$ tabulated is the sum of the mass of the nucleus and the mass of Z electrons (see the discussion preceding eq. 3-10).

The following table is based on the 1960–1961 agreements of the International Union of Pure and Applied Physics and the International Union of Pure and Applied Chemistry to adopt C^{12} as the standard mass. Prior to that time, the standard masses in the chemical scale of atomic weights (natural oxygen mass set equal to 16.00000) and in the physical scale of atomic masses ($O^{16} = 16.000000$) were slightly different.

In this table the mass of the nuclide C^{12} is set equal to exactly 12.000000. The "Atomic Weight" column [1] gives the average mass of the naturally occurring isotopes of each element, weighted according to their relative abundances. This column differs by only about 40 parts per million from the pre-1961 atomic weights. (The title "Atomic Weights" is repeated here because the technical literature invariably uses "atomic weight" or "molecular weight" to refer to this average isotopic mass, but "Chemical Atomic Mass," or "Average Mass of Element," or another term would be more appropriate.)

The "Nuclide Mass" column [2] gives the masses of all of the naturally occurring nuclides with $Z \leq 20$, of the most abundant of the isotopes of each Z with $Z > 20$, and a few other nuclides. Only about 40% of the stable nuclides are listed in this table. Radioactive nuclides are designated by an asterisk, and only a small fraction of the hundreds of known radioactive nuclides are included here. Nuclide masses appearing in this column are about 318 parts per million smaller than comparable numbers on pre-1960 scale, which used O^{16} as the standard mass.

Relative abundances are from the compilation by G. L. Trigg.[3]

[1] *Chemical and Engineering News*, November 20, 1961, p. 43. By permission of The International Union of Pure and Applied Chemistry and Butterworths Scientific Publications.

[2] L. A. König, J. H. E. Mattauch, and A. H. Wapstra, *Nuclear Physics*, **31**, 18–42 (1962).

[3] *American Institute of Physics Handbook*, McGraw-Hill, New York (1957).

Z	Symbol	Atomic Weight	Element	A	Relative Abundance, %	Nuclide Mass
0	n		Neutron	1		1.008665
1	H	1.00797	Hydrogen	1	99.985	1.007825
				2	0.015	2.014102
				3 *		3.016049
2	He	4.0026	Helium	3	0.00015	3.016030
				4	~100	4.002604
3	Li	6.939	Lithium	6	7.52	6.015126
				7	92.48	7.016005
4	Be	9.0122	Beryllium	8 *		8.005308
				9	100	9.012186
5	B	10.811	Boron	10	18.7	10.012939
				11	81.3	11.009305
6	C	12.01115	Carbon	12	98.892	12.0000000
				13	1.108	13.003354
				14 *		14.003242
7	N	14.0067	Nitrogen	13 *		13.005739
				14	99.635	14.003074
				15	0.365	15.000108
8	O	15.9994	Oxygen	16	99.759	15.994915
				17	0.037	16.999133
				18	0.204	17.999160
9	F	18.9984	Fluorine	19	100	18.998405
10	Ne	20.183	Neon	20	90.92	19.992440
				21	0.257	20.993849
				22	8.82	21.991384
11	Na	22.9898	Sodium	22 *		21.994435
				23	100	22.989773
12	Mg	24.312	Magnesium	24	78.60	23.985045
				25	10.11	24.985840
				26	11.29	25.982591
13	Al	26.9815	Aluminum	27	100	26.981535
14	Si	28.086	Silicon	28	92.27	27.976927
				29	4.68	28.976491
				30	3.05	29.973761
15	P	30.9738	Phosphorus	31	100	30.973763
16	S	32.064	Sulfur	32	95.018	31.972074
				33	0.750	32.971460
				34	4.215	33.967864
				36	0.017	35.967091
17	Cl	35.453	Chlorine	35	75.4	34.968854
				37	24.6	36.965896

Z	Symbol	Atomic Weight	Element	A	Relative Abundance, %	Nuclide Mass
18	Ar	39.948	Argon	36	0.337	35.967548
				38	0.063	37.962724
				40	99.6	39.962384
19	K	39.102	Potassium	39	93.08	38.963714
				40 *	0.0119	39.964008
				41	6.91	40.961835
20	Ca	40.08	Calcium	40	96.97	39.962589
				42	0.64	41.958628
				43	0.145	42.958780
				44	2.06	43.955490
				46	0.0033	45.953689
				48	0.0185	47.952363
21	Sc	44.956	Scandium	45	100	44.955919
22	Ti	47.90	Titanium	48	73.45	47.947948
23	V	50.942	Vanadium	51	99.76	50.943978
24	Cr	51.996	Chromium	52	83.76	51.940514
25	Mn	54.9380	Manganese	55	100	54.938054
26	Fe	55.847	Iron	56	91.68	55.934932
27	Co	58.9332	Cobalt	59	100	58.93319
28	Ni	58.71	Nickel	58	67.76	57.93534
29	Cu	63.54	Copper	63	69.1	62.92959
30	Zn	65.37	Zinc	64	48.89	63.92914
31	Ga	69.72	Gallium	69	60.2	68.92568
32	Ge	72.59	Germanium	74	36.74	73.92115
33	As	74.9216	Arsenic	75	100	74.92158
34	Se	78.96	Selenium	80	49.82	79.91651
35	Br	79.909	Bromine	79	50.52	78.91835
36	Kr	83.80	Krypton	84	56.90	83.91150
37	Rb	85.47	Rubidium	85	72.15	84.91171
38	Sr	87.62	Strontium	88	82.56	87.90561
39	Y	88.905	Yttrium	89	100	88.90543
40	Zr	91.22	Zirconium	90	51.46	89.90432
41	Nb	92.906	Niobium	93	100	92.90602
42	Mo	95.94	Molybdenum	98	23.75	97.90551
43	Tc		Technetium	98 *		97.90730
44	Ru	101.07	Ruthenium	102	31.3	101.90372
45	Rh	102.905	Rhodium	103	100	102.90480
46	Pd	106.4	Palladium	106	27.2	105.90320
47	Ag	107.870	Silver	107	51.35	106.90497
48	Cd	112.40	Cadmium	114	28.86	113.90357
49	In	114.82	Indium	115	95.77	114.90407

Z	Sym-bol	Atomic Weight	Element	A	Relative Abundance, %	Nuclide Mass
50	Sn	118.69	Tin	120	32.97	119.90213
51	Sb	121.75	Antimony	121	57.25	120.90375
52	Te	127.60	Tellurium	130	34.49	129.90670
53	I	126.9044	Iodine	127	100	126.90435
54	Xe	131.30	Xenon	132	26.89	131.90416
55	Cs	132.905	Cesium	133	100	132.90509
56	Ba	137.34	Barium	138	71.66	137.90501
57	La	138.91	Lanthanum	139	99.911	138.90606
58	Ce	140.12	Cerium	140	88.48	139.90528
59	Pr	140.907	Praseodymium	141	100	140.90739
60	Nd	144.24	Neodymium	142	27.13	141.90748
61	Pm		Promethium	145 *		144.91231
62	Sm	150.35	Samarium	152	26 63	151.91949
63	Eu	151.96	Europium	153	52.23	152.92086
64	Gd	157.25	Gadolinium	158	24.87	157.92410
65	Tb	158.924	Terbium	159	100	158.92495
66	Dy	162.50	Dysprosium	164	28.18	163.92883
67	Ho	164.930	Holmium	165	100	164.93030
68	Er	167.26	Erbium	166	33.41	165.93040
69	Tm	168.934	Thulium	169	100	168.93435
70	Yb	173.04	Ytterbium	174	31.84	173.93902
71	Lu	174.97	Lutetium	175	97.40	174.94089
72	Hf	178.49	Hafnium	180	35.44	179.94681
73	Ta	180.948	Tantalum	181	100	180.94798
74	W	183.85	Tungsten	184	30.6	183.95099
75	Re	186.2	Rhenium	187	62.93	186.95596
76	Os	190.2	Osmium	192	41.0	191.96141
77	Ir	192.2	Iridium	193	61.5	192.96328
78	Pt	195.09	Platinum	195	33.7	194.96482
				198	7.23	197.96753
79	Au	196.967	Gold	197	100	196.96655
80	Hg	200.59	Mercury	202	29.80	201.97063
81	Tl	204.37	Thallium	205	70.50	204.97446
82	Pb	207.19	Lead	208	52.3	207.97664
83	Bi	208.980	Bismuth	209	100	208.98042
84	Po		Polonium	210 *		209.98287
85	At		Astatine	211 *		210.98750
86	Em		Emanation	211 *		210.99060
87	Fr		Francium	221 *		221.01418
88	Ra		Radium	226 *		226.02536
89	Ac		Actinium	225 *		225.02314

Z	Symbol	Atomic Weight	Element	A	Relative Abundance, %	Nuclide Mass
90	Th	232.038	Thorium	232 *	100	232.03821
91	Pa		Protactinium	231 *		231.03594
92	U	238.03	Uranium	233 *		233.03950
				235 *	0.715	235.04393
				238 *	99.28	238.05076
93	Np		Neptunium	239 *		239.05294
94	Pu		Plutonium	239 *		239.05216
95	Am		Americium	243 *		243.06138
96	Cm		Curium	245 *		245.06534
97	Bk		Berkelium	248 *		248.07305
98	Cf		Californium	249 *		249.07470
99	Es		Einsteinium	254 *		254.08811
100	Fm		Fermium	252 *		252.08265
101	Md		Mendelevium	255 *		255.09057
102	No		Nobelium	254 *		
103	Lw		Lawrencium	257 *		

Appendix D
Conversion of mksa to cgs Units

The units throughout this book are "rationalized mks ampere units." The only exceptions to this statement are the frequent use of the electron volt as an alternative energy unit instead of the joule and of angstroms instead of meters. The reader who is unfamiliar with the mks system should study an elementary electricity and magnetism text that uses these units.*

The mathematical expressions and problems in this book can be used with unrationalized cgs units if the following operations are performed: (1) Convert the individual given quantities like length and charge to either esu or emu by the table below. (2) If ϵ_0 appears in the expression, replace it by $1/4\pi$ and use esu. (3) If μ_0 appears in the expression, replace it by $1/4\pi$ and use emu. The reason for operations 2 and 3 is that formulas based on Coulomb's law or Gauss' law have different forms in the different systems of units. These operations will be illustrated by two examples.

Example 1. Problem 1-8. The required e/m can be calculated from eq. 1-9:

$$\frac{e}{m} = \frac{V_d^2}{2V_0 d^2 \mathcal{B}^2}$$

This expression does not contain ϵ_0 or μ_0, and therefore we need only convert the individual quantities. We shall use the emu system (we could use esu).

$$V_d = 184 \text{ volts} = 184 \times 10^8 \text{ abvolts}$$

$$V_0 = 300 \text{ volts} = 300 \times 10^8 \text{ abvolts}$$

$$d = 0.015 \text{ m.} = 1.5 \text{ cm}$$

$$\mathcal{B} = 0.0012 \text{ weber/m.}^2 = 12 \text{ gauss}$$

$$\frac{e}{m} = \frac{(1.84)^2 \times 10^{20}}{6 \times 10^{10} \times (1.5)^2 \times (12)^2} = 1.74 \times 10^7 \text{ abcoulombs/gram}$$

---

* For example, D. Halliday and R. Resnick, *Physics for Students of Science and Engineering*, Part II, Wiley, New York, 2nd Ed., 1962.

This is the value measured in the experiment. The accepted value is 1.759×10^7 abcoulombs/gram.

Example 2. Find the energy levels of the hydrogen atom in ergs. The required E_n can be calculated from eq. 6-2:

$$E_n = - \frac{e^4 m}{n^2 h^2 8\epsilon_0^2} \text{ joules}$$

We must replace ϵ_0 by $1/4\pi$ and use the esu system:

$$E_n = - \frac{2\pi^2 e^4 m}{n^2 h^2}$$

$$e = 1.602 \times 10^{-19} \text{ coulomb} = 4.80 \times 10^{-10} \text{ statcoulomb}$$

$$m = 9.11 \times 10^{-31} \text{ kg} = 9.11 \times 10^{-28} \text{ gram}$$

$$h = 6.62 \times 10^{-34} \text{ joule sec} = 6.62 \times 10^{-27} \text{ erg sec}$$

$$E_n = - \frac{2\pi^2 \times (4.8)^4 \times 9.11 \times 10^{-68}}{n^2 \times (6.62)^2 \times 10^{-54}}$$

$$E_n = - \frac{2.18 \times 10^{-11}}{n^2} \text{ ergs}$$

We can convert back to electron volts as a check. From the definition of the electron volt,

$$1 \text{ e.V.} = (4.80 \times 10^{-10} \text{ statcoulomb})(\tfrac{1}{300} \text{ statvolt}) = 1.60 \times 10^{-12} \text{ erg}$$

Hence

$$E_n = - \frac{2.18 \times 10^{-11}}{n^2 \times 1.60 \times 10^{-12}} = - \frac{13.6}{n^2} \text{ e.V.}$$

CONVERSION TABLE

Equal signs are understood across each row. Always make sure that the conversion factor is being applied in the correct direction by first converting some unit for which the answer is known (e.g., length).

The factors of 3 in the table arise from the velocity of light, which has been set equal to 3×10^8 m./sec. In very precise work each factor of 3 should be replaced by a factor of 2.99793. The factors of 4 are exact. For example, the exact value of ϵ_0 is $[4\pi(2.99793)^2]^{-1} \times 10^{-9} = 8.8542 \times 10^{-12}$ farad/m.

			cgs Units	
Quantity and Symbol		mksa Unit	emu	esu
Time	t	1 second (sec)	1 sec	1 sec
Length	d	1 meter (m.)	100 cm	100 cm
Mass	M	1 kilogram (kg)	1000 gm	1000 gm
Energy	E	1 joule	10^7 ergs	10^7 ergs
Force	F	1 newton	10^5 dynes	10^5 dynes
Charge	q	1 coulomb	$\frac{1}{10}$ abcoulomb	3×10^9 statcoulombs
Current	I	1 ampere (amp)	$\frac{1}{10}$ abamp	3×10^9 statamps
Potential	V	1 volt	10^8 abvolts	$\frac{1}{300}$ statvolt
Electric field intensity	$\mathcal{E}$	1 volt/m.	10^6 abvolts/cm	$\frac{1}{3} \times 10^{-4}$ statvolt/cm
Magnetic induction	$\mathcal{B}$	1 weber/m.^{2}	10,000 gauss	$\frac{1}{3} \times 10^{-6}$ statvolt sec/cm^2
Magnetic field intensity	$\mathcal{H}$	1 amp-turn/m.	$4\pi \times 10^{-3}$ oersted	$12\pi \times 10^7$ statamps/cm
Magnetization		1 weber/m.^{2}	$10^4/4\pi$ gauss	$\frac{1}{12\pi} \times 10^{-6}$ statvolt sec/cm^2
Electrical permittivity of vacuum	ϵ_0	$\frac{1}{36\pi} \times 10^{-9}$ farad/m.		Unity
Magnetic permittivity of vacuum	μ_0	$4\pi \times 10^{-7}$ henry/m.	Unity	

Appendix E
Pulse Spectra and
the Indeterminacy Principle

Equation 4-24 at the end of Sec. 4-12 was discussed in terms of a discrete frequency spectrum in order to show as simply as possible that a "spread" of frequencies was required to obtain resolution in time. The $\Delta E \, \Delta t \geqslant h$ indeterminacy, which followed from that equation, can be treated more satisfactorily by considering *continuous* frequency distributions, and we shall undertake such a treatment here. We shall proceed by first finding the "frequency spectrum" of a typical pulse that might develop in a radar or communication system. It is illustrated in Fig. E-1 and is a rectangular voltage pulse of duration 2τ and of height V_0. That is, we shall find the frequencies and amplitudes of ordinary sinusoidal oscillations such that when they are superimposed they give the $V(t)$ in Fig. E-1. This distribution will be called $G(\omega)$, where G is the amplitude and $\omega = 2\pi\nu$ is the "angular frequency." $G(\omega)$ is called the *spectrum* of the pulse:

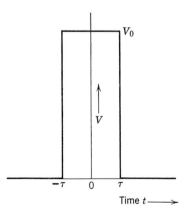

Fig. E-1. Voltage as a function of time for a rectangular pulse of duration 2τ.

597

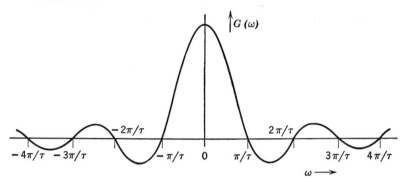

Fig. E-2. Spectrum of the pulse of Fig. E-1.

The spectrum is found by the use of the Fourier integral theorem. The pulse of Fig. E-1 is an "even function" or symmetrical function of t; that is, $V(t)$ equals $V(-t)$. For such functions the Fourier integral theorem takes on a slightly simpler form than the general form. The simpler form for $G(\omega)$ applicable here is

$$G(\omega) = (2/\pi)^{\frac{1}{2}} \int_0^\infty V(t) \cos \omega t \, dt \qquad \text{(E-1)}$$

The theorem states that

$$V(t) = (2/\pi)^{\frac{1}{2}} \int_0^\infty G(\omega) \cos \omega t \, d\omega \qquad \text{(E-2)}$$

In other words, the pulse can be "recovered" from the function $G(\omega)$.

This may not appear to be very helpful, since we seem to be going around in circles. In electric-circuit theory, it *is* very helpful, however, since it permits the prediction of the distortion produced by an amplifier or other circuit with a finite "pass band." The gain of such a circuit is not constant at all frequencies, but in general the gain is some function $A(\omega)$.* In order to see what distortion of the pulse this produces, all we need to do is to multiply $A(\omega)$ by $G(\omega)$ and then find the new (distorted) $V'(t)$:

$$V'(t) = (2/\pi)^{\frac{1}{2}} \int_0^\infty G(\omega) A(\omega) \cos \omega t \, d\omega \qquad \text{(E-3)}$$

---

* In general there will also be a phase shift $\phi(\omega)$. In many practical systems the phase shift is proportional to frequency: $\phi(\omega) = \omega t_0$, where t_0 is a constant. The analysis given here applies to such a system if the term $\cos \omega t$ is replaced by $\cos (\omega t - \omega t_0)$. The effect of such a linear phase shift is simply to delay the output pulse by a time t_0.

For the pulse in our example, $V(t) = 0$ except when $-\tau < t < \tau$ and equals V_0 within this interval:

$$G(\omega) = (2/\pi)^{\frac{1}{2}} \int_0^{\tau} V_0 \cos \omega t \, dt = V_0 \, (2/\pi)^{\frac{1}{2}} \frac{\sin \omega \tau}{\omega} \qquad \text{(E-4)}$$

This is plotted in Fig. E-2. It is noteworthy that a spectrum of frequencies from $-\infty$ to $+\infty$ is contained in the pulse.

A rather good approximation to the pulse could be obtained by cutting off this spectrum at $\pm \pi/\tau$. In order to see what the pulse would be after passing through an amplifier with unity gain for $-\pi/\tau < \omega < \pi/\tau$ and zero gain for other frequencies, we write

$$V'(t) = (2/\pi)^{\frac{1}{2}} \int_0^{\pi/\tau} G(\omega) \cos \omega t \, d\omega$$

$$= \frac{2V_0}{\pi} \int_0^{\pi/\tau} \frac{\sin \omega \tau \cos \omega t}{\omega} \, d\omega \qquad \text{(E-5)}$$

The resulting $V'(t)$ is plotted in Fig. E-3.

A radar pulse or wave packet consists of a function like that of Fig. E-1 modulating a sinusoidal oscillation at high frequency $\nu_0 = \omega_0/2\pi$, the "carrier" frequency. The foregoing theory then becomes

$$V(t) = V_0 \cos \omega_0 t \qquad \text{for } -\tau < t < \tau$$

$$V(t) = 0 \qquad \text{for } t < -\tau \quad \text{or} \quad t > \tau$$

Since this is still an even function we can apply the above theory and obtain

$$G(\omega) = V_0 \, (2/\pi)^{\frac{1}{2}} \frac{\sin (\omega - \omega_0)\tau}{\omega - \omega_0} \qquad \text{(E-6)}$$

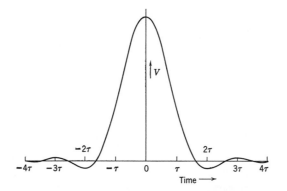

Fig. E-3. Pulse of Fig. E-1 after passing through an amplifier with a pass band of $1/\tau$ cycles.

The only effect of introducing the carrier frequency oscillation has been to shift the center of the frequency spectrum from $\omega = 0$ to $\omega = \omega_0$; the width and shape are unchanged.

Let us now examine Fig. E-3 to find what spread of frequencies is required to fix the time a pulse passes a certain point (or to fix the time of emission) within a precision Δt. The angular frequency spread is $\Delta\omega = 2\pi/\tau$, and the frequency spread is $\Delta\nu = 1/\tau$ (the bandwidth of the amplifier). The time of arrival of this pulse can be determined only to a precision Δt of the order of τ, since it extends in time to this extent (Fig. E-3). Hence

$$\Delta\nu \, \Delta t = (1/\tau)\tau = 1 \tag{E-7}$$

This means that if we wish a sharp enough pulse so that its flight can be timed to a microsecond, we must use amplifiers and transmission circuitry that will pass frequencies in a 1-megacycle pass band.

When applied to the wave packet for a photon or electron, the result expressed in eq. E-7 leads to the Indeterminacy Principle, $\Delta E \, \Delta t \geq h$. The form of the wave packet that gives the highest precision in the simultaneous determination of $E = h\nu$ and of t is the "Gaussian" form illustrated in Fig. E-4a. Its spectrum is illustrated in Fig. E-4b. The Δt and $\Delta\nu$ identified on the figure are such that their product is unity.

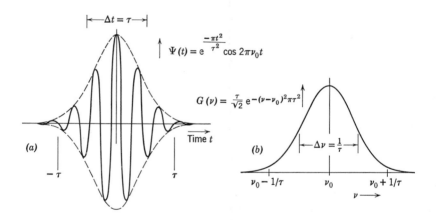

Fig. E-4. A wave packet of the Gaussian form is shown in (a) and its spectrum of frequencies in (b). There should be many more oscillations of the cosine term in (a), since $\nu_0 \gg 1/\tau$ for the usual wave packet.

Appendix F
Energy, Momentum, and
Wave Functions

Quantum mechanics provides predictions of the values of any quantity that can be measured in an experiment. It recognizes no obligation to make predictions about experiments that cannot be performed, but it provides information about any *observable* quantity. The general procedure is as follows: *To each observable* (such as the energy) *there corresponds an operation on* Ψ. If this operation is performed the result is the product of Ψ and the desired observable quantity. (Throughout the following analysis we shall assume that Ψ has been normalized.)

We shall illustrate the procedure by the example of the energy E. The "operator" to determine E is $\dfrac{-h}{2\pi i}\dfrac{\partial}{\partial t}$. The operation on Ψ is

$$\frac{-h}{2\pi i}\frac{\partial \Psi}{\partial t} = E\Psi \tag{F-1}$$

If we wish to compute E from a known Ψ, we can multiply both sides of this equation by Ψ*, the complex conjugate of Ψ, and integrate over all space:

$$\iiint \Psi^* \left(\frac{-h}{2\pi i}\frac{\partial \Psi}{\partial t} \right) dv = \iiint \Psi^* E \Psi \, dv \tag{F-2}$$

Here dv is the element of volume in the coordinates being used; it is $dx\,dy\,dz$ in rectangular coordinates (or simply dx for one-dimensional problems) and $r^2 \sin\theta\,dr\,d\theta\,d\phi$ in spherical polar coordinates. Since E is a number (not a function of the coordinates) and since Ψ has been normalized, this equation is simply

$$E = \iiint \Psi^* \left(\frac{-h}{2\pi i}\frac{\partial \Psi}{\partial t} \right) dv \tag{F-3}$$

In order to illustrate the technique of using equations like eq. F-3, we shall insert into this equation the wave function

$$\Psi = \psi(x)e^{(-2\pi iE/h)t}$$

which is the combination of eqs. 5-3 and 5-5. Partial differentiation of this Ψ with respect to t gives $(-2\pi iE/h)\Psi$; and, since Ψ is normalized, we find that eq. F-3 reduces to the identity $E = E$. Thus we have learned nothing we did not already know, but we have seen how the observable quantity (in our example, the total energy) can be obtained when the wave function is known.

The operator for the linear momentum * p_x is $\dfrac{h}{2\pi i}\dfrac{\partial}{\partial x}$ and for the z component of the angular momentum m_z is $\dfrac{h}{2\pi i}\dfrac{\partial}{\partial \phi}$. The procedure for finding p_x or m_z is just the same as the foregoing procedure for E: We perform the indicated operation, multiply by Ψ^*, and integrate over all space.

It is interesting to explore the connection between the operator for p_x and the Schrödinger equation. First we note that to compute the square of an observable quantity, the operation corresponding to that quantity should be performed twice. Thus the operator for $p_x{}^2$ is $\dfrac{-h^2}{4\pi^2}\dfrac{\partial^2}{\partial x^2}$. The operator for the potential energy is simply P, and the operation consists of multiplying Ψ by P. Then we note, as we might guess, that the operator for the total energy can be written as the sum of kinetic and potential energy terms, $p^2/2m + P$, operating on Ψ. This operation gives

$$\frac{-h^2}{8\pi^2m}\left(\frac{\partial^2}{\partial x^2} + \frac{\partial^2}{\partial y^2} + \frac{\partial^2}{\partial z^2}\right)\Psi + P\Psi = E\Psi \tag{F-4}$$

When this equation is compared with eq. F-1, it is apparent that we have produced the Schrödinger equation, eq. 5-1:

$$\frac{h^2}{8\pi^2m}\left(\frac{\partial^2\Psi}{\partial x^2} + \frac{\partial^2\Psi}{\partial y^2} + \frac{\partial^2\Psi}{\partial z^2}\right) - P\Psi = \frac{h}{2\pi i}\frac{\partial\Psi}{\partial t} \tag{5-1}$$

This paragraph is in no sense a rigorous derivation of the Schrödinger

---

* Since $\partial\Psi/\partial x$ is therefore proportional to the momentum, it should now be clear why we insisted in eq. 5-8 that $\partial\Psi/\partial x$ be continuous. A discontinuity in the momentum would mean a failure of the law of conservation of momentum (except in the idealized case in which the potential energy is infinite).

equation from postulates about operators, but it does show that the operator approach and the Schrödinger equation are consistent and connected.

Equations like eq. F-1 and eq. F-4 are usually used to find Ψ, rather than to find E when Ψ is known. Such equations generally have "well-behaved" solutions (that is, solutions satisfying eqs. 5-6, 5-7, and 5-8) only for certain values of E. These values are called *eigenvalues*, and the corresponding ψ's are called *eigenfunctions*. Examples are the $\psi_0, \psi_1, \cdots$, eigenfunctions for the harmonic oscillator with eigenvalues $E_0, E_1, \cdots$. (Refer to Sec. 5-2 for the difference between Ψ and ψ.)

The only result possible for a well-performed experiment to measure E is one or another of the eigenvalues E_n. But if we make another measurement of E on the same system we may find another one of the eigenvalues. For example, a hydrogen atom is usually in its ground state ($n = 1$), and therefore a measurement would give $E = E_1$. But it may be in an excited state part of the time, in which case the value E_2, or E_3, etc., might be found. When a system may be in any one of a number of states, the wave function is

$$\Psi = a_0\Psi_0 + a_1\Psi_1 + a_2\Psi_2 + \cdots \tag{F-5}$$

where Ψ_n is the eigenfunction for the eigenvalue E_n, the a_n's are constants, and the ground state is $n = 0$. If, for example, the system spends one-half of the time in the state E_1 and one-half in E_2, then

$$\Psi = 2^{-\frac{1}{2}}(\Psi_1 + \Psi_2) \tag{F-6}$$

The equality of the two coefficients makes it equally likely that the system will be found in either state. The coefficient in eq. F-6 has been chosen to normalize Ψ (assumed here to be a function only of x):

$$\int_{-\infty}^{\infty} \Psi\Psi^* \, dx = \frac{1}{2}\int_{-\infty}^{\infty} (\Psi_1 + \Psi_2)(\Psi_1{}^* + \Psi_2{}^*) \, dx = \frac{1}{2}\int_{-\infty}^{\infty} \Psi_1\Psi_1{}^* \, dx$$

$$+ \frac{1}{2}\int_{-\infty}^{\infty} \Psi_2\Psi_2{}^* \, dx + \int_{-\infty}^{\infty} \Psi_1\Psi_2{}^* \, dx + \int_{-\infty}^{\infty} \Psi_2\Psi_1{}^* \, dx$$

Each of the two integrals containing only Ψ_1 or only Ψ_2 equals unity, since Ψ_1 and Ψ_2 are normalized. Each of the last two integrals equals zero, since it is a general property of solutions of the Schrödinger equation that

$$\int_{-\infty}^{\infty} \Psi_n\Psi_m{}^* \, dx = \int_{-\infty}^{\infty} \psi_n\psi_m{}^* \, dx = 0 \tag{F-7}$$

unless $n = m$ (see prob. F-1).

The average value $\bar{E}$ of the energy for a large number of measurements on the same system is frequently of interest. To find $\bar{E}$ for a one-dimensional problem, we use the operator of eq. F-4

$$\frac{-h^2}{8\pi^2 m}\frac{\partial^2}{\partial x^2} + P$$

and the procedure used to obtain eq. F-3. (Compare the procedure for finding $\bar{x}$ in Sec. 5-7.)

$$\bar{E} = \int_{-\infty}^{\infty} \Psi^* \left\{ \frac{-h^2}{8\pi^2 m}\frac{\partial^2 \Psi}{\partial x^2} + P\Psi \right\} dx \qquad \text{(F-8)}$$

or

$$\bar{E} = \int_{-\infty}^{\infty} \psi^* \left\{ \frac{-h^2}{8\pi^2 m}\frac{d^2 \psi}{dx^2} + P\psi \right\} dx \qquad \text{(F-9)}$$

If this procedure is applied to a wave function like eq. F-5, the result will not usually be that $\bar{E}$ equals one of the E_n's. ($\bar{E}$ equals E_n only if $a_n = 1$ and all the other a's equal zero, i.e., if the system is certainly in the nth state.) But any single measurement of E can yield only one of the eigenvalues E_n. A result precisely similar to eq. F-8 or eq. F-9 is obtained for the average momentum p_x by substituting the momentum operator for the energy operator.

There is a very useful method of obtaining the ground state (lowest-energy eigenvalue) of a system by the use of the energy operator and the concept of average energy. It is based on a mathematical procedure known as the variational method. The principle of this method is that the correct ψ for the ground state is the ψ that makes the $\bar{E}$ of eq. F-9 a minimum. A proof of this principle will now be sketched for a wave function that is a function only of x.

Let us guess a wave function ψ, called the "trial" wave function, for a particular problem. Since this was only guessed, it is not one of the solutions ψ_n of the Schrödinger equation for this problem. But *any* function satisfying the boundary conditions for the problem can be expanded in an infinite series of the ψ_n's (just as any physical function can be expanded in a Taylor series). We expand our trial function in this way:

$$\psi = a_0\psi_0 + a_1\psi_1 + a_2\psi_2 + \cdots \qquad \text{(F-10)}$$

The average energy $\bar{E}$ is found by substituting eq. F-10 into eq. F-9. The resulting very complicated expression can be greatly simplified by using eq. 5-4 and then eq. F-7 with the outcome:

$$\bar{E} = |a_0|^2 E_0 + |a_1|^2 E_1 + |a_2|^2 E_2 + \cdots \qquad \text{(F-11)}$$

Thus $\bar{E}$ is a weighted average of the E_n's, with weighting factors $|a_n|^2$. The sum of these factors is

$$\Sigma |a_n|^2 = 1 \qquad \text{(F-12)}$$

which can be proved as a generalization of the foregoing proof that the Ψ in eq. F-6 was normalized (see prob. F-3).

We now show that $\bar{E}$ for any ψ (other than ψ_0) is always greater than E_0. We divide eq. F-11 by E_0

$$\frac{\bar{E}}{E_0} = |a_0|^2 + |a_1|^2 \frac{E_1}{E_0} + |a_2|^2 \frac{E_2}{E_0} + \cdots \qquad \text{(F-13)}$$

Since E_0 is the lowest of the E_n's, each term on the right is greater than or equal to the similar term in

$$|a_0|^2 + |a_1|^2 + |a_2|^2 + \cdots = \Sigma |a_n|^2 = 1$$

Therefore the right side of eq. F-13 is $\geqq 1$, and

$$\bar{E} \geqq E_0 \qquad \text{(F-14)}$$

Thus any $\bar{E}$ we calculate with a trial function is always greater than or equal to the ground-state energy E_0.

Furthermore, if our trial function happened to have been guessed correctly, a_0 equals 1, all the other a_n's equal zero, and $\bar{E} = E_0$. Hence we know that the trial function is a better approximation to the actual ground-state wave function if it gives a lower $\bar{E}$ than some other trial function. In practice, we guess a form for $\psi(x)$ that satisfies the boundary conditions and contains a few parameters, and then we vary these parameters until the integral (eq. F-9) is a minimum. Thus we might guess that ψ equals $e^{-ax^2} \cos bx$ for a particular problem and vary a and b until $\bar{E}$ is a minimum. This may not be the correct wave function of the ground state, but it will be very close to it if we have made a wise guess. If we wish to do better we can try again with a different form for ψ or with a larger number of adjustable parameters.

Our principal interest in this method is that it permits sketching with fair accuracy the wave function of the ground state of a system, without the use of any mathematics. First, however, we must work eq. F-9 into a different form:

$$\bar{E} = \int_{-\infty}^{\infty} \frac{-h^2}{8\pi^2 m} \psi^* \frac{d^2\psi}{dx^2} \, dx + \int_{-\infty}^{\infty} P\psi^*\psi \, dx \qquad \text{(F-15)}$$

The second integral is just the average value of the potential energy $\bar{P}$, and therefore the first integral must be $\bar{K}$ (since $\bar{E} = \bar{K} + \bar{P}$). $\bar{K}$ can be put into a more suitable form by integration by parts. The general procedure of integration by parts is, of course:

$$\int_{v_1}^{v_2} u \, dv = uv \Big]_{v_1}^{v_2} - \int_{v_1}^{v_2} v \, du$$

We take $\psi^* = u$ and $(d^2\psi/dx^2) \, dx = dv$. Then $v = d\psi/dx$ and $du = (d\psi^*/dx) \, dx$. Hence

$$\int_{-\infty}^{\infty} \psi^* \frac{d^2\psi}{dx^2} \, dx = \psi^* \frac{d\psi}{dx} \Big]_{x=-\infty}^{\infty} - \int_{-\infty}^{\infty} \frac{d\psi}{dx} \frac{d\psi^*}{dx} \, dx$$

The condition eq. 5-6 ensures that $\psi^*(d\psi/dx) = 0$ at $x = \pm\infty$. Also, the operation $d\psi/dx$ does not involve i, and so $(d\psi^*/dx) = (d\psi/dx)^*$, and therefore $\dfrac{d\psi}{dx} \dfrac{d\psi^*}{dx} = \left| \dfrac{d\psi}{dx} \right|^2$. Thus

$$\int_{-\infty}^{\infty} \psi^* \frac{d^2\psi}{dx^2} \, dx = -\int_{-\infty}^{\infty} \left| \frac{d\psi}{dx} \right|^2 dx$$

When we put this result into eq. F-15, we obtain

$$\bar{E} = \int_{-\infty}^{\infty} \frac{h^2}{8\pi^2 m} \left| \frac{d\psi}{dx} \right|^2 dx + \int_{-\infty}^{\infty} P\psi^*\psi \, dx \qquad \text{(F-16)}$$

The kinetic energy K is therefore proportional to $|d\psi/dx|^2$. In order to obtain the ground state ψ, we therefore minimize the sum of the average P and the average K (proportional to $|d\psi/dx|^2$), while always keeping

$$\int_{-\infty}^{\infty} \psi^*\psi \, dx = 1$$

Illustrations of this method of finding ground state wave functions are presented at the end of Sec. 5-7 and in probs. 5-61, 5-62, and 5-63. The concepts of average energies $\bar{K}$ and $\bar{P}$ and of minimizing their sum to obtain the ground state wave function of a system are also used at many other points in our work, such as in the discussion of molecular binding and in the exchange-force explanation of ferromagnetism.

PROBLEMS

F-1. Prove eq. F-7 for the special case of ψ_1 and ψ_2 from eq. 5-24.

F-2. Carry out the procedure described just preceding eq. F-8 to derive that equation.

F-3. Show that eq. F-12 follows from the fact that ψ and the ψ_n's in eq. F-10 are normalized.

F-4. Prove eq. F-11.

F-5.* The three wave functions of the hydrogen atom with $n = 2, l = 1$ are:

$$\psi_1 = A 2^{-\frac{1}{2}} e^{i\varphi} \sin \theta \qquad (m_l = 1)$$

$$\psi_0 = A \cos \theta \qquad (m_l = 0)$$

$$\psi_{-1} = A 2^{-\frac{1}{2}} e^{-i\varphi} \sin \theta \qquad (m_l = -1)$$

in which $A = \pi^{-\frac{1}{2}}(2\rho)^{-\frac{5}{2}} r e^{-r/2\rho}$. Calculate the z component m_z of the angular momentum for each state. *Hint:* The problem can be solved without executing any of the integrations by noting that $(\partial/\partial\varphi)e^{i\varphi} = i e^{i\varphi}$ and that $\iiint \psi^*\psi \, dv = 1$.

Appendix G
Three-Dimensional
Square Well

The three-dimensional analogue of the square well with infinitely high sides of Sec. 5-4a consists of a region in space where the potential energy P is zero surrounded by abrupt, infinitely high potential barriers. If the shape of this region is a sphere or a rectangular parallelepiped, the Schrödinger equation can be readily solved.

The cube, a special case of the rectangular parallelepiped, is the easiest geometry, and the solution for this geometry illustrates the nature of the solutions for all such wells. We shall consider a cube with edge length $2a$ and with the origin of rectangular coordinates at the center. By the arguments of Sec. 5-4a and 5-4b, we know that $\Psi = 0$ at $x = \pm a$, $y = \pm a$, and $z = \pm a$. We can apply the same technique of separating the variables and thereby obtaining ordinary differential equations that was used in Sec. 5-2. Here we let

$$\Psi(x, y, z, t) = \psi(x)\xi(y)\chi(z)\phi(t) \tag{G-1}$$

and insert this Ψ into eq. 5-1. After carrying out the indicated differentiations and dividing by Ψ, we obtain

$$\frac{h^2}{8\pi^2 m}\left(\frac{1}{\psi}\frac{d^2\psi}{dx^2} + \frac{1}{\xi}\frac{d^2\xi}{dy^2} + \frac{1}{\chi}\frac{d^2\chi}{dz^2}\right) - P = \frac{h}{2\pi i \phi}\frac{d\phi}{dt} \tag{G-2}$$

Since the left side does not contain the time and the right side does not contain the space variables, each side must equal the same constant $-E$, and eq. 5-5 gives $\phi(t)$.

In our problem, $P = 0$, and hence

$$\frac{h^2}{8\pi^2 m}\left(\frac{1}{\psi}\frac{d^2\psi}{dx^2} + \frac{1}{\xi}\frac{d^2\xi}{dy^2} + \frac{1}{\chi}\frac{d^2\chi}{dz^2}\right) + E = 0$$

Now $(1/\psi)(d^2\psi/dx^2)$ can be a function only of x or a constant, and similarly for the other terms. Thus again the only possibility is that each term equals a constant. Therefore

$$\frac{h^2}{8\pi^2 m}\frac{d^2\psi}{dx^2} + E_1\psi = 0 \tag{G-3}$$

and similar equations for y and z, with $E_1 + E_2 + E_3 = E$. We can write the solutions for ψ by following the procedure of Sec. 5-4a, and we obtain

$$\psi = A\cos\beta_1 x \qquad (n_1 = 1, 3, 5, \cdots)$$

or $\tag{G-4}$

$$\psi = A\sin\beta_1 x \qquad (n_1 = 2, 4, 6, \cdots)$$

in which $\beta_1 = n_1\pi/2a = (2\pi/h)(2mE_1)^{\frac{1}{2}}$, and A is to be determined by normalization. These expressions satisfy both eq. G-3 and the boundary conditions that $\psi = 0$ at the edges of the well. Thus

$$E_1 = \frac{n_1{}^2 h^2}{32ma^2} \tag{G-5}$$

as in eq. 5-26.

The solutions for ξ and χ are evidently similar but with n_1 and E_1 replaced by quantum numbers n_2 and n_3 and energies E_2 and E_3. The total energy E (which is also the kinetic energy since $P = 0$) is therefore

$$E = E_1 + E_2 + E_3 = (h^2/32ma^2)(n_1{}^2 + n_2{}^2 + n_3{}^2) \tag{G-6}$$

There are actually *two* different quantum states for each combination of integers n_1, n_2, n_3, one with spin $+\frac{1}{2}$ and one with spin $-\frac{1}{2}$. On the other hand, changing n_1 to $-n_1$ does *not* change the wave function, and hence in counting quantum states we should include only positive values of the quantum numbers.

Our principal application of this analysis of the three-dimensional square well is to determine how closely packed quantum states can be. For example, in Sec. 9-4 we need to know how many quantum states per unit energy and per unit volume there are for electrons confined to a metal but free to move inside the metal. We shall define this "density of states" by letting $S(E)\,dE$ be the number of quantum states per unit volume between E and $E + dE$. Clearly $S(E)$ according to eq. G-6 is not a continuous function, since the n's take on only integer values; but fortunately we need $S(E)$ only for values of E far in excess of the ground state $(n_1 = 1 = n_2 = n_3)$ energy, and for such E's, $S(E)$ is practically continuous.

Our problem can be visualized by studying Fig. G-1, in which the values of n_1, n_2, and n_3 are plotted in "quantum number space." Each

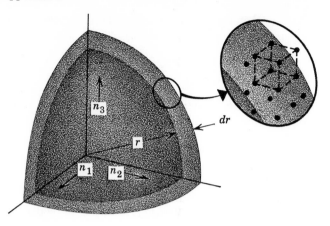

Fig. G-1. Quantum number space. Each dot represents a particular combination of integer values of n_1, n_2, and n_3, and therefore each dot represents a pair of quantum states. All the states in the spherical shell between r and $r + dr$ have approximately the same energy.

point corresponds to a particular combination of integers and therefore to a particular pair of quantum states (with $\pm \frac{1}{2}$ spin). Thus there are two quantum states per unit volume in this quantum number space, since there is one point (one combination of integers) for each little cube with unit length of sides. The number of quantum states $S(r)\, dr$ between r and $r + dr$ is thus the number per unit volume (2) times the volume between r and $r + dr$ (the volume of one octant of a spherical shell, $4\pi r^2\, dr/8$), or

$$S(r)\, dr = \pi r^2\, dr \qquad (G\text{-}7)$$

The factor "1/8" enters since negative n's do not give different wave functions from positive n's, and hence in counting quantum states we count only in the all-positive octant of quantum number space.

Now each value of r in Fig. G-1 corresponds to a value of E, since $r^2 = n_1{}^2 + n_2{}^2 + n_3{}^2$, and eq. G-6 can thus be expressed as

$$E = (h^2/32ma^2)r^2 \quad \text{or} \quad r = (2^{5/2}m^{1/2}a/h)E^{1/2} \qquad (G\text{-}8)$$

We can obtain $S(E)\, dE$ by using eq. G-8 to replace both r and dr in eq. G-7 by their values in terms of E:

$$S(E)\, dE = \pi(32ma^2E/h^2)(2^{3/2}m^{1/2}a/hE^{1/2})\, dE$$

$$= (\pi 2^{13/2}m^{3/2}a^3/h^3)E^{1/2}\, dE$$

This is the density of states in a cube in real (x, y, z) space of volume $(2a)^3$. Thus the density of states per unit volume is

$$S(E)\ dE = (2^{7/2}m^{3/2}\pi/h^3)E^{1/2}\ dE \qquad \text{(G-9)}$$

Of course the cube dimensions no longer appear, which suggests that the assumed size and shape of the square well are of no consequence in the result. Equation G-9 is, in fact, a general expression for the density of states. One of its applications is to conduction band electrons in solids, and it appears as eq. 9-6 in Sec. 9-4.

It is also of interest to examine the packing of quantum states in *momentum* space. A line from the origin to a general point in this space represents the magnitude and direction of the vector momentum associated with a quantum state. We shall work first with the wave functions developed above in eqs. G-4. These wave functions describe *standing* waves, that is, electron waves reflected back and forth at the edges of the well and interfering constructively. Strictly speaking, such waves possess no net momentum since there is just as much electron "motion" in any one direction as in the opposite direction. But it is helpful to interpret the packing of quantum states in terms of p^2, the square of the magnitude of the momentum associated with the back-and-forth motion. Since $P = 0$ in our well, $E = K = p^2/2m$, and since $p^2 = p_x^2 + p_y^2 + p_z^2$, we have

$$E = E_1 + E_2 + E_3 = (1/2m)(p_x^2 + p_y^2 + p_z^2)$$

This equation gives us, in fact, three equations of the form

$$E_1 = (1/2m)p_x^2 \qquad \text{(G-10)}$$

since the x, y, and z motions are independent. That is, we can vary p_x while keeping p_y and p_z constant (and therefore keeping E_2 and E_3 constant), and thus E_1 and p_x must be related by eq. G-10. Then by using eq. G-5 in eq. G-10 we find the relation between each component of momentum and the related quantum number:

$$p_x = +(2mE_1)^{1/2} = n_1(h/4a) \qquad \text{(G-11)}$$

and similarly for the y and z components. (We could as well have used the operator for p_x^2, namely $(-h^2/4\pi^2)(\partial^2/\partial x^2)$, from Appendix F in order to obtain this relation.) Thus the momentum magnitude p_x is proportional to n_1, and momentum space and quantum number space are identical except for the scale factor $h/4a$. The quantum numbers take on only positive values (negative values do not correspond to different standing wave states), and therefore p_x, p_y, and p_z (the momentum magnitudes) take on only positive values. Since there are two quantum states for each unit volume in quantum number space, there are two quantum states for each volume $(h/4a)^3$ in momentum space. The result of the preceding paragraph pertains to a volume $(2a)^3$ in coordinate space. Thus a volume $(h/4a)^3(2a)^3 = (h/2)^3$ of the six-dimensional momentum-coordinate space contains two quantum states. Stated slightly differently, there are two quantum states (differing in spin) in each volume

$$\Delta p_x \, \Delta p_y \, \Delta p_z \, \Delta x \, \Delta y \, \Delta z = (h/2)^3 \qquad \text{(G-12)}$$

of momentum-coordinate space. This statement is used in Secs. 6-3, 7-2, and others in connection with the implications of the Exclusion Principle.

All of this analysis of density of quantum states could have been developed for *traveling* waves, instead of for standing waves.* For traveling waves, unlike standing waves, there is a difference between plus and minus momentum components, since they represent waves traveling in opposite directions. The analysis for traveling waves is a little more complicated, since it cannot build so directly upon the work of Chapter 5. The result is that there are two traveling wave states in each volume

$$\Delta p_x \, \Delta p_y \, \Delta p_z \, \Delta x \, \Delta y \, \Delta z = h^3 \qquad \text{(G-13)}$$

It should be noted that eqs. G-12 and G-13 lead to the same density of states $S(E)$, namely, eq. G-9. This comes about because the entire spherical shell (both $+$ and $-$ momenta) of momentum space is used to go from eq. G-13 to $S(E)$, whereas only the all-positive octant is summed to go from eq. G-12 to $S(E)$. (See probs. G-5 and G-6.) For a systematic treatment of traveling waves, see C. Kittel, *Introduction to Solid State Physics*, Wiley, New York, 2nd Ed., 1956, pp. 243–250.

PROBLEMS

G-1.* Let the origin of coordinates be at the corner of the cube for the three-dimensional square well (instead of at the center, as in the text). Write the general solution $\Psi(x, y, z, t)$ in terms of quantum numbers n_1, n_2, and n_3.

G-2.* Normalize the wave function of prob. G-1.

G-3. Consider the classical vibration of a square drumhead, a membrane held rigidly at $x = \pm a$ and $y = \pm a$. Assume that a wave equation like the Schrödinger equation holds for small amplitude vibrations. Write a general expression for the z displacement of the drumhead as a function of x, y, and t. Do not attempt to work out the values of the constants, but show that (unlike a violin string) the natural frequencies of vibration are not merely a series of harmonics of a fundamental.

---

* Momentum is, in fact, satisfactorily defined *only* for traveling waves.

G-4.* Use the methods of Appendix F to find p_x^2 from $\Psi(x, y, z, t)$ for the three-dimensional square well.

G-5. Derive eq. G-9 from eq. G-12. *Hint:* Write E in terms of p and proceed to transform variables as in the r-to-E transformation in the text.

G-6. Derive eq. G-9 from eq. G-13. *Hint:* Same as in prob. G-5.

Appendix H
Answers to Problems
Marked by Asterisks

Only the numerical part of the answer is given here. Analysis, explanation, and answers to specific questions are parts of the answers to many problems.

Chapter 1

1-1. 1.684×10^{-19} coulomb; ($a = 1.665 \times 10^{-6}$ m.).
1-3. 4.6×10^{-5} coulomb/kg. **1-5.** 19 m./sec; 35°.
1-8. 1.74×10^{11} coulombs/kg. **1-9.** $v = 5.93 \times 10^5 \, V_0^{\frac{1}{2}}$ m./sec.
1-11. 1.21 amp; 0.028 amp. **1-12.** 1.24×10^{36}.
1-16. 0.511 M.e.V.; 938 M.e.V. **1-19.** 0.412; 0.863; 0.9957.
1-21. 1.156; 11.3%. **1-24.** 0.164 m.; 0.095 m.
1-26. 3.5×10^7 sec^{-1}; 1.95×10^7 sec^{-1}. **1-28.** 0.9793; 4.91.
1-30. 1.49×10^5 sec^{-1}; 4.77×10^5 sec^{-1}.

Chapter 2

2-1. 477 m./sec. **2-3.** 3.8 μvolts; 26. **2-5.** 1.6×10^{-4}. **2-8.** N.
2-9. $Nv^2(2/\pi)^{\frac{1}{2}}(M/kT)^{\frac{3}{2}}e^{-Mv^2/2kT}$. **2-11.** $2kT$. **2-13.** 0.68; 0.26.
2-16. 0.002% for H$_2$; 0.03% for N$_2$. **2-18.** 62 m. **2-19.** 1.9×10^{-19} m.
2-20. 2.9×10^{-6} m.

Chapter 3

3-1. 4.08 e.V./molecule. **3-4.** 1.00206. **3-5.** 3.1×10^{-14} m.
3-7. 10.4 M.e.V. **3-9.** 2.4×10^{-3} Å. **3-10.** 0.99939; 0.9962.
3-12. 0.119 weber/m.^{2} **3-16.** 5.1 years.
3-20. 31.972071 amu; 15.994912 amu. **3-23.** 1.46×10^{-8} amu.
3-26. $Mgh_1/3$; $8Mgh_1/3$. **3-28.** 0.094 M.e.V. **3-30.** 22.4 M.e.V.
3-32. 6.46 M.e.V./nucleon.

Chapter 4

4-2. 2.48 e.V.; 24,800 e.V. **4-3.** 2750 Å; 4960 Å. **4-5.** 0.71 volt; -0.79 volt.
4-9. 6.63×10^{-28} joule; 1.51×10^{22}. **4-11.** 10 watts/Å; about 1000 times as large.
4-13. 10.20 e.V.; 3×10^4; 4.5 μwatts. **4-15.** 2.82 Å.
4-18. 6.15×10^{26} molecules/kilomole. **4-19.** $2^{-\frac{1}{2}}d_0$; $3^{-\frac{1}{2}}d_0$.
4-22. 1.08×10^{-5}. **4-23.** $\sim$28% of the distance from grid to anode.
4-24. 10.2 e.V.

4-28. 4000 volts in transmission Laue method with film close to crystal; 2800 volts in back-reflection Laue method.

4-29. 0.276 Å.   **4-33.** 67.5°.   **4-34.** $v = 3c/5$.   **4-37.** 0.23°.

4-40. $2.9 \times 10^7/T$ Å.   **4-43.** 236 watts.   **4-44.** 1392°C.   **4-48.** 0.30 mm.

4-49. 1.45×10^{-16} newtons.   **4-51.** 13 e.V.   **4-53.** 6.63×10^{-32} m.

4-55. 5.18×10^{-15} m.   **4-56.** 2.72×10^3 m./sec; 1.45 Å.   **4-57.** 9080 rpm.

4-59. $\Delta x > \sim 2.2 \times 10^{-27}$ m.   **4-62.** 10^8 sec^{-1}; 1.67×10^{-7}.

4-63. $< \sim 5$ μsec; $> \sim 200$ kilocycles/sec.

Chapter 5

5-3. $b^2 = \int_{-\infty}^{\infty} \int_{-\infty}^{\infty} \int_{-\infty}^{\infty} |\psi_1|^2 \, dx \, dy \, dz;$

$b^2 = \int_0^{\infty} \int_0^{\pi} \int_0^{2\pi} |\psi_1|^2 r^2 \sin\theta \, dr \, d\theta \, d\varphi;$

$b^2 = 4\pi \int_0^{\infty} |\psi_1|^2 r^2 \, dr.$

5-12. $\psi = x_0^{-1/2} \sin\beta x$; $\beta = n\pi/2x_0$; $n = 1, 2, 3, \cdots$.

5-15. 4.18 e.V.; 16.71 e.V.; 37.60 e.V.

5-16. $5.49 \times 10^{-61} n^2$ joule; 1.35×10^{30}; 1.48×10^{-30} joule.

5-17. 4.18 e.V.; 8.2×10^{-6}.   **5-18.** 5.73 M.e.V.; 6.1×10^{-3}.

5-23. 0.36; 8.7×10^{-20}.   **5-26.** $\Delta p_x = 2\pi\sqrt{2m(P_0 - E)}$; $\Delta K = 4\pi^2(P_0 - E)$.

5-28. 0.36; 8.5×10^{-15}.   **5-33.** $P_0^2/16E^2$; $1 - 4(1 - P_0/E)$.   **5-37.** 19 Å.

5-38. 1.53 e.V.

5-42. $\mathcal{P}(x) = \pi^{-1}(x_0^2 - x^2)^{-1/2}$; $\mathcal{P}(0) = 1/\pi x_0$;

$\mathcal{P}(x_0) = \infty$; $\mathcal{P}(0.8x_0) = 5/3\pi x_0$.

5-43. 3.52×10^{13} sec^{-1}; 0.0728 e.V.; 0.2184 e.V.

5-44. 0.159 sec^{-1}; 9.48×10^{32}; 1.05×10^{-34} joule.   **5-48.** $N = n - 1$; $N = n$.

5-52. $\Delta p \cong 0.224 \, (Mh\nu_0)^{1/2}$; $\Delta x \cong 4.47 \, (h/M\nu_0)^{1/2}$; ~ 40.

5-55. $(x_0/\pi)[(\pi^2/3) - 2]^{1/2}$.   **5-56.** $x_0/3^{1/2}$.   **5-57.** $1/2a$; $1/a2^{1/2}$.

5-58. $h\nu_0/4$; $3h\nu_0/4$.   **5-60.** 0; $n^2h^2/32mx_0^2$.

Chapter 6

6-3. 10.20 e.V.; 1216 Å.   **6-4.** $r = \rho$.   **6-7.** $2^{-5/2}\rho^{-3/2}\pi^{-1/2}$.   **6-9.** 5ρ.

6-11. $-e^4m/h^24\epsilon_0^2$.   **6-12.** $e^4m/h^28\epsilon_0^2$.

6-13. $e^4m/n^2h^28\epsilon_0^2$; $e^2m/nh2\epsilon_0$; $2n\epsilon_0h^2/e^2m$; n.

6-14. 5.47×10^5 m./sec; 1.00000166.   **6-15.** $e/4\pi^{3/2}\epsilon_0^{1/2}m^{1/2}r_c^{3/2}$.

6-16. $e/4\pi^{3/2}\epsilon_0^{1/2}m^{1/2}(\bar{r})^{3/2}$.   **6-21.** $N_2/N_1 = 3 \times 10^{-17}$.   **6-23.** $-e^4m/24h^2\epsilon_0^2$; 2ρ.

6-26. 54.4 volts; 122.4 volts.   **6-27.** 5.06×10^{-6}.   **6-28.** -3.49 e.V.

6-35. 2.10 volts.   **6-36.** 0.00214 e.V.   **6-38.** 9.26×10^{-5} e.V.; 0.26 Å.

6-39. 1.12×10^{10} cycles/sec.   **6-40.** $-2N_1\mathfrak{M}_B\mathfrak{B}/kT$.   **6-42.** 1.226.

Chapter 7

7-3. 0.18 e.V.   **7-4.** 22.   **7-6.** 4.0 e.V.; 6.8 e.V.   **7-7.** 10^{-240}.

7-10. 13.6 e.V.   **7-11.** -54.4 e.V.   **7-24.** 0.0545%; decrease of 0.00741 e.V.

7-26. $E(H_2) = (2/3^{1/2})E(HD) = 2^{1/2}E(D_2)$; 0.078 e.V.   **7-28.** 2.6×10^{-10}.

7-32. 11 e.V.   **7-36.** 1.30 Å.

7-37. 4.55×10^{-48} m.^{2}/kg; 0.0152 e.V., 0.0456 e.V., 0.0912 e.V., and 0.1521 e.V. above $- 4.48$ e.V.

7-39. 2.71 in $J = 1$ for each in $J = 0$.   **7-42.** $\nu_c = (1/2\pi)(2E/B)^{1/2} = \nu_w$.

Chapter 8

8-1. 23.1.   **8-2.** 7.64 e.V.   **8-4.** 9.1.

8-11. $E = E_c + 3.8(R - R_0)^2$ e.V., where R and R_0 are in Å.

8-16. $(d^2E/dR^2)_{R=R_0} = 59$ joules/m.^{2} per ion; 4.8×10^{12} sec^{-1}; 62 μ.

8-18. 3.04 Å.   **8-22.** $3^{1/2}2^{-1}a$; $2^{-1/2}a$.

8-23. On a cube face, a distance $5^{1/2} a/4$ from each of two adjacent cube corners.

8-31. $9h^2/32mR^2$.

Chapter 9

9-2. 3×10^{-8} e.V.; $\sim 3 \times 10^9$ volts/m.   **9-7.** $\delta = 1.1kT$.

9-9. $0.018; 0.00033; 4.6 \times 10^{-18}$.   **9-10.** 1.8% error; 0.0045% error.

9-15. 3.16 e.V.; 7.06 e.V.   **9-16.** 7.52 e.V.   **9-17.** 82,000°K.

9-18. $m^*/m = 1.47$.   **9-20.** 0.024 e.V.

9-25. 7.1×10^{12} sec^{-1}; 1.8×10^6 sec^{-1}; 339°K; 8.6×10^{-5}°K.

9-27. $3.09k$; $3.03k$.   **9-28.** 2.55×10^4 joules/kilomole; 1.78×10^4 joules/kilomole.

9-29. $(12\pi^4R/5)(T/\Theta)^3$.   **9-32.** $0.0604R = 500$ joules/kilomole.

9-33. 0.80 joule/kilomole, compared to 0.26 joule/kilomole for lattice.

9-42. $5.93 \times 10^5 E_0^{1/2}$ m./sec; 1.58×10^6 m./sec.   **9-43.** 320 Å; 390 Å.

9-44. 0.32 m./sec; 2×10^{-7}.   **9-52.** 6.6 μvolts.

9-53. -0.74×10^{-10} volt m.^{3}/amp weber.   **9-54.** $4.4kT_c$; $5.1kT_c$; $4.5kT_c$.

9-57. $4.5 \times 10^{-3}N\mathfrak{M}_B$.   **9-58.** $eh/4\pi m$ for $m_l = 1$.

9-61. 3.9 webers/m.^{2}; 3.2 webers/m.^{2}; 2.1 webers/m.^{2}

9-64. $3\mathfrak{M}_B$; 0.43 weber/m.^{2} **9-65.** $14\mathfrak{M}_B$.

Chapter 10

10-1. $\sim 50d_0$.   **10-6.** 625 mhos/m.   **10-11.** $20d_0$.

10-12. 1.30 μ; 63 seconds of arc.   **10-16.** 10^{-303}.   **10-17.** 0.10 sec.

10-19. Net jump probability is $(a^2\nu_0q\mathcal{E}/kT)e^{-\Delta E/kT}$; $\mu = (a^2\nu_0q/kT)e^{-\Delta E/kT}$.

10-21. $E_g > 3$ e.V.   **10-22.** ~ 500 Å.   **10-26.** Blue-green; bluish-purple.

10-28. 10^7 holes, 10^7 electrons; 1.6×10^{-12} coulomb.   **10-29.** 0.053 volt.

10-33. 1.7×10^{-25} kg m./sec; 1.1×10^{-27} kg m./sec.   **10-39.** 0.39 e.V.

10-43. Violet (Ag); blue-green (Cu); orange-red (Mn).

10-48. $152,000; 9.1 \times 10^{-4}$ Å; 1.31×10^{-7}.

Chapter 11

11-3. $0.82kT$ below it.   **11-5.** 2.0 mhos/m.; 4.6×10^{-4} mho/m.

11-8. 0.098 e.V. below bottom of conduction band.

11-10. 0.16 e.V. above top of valence band.

11-11. 0.040 e.V. below bottom of conduction band.

11-13. 0.16 e.V. below bottom of conduction band.   **11-14.** 3000 mhos/m.

11-17. $N_d = 2.34N_ce^{-E_g/2kT}$.   **11-19.** $E_0 = E_d$; $E_g - E_d = 0.015$ e.V.

11-21. 580, 300, 5000, and 61,000 mhos/m.

11-23. 0.33 m.^{2}/volt sec; 1.9×10^{21} m.$^{-3}$; 1.8×10^{21} m.$^{-3}$; 4.1×10^{-8}

11-25. -2.1.   **11-26.** $E_0 = E_g - 1.17$ e.V.; $\sigma = 0.017$ mho/m.

11-28. $N_p \sim e^{-E_g/kT}$.   **11-30.** $2N_i = 4.5 \times 10^{19}$ m.$^{-3}$.

11-32. $d_n = 3.1 \times 10^{-7}$ m.; $d_p \ll d_n$.   **11-34.** $d_n \sim V^{1/2}$.

11-37. $\rho = 2.6 \times 10^8$; $\rho = 10^{2520/T}$ or $d\rho/\rho = -19.4\,(dT/T)$.

11-38. $26/I$(ma) ohms.

11-39. 0.179 and 0.238 volt; 0.184 and 0.288 volt (corrected).

11-40. 0.010 m.^{2}/sec.　　**11-41.** $0.12\ \mu$amp.　　**11-44.** $(I_p/I_n) = L_n\sigma_p/L_p\sigma_n$.
11-48. 0.47 volt.　　**11-49.** 0.72 m.^{2}; 1.2×10^{-6} m.^{2}　　**11-51.** 0.22 volt.
11-52. 2.5×10^{25} m.$^{-3}$　　**11-54.** $29\ \mu\mu$farad.　　**11-56.** 4×10^{-8} sec.
11-60. $49;980$.　　**11-62.** $\sim 1/(1 - a)$; 50 or 17 db.　　**11-64.** $53\ \mu$sec.
11-68. 0.028 e.V.; 0.080 joule/coul.

Chapter 12

12-2. $(2.12 \times 10^{13}/\epsilon T^2)e^{-e\phi/kT}$.　　**12-3.** 2890 amp/m.^{2}　　**12-5.** 4.43 volts.
12-6. 1.7.　　**12-7.** 10^5 volts/m.; 6.0×10^{-8} m.; 0.012 volt; 1.085.
12-12. -2.9 volts.　　**12-16.** 0.0044 amp/watt; 0.25 amp/watt.
12-19. Increases a factor of 2.71; -0.086 e.V.　　**12-23.** $e(V_0 + 4.6)$ joules.
12-25. 210 amp/m.^{2}　　**12-26.** V_0/V_c; 27%; 67%.　　**12-27.** 0.87 watt; 18%.
12-29. $\sim 5\ \mu$.　　**12-30.** 6.2×10^4 amp/m.^{2}; $0.062\ \mu$amp.
12-32. 9.4×10^{-25} kg m./sec; 1.6×10^{-27} kg m./sec.　　**12-33.** $0.072\ \mu$amp.
12-40. $\sim 2 \times 10^{-15}$ sec.　　**12-43.** $3.22 \times 10^{22}p$ atoms/m.^{3}　　**12-44.** $10\ \mu$amp.
12-45. 2.41 e.V.　　**12-46.** 4.0×10^{20} m.$^{-3}$ sec^{-1}.　　**12-49.** 0.20 milliamp.
12-51. 10.4 e.V.; 19.6 e.V.; 41 e.V.; 49 e.V.　　**12-53.** $\sim 8 \times 10^{-8}$; 1.7×10^{-26} m.^{2}
12-57. $\sim 5 \times 10^{-17}$ sec; $\sim 10^{-14}$ sec.　　**12-60.** $(v_2/v_1) = (eV_2 + K_0)^{1/2}(eV_1 + K_0)^{-1/2}$.
12-64. 0.037 Å; 6 Å.

Chapter 13

13-1. 0.086 M.e.V.; 4.79 M.e.V.　　**13-3.** $\tau/0.693$.
13-4. $n_p = n_0e^{-\lambda_p t}$; $n_d = (\lambda_p n_0)(\lambda_d - \lambda_p)^{-1}(e^{-\lambda_p t} - e^{-\lambda_d t})$.　　**13-7.** 28.2 M.e.V.
13-8. 4.3×10^{-14} m.　　**13-10.** 2.3×10^{17} kg/m.^{3}
13-11. 2.9×10^{-12}; 2.9×10^{-6}.　　**13-14.** 5.7 M.e.V.　　**13-16.** 4.28×10^7 ℬ sec^{-1}.
13-17. 0.783 M.e.V.　　**13-20.** 0.0475 m.　　**13-22.** $E_n = -6.80/n^2$ e.V.
13-25. 98.3%.　　**13-26.** 0.0124 Å.　　**13-31.** 2×10^5 atoms.
13-33. 2.1×10^{-4} m./sec.
13-37. $_zM^A = 923.8A - 0.78Z + 17.8A^{2/3} + 0.71Z(Z - 1)A^{-1/3}$
$\qquad + 23.6A^{-1}(A - 2Z)^2 \mp 132A^{-1}$ M.e.V.
13-38. 29.2.　　**13-43.** 0.47 M.e.V.　　**13-45.** 0.0032.　　**13-46.** 0.7 barn.
13-49. 10^{-12}.　　**13-51.** $5.24, 5.23, 5.22,$ and 5.18 M.e.V.　　**13-53.** 3.8 M.e.V.
13-55. $180°$.　　**13-57.** $179.73°$.　　**13-59.** 4.1 M.e.V.

Chapter 14

14-1. $E \sim x^{-1}$; $\sim 1.2 \times 10^{-14}$ m.　　**14-2.** 1.260; ~ 176 M.e.V.
14-7. 1.35×10^{-3} m.　　**14-13.** $-15.7, -173.8, 362.2, 9.1,$ and 181.8 M.e.V.
14-16. 0.284; 0.889; 1.000.　　**14-17.** 120; 2×10^{-4} sec.
14-19. 1.57×10^{17} sec^{-1}; 5000 kilowatts; 4×10^7 curies.　　**14-20.** 0.55%.
14-22. 4 barns.　　**14-24.** 0.9957.　　**14-26.** 1.47×10^{20}; 16 curies.
14-28. Eq. 14-10, 3.52 M.e.V./amu; eq. 14-1, 0.85 M.e.V./amu.　　**14-29.** 65 kg.
14-32. 0.7 M.e.V.; 8×10^9 °K.　　**14-34.** 6.3×10^{-4} m.　　**14-36.** 3.5.
14-39. 0.34 M.e.V.　　**14-41.** 3.3×10^5; 5.3×10^{-11} coulomb; 0.53 millivolt.
14-49. $\theta = 0$; $\theta = 41.2°$.　　**14-51.** 1.88×10^3 watts/kg; 1.41×10^5 watts/kg.
14-53. 4.5×10^{-10} curie; 0.045 curie/kg; 6.2×10^{-10}.

Appendices

F-5. $h/2\pi$; 0; $-h/2\pi$.
G-1. $\psi = A \sin\beta_1 x \sin\beta_2 y \sin\beta_3 z$, where $\beta_1 = n_1\pi/2a = (2\pi/h)(2mE_1)^{1/2}$, etc.
G-2. $A = a^{-3/2}$.　　**G-4.** $2mE_1$.

Index